Lecture Notes in Production Engineering

More information about this series at http://www.springer.com/series/10642

Christian Niemann-Delius
Editor

Proceedings of the 12th International Symposium Continuous Surface Mining - Aachen 2014

Editor
Christian Niemann-Delius
Department of Mining Engineering III
RWTH Aachen University
Aachen
Germany

ISSN 2194-0525 ISSN 2194-0533 (electronic)
ISBN 978-3-319-12300-4 ISBN 978-3-319-12301-1 (eBook)
DOI 10.1007/978-3-319-12301-1

Library of Congress Control Number: 2014950638

Springer Cham Heidelberg New York Dordrecht London

Printed on acid-free paper

Springer is part of Springer Science+Business Media (www.springer.com)

Welcome to the ISCSM 2014

Despite the fact that continuous mining systems are operated in all five continents and in unconsolidated as well as consolidated material, the technique is still associated with lignite mining in Central Europe. The first International Symposium was, however, held in Edmonton, Canada in 1986 at a time when continuous systems marked the first steps of the fast expanding oil sand mining. With the expansion of elements of continuous systems into hard rock mines the development of the technique now can be found in all kind of mines.

The scope of the ISCSM is to reference the state of the art as well as inform about research results and trends in continuous surface mining. It is jointly organized by Aachen University and Bergakademie Freiberg and in partnership with DEBRIV. The event takes place every three years alternating between the two universities and international locations. The ISCSM 2014 Call for Papers attracted more than 65 abstracts from 15 countries.

The papers that have been accepted will be published in hard cover by the publishing house Springer in the "Lecture Notes in Production Engineering"-Series as part of the conference package. This is an innovation insofar as the selection and evaluation of papers is now done in a standardized peer review process. The symposium committee apologies for the burden imposed by the online conference system OCS especially on the authors from outside the academic structures. In engineering evaluation of conference papers and publications in journals traditionally takes place by praxis rather than in academic circles. The symposium committee, however, felt obliged to comply with the urge of university management worldwide towards the peer reviewed system - common in natural science.

After the opening session the symposium continues in three parallel ones grouped into seven issues to allow the audience to focus on distinct topics. Apart from the traditional mining in unconsolidated material (1), drilling, dewatering, slope stability and soil mechanics (2), deposit and mine simulation in computerized mine planning (3), control of operations, organization and management (4) as well as environmental questions (5) and the now established praxis in consolidated material for example IPCC (6) will be dealt with. An individual session has been added to address legislation and mine administration (7).

Regardless the effort to comply with university standards, the symposium apart from scientific research specifically focuses on the state of the art in the industry. Here, the event will present ample opportunities for discussions with national and international professionals and features a poster presentation. The field trip to the famous Hambach Mine will give an insight into the state of the art in mining technology and consideration of environmental aspects in the Rhenish lignite district.

At this point, I would like to thank all speakers, companies, institutions and supporters contributing to the success of the conference. This foremost goes to DEBRIV for its active partnership in the preparation of the upcoming event.

We hope that you will be able to join us at the Symposium and look forward to extending a warm welcome to all delegates and affiliates.

September 2014 Christian Niemann-Delius

Organization

Reviews by Program Committee

Agioutantis, Prof. Zacharias
Arnold, Ingolf
Asenbaum, Dr. Peter
Bui, Prof. Xuan-Nam
Chakravarty, Prof. Debashish
Czaja, Prof. Pjotr
Dirner, Prof. Vojtech
Günther, Andreas
Hardygóra, Prof. Monika
Kuhnke, Claus
Kulik, Dr. Lars
Lazar, Prof. Maria
Moser, Prof. Peter
Niemann-Delius, Prof. Christian
Panagiotou, Prof. George
Paraszczak, Prof. Jacek
Pavlovic, Prof. Vlada
Rakishev, Prof. Bajan
Singh, Prof. Pradeep K.
Vossen, Dr. Peter

Organizing Committee

Dr.-Ing. Peter Vossen
Kerstin Schroeder M.Sc.
Dipl.-Ing. Mirjam Rosenkranz
Elisa Kuhnke B.Sc.

Contents

Continuous Surface Mining Equipment and Mining Systems

Drilling, Dewatering, Slope Stability and Soil Mechanics

Deposit and Mine Simulation, Computerized Mine Planning

Control of Operations, Organization and Management

Continuous Mining and Transportation in Consolidated Rock

Mining Legislation and Administration

Conti® MegaPipe – A New Dimension in Closed-Trough Belt Technology

Thomas Neumann and Andrey Minkin

ContiTech Transportbandsysteme GmbH,
PO Box 1169,
37154 Northeim, Germany

ContiTech is adding a new dimension to closed-trough belt technology. Enlarging the maximum diameter of the closed-trough belt to max. 900 mm will considerably expand the number of uses to which it can be put. It will give rise to completely new application areas for the transport of bulk material. Considerable reductions in fixed costs and CO_2 emissions are possible, particularly for solid rock digging in surface mining. For all other industrial applications, this innovation will double the volumetric flow rate!

1 Challenges Facing Closed-Trough Belt Technology

Is it possible to transport large volumetric flow rates over complicated routes without generating high noise levels? Can the transported material be protected, at the same time, from environmental influences such as rain and wind? Is steep conveyance possible? Can all of this be done, even under extreme environmental conditions such as very low temperatures?

These, and many other questions from our customers, encouraged us to investigate the current state of closed-trough belt technology and compare this with current opportunities. In doing so, we drew on a rich collection of closed-trough belts that have already been delivered. Under the name "HS-Rollgurt," Conti closed-trough belts have become a technological standard in a large number of industrial sectors.

But time presses on, yet the further development of drive technology makes progress possible, particularly in light of EU environmental legislation grounded in the new requirements to lower CO_2 emissions.

The technical pre-requisites, timing resources, and naturally the motivation of excellently trained application engineers and developers form the basis for successful implementation of the idea.

1.1 High Inclined Angles and Small Curve Radii

Closed-trough belts generally allow the curve to be designed in many different ways. They can thus be made to fit in very well with existing natural or even

C. Niemann-Delius. (ed.), *Proc. of the 12th Int. Symp Continuous Surface Mining - Aachen 2014*,
Lecture Notes in Production Engineering, DOI: 10.1007/978-3-319-12301-1_1

urban profiles. The low noise levels and the fact that the transported material is enclosed frequently make the closed-trough belt a good choice.

Steep conveyance is a special form of flow guidance. A large number of special steep conveyor belt solutions as well as vertical conveyor belts are currently in use. This means that inclines of up to 20° can be easily handled using standard troughed belt systems. Depending on the characteristics of the bulk materials, inclines of up to 35° can be created using closed-trough belts.

A particular challenge in the area of steep conveyance arises in solid rock digging. Nowadays, heavy-duty trucks are mainly used here; these transport the material from the surface mine at an incline of up to 7°. In addition to enormous energy consumption, other characteristics of this mode of transport are high operating costs, high expenditures for servicing and a very high demand for qualified personnel. Many mines are located in inaccessible areas where labor is scant. The attractiveness of such projects is thus also very low.

Mining companies are showing increasing interest in finding alternative means of transporting ore or rubble. By switching over from discontinuous conveyor technology to continuous conveyor belt technology, it is possible to achieve considerable gains in efficiency, an improvement in the CO_2 footprint, and improved process reliability.

Software tools are very helpful in convincing people to switch over. They take actual data compiled on the current mine conditions and compare conveyor technologies. This allows for comparison of CO_2 footprints as well as investment and operating costs. This then provide an excellent basis for making a decision.

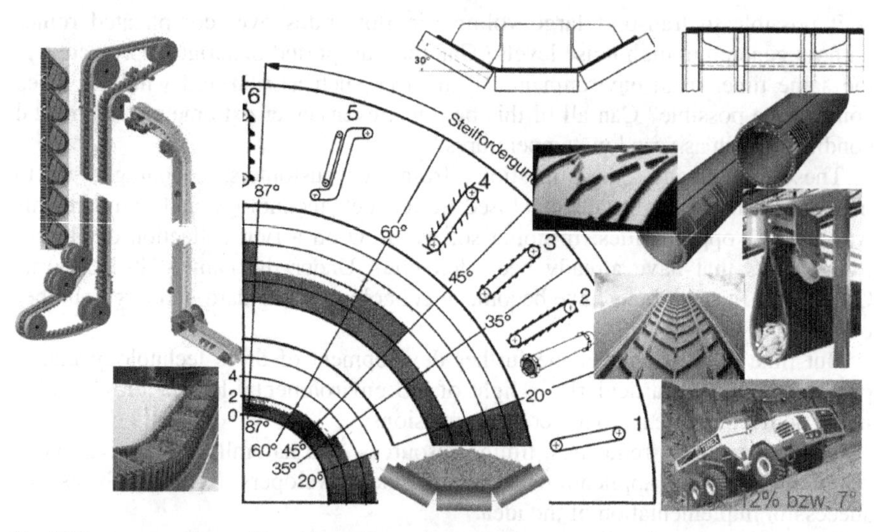

Fig. 1 The opportunities offered by different belt solutions as a function of conveyor angle

1.2 High Mass Flows

How to increase efficiency poses a constant challenge for all customers. Closed-trough belts allow a considerable improvement here thanks to their high speed, optimum routability and expanded conveyor diameters. Conti engineers have an excellent pool of knowledge at their disposal that enables them to superbly adjust the belts to be delivered to meet customer requirements. This is where we use our extremely extensive wealth of experience, combined with modern calculation procedures and technological expertise, to achieve optimum results.

High mass flows pose new requirements for the transmission of higher tractive forces, optimization of the forming forces, and belt designs.

1.3 FEM and DEM Method for Designing a Closed-trough Belt

Designing closed-trough belts is a time-consuming process. Numerous factors have to be taken into consideration. This is why specific calculation programs developed by Conti engineers for this very purpose as well as other known software are used for DEM and FEM analysis. Test samples are then created to check the calculation results on special test stands in the technical center. The result is an optimally designed closed-trough belt that exactly meets the requirements of the individual project.

2 Expanding the Technical Opportunities with Conti® MegaPipe Technology

In developing the Conti® MegaPipe, the ContiTech Conveyor Belt Group has opened up a whole new dimension in closed-trough belting. Figure 2 shows the structure of a MegaPipe with one textile tensile member and one steel cord tensile member. The rigidity of the MegaPipe in a transverse direction and the shape of the overlapping area play a decisive role in ensuring that the pipe functions safely and reliably. As is the case with conventional closed-trough belts, the belt must be designed in such a way as to ensure optimum transverse rigidity.

If this is not the case, the belt edges of the overlapping area can crumble (transverse rigidity is too low) or the system's energy consumption will be too (in some cases, the closed-trough belt risks opening up between idler roller stations). The shape of the overlapping area must securely seal the material in the closed-trough belt.

As shown in Figure 2, optimum transverse rigidity of a steel cord MegaPipe is achieved by means of a system consisting of several textile and steel breakers that have been attached at a particular distance from the neutral phase (this is frequently the central level of the cord) in the cross-section of a closed-trough belt. In the case of textile EP MegaPipes, optimum transverse rigidity is ensured by maintaining a particular distance between textile EP layers and by the layout of a steel breaker.

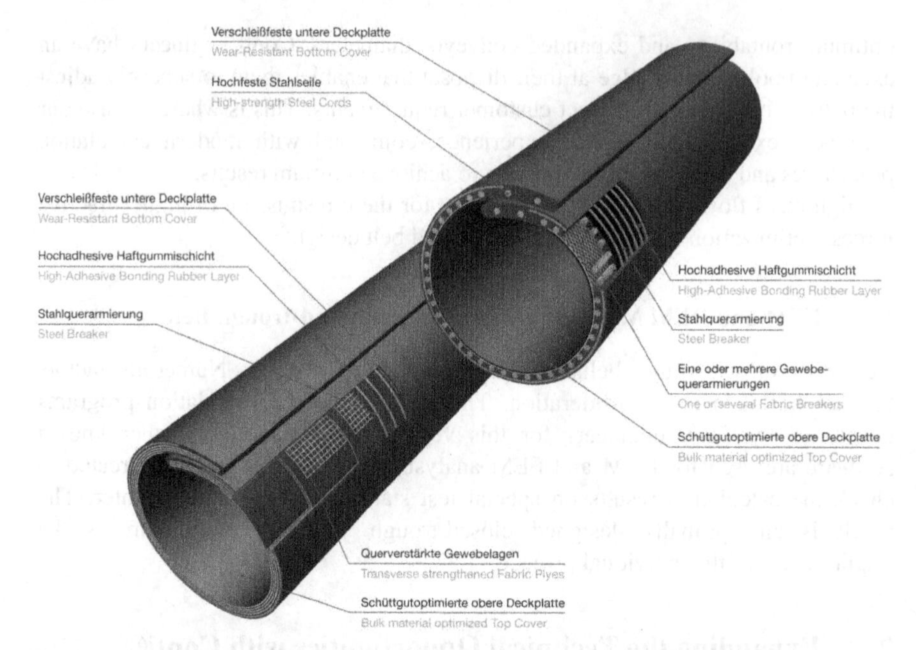

Fig. 2 Structure of a Conti® MegaPipe with one textile and one steel cord tensile member

2.1 *Increasing the Volumetric Flow Rate*

Whereas conventional models do not exceed an outer diameter of 700 mm, the Conti®MegaPipe can have an outer diameter of as much as 900 mm. The larger diameter offers two distinct advantages. Firstly, the larger belt width of up to 3,200 mm and the higher conveyor speed enable a significant increase in the capacity of the conveyor system. This allows for an increase in the flow rate of up to 100 percent.

2.2 *Increased Grain Size*

The Conti®MegaPipe can transport coarse-grained material. The Conti®MegaPipe can handle material with a grain size up to 30 percent larger than what conventional systems can deal with. This facilitates transportation of large chunks of material with a grain size of up to 350 mm. This makes it possible to use the Conti®MegaPipe directly after a primary crusher.

2.3 *Adapting to the Profile of the Surface Mine*

Solid rock digging, in particular, involves a very steep angle of approach. For this reason, the use of continuous conveyor devices has only been possible so far with considerable additional investments. It is then necessary to create intricate approach sections if belt systems are used whose conveying angle is limited by the bulk material being transported. The Conti®MegaPipe can be adapted exactly to the existing route profile.

2.4 Long-Distance Transport

Conti®MegaPipe closed-trough belts are also suitable for very long transport routes. The absence of transfer points, and the enormous adaptability offer an interesting alternative here to discontinuous transport involving heavy-duty trucks or even rail transport. In the latter case, neither the time-consuming loading and unloading processes nor the complete technical and HR infrastructure for rail transport are required.

This mode of transport offers optimum protection to the transported material. And quantity losses are minimal.

2.5 Flow Rates within Companies

The opportunity to realize 3D routes makes the use of closed-trough belts within companies very interesting. Conti®MegaPipe expands the range of uses here in relation to volumetric flow rates. Loss of material at transfer points and scrapers is now a thing of the past.

Table 1 Overview of belt width, outer diameter, volumetric flow rate, max. grain size, and conveyor speed

Outer diameter	Belt width	Volumetric flow rate (for v=1 m/s and filling level of 75 percent)	Recommended conveyor speed	Max. grain size
mm	mm	m^3/h	m/s	mm
780	2,800	1,100	5.8	
830	3,000	1,250	6.0	250…350
890	3,200	1,450	6.5	

3 Advantages for the User

3.1 System Technology

Closed-trough belts make possiblevery tight curve radii and high inclined angles. This enables the belt's route to be optimally adapted to the terrain.

The belt features a very compact design. This makes it ideal for use in underground mining and tunnel construction.

The absence of transfer points considerably cuts costs. Even the material losses, which are difficult to avoid here, and the HR outlay for manual removal of contamination on the transport route are a thing of the past.

Thanks to standardized and low-wear components, the servicing and cleaning outlay is greatly reduced.

Thanks to the closed-trough belt concept, very steep conveyor angles of up to 35° are possible.

Conti closed-trough belts can be designed for high conveyor speeds.

3.2 Product

High conveyor strengths make large distances between axes possible, even in the case of larger diameters and inclines.

The short clamping distances result from the low-stretch reinforcements.

High straight-line stability, and optimum sealing of the overlapping zone are hallmark features of Conti®MegaPipe closed-trough belts. A high level of experience and accuracy in designing tensile members and belt material is required to achieve this very characteristic. This characteristic is particularly important in the event of adverse external circumstances.

Thanks to the company's decades of experience, and the extreme reliability of the belt design, a long service life is guaranteed.

The particular dimensions and characteristics of the Conti®MegaPipe allow the DIRECT transport of broken ore away from the PRIMARY CRUSHER. This means that transportation by heavy-duty truck can be minimized.

3.3 Protecting the Environment and the Transported Material

Conti® MegaPipe technology also makes it possible to dispense with discontinuous material transport by means of heavy-duty trucks. This considerably reduces energy consumption , resulting in a significant reduction in CO_2 emissions. ContiTech supports customers in implementing this universal goal. This contributes toward implementation of the ContiTech "Green Engineering" key principle.

Closed-trough belts protect both the environment and operatives from dangerous, contaminated, dusty, and malodorous materials (such as chemicals, waste, ash, rubble, etc.). What is more, the material transported is not contaminated by outside influences.

Instead, it is protected from environmental influences such as wind, snow, and rain.

4 Conti® MegaPipe in Solid Rock Digging

The feasibility study focuses on an existing solid rock mine. The role of the closed-trough belt is to transport the metal ore from the bottom of the mine to the upper level. Heavy-duty trucks are used only for haulage from the extraction equipment to the central loading point of the Conti® MegaPipe with primary crusher.

4.1 Mining Parameters

Conveyor height: 500 m
Conveyor length: 1,000 m
Mass flow: 4,000 t/h
Inclined angle: 30°
Grain size: Rubble or ore with a grain size up to max. 300 mm

4.2 Main Parameters of the Belt System with Conti® MegaPipe

Conveyor belt: Conti® MegaPipe 3000 St7400 11F:11SS DIN-X,
Belt thickness: approx. 35 mm; width: 3,000 mm, belt weight: 211 kg/m, nominal strength: 7,400 N/mm
Outer diameter (width across flats of the hexagonal idlers): 830 mm
Idler spacing: 1,500 mm
Fictitious friction coefficient DIN-f (worst case): 0.08
Necessary drive power at the head (worst case): 7,800 kW
Installed drive power at the head: 2 × 5,000 kW
Drive pulley diameter: 2,200 mm
Conveyor speed: 3 m/s
Filling level: approx. 50 percent

4.3 View of the Energy Saving / CO_2 Reduction

The comparison juxtaposes the expected energy consumption of the conveyor belt system with realization of the same conveyor quantity (4,000 t/h) using heavy-duty trucks.

4.3.1 Conti® MegaPipe Consumption Calculation

- Required drive power for the closed-trough belt system: 7,800 kW
- Reduction to be expected with the belt retracted: approx. 9 percent
- Expected power consumption of the belt system in full-load operation: **7,100 kWh**

4.3.2 Heavy-Duty Truck – Consumption Calculation

- Power class of the heavy-duty truck: 200 t
- Cycles per hour: 2
- Required number of heavy-duty trucks: 10
- Diesel consumption: 200 l / h
- Energy content: 9,800 kWh/m^3 = 9.8 kWh/l (source: Wikipedia "Fuel")
- Total diesel consumption of the heavy-duty truck fleet: 4,000 l/h
- Total energy consumption of the heavy-duty truck fleet: **19,600 kWh**

This rough examination shows a considerable potential for reducing energy consumption and CO_2 emissions. The belt system consumes less than half as much transport energy as heavy-duty truck transport.

This does not take into account further cost reductions arising from reduced operating, HR, and servicing expense as well as profits from higher process reliability, etc..

4.4 Cost Advantage Resulting from Doing without a Secondary Crusher Thanks to a Conti® MegaPipe

There is a limit to the grain diameter size in the case of closed-trough belts. In general, the internal closed-trough belt diameter must be twice or three times the max. material grain size in order to cope with this material. In standard closed-trough belts, a secondary crusher is therefore required immediately after the primary crusher. It reduces the grain size and thus make the material "transportable" for a closed-trough belt (see Figure 3).

A simple cost-effectiveness calculation shows the following: wearing parts must be replaced 14 times a year or even more often. If the costs for wearing parts over a 25-year period (the usual service life of a secondary crusher) are added to its cost of purchase, it becomes obvious that using a Conti®MegaPipe and the associated elimination of the secondary crushers can mean that getting around the cost of an eight-figure investment!

Even if use of a secondary crusher is unavoidable in the technological chain, a MegaPipe allows it to be used flexibly, i.e. the secondary crusher can be installed after both the primary crusher as well as the MegaPipe discharge point. This means that a mine operator has more latitude in the positioning of a secondary crusher in the technology chain.

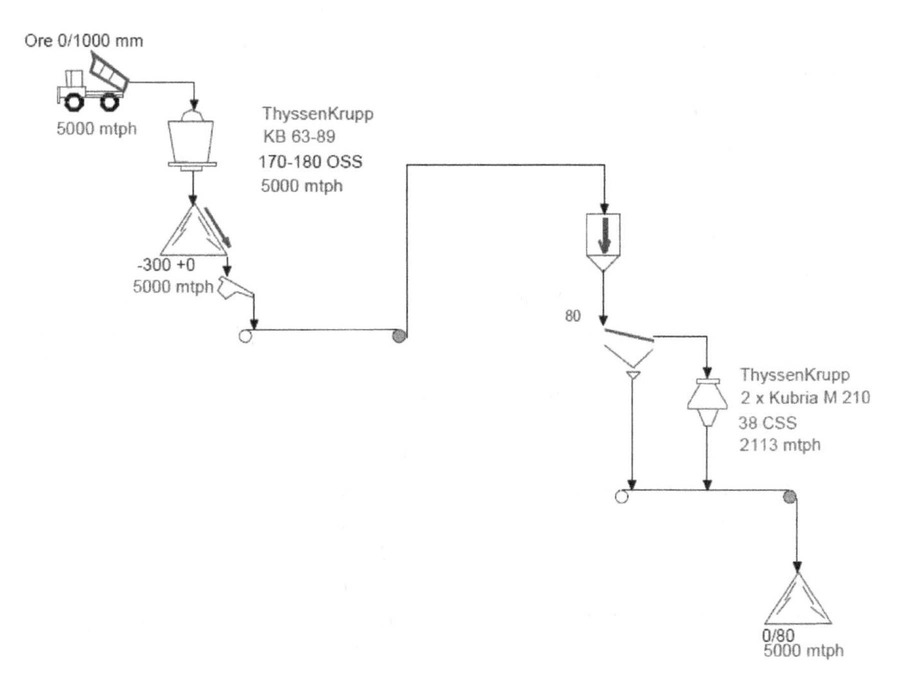

Fig. 3 Typical flow chart from primary and secondary crusher (source: ThyssenKrupp)

5 Conti®MegaPipe for Large Mass Flows

The second feasibility study focuses on a long, curved conveyor route. The parameters are taken from a real project [3], and describe the project conditions. Conti®MegaPipes satisfy each and every one of these conditions. Even under tough operating conditions, the advantages of this technology are utilized fully.

5.1 Parameters of the Transport Route

Conveyor profile: 22 horizontal curves and 45 vertical curves
Conveyor length: 3,414 m
Mass flow: 4,500 t/h
Inclined angle: max. 11°
Temperature range: -40°C … +40°C

5.2 Main Parameters of the Belt System with CONTI® MegaPipe

Conveyor belt: Conti®MegaPipe 2250 S-K2 11F:11SS Conti Extra, 3,000 mm wide

Outer diameter (width across flats of the hexagonal idlers): 830 mm
Minimum curve radius: $R_{min} \geq 500$ m
Installed drive power: 5,000 kW
Belt speed: 4.2 m/s

6 Summary

Closed-trough belts are a reliable technology for continuous conveyor technology. The advantages they offer come into play in very demanding tasks in particular. Conti®MegaPipes expand the range of application possibilities considerably. They open up an entirely new dimension in conveyor technology for mining .

The Conti®MegaPipe combines the well-known advantages of a closed-trough belt -- such as environmentally compatibility, enclosed transport of materials, small curve radii, high incline angles of up to 35°, elimination of transfer points -- with new characteristics such as the possibility of material grain sizes of up to 350 mm, high conveyor speeds, and high mass flows.

This presentation outlines how continuous closed-trough belt technology is used in solid rock mining. It has considerable potential to reduce energy consumption, and CO_2 emissions, with a resulting reduction in the cost of transporting ore and rubble from the mine.

Thanks to the possible omission of a secondary crusher, eight-figure investments can be avoided. In addition, a mine operator has more latitude with regard to the positioning of a secondary crusher in the technology chain.

This technology is also an interesting solution for large flow rates over complicated routes and relatively long distances.

References

[1] Minkin, A., Dilefeld, M., Fischer, W.: Schlauchgurtförderer für die Kraftwerksentsorgung ("tube belt conveyors for power plant waste disposal"). Schüttgut 18(6) (2012)
[2] Wikipedia – http://de.wikipedia.org/wiki/Kraftstoff
[3] Minkin, A., Jungk, A., Hontscha, T.: Belt Replacement at a Long Distance Pipe Conveyor. Bulk Solids Handling 32(6) (2012)
[4] Neumann, T.: " Трубчатые конвейерные ленты ContiTech — превосходное транспортное решение!" ("ContiTech closed-trough belts – an outstanding transport solution.") UGOL (March 2013)

Belt Positioning and Skewing Prevention in Lignite Mining Using Long-Wavelength Infrared Cameras

Karl Nienhaus[1], Manuel Warcholik[1], Christoph Büschgens[1], and Dietmar Müller[2]

[1] Institute for Mining and Metallugical Machinery (IMR), RWTH Aachen University, Germany
[2] Mitteldeutsche Braunkohlengesellschaft mbH (MIBRAG), Germany

Abstract. In open pit mines, one of the key aspects of automation is the transport of the mined minerals. Due to their economic benefit, conveyor belts are used to transport masses continuously. High availabilities and safe working conditions can be achieved by automating the process. Especially the feeding of the conveyor system has a strong automation potential. At the moment, the positioning of transfer booms or transfer chutes is done manually by a machine operator. The decisions of the operator are driven by his experience and his visual perception. The visual perception is often limited, e.g. by low light (in the evenings or at night), blinding by the sun or other environmental conditions such as dust or rain which can be present in open pit mining. This can lead to sub-optimal loading of the material. In case of off-centered loading, the conveyor belt can start to drift away from a centered position, which is called skewing. This skewing results in increased wear and tear of the system leading to breakdowns or increased maintenance times. Additionally, the transport capacity of the conveyor belt decreases because less area can be loaded with material.

To detect skewing on conveyor belts, various systems are available on the market. The most common types of detection systems are mechanical components like switches and guide roller constructions. A second group of sensors used for skewing detection are supersonic systems. Both, mechanical and supersonic systems are not able to prevent belt skewing. These systems only detect an existing misalignment.

To be able to prevent belt misalignments, a system based on long-wavelength infrared (LWIR) cameras which monitor the loading process as well as the position of the conveyor belt is proposed by the authors. The obtained data is correlated and analyzed with regard to the position of the belt when entering the loading station, the loading process itself and the belt position when leaving the station. This data can either be used for automation or if visualized to the operator, as an assistance system. This information can be used to optimize the loading process in order to prevent skewing, or to counteract already existing drift of the conveyor belt to prevent damages on the conveyor belt or other systems.

C. Niemann-Delius. (ed.), *Proc. of the 12th Int. Symp Continuous Surface Mining - Aachen 2014*,
Lecture Notes in Production Engineering, DOI: 10.1007/978-3-319-12301-1_2

1 Conveyor Systems in Open Pit Mines

Conveyor belts have been used in German open pit mines since the 1950s. Due to the comparably low operational costs and the possibility of continuous transportation, the conveyor belt quickly replaced train conveyors [Hartmann 2002][Stoll 2009]. Furthermore, in comparison to train conveyors the new belt conveyors could be applied to higher inclinations [Kunze 2002].

A conveyor belt system usually consists of two major conveyor drums of which at least one is a drive pulley (Figure 1). These drums are situated on the deflection points of the conveyor. Moreover, segments which are built in transfer direction are situated between driving (4) and return station (1). These segments (2) are used to support the belt and prevent it from sagging. In smaller mines the material is loaded onto the belt directly from the excavator. In larger mines the raw material is charged with a feeding chute (3)[Stoll 2009].

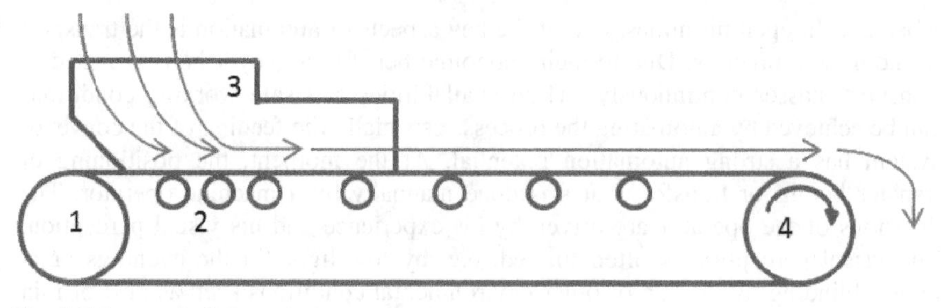

Fig. 1 Scheme of a belt conveyor

Since the belt has to meet all requirements regarding lifetime and functionality, the belt itself is the limiting part of the belt conveyor. Wear and tear of the conveyor belt is the limiting criterion related to economics. Hence, poorly operated systems will waste a significant amount of money.

Literature depicts various reasons for belt misalignments. In this paper, the focus is on misalignment caused by the loading unit. Normally, the loading unit operator ensures that the belt stays aligned on the conveyor. Belt misalignment leads to massive damages on the belt's edges and can also lead to material loss or material migration to the belt underneath. As a result, disturbances in the sequence of operation can occur.

A hopper car as shown in Figure 2 can be driven on the loading unit independently. By positioning the hopper car, one can influence the alignment of the conveyor belt and the position of the feeder boom (Figure 3).

Nowadays, the loading unit driver adjusts the feeding position manually. Therefore, he relies on experience as well as sight and sound. Due to harsh conditions in the open pit mine appropriate positioning is not always possible. If the feeder position is not centered, the weight and the impulse of the bulk material hitting the conveyor belt during charging force the belt to misalign. The misalignment increases with raising deviation of the feeder position from the middle [Seeliger 2002].

Fig. 2 Hopper car

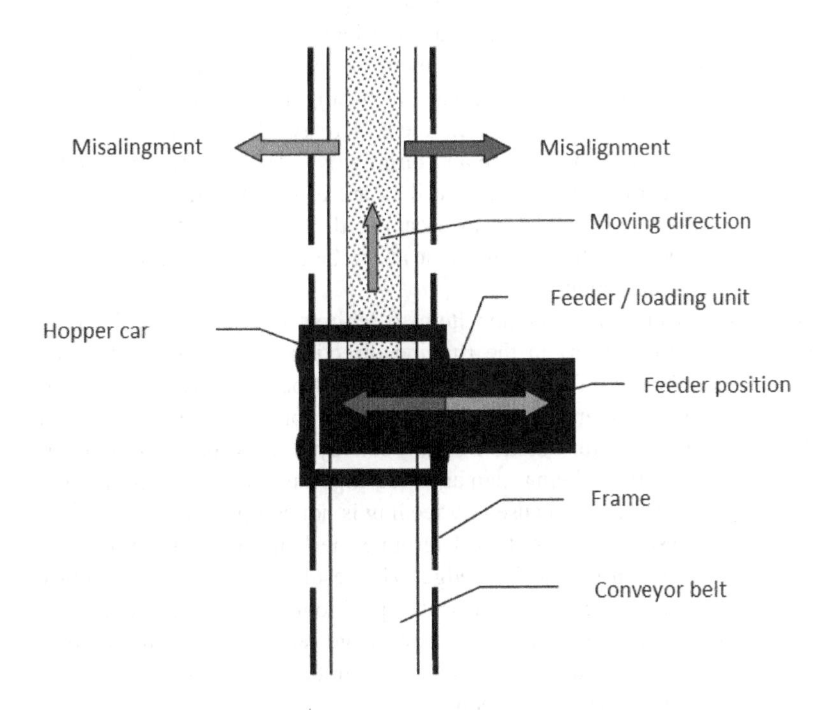

Fig. 3 Misalignment caused by loading unit and hopper car [Seeliger 2002]

During production, the hopper car is moved depending on the loading unit's position. Moving the hopper car is necessary if the maximal charging range is reached. Changes in the feeder position have to be corrected in order to prevent misalignments.

Besides the positioning of the boom which is done by the loading unit driver, there are a number of material parameters influencing misalignments on the conveyor belt. Changes in density and different rebound properties at the impact plate as well as water content of the material can considerably influence the position of the belt [Seeliger 2002].

To detect misalignments during production various systems were developed. These systems will be introduced in the following.

2 Existing Systems for Skewing Detection/Prevention

Systems for skewing detection can be divided into the following groups:

- Mechanical systems
- Ultrasonic systems
- Optical systems

Mechanical systems cover by far the biggest group of detection systems. Devices used for skewing detection are mechanical switches which are mounted next to the belt. In case of a misalignment the switch is pulled. Modern systems have got two phases. In the first phase of misalignment an alert is triggered, but the belt is still running. If the misalignment increases the mechanical system stops the belt to prevent it from damage and turning over. Another belt tracking method is the employment of rollers. These rollers are installed in recurring distances to guide the belt and prevent it from escaping the slide way. Using guide rollers causes stress and, therefore, higher tear at the belt edges. However, this measure does prevent misalignment.

A tracking roller is a common item to redirect the belt passively. If the belt misaligns, the roller turns in the running direction guiding the belt back to the center. This is manly influenced by the weight of belt and material as well as the friction between belt and tracking roller. This roller is usually installed right before driving- and return station of the conveyor to ensure a straight turn and prevent the belt from being damaged during the turning process. From this follows, that misalignment caused by feeding is not compensated.

Ultrasonic systems are used to determine the belt's position by detecting the distance between sensor and belt edge. The results of this measurement can be influenced negatively, due to poor sensor positioning. Furthermore, a damage of the belt edges leads to mistakes. Damaged edges are not reflecting the ultrasonic waves to the sensor, but scatter and dampen them. Therefore, no signals are detected and false information is generated.

Optical systems comprise the last group of belt tracking systems. Due to modern camera and video processing technologies, belt misalignment can be detected with ease. Unfortunately, optical cameras are strongly influenced by environmental conditions. Especially dust, rain and poor light conditions can cause false results or make it impossible to measure the belt track.

3 Principles of Infrared Thermography

Infrared radiation is part of the electromagnetic spectrum located between the visual light and microwave/radio radiation with a wavelength between 0.78 and 1000 μm (see Figure 4). All bodies with a temperature above absolute zero (0 K or -273.15°C) emit electromagnetic radiation due to molecular movements. The wavelength of this emitted radiation is dependent on the temperature of the object and shortens the higher the temperature of the object gets. For ambient temperatures, the maximum of this emitted radiation is located inside the long-wavelength infrared (LWIR or LIR) spectrum between 8 and 12 μm while for higher temperatures, the spectral emission shifts towards the visual light, according to Wien's displacement law. [Schuster 2009] The radiation is detected by infrared-sensitive sensor arrays and visualized in images, similar to digital cameras. Therefore, detection of this radiation is also referred to as thermography or thermal imaging since it (mainly) mirrors the temperature of observed objects. The images contain information of relative temperature differences, with the coldest part of the image shown in black while the hottest point inside the image is white. [Nienhaus 2010]

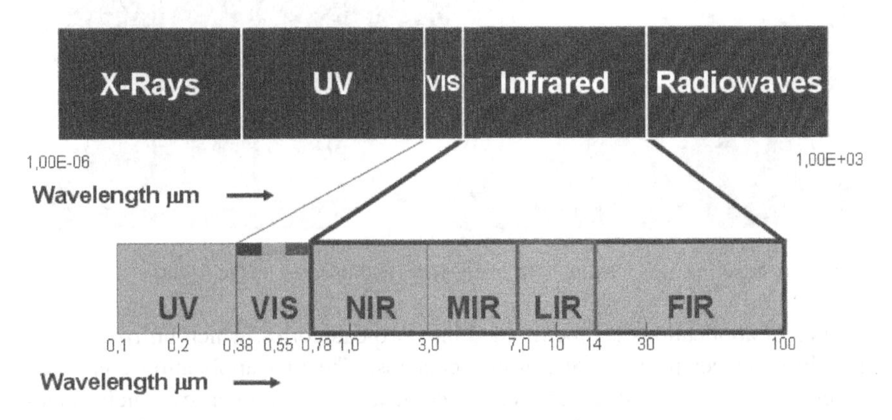

Fig. 4 Allocation of infrared radiation in the electromagnetic spectrum

Compared to visual light detection in digital cameras, the main advantage of thermal imaging is the robustness towards environmental influences such as bright light and small particles in the surrounding atmosphere (like dust or water particles). Small particles are penetrated by long-wavelength infrared radiation and are thereby no source of influence for the resulting image. The advantages of infrared imaging can be seen in Figure 5. This figure shows the sight of a visual light camera and an infrared camera onto the feeding process in the hopper car. While the visual camera is blinded by the sun and sight ends at the exit of the hopper car, it is possible to see the loaded, outgoing belt over a long distance in the infrared image. Additionally, the contrast between the material and the hopper car parts is almost non-existent in the visual light image, which makes it difficult to see the edges of the material. In the infrared image, a clear differentiation between material, hopper car and conveyor belt is possible. Due to these advantages, usage of infrared cameras is highly recommended in the mining industry, compared to other imaging methods. [Mavroudis 2011]

During the last decade, the progress in modern electronics and the corresponding miniaturization of devices have led to a rapid cost decay for infrared sensors, which eliminates the main disadvantage of thermographic cameras. Remaining disadvantages compared to digital cameras are the reduced (geometrical) resolution and availability of complementary equipment e.g. lenses for customization of aperture angles. Modern thermographic cameras for industrial usage have a resolution of about 640x480 pixels and a noise-equivalent temperature difference of less than 50 mK. For mining applications, the use of robust sensors able to withstand vibration and shock is necessary. [Mavroudis 2011]

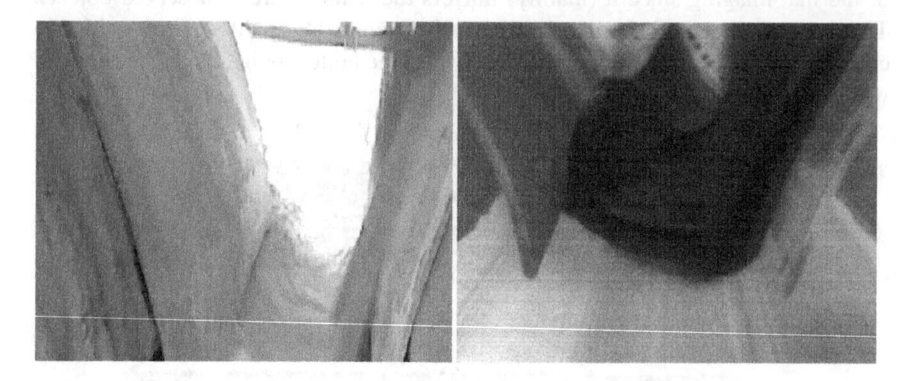

Fig. 5 Comparison between visual light image (left) and infrared image (right)

Fields of applications can be divided into two categories, which utilize the two main different perspectives of infrared cameras. The first application comprises the detection of thermal differences to detect features and properties which cannot be seen in images from visual cameras. As an example, these thermal features can be used for an indirect differentiation of material, depending on the temperature or thermal conductivity. The second application is image acquisition under rough conditions, which would render visual cameras useless. The independence on external light sources makes night vision devices based on infrared possible, while the insensitiveness towards dusty or foggy environments lead to applications in mining, e.g. boundary layer detection for shearer loader automation underground [Nienhaus 2010].

4 Application of Thermal Cameras for Skewing Detection and Prevention

In order to overcome the difficulties of current applications for belt skewing detection, imaging systems can be used to obtain a view of the position of the belt. By detecting the position of material on the belt and the position of the belt in comparison to the frame of the conveyor system, skewing can be detected and, hence, prevented. While observing the loading process, misalignment during the loading can be detected as well. As already stated, the main disadvantage of visual cameras is the vulnerability to all environmental influences present in an open pit mine. Therefore, infrared cameras were chosen for the given task.

The positions of the installed infrared cameras were chosen to cover the three relevant steps of the process: the incoming belt position, the position of loading and the outgoing belt (see Fig. 6). The incoming belt is observed to detect potential misalignment already present before loading, while the other two cameras are used to monitor the loading process and its influence on the belt position.

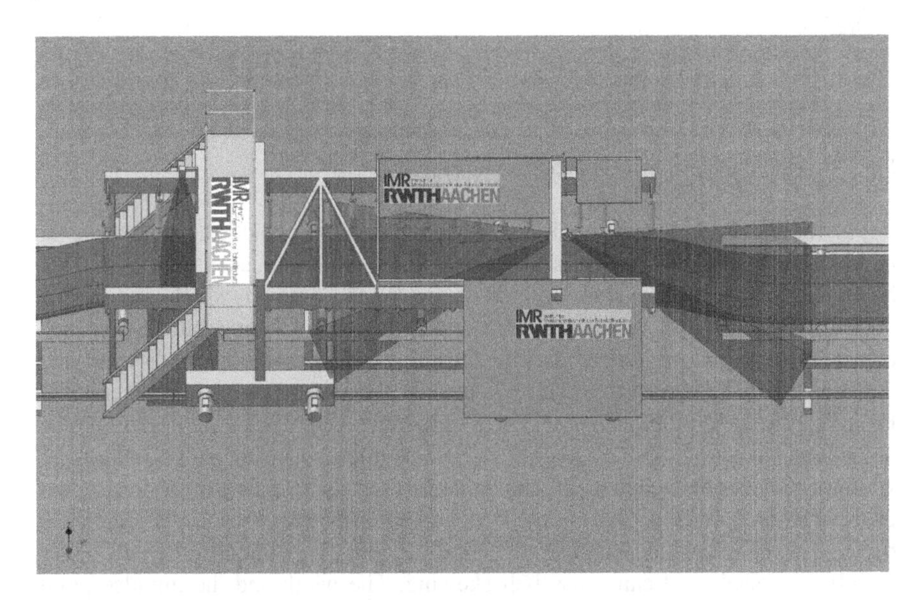

Fig. 6 Position of the thermal cameras on the hopper car

With the aim of preventing belt skewing, the images are processed in order to find relevant features capable of describing and analyzing the loading process. Following, two example pictures of the infrared cameras are depicted after algorithms were applied. Additionally, a short description of the used algorithms and key points inside the images are given.

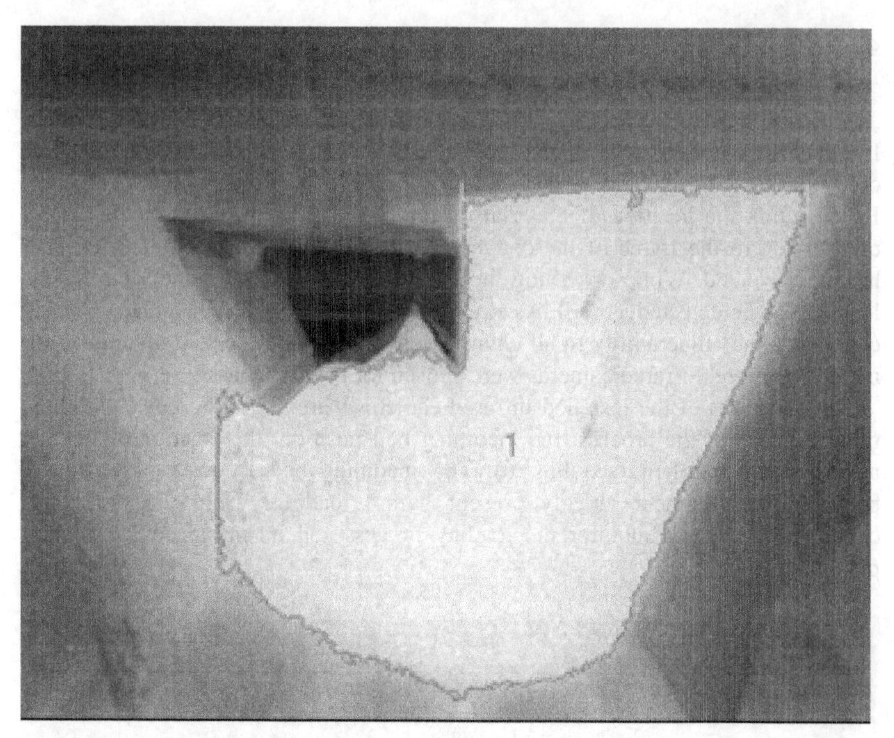

Fig. 7 Processed thermal image of the loading process

For the different camera positions, different features inside the image are of interest. For the camera which is monitoring the loading process (see Figure 7), a blob detection algorithm was applied to detect the area in which material is present. This blob detection is used to find the edges of an arbitrarily shaped object inside an image. In this case, this object is the material which is loaded into the hopper car. The result of the blob detection is the orange edge inside Figure 7. From this detected blob, different points of interest were derived to determine the amount and position of the loaded material. On the upper part of the image, the width of the material is detected, which can be used for an estimate on the amount of the material. Additionally, on the belt's surface, the impact point is identified, because in this point, the impulse of the falling mass is applied to the belt. This impulse, if misaligned can cause belt skewing. The width and the impulse point are depicted in green inside the image.

For the other two cameras, different points of interest were chosen. For the determination of the position of the belt, the edges of the belt in relation to the frame and the position of material on the belt are of interest. In order to detect this features inside the image (see Figure 8), line detection algorithms were applied for vertical lines. There are multiple lines which can be detected inside an image. On the incoming side of the system, there are two lines for the edge of the belt and two lines for the beginning of the frame. On the outgoing side, there are two additional lines on the edge of the loading. The edges of these lines are highlighted in red, while numbers on the line correspond to the amount of pixels from one edge to the next and can be transformed into geometrical distances. If those edges are compared to the middle of the image, skewing of the belt can be detected. Analyzing the width of material enables estimation on the amount of loaded material.

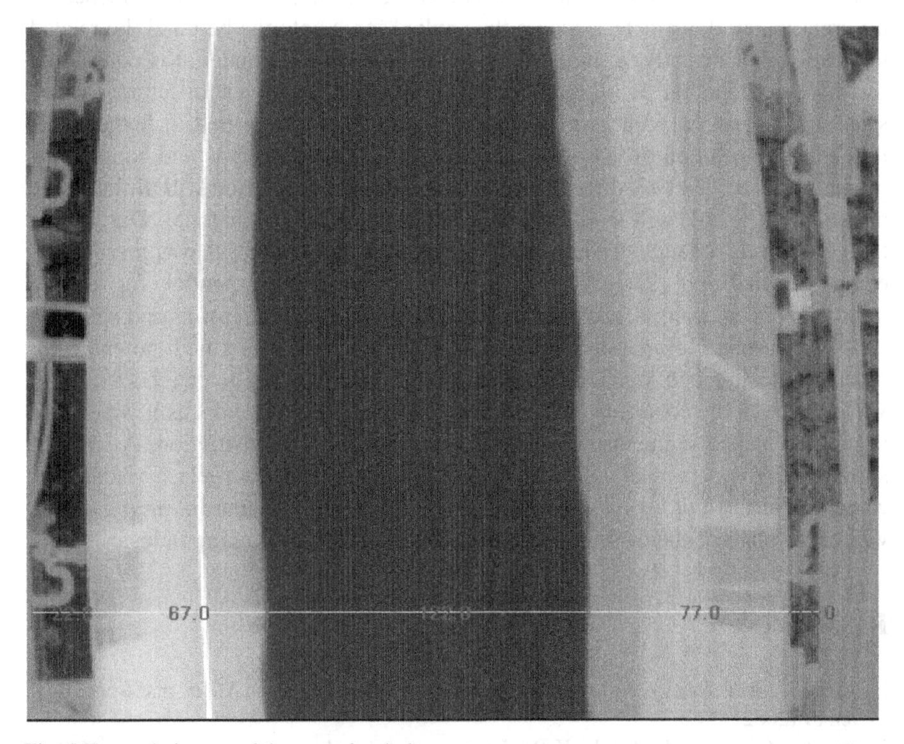

Fig. 8 Example image of the outgoing belt

In a second step, after the algorithms have processed the individual camera data, their results are correlated. Due to the different positions, there is a short time delay between the correlated images as the belt needs to pass through the hopper car. This time delay is constant and depends on the belt velocity. By correlating the three corresponding images measures to counteract skewing can be applied by changing the position of loading.

5 Summary and Outlook

The test installation with thermal cameras showed the general applicability of infrared cameras for the detection of belt skewing and off-centered loading. All important features of the loading process can be seen in the infrared images which are insensitive to the environmental influences. This makes infrared imaging superior to imaging with visual light cameras. Additionally, one of the main disadvantages of ultrasound sensors, the influence of wear on the belt edges is not only irrelevant for the detection of belt skewing with infrared, but can also be detected. This could be used for improved maintenance schedules to reduce the risk of major belt damages.

Next steps during this project include the determination of the influence of different material characteristics on the applied force onto the belt and the related belt skewing. Currently, only the positioning of the loaded material is considered for the determination of a misalignment in loading. However, different particle sizes of the material will result in different impulses on the belt and, therefore, influence the position of the belt in different ways. Also the type and consistency of material can lead to differences in belt skewing, e.g. rain will fluidize the material while frost will lead to bigger particles and harder material. During this test phase, only a small subset of possible material constellations was present and could be analyzed. Thereby, there might be a huge potential for further improvements when material changes due to the influence of seasons and weather.

The robustness in imaging with infrared technology shows a high potential for other applications in open pit mines. The thermal detection can be used for surveillance of the condition of support segments of the conveyors as it is possible to detect blocked segments which will heat due to increased friction. Assistance systems for the personnel can be installed for supervision and replace visual light cameras. Because of all its advantages, it is to be expected that infrared cameras will be the state-of-the-art imaging tool in the future for the mining industry.

References

Hartman, H., Mutmansky, J.: Introductory mining engineering. John Wiley and Sons, New Jersey (2002)

Kunze, G., Göhring, H., Jacob, K.B.: Erdbau- und Tagebaumaschinen. Vieweg Verlag, Braunschweig/Wiesbaden (2002)

Stoll, R.D., Niemann-Delius, C., Drebenstedt, C., Müllensiefen, K.: "Der Braunkohlentagebau: Bedeutung, Planung, Betrieb, Technik. Umwelt". Springer, Heidelberg (2009)

Seeliger, A., Schwoon, O., Müller, B.: Automatisierung von Belade- und Bandschleifenwagen. Interner Bericht Institut für Bergwerks- und Hüttenmaschinenkunde RWTH Aachen (2002)

Nienhaus, K., Mavroudis, F., Warcholik, M.: Coal Bed Boundary Detection using Infrared Technology for Longwall Shearer Automation. In: Twentieth International Symposium on Mine Planning and Equipment Selection, MPES (2010)

Schuster, N., Kolobrodov, V.: Infrarotthermographie. Wiley-VCH (2009)

Mavroudis, F.: Infrarotsensorik zur Grenzschichterkennung. Zillekens Verlag, Stolberg (2011)

Maintenance of Belt Conveyor Systems in Poland – An Overview

Radoslaw Zimroz[1,2], Monika Hardygóra[1,2], and Ryszard Blazej[2]

[1] KGHM CUPRUM Ltd CBR Sikorskiego 2-8, 53-659 Wrocław, Poland
[2] Machinery Systems Division, Wroclaw University of Technology, Na Grobli 15, 50-421 Wroclaw, Poland
{monika.hardygora,radoslaw.zimroz,ryszard.blazej}@pwr.wroc.pl

Abstract. The tendency to increase efficiency and safety in mining industry leads to quick development of monitoring, diagnostic and management systems. It can cover different aspect of mining activities. In this paper recent achievement in Polish mining industry related to machinery systems will be discussed. Selected solutions deployed by research teams from Wroclaw University of Technology in both opencast and underground mining companies in Poland will be briefly presented and discussed. The purpose of this paper is to provide current state of the research and practices focused on maintenance of belt conveyors. In particular, among others, following projects will be overviewed :i) Online monitoring system for drive units developed for lignite mine, ii) Portable monitoring system for conveyor belt with steel cords developed for lignite mine, iii), Diag Manager – maintenance system for belt conveyor transport network, iv) Laboratory and in situ research on belt conveyor technology (belts, joints, idlers resistance to motion, energy efficiency, etc). The Authors will try to summarize challenges in this field and point out future work at the end of the paper

Keywords: belt conveyor, condition based maintenance, nondestructive testing, condition monitoring.

1 Introduction

Nowadays every company should compete with other players on the market in terms of efficiency, quality of products, production cost etc. Any troubles (breakdowns) with often complicated, unique machinery systems used in the company is very undesirable due to financial loses as well as safety of the staff and environmental issues. It was reported by Babiarz and Dudek [1] that due to unexpected breakdowns several heavy duty machines in Polish opencast lignite mining was destroyed completely and other required significant investment to rehabilitate machine or used mechanical systems. Similar problem might be find in underground mining (hard coal, copper). What kind of reasons have been occurred in mentioned cases? The most spectacular have been related to basic machines in lignite mine industry as bucket wheel excavator (planetary gearbox failure – several

C. Niemann-Delius. (ed.), *Proc. of the 12th Int. Symp Continuous Surface Mining - Aachen 2014*, Lecture Notes in Production Engineering, DOI: 10.1007/978-3-319-12301-1_3

month of production breakdown, BWE boom failure, burning of machine due to belt ignition, etc).

In this paper we would like to be focused on belt conveyors in both opencast and underground mines. Critical problems for this mechanical system are related to belt (longitudinal intersection, joint failure), pulleys (damages of coating, bearings, shaft) and other element of drive units (damages of electric motor, coupling, gearbox). Due to mechanical contact of pulley' coating and belt, it has sometimes happen that temperature is increasing and belt ignition might initiate. This is very dramatic situation for underground mines and usually results in serious consequences in terms of organization of production in the mine, investigation by local mining authority, financial losses for mine and managers, unfortunately it might me also affect safety of mining staff.

Any failure of the machines in the mine might be very difficult to repair due to sparse parts and experienced staff availability, environmental factors etc. To avoid such situations, to maintain effectively machines and to manage of maintenance process, many project have been launched in last decade in Polish mining industry. The purpose of this paper is to list and review them as well as to synthesize current effort to project what would be important in the near future in Poland.

A current state of the research and practices focused on maintenance of belt conveyors will be discussed, in particular, among others, following projects will be overviewed: i) Online monitoring system for drive units developed for lignite mine(2010), ii) Portable monitoring system for conveyor belt with steel cords developed for lignite mine(2013), iii), Diag Manager – maintenance system for belt conveyor transport network (2011 and 2013), iv) Laboratory and in situ research on belt conveyor technology (belts, joints, idlers resistance to motion, energy efficiency, etc). Mentioned industrial project have been deployed by research teams from Wroclaw University of Technology, to provide whole picture regarding situation in Poland, other important implementation in both opencast and underground mining companies in Poland will be recalled as well.

2 Belt Conveyor and Belt Conveyor System

To realize how important problem is discussed in the paper a brief description of conveyor and example of conveyor based transportation system in the mine will be presented. Fig 1a presents simplified diagram of a belt conveyor, which consists of: a drive unit(s), a belt loop, idlers and some auxiliary elements. From the very first view it doesn't look so complicated. However, the problem with maintenance of belts, drive units, idlers etc is related to specificity of mining industry. Today over 500 km of conveyor belts are installed in 4 lignite surface mines in Poland, even more conveyor belts are installed in underground mines. In this paper research is focused on belts and drive units. Different loading, harsh environmental conditions, large dimensions and weight of conveyor components, high power required to transport materials etc require innovative approaches for belt conveyor maintenance management. It is important to notice, that elements of conveyor-based system are located on large area of mine (so called spatially distributed system) that requires also specific logistic procedures (Fig2). Taking into account all these factors and different detailed problems of given conveyor component, dedicated systems have to be developed, and finally information from them should be integrated to make a decision on maintenance action.

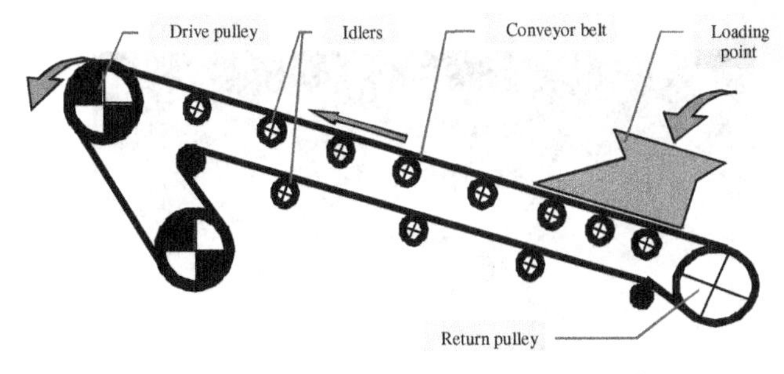

Fig. 1 Simplified diagram of a belt conveyor

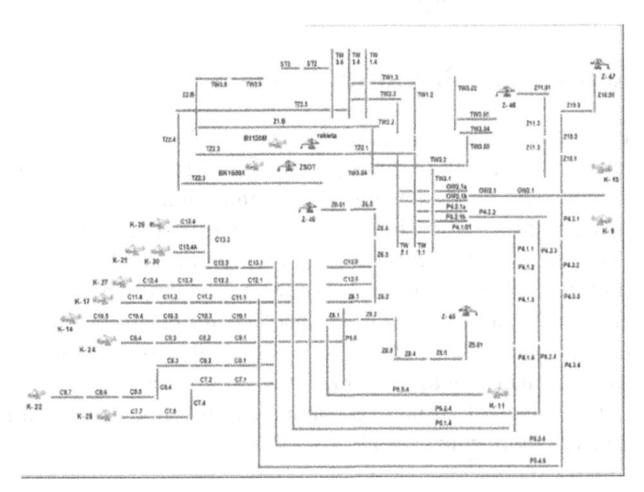

Fig. 2 Example of belt conveyor based transportation system for one of lignite mine in Poland

3 Belt Conveyor Laboratory

Transportation system in the mine is a key component in production chain in the mine. There is very undesired by the miners to do any "passive" experiments that might affect production efficiency. Any "active" experiments (introducing damage) to test efficiency of monitoring system are not possible due to financial, organizational and safety reasons. In such a case the only possible solution is to play with test rig. Such a conveyor for testing was built several years ago in our laboratory to test diagnostic systems. The belt is 17m length, 40 cm width, conveyor is equipped with speed control system, unfortunately, there is no load applied to the conveyor (no material stream). It should be highlighted that it cannot replace final test in situ, however it significantly decreased cost of testing belt diagnostic systems, vibration monitoring system, infrared thermography based diagnostic procedures. Other equipments to identify parameters of belts or their joints have been described in [2]

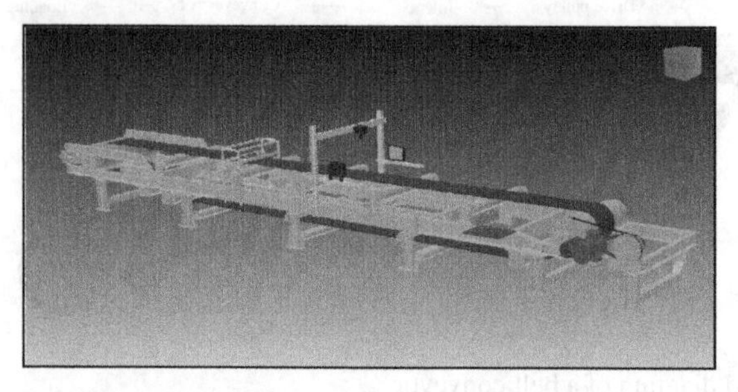

Fig. 3 A test rig used for monitoring/diagnostic oriented research

4 Monitoring and Diagnostic Systems for Belt Conveyor

In this chapter following subsections will discuss: i) Online monitoring system for drive units developed for lignite mine (2010), ii) Portable monitoring system for conveyor belt with steel cords developed for lignite mine (2013), iii), Diag Manager – maintenance system for belt conveyor transport network developed for underground copper ore mine (2011 and 2013). Again, number of element, their different importance in the system, motivates us to develop different solution: portable diagnostic system that has to be installed for each measurement and deinstalled after that and online, stationary system. Reasonable decision making should be based on many factors, including average lifetime, loading condition, damage type, importance of monitored component in whole chain and finally cost of single stationary monitoring system

4.1 Diagnostic System for Conveyor Belts

A key difference between condition evaluation of belts and other object is that length of the belt could reach several kilometers and the belt is moving with speed equals approx. several m/s. Moreover, number, size and density (along belt length and width) of damage could be very different and non-uniformly distributed. It is due to the fact that belt' loop consists of sections with different length. So in other words there is no other option for using automatic monitoring and diagnostic system. However, specific practices in belt operation requires dedicated solution that need to be implemented in the system ("beginning" of the loop, differentiation between damage and joint, influence of belt speed, etc.). All these issues have been taken into account in system presented in Fig 4 – hardware and software part are presented, respectively [3, 4].

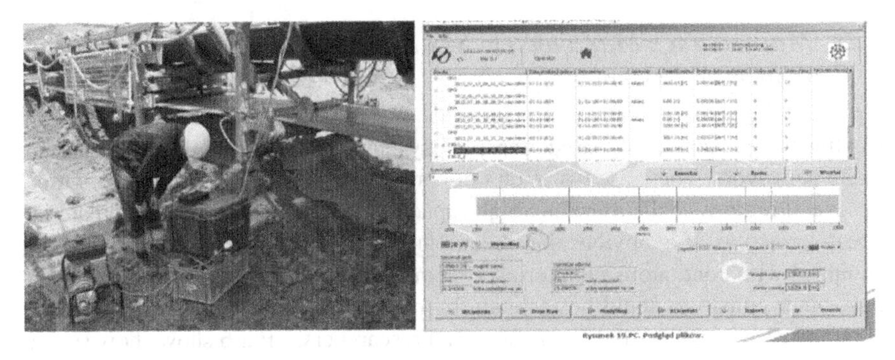

Fig. 4 Diagnostic system in operation: a) during measurement in the mine, b) view on main window of developed software

Fig5 shows example of results obtained for conveyor operating in lignite mine. Fig5a shows spatial distribution of detected damages (along width and length of the belt, note that joints are not visualized here), while Fig 5b-d aggregated results describing number of damages vs belt width, length and finally sum of amplitudes of detected damages vs belt width.

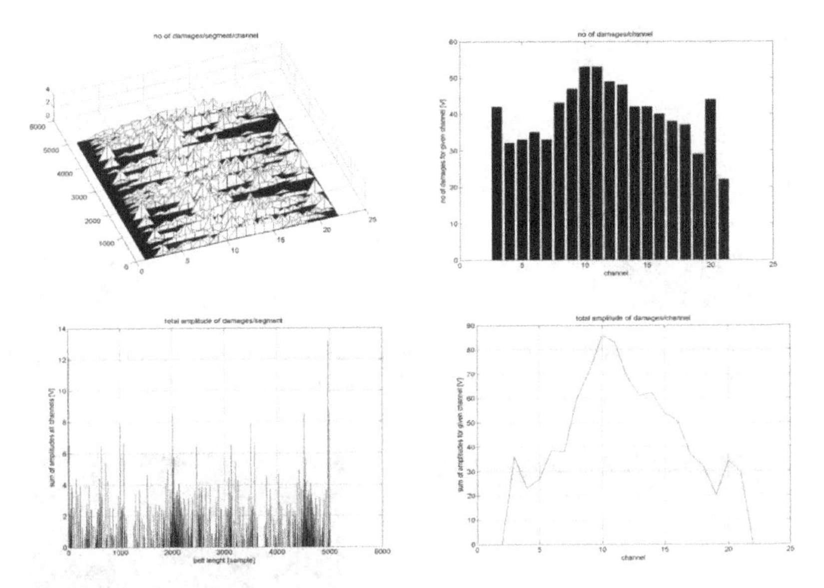

Fig. 5 Visualisation of diagnostic results: a) spatial distribution of detected damages, b) histogram no of damages per channel ,c) distribution of damages vs. belt' length, d)distribution of amplitudes of damages per channel,

From Fig 5 one might extract knowledge about size number and distribution of damages that should be basis for decision making in modern maintenance management. More information about system could be found in [3,4]

4.2 Maintenance Management System for Underground Belt Conveyor System

Continues belt conveyor based transport system form of complex structure of many conveyors. There are two group of conveyors: low capacity conveyors used in several regions of the mine and high capacity conveyors used for main material stream transport. It is obvious that second group is more important and should be equipped with continuous monitoring systems. Due to financial reasons and lower risk, other conveyor could be maintain by periodic diagnostic inspection. It was common conclusion made by engineers and researchers. Fig 6 shows how such a measurement is performed. Installation of the sensors, launching the software and data acquisition takes no more than a few minutes. The software installed on the laptop (fulfilling military requirements due to harsh environment in underground mine) automatically collect and process the data (Fig 7) and finally provides information on machine condition. The portable data acquisition system is also providing a data package for data base – kind of maintenance management software that is used in the office to prepare reports, to plan maintenance action and so on (Fig 8). More information are provided by [5,6]

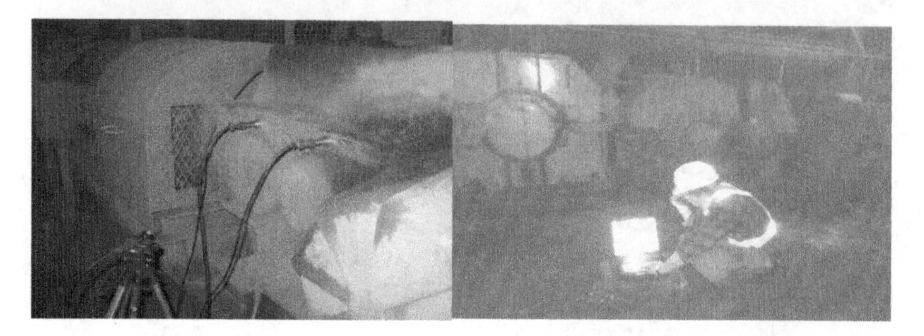

Fig. 6 Diagnostic data acquisition: a) view on sensors location (tacho probe and 2 accelerometers), b) laptop with appropriate software for data acquisition and diagnostics

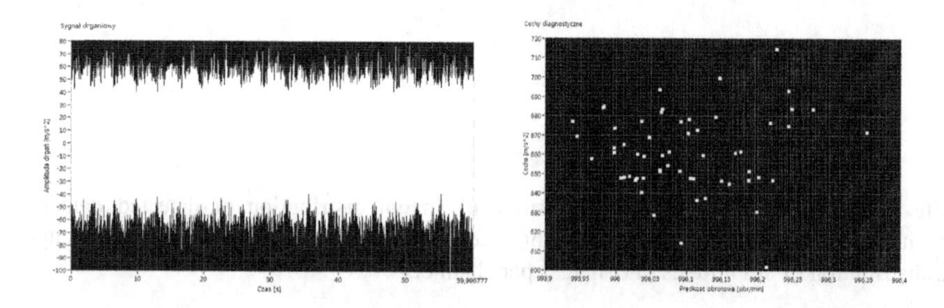

Fig. 7 Visualisation of measurement and feature extraction: a) raw vibration signal (60s), b) features extracted after signal segmentation – 60 features with reference to operating cond

•

Fig. 8 Diag Manager II – software for maintenance management of belt conveyor based transportation system

4.3 Condition Monitoring System for Lignite Belt Conveyor

As it was said, in the strategic part of transportation system (i.e. lignite to the power plant or main transport line directly to mining shaft), there is a serious need to assure very high reliability of belt conveyors. Due to tendency to make the conveyor operation automatic, a dedicated condition monitoring system have been developed and installed in the mine. Fig 9 show a picture of monitored drive station and scheme of belt conveyor with marked key components to be monitor. The system collect different type of physical variables (temperatures, currents, rotational shaft speed and vibration), process them, diagnose condition of engine, gearbox and pulleys components. The system follow modular/layer structure and consists of sensors, wire transmission to the Data Acquisition Unit (DAU, Fig 10a), WiFi transmission from DAU to Central Unit (CU), processing layer at CU (Fig 10b, 11a) and finally visualization, reporting and publishing (Web) layer (Fig 11b). More details about the system could be found in [7]

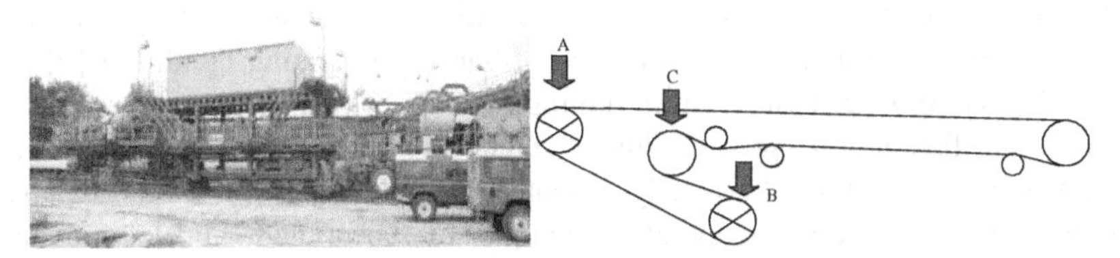

Fig. 9 a)View on monitored drive station, b) scheme of belt conveyor with marked key components to be monitor

Fig. 10 Hardware components developed for the project:a) data acquisition unit DAU, b) central unit CU with computer for control, visualization and web-acces services

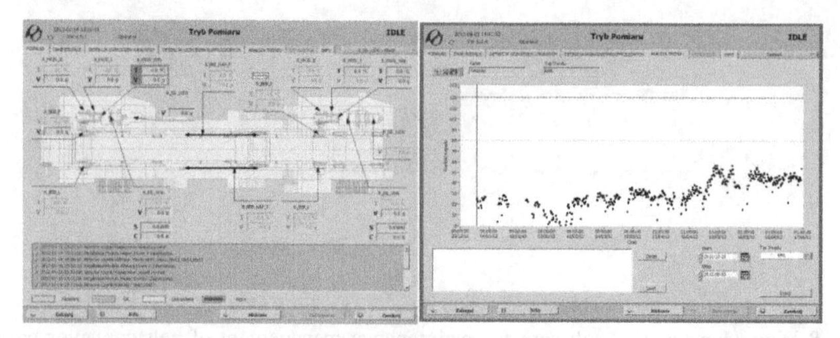

Fig. 11 Visualisation of diagnostic results: a) main window of user interface with all channels providing information about current condition, b) long term trend analysis

5 Trends in Monitoring and Diagnostic Systems for Belt Conveyor

Development of sensor and measurement technologies as well as data transmission and processing techniques make possible to apply very advanced, multichannel monitoring systems for maintenance purposes. Moreover, apart from measurement, it is often require to combine knowledge from measurements with knowledge about failure analysis, human factors influence, environmental impact and so on. All these information are nowadays available in modern corporations in various data bases and information systems. In authors belief, the future of maintenance management will be based on data/information and knowledge fusion. Problem of mining machines is that they operates in very harsh conditions, and problem of maintenance seems to be very complicated [8]. It is due to number of factors that should be taken into account. Additional problem that was recently deeply studied by Bartelmus is influence of varying operating condition [9] and unique degradation scenarios [10] in mining industry. This problem is partially related to data fusion concept suggested by Galar [11], because apart from current diagnostic features values, knowledge about operational condition should be taken into account Also history of operation (number of startup and shut down events, repair action etc.) might significantly modify results of condition assessment and prognosis. Research team lead by Barszcz [12] pointed out that due to harsh conditions, complexity of machines and influence of external factors, data collecting by automatic systems should be carefully interpreted and validated. It appears that many data acquired by condition monitoring systems doesn't contain any diagnostic information due to, for example, damaged sensor. Finally, the last key issue and in fact serious challenge for diagnostic is problem of decision making based on collected data. Decision scheme commonly used in monitoring system follow the "if then else" rule, that basically mean constant thresholds for warning or alarm [13-15]. It has been proved among others by Bartelmus [9] that in machines operated in time varying conditions values of thresholds should be load dependent and might be linear or nonlinear function of load. Mentioned challenges (feature-operating condition dependencies, validation, load dependent threshold, data fusion) seems to be universal problems both for rotating machines as well as belts condition monitoring. It should be noticed that there are other commercial solutions that also are used in mining companies ([16-18]) as well as other research institutes. As far as we know, at least in systems operated in mining companies, mentioned problems were not solved in systematic way. Thresholds are "adapted" according to engineering experience (what if machine is unique?), data fusion is not performed, varying load is not taken into account. A final conclusion could be formulated in optimistic manner: there is a lot of work to do with monitoring and diagnostic in mining industry from research perspective and fortunately, there are more and more mining companies that want to invest money in Condition Monitoring.

6 Conclusion

In the paper a review of last developments of condition based maintenance solutions dedicated to belt conveyors is presented. They have been prepared by members of Machinery Systems Division and prototypes were installed in opencast and underground mine. A brief description of basic functions implemented in these system is provided in appropriate sections. Finally, several challenges for future work are pointed out. They seem to be complicated task and according to authors knowledge many research teams work on them. Finally in the paper we point out several challenges that come out from many years of experience with mining machines, monitoring and diagnostics. Unfortunately they seem to be classical ones: experimental data validation, data processing, modeling, analysis, decision making, information processing and visualization.

Acknowledgements. This work is partially supported by Applied Research Programme (PBS): "Intelligent system for automated testing and continuous diagnosis of the conveyor belt" (2012-2015) – R. Blazej and KGHM Cuprum – M Hardygora and R. Zimroz

References

[1] Babiarz, S., Dudek, D.: Kronika awarii i katastrof maszyn podstawowych w polskim górnictwie odkrywkowym Oficyna Wydawnicza Politechniki Wroclawskiej (2007)
[2] Błażej, R., Jurdziak, L., Zimroz, R., Hardygóra, M., Kawalec, W.: Investigations of conveyor belts condition in the Institute of Mining Engineering at Wroclaw University of Technology. In: 23rd World Mining Congress, Montreal, Canada, August 11-15, pp. 1–9. Canadian Inst. of Mining, Metallurgy and Petroleum (2013)
[3] Błażej, R., Jurdziak, L., Zimroz, R.: Novel approaches for processing of multichannels NDT signals for damage detection in conveyor belts with steel cords. In: Key Engineering Materials, vol. 569/570, pp. 978–985 (2013)

[4] Błażej, R., Zimroz, R., Jurdziak, L., Hardygóra, M., Kawalec, W.: Conveyor belt condition evaluation via non-destructive testing techniques Mine planning and equipment selection. In: Drebenstedt, C., Singhal, R. (eds.) Proceedings of the 22nd MPES Conference 2013, Dresden, Germany, October 14-19, vol. 2, pp. S.1119–S.1126. Springer (2014)

[5] Zimroz, R., Król, R., Hardygóra, M., Górniak-Zimroz, J., Bartelmus, W., Gładysiewicz, L., Biernat, S.: A maintenance strategy for drive units used in belt conveyors network. In: Eskikaya, Ş. (ed.) 22nd World Mining Congress & Expo, Istanbul, September 11-16, vol. 1, Aydoğdu Of set, Ankara (2011)

[6] Stefaniak, P., Zimroz, R., Krol, R., Gorniak-Zimroz, J., Bartelmus, W., Hardygora, M.: Some remarks on using condition monitoring for spatially distributed mechanical system belt conveyor network in underground mine – A case study. In: Fakhfakh, T., Bartelmus, W., Chaari, F., Zimroz, R., Haddar, M. (eds.) Condition Monitoring of Machinery in Non-Stationary Operations, vol. 110, pp. 497–507. Springer, Heidelberg (2012), http://dx.doi.org/10.1007/978-3-642-28768-8_51

[7] Zimroz, R., Król, R., Hardygóra, M., Bartelmus, W., Gładysiewicz, L.: Condition monitoring system for drive units in belt conveyor. Transport & Logistics (Belgrade) 7, 342–346

[8] Hardygóra, M., Bartelmus, W., Zimroz, R., Król, R., Błażej, R.: Maintenance, diagnostics and safety of belt conveyors in the operations. Transport & Logistics (Belgrade) 6, 351–354

[9] Bartelmus, W., Zimroz, R.: A new feature for monitoring the condition of gearboxes in non-stationary operating conditions. Mechanical Systems and Signal Processing 23, 1528–1534 (2009)

[10] Bartelmus, W.: Gearbox damage process. Journal of Physics. Conference Series. 305(1), 1–9 (2011)

[11] Galar, D., Gustafson, A., Tormos, B., Berges, L.: Maintenance Decision Making based on different types of data fusion. Eksploatacja i Niezawodnosc – Maintenance and Reliability 14(2), 135–144 (2012)

[12] Jablonski, A., Barszcz, T.: Validation of vibration measurements for heavy duty machinery diagnostics. Mechanical Systems and Signal Processing 38(1), 248–263

[13] Cempel, C.: Limit value in practice of vibration diagnosis. Mechanical Systems and Signal Processing 4(6) (1990)

[14] Brooks, R., Thorpe, R., Wilson, J.: A new method for defining and managing process alarms and for correcting process operation when an alarm occurs. Journal of Hazardous Materials 115 (2004)

[15] Jablonski, A., Barszcz, T., Bielecka, M., Breuhaus, P.: Modeling of probability distribution functions forautomatic threshold calculation in condition monitoring systems. Measurement 46(1), 727–738 (2013)

[16] http://famur.com.pl/ - description of FAMAC VIBRO system

[17] http://www.pruftechnik.com.pl/, description of diagnostic systems from Pruftechnik

[18] http://www.ec-systems.pl/pl/, description of diagnostic solutions from EC Systems

Novel Techniques of Diagnostic Data Processing for Belt Conveyor Maintenance

Radoslaw Zimroz[1,2], Paweł K. Stefaniak[1], Walter Bartelmus[1], and Monika Hardygóra[1,2]

[1] Diagnostics and Vibro-Acoustics Science Laboratory, Wroclaw University of Technology, Na Grobli 15, 50-421 Wroclaw, Poland
{radoslaw.zimroz,pawel.stefaniak,walter.bartelmus, monika.hardygora}@pwr.wroc.pl

[2] KGHM Cuprum Ltd., Research & Development Centre, Sikorskiego 2-8, 53-659 Wroclaw, Poland

Abstract. In the paper a new diagnostic approach for gearbox used in belt conveyors will be discussed. The purpose of the work is to provide novel view on diagnostic data processing in the context of detection of changes in condition for population of gearboxes used in belt conveyor network. The idea will be presented by examples: a data base of diagnostic features collected during last 3 years (real data from conveyors operating in mining company) will be used for illustration.

The method takes advantage from recent results of research carried out by authors and other researchers related to different types of gearboxes used in mining and other machines, (i.e. belt conveyors, bucket wheel excavators, coal shearers, wind turbines and helicopters). A serious dependency between diagnostic features and operational conditions (speed/load) it is shown in mentioned works.

A novel research hypothesis has been formulated that behavior of machine in bad condition is unstable and it is more visible for heavy loaded machine.

It results with diagnostic data set with higher data dispersion than for healthy one. In the paper we will prove that feature load dependency and data dispersion might be a basis for novel approach for condition monitoring of gearboxes used in belt conveyors. An advantage of such approach is its simplicity and strong physical background.

Keywords: belt conveyor, diagnostics, novel features, data processing.

1 Introduction

Fundamental basis for decision making in condition monitoring is that given set of diagnostic parameters (values of vibrations, temperatures, pressures etc.) is dependent on change of machine condition and independent from the rest of possible factors. It is well-known nowadays, many of machines operate under time varying load conditions (TVLC) and classical parameters (called features) might be (in fact they are) load/speed dependent [1,2,4]. It means that our diagnostic decision

C. Niemann-Delius. (ed.), *Proc. of the 12th Int. Symp Continuous Surface Mining - Aachen 2014*, Lecture Notes in Production Engineering, DOI: 10.1007/978-3-319-12301-1_4

should be function of two groups of variables *D=f(operating conditions descriptors, machine condition descriptors)* and reasoning process become complicated. Many researchers try to collect classical data and process them using advanced data mining techniques to cancel out influence of TVLC. For example there are papers where application of PCA [3,4] is advised. Also neural-network-based approaches are used. Search for load independent, fault sensitive features is difficult direction of research and still seems to be a challenging issue. Other idea, proposed by [1] was to extend a bit classical approach and to take into account second parameter(s) influencing decisions. It has appeared that simple, classical features are very complicated to interpret alone for TVLC case, but combining with descriptors of TVLC become again very intuitive and simple.

Following based on assumptions of [1] we would like to propose a procedure that help us to understand the data we acquired from many gearboxes used in belt conveyor drives. They operate under different external load, they are in unknown condition and problem of proper decision making (accept to continue further operating or stop the machine?) is still open. Unfortunately, actual condition of machines, pre-defined decision threshold and methodology how to deal with such problem hadn't been precisely defined until now, thus we didn't have any key or reference during analysis. Due to relatively short observation (approx. 1 min) and actually approximately constant loading conditions during 1 min of measurement, the concept proposed by [1] cannot be applied directly. However, knowledge from that paper, number of objects and some new ideas allow us to propose a method to estimate what group of machines is in good, warning or bad condition, respectively. The paper is organized as follows: a short review on the subject and current practice in the mine will be described; based on preliminary analysis, some remark and assumption will be formulated and novel procedure will be presented and discussed; finally, application of the method will be provided.

2 Problem Definition

Machine condition monitoring means collecting signals (vibro, temperature, current, etc.) and compare actual values with threshold. Progress in sensors data transmission/storage technologies make possible quick development of CM systems. Surprisingly, it appeared that there are just a few papers discuss a problem of decision threshold estimation in context of condition monitoring [9-11]. The ideal situation is when two data sets (good/bad conditions) are available. Then using statistical data modeling or neural networks so called "classification system" might be trained and validated using appropriate data sets. Unfortunately, in our case, there are several facts that make this problem very different. First, three (not two) classes are expected (good/warning/bad). Second, there are many data in unknown and very different (from very good to very bad) conditions' cases. Third, we have a large dataset but for many objects, for certain gearbox there are only 60 samples of diagnostic feature. All these data are stored in kind of decision support system developed for maintenance management which has been used nowadays by the maintenance staff [7]. Computerised systems for maintenance management is a popular way for machinery management [5,6]. Values of threshold established in the system are probably too high, and they don't depend on operating conditions (load) that might be source of misleading decisions.

3 The Approach Used in the Industry and Recent Research-Based Solutions -A Discussion

A condition based maintenance approach were applied for discussed system in 2009. Since then many measurement have been done (Fig. 1a). A key problem is to establish two values of threshold: for warning and alarm levels. Currently, in diagnostic Decision Support System (DSS) constant values of thresholds are used as presented in Fig. 1b [7].

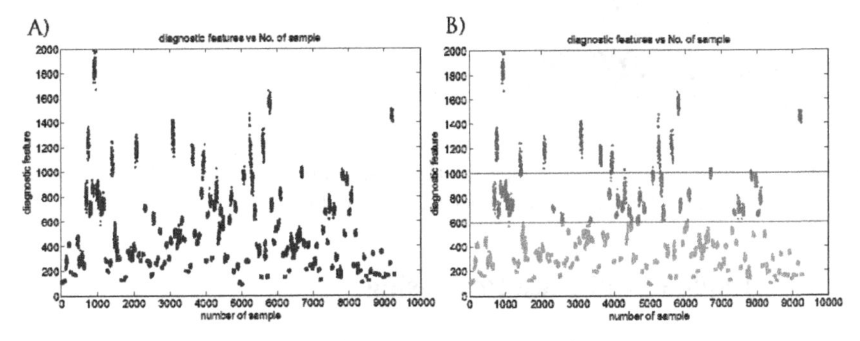

Fig. 1 Input dataset and thresholds used by the DSS in the mine

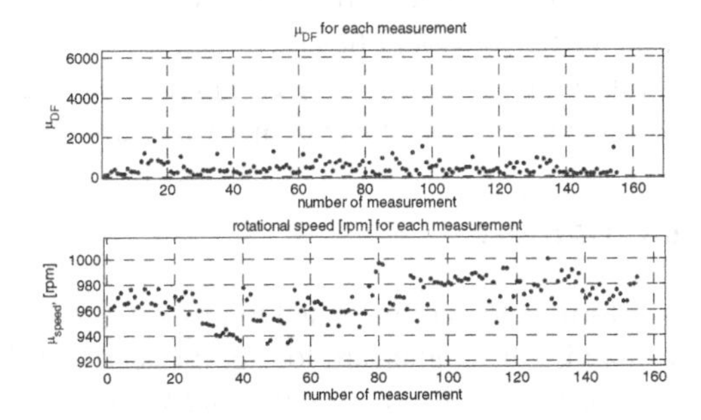

Fig. 2 Mean values of diagnostic features and speeds for all data sets (for each gearbox)

From the literature [1,2,13] it is known fact that one should expect two phenomena in such data set: i) operating conditions expressed here as mean value of instantaneous speed should be time varying, ii) features should be somehow correlated with speed variation (should depend on speed). At least first condition is clearly visible from Fig. 2: speed varies from 930 to nearly 1000rpm.

It means that one should analyze diagnostic data with reference to operating condition (speed here). It can be considered as the simplest form of data fusion as recommended in [4]. Fig. 3 shows representation of data in 2D space (2a) and visualization of decision boundaries according to Fig. 1 (thresholds does not depend on speed). However, it is known that in such a case due to feature-speed dependency, values of thresholds should not be constant, they should be adaptively calculated for actual speed value. Unfortunately, it requires kind of training process for extraction of knowledge about properties of bad and good datasets.

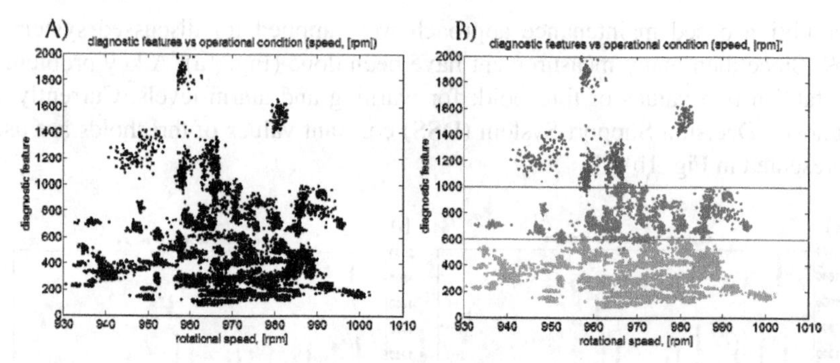

Fig. 3 Representation of data in 2D space feature –speed (as recommended in [1]): a) raw data, b) visualization of decision boundaries according to Fig. 1

4 Methodology of Data Processing

Before a novel approach will be presented, a few figures with preliminary analysis will be discussed. In fact, these observations have been a basis to define new approach presented in the paper. It should be noted, that it is strongly linked to previous works lead by Bartelmus and Zimroz [1,2,13] related to diagnostics of gears used in bucket wheel excavator, belt conveyor and wind turbine.

4.1 Preliminary Analysis

A first remark we will formulate here is related to properties of data clouds. Some of them are very scattered, while other are located in nearly in the same coordinates on feature-speed map. What is a reason of that? We believe, that it could be related with technical condition of machine, and kind of its "stability" of operation. Machine in good condition should manifest following properties: small value of feature (classic approach) and small scatter. There are two scenarios for degradation: machine could increase its feature according to wear development (and scatter but not so much) or could increase its scatter more than value of the feature. Fig. 4 shows mean values of data clouds (as blue *) and values of max and min values of features (represented as red circle). We propose to introduce simple parameter as a measure of data cloud scatter, namely *Max-Min = max(DF)*

– *min(DF)*. *Max-Min* values are shown in bottom subplot, Fig. 4. Next, let's have a look in Fig. 5, where *Max-Min* is plotted vs. diagnostic feature DF (left subplot) and speed (right subplot), respectively. This figure will be a basis for the method proposed in the next section.

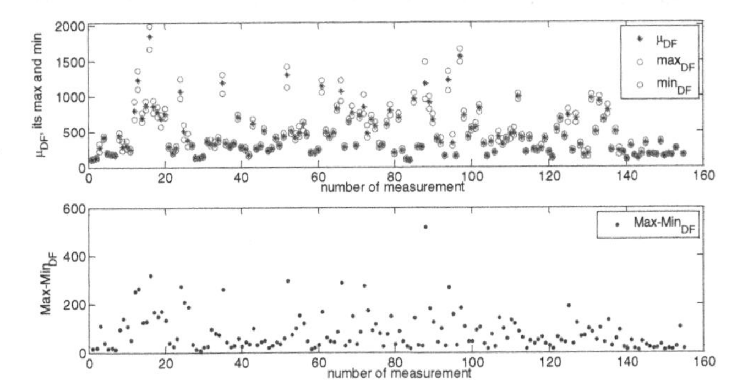

Fig. 4 Mean values and scatter (*Max-Min*) of features for each measurement

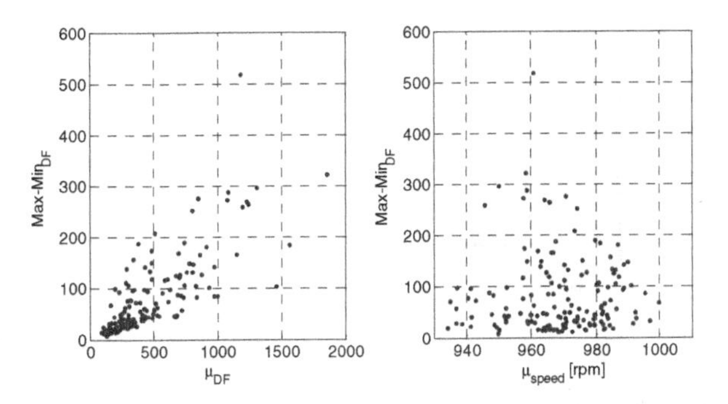

Fig. 5 a) Relationship between dispersion of features and their mean value for each measurement b) Relationship between dispersion of features and mean speed

Detailed analysis of Fig. 5 will allow us to notice two characteristic areas (with small features and scatter and big features and scatter) corresponding to two machine conditions, moreover, right subplot show how scatter depends on speed for healthy and worn machine in very similar way as in [1,13]

4.2 *Proposal of a New Approach*

In this section a proposal of complete procedure describing how to process data to estimate decision boundaries will be presented. As it was said, the procedure is based on experience with mining machines, previous publications and some observation related to this specific data set. It might be possible, that this approach is valid for very specific data from mining machines only. The FlowChart of the method is presented in Fig. 6. According to this FlowChart, first one should check basic properties of data, their variability, *feature-speed relationship* and so on. If dataset doesn't reveal mentioned properties, there is no need (and even more - no possibility!) to use our approach. Next step by step we try to understand our data and introduce constraints related to scatter, values of features, scatter-speed relation and we visualize iteratively results using 3 colors in feature-speed plane for validation.

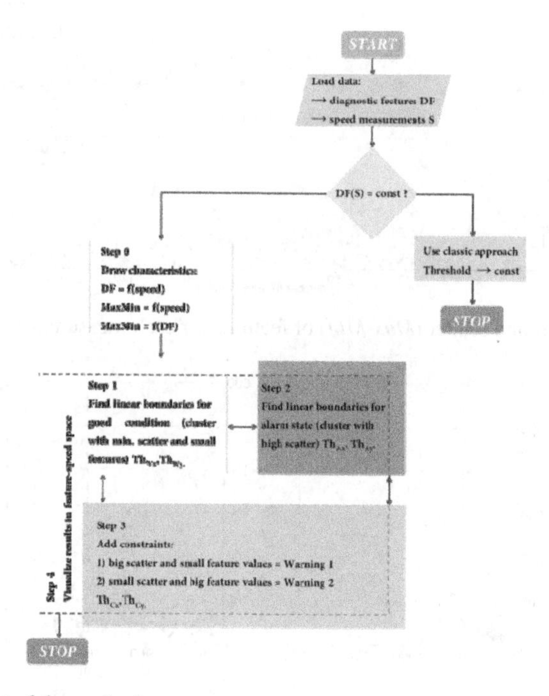

Fig. 6 FlowChart of the method

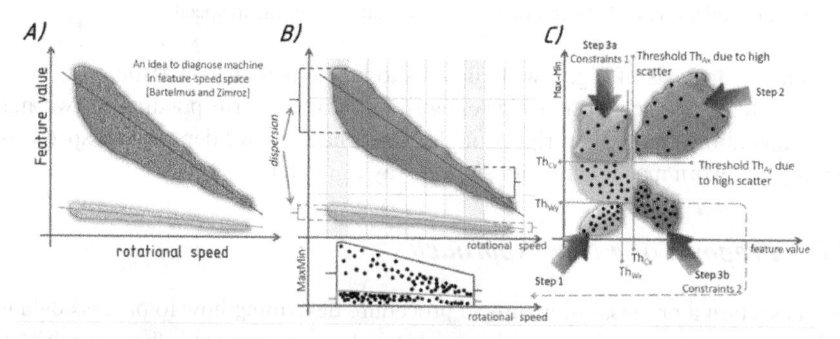

Fig. 7 Supplementary explanation to the FlowChart

As it was said, machine in good condition should manifest following properties: small values of features (classic approach) and small scatter. In Step 1 we identify a cluster of data (representing set of machines) with such properties.

Second Step is related to preliminary identification of dataset with very big scatter and big values of diagnostic features (it is done based on *scatter – speed relation* analysis). Visualisations after first iteration might not be so clear due to simple decisions we made. Now is time to introduce some extra constraints that are based on knowledge about machine. It is obvious that properly loaded (this is a reason we need feature-speed plot!) machine with very small features- cannot be considered as damaged. First constraints (Step 3a) is as follow: discriminate data with big scatter into: small and other features. It should be done by visual inspection of data, in the worst case the highest value from pointed out 'good condition' dataset could be used. The last step (Step3b) is related to understanding properties of data from class "warning" that is between good and bad condition (plotted in yellow). In fact it consist in 3 subclasses: medium features and scatter, small features/big scatter and small scatter/big features. They represent very different scenarios of degradation and in our opinion should not be treated in the same way [12, 13]. More explanation will be presented during application results discussing.

Now we will present and discuss results obtained when applied the procedure to raw data presented in Fig. 1. Follow the FlowChart we provide illustration of each step. In Fig. 8 results for step 1 and 2 are presented. Identification of good condition dataset was obtained using scatter-feature-value analysis.

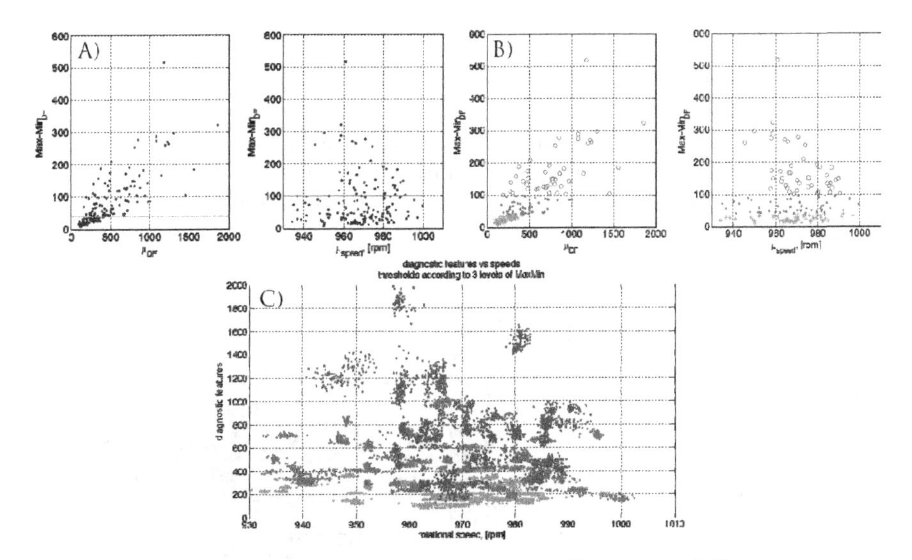

Fig. 8 Performing of step 1 and 2: a selection of small scatter (a left) and big scatter (a right), b, c) visualization green-good, magenta-warning, red-bad condition.

A value of threshold was set Th_{Wy} = *40* based on similarity (most closest) of cluster for good condition. From scatter-speed plot it has been found that there are mixture of two very different datasets: some of data create of patter significantly dependent of speed, while other rather not. A threshold was set Th_{Cy} = *100,*

Fig. 8b and c visualize our decisions. From Fig. 8c one might conclude that values of thresholds are not really good due to number of clusters classified as bad, but in fact they are located in the middle of warning and very close to good condition. Fig. 9 shows results for the first of constraints applied to the data. Simple linear discrimination for Th_{Cx} = *650* significantly improved the plot 9b. Fig. 10 presents a decomposition of warning class into 3 subclasses. Fig. 10b shows that data plotted in black (small scatter/big feature values) should be rather considered as bad condition. Fig. 11 shows final picture - results of decision boundaries estimation. It should be noted that our warning and alarm levels are smaller than currently used in the mining DSS that means we are able to recognize incorrect operation of drive unit's component earlier.

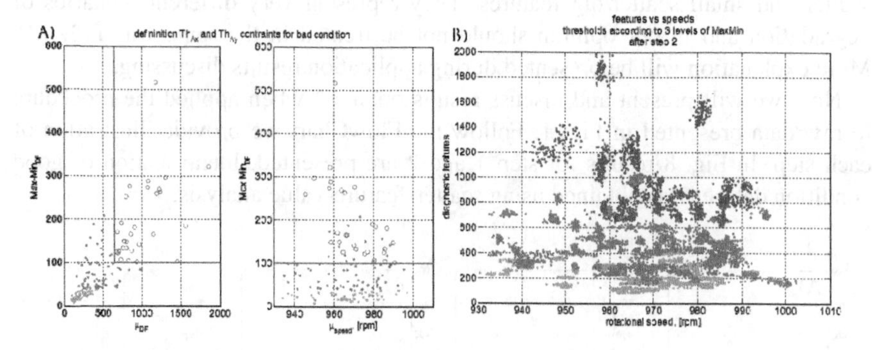

Fig. 9 Performing of step 3a: a) adding constraints related to high scatter and small values of features (a) left), note overlapping between bad and warning in Fig 9a) right; Fig9 b) visualization of feature-speed: green-good, magenta-warning, red, -bad condition.

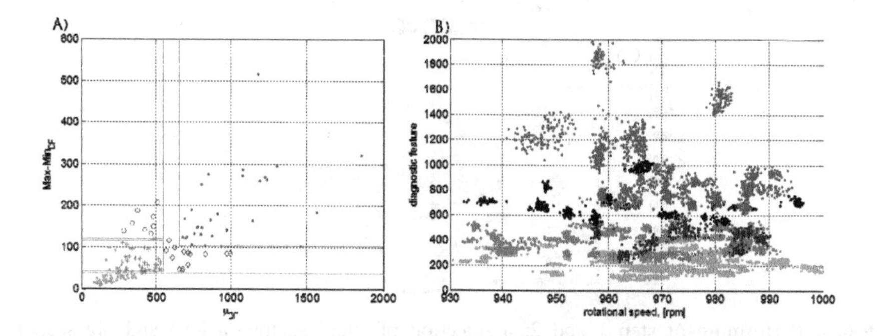

Fig. 10 Performing of step 3b: a) adding constraints related decomposition of warning class into 3 subclasses (fig a) left), b) visualization green-good, magenta-warning, red-bad condition, blue – small features/big scatter, black-small scatter/big features.

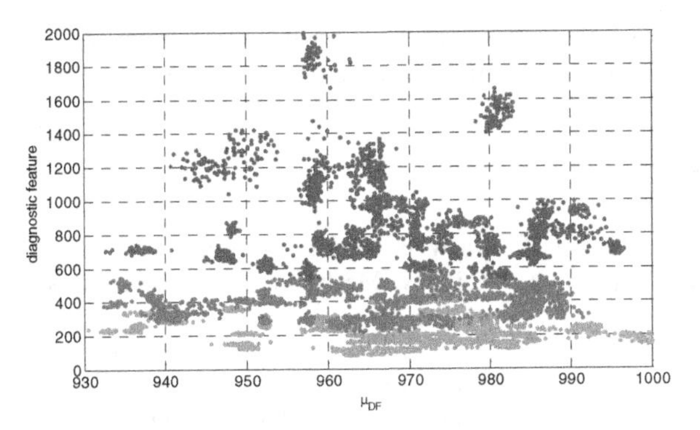

Fig. 11 Final plot: visualization of data divided into 3 subclasses

5 Conclusions

The main aim of this work was to develop a procedure for finding of new decision boundaries (thresholds) to increase effectiveness of diagnostic decision support system (DSS) used in copper mine for belt conveyors maintenance.

Unfortunately, the belt conveyors are particularly difficult kind of machines to diagnose (for various reasons), thus detailed analysis including: exploration of the relationships among diagnostic data, relating them to the machines' behaviors and taking into account factors of degradation are needed. Currently, authors simultaneously are working on several ideas regarding modeling of degradation processes for belt conveyors.

The novel approach presented in this paper introduces a new parameter namely *Max-Min* to determine a decision threshold. *Max-Min* is measure of scatter of data cloud from single diagnostic measurement. Multiple visualizations of *Max-Min* vs. *external load* or vs. *mean of diagnostic features* allow to divide dataset into two main clusters: (a) good condition characterized by data clouds with small values of features and small scatters, (b) bad condition - data clouds reveal high values of features and big scatters. Detailed analysis of obtained clusters showed that data depend on speed, much more for bad condition than for other classes. It follow the idea proposed in [1].

Moreover, for a large population of data clouds no above mentioned properties has been identified. It has been assumed that they represent very different scenarios of degradation and in our opinion should not be treated in the same way. Therefore, we decompose that intermediate condition, so called warning state into 3 subclasses: (a) small scatter/big feature values, (b) big scatter/small feature values and (c) medium scatter/feature values. After further analysis we decided to classify "small scatter/big feature values" as a bad condition. We believe that these different behaviors of machines are strong dependent on: (a) operating parameters (rotational speed, external load), (b) design properties (technical configuration, modulus of elasticity etc.) and (c) degree of wear (e.g. pitting, scuffing). What's more, perhaps the ratio of scatter value to feature value (or similar) should be used to estimate failure risk index in reliability analysis or prediction of further degradation in belt conveyor case.

Acknowledgements. This work is partially supported by the statutory grant No. S300973 (P. Stefaniak). M. Hardygora and R. Zimroz are supported by KGHM Cuprum Ltd.

References

1. Bartelmus, W., Zimroz, R.: A new feature for monitoring the condition of gearboxes in non-stationary operating conditions. Mech. Syst. and Signal Proc. 23, 1528–1534 (2009)
2. Zimroz, R., Bartelmus, W., Barszcz, T., Urbanek, J.: Diagnostics of bearings in presence of strong operating conditions non-stationarity-A procedure of load-dependent features processing with application to wind turbine bearings. Mechanical Systems and Signal Processing (2013), doi:10.1016/j.ymssp.2013.09.010
3. Bellino, A., Fasana, A., Garibaldi, L., Marchesiello, S.: PCA-based detection of damage in time-varying systems. Mechanical Systems and Signal Processing 24(7), 2250–2260 (2010)
4. Zimroz, R., Bartkowiak, A.: Two simple multivariate procedures for monitoring planetary gearboxes in non-stationary operating conditions. Mechanical Systems and Signal Processing 38(1), 237–247 (2013)
5. Kacprzak, M., Kulinowski, P., Wedrychowicz, D.: Computerized information system used for management of mining belt conveyors operation. Eksploatacja i Niezawodnosc – Maintenance and Reliability 50(2), 81–93 (2011)
6. Lodewijks, G.: Strategies for Automated Maintenance of Belt Conveyor Systems. Bulk Solids Handling 24(1), 16–22 (2004)
7. Zimroz, R., Krol, R., Hardygora, M., Gorniak-Zimroz, J., Bartelmus, W., Gladysiewicz, L., Biernat, S.: A maintenance strategy for drive units used in belt conveyors network. In: Eskikaya, Ş. (ed.) 22nd World Mining Congress & Expo, Istanbul, September 11-16, vol. 1, pp. 433–440. Aydoğdu of set, Ankara (2011)
8. Galar, D., Gustafson, A., Tormos, B., Berges, L.: Maintenance Decision Making based on Different types of Data Fusion. Eksploatacja i Niezawodnosc – Maintenance and Reliability 14(2), 135–144 (2012)
9. Cempel, C.: Limit value in practice of vibration diagnosis. Mechanical Systems and Signal Processing 4(6) (1990)
10. Brooks, R., Thorpe, R., Wilson, J.: A new method for defining and managing process alarms and for correcting process operation when an alarm occurs. Journal of Hazardous Materials 115 (2004)
11. Jablonski, A., Barszcz, T., Bielecka, M., Breuhaus, P.: Modeling of probability distribution functions for automatic threshold calculation in condition monitoring systems. Measurement 46(1), 727–738 (2013)
12. Bartelmus, W.: Gearbox damage process. J. Phys.: Conf. Ser. 305(1), paper no 012029 (2011), doi:10.1088/1742-6596
13. Bartelmus, W., Chaari, F., Zimroz, R., Haddar, M.: Modelling of gearbox dynamics under time-varying nonstationary load for distributed fault detection and diagnosis. European Journal of Mechanics. A, Solids 29(4), S.637–S.646 (2010)

Development of Inclined Conveyor Hard Rock Transportation Technology by the Cyclical-and-Continuous Method

N. Kuchersky[1], Vasiliy Shelepov[2], Illia Gumenik[3], and A. Lozhnikov[3]

[1] Navoy
[2] SE Institute UkrNDIproekt, Kiev
[3] Department of open cast mine,
National Mining University, Dnepropetrovsk, Ukraine

Abstract. The article is devoted to improving the efficiency of rock mass transportation in open-cast mining steep deposits. Implementation of cyclic-and-continuous transport chart on these deposits is analysed. The conducted researches consist improvement of hard rocks and ore transportation chart that combines trucks, conveyor and railway. It is based on the design developed of the steeply inclined conveyor with a pressure belt. As a result of the proposed technology application is reduction of internal quarry mileage of trucks and the maximum output of the conveyor.

Keywords: Hard rock, Cyclical-and-Continuous Method, Inclined Conveyor, Transportation System, Management Information System.

The development of steep deposits depends on increases of the depth of the quarries. As a result the volumes and the distance of transportation of the mountain mass increase. Neither type of the quarry transport has no totality economic advantages which would allow to use it without other types of the transport one quarry of large depth. That why using of two and more types of the transport in independent or combined objectively unavoidable.

The more promising is a combined motor-conveyer transport, i.e. cyclic-and-continuous transport (CCT), which economic expediency increases with depth of the quarry and the rock mass. The challenges for implementation of CCT on working pit are placement of the conveyer systems on the formed slopes and non-coordination of the cyclic and the continuous parts of an integral transport system.

Increasing of effectiveness CCT consist in using the steeply inclined lifting conveyer that reduces up to the minimum the volumes of mining permanent works when preparing the route for on the formed pit walls. Unfortunately, application of the steeply inclined conveyers SIC in the quarries of the CIS countries was limited only with theoretic researches. Results of this researches were applied in the quarry "Muruntau" of Navoi Mining-and-Metallurgical Industrial Complex.

C. Niemann-Delius. (ed.), *Proc. of the 12th Int. Symp Continuous Surface Mining - Aachen 2014*,
Lecture Notes in Production Engineering, DOI: 10.1007/978-3-319-12301-1_5

Aim of researches consist in development of efficient CCT for the deep quarry on the base of resource-saving cyclic-and-continuous transport with steeply inclined conveyer (SIC).

For realize this idea the Navoi MMIC has adopted and fulfilled a program with main specialists from Ukraine. The main components of CCT-ore: CRS - crushing-reloading station (point); SIC – steeply inclined conveyer, LSC - loading-storage complex; ASMODU – automated system of monitoring and operative-dispatcher control (fig. 1, 2).

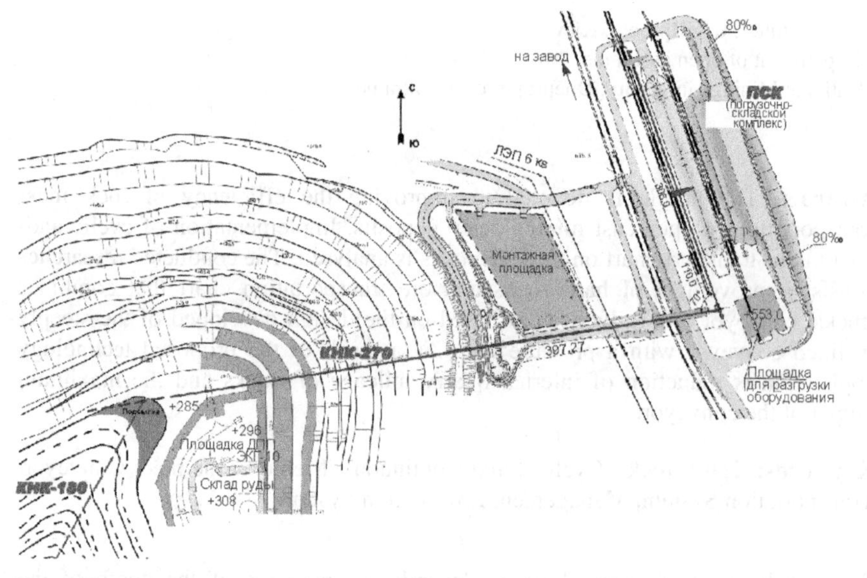

Fig. 1 Situational layout of CCT with steeply inclined conveyer SIC-270 and direct flow loading into the railway transport

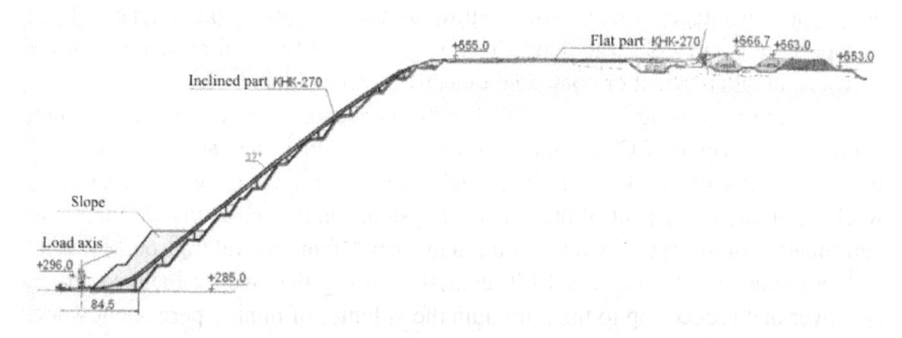

Fig. 2 Profile of continuous transport with SIC-270

This innovative technology of the cyclic-and-continuous transport was created with joint efforts of Navoi MMIC, designers, developers, scientists and machine-builders whose collectives are in Ukraine, Uzbekistan and Russia. Thus, it was confirmed again that the most significant results are achieved on the junction of the sciences when integrating the efforts of scientists, developers, specialists of machine-building and operating enterprises.

In the process of the project development, manufacturing, installation and commissioning of the steeply inclined conveyer SIC-270 it had been set to the developed design of the steeply inclined conveyer with a rider belt is operable, has high reliability, ensures stable reception, lifting at angle of 37 degrees at the height of 270 m and unloading of the crushed rock mass with capacity of more than 3500 t/h (fig. 3)

Fig. 3 Inclined lifting conveyer

The created steeply inclined conveyer SIC-270 satisfies the conditions of operation in the quarry, is provided with the servicing facilities, is recognized to be successful solution and is recommended for use and. This conveyor can combine in the unique module the lifting and the transferring functions that is its feature in the quarry "Muruntau" (fig. 2). The route of the steeply inclined conveyer to be placed perpendicularly to the berms of the ledges at general angle of the inclination of the pit wall.

The developed design layout of pivot joint of the sections between themselves and the supports on the ledges along the entire length of the steep part without discontinuities for compensation of temperature deformation of supporting metal structures confirmed its operability and reliability.

Slope on the area of placing SIC-270 is sufficiently stable, the possibility of appearance of large deformations being able to result in emergency situations is practically absent, and the design of the foundation structures for supports (know how) reduces the possibility of local collapses of the ledges with deformation of support of the steep part of SIC-270. The safety factors of mountain massifs from their additional loading with steeply inclined conveyer are 1.3...1.95.

The possibility of installation on the pit wall of large components (length 51 m, cross-section 18 m^2 and weight 84 tons) of steep part of SIC-270 is proved with the consecutive lowering of the sections on the guides stacked on the upper structure of supporting metal structures.

The screw-tooth crusher confirmed its operability on the rocky mountain mass of the quarry "Muruntau". For the increase of its capacity up to the design values and the durability of its operating elements the grate bars for the preliminary screening before the crushing are provided.

In the process of installation and commissioning of the loading-storage complex on the surface of the quarry "Muruntau" it has been set to that self-propelled rocky loader-stacker ensures the continuous direct flow loading of the ore into the dump cars or to the storage. In result, the continuous delivery of the ore from the bottom of the quarry to the plant without unnecessary transshipment on the surface is formed (fig. 4). It was determined that the combined capacity of operative and motor transport warehouse provides six days stock of the ore for its shipment to the plant.

a) b)

Fig. 4 Self-propelled rocky loader-stacker: a) loading to the temporary warehouse; b) loading to the railway dump cars

Installation of the additional conveyer line gave to the CCT system the additional flexibility which allows to adapt it to the current prevalence of the volumes of mixed flows of the ore or overburden which is formed in the quarry.

At the installation and commissioning of ASMODU (automated system of monitoring and operative-dispatcher control) of the continuous transport for the ore of the quarry "Muruntau" was determined that the multi-level computer-aided system of control Automated Control System "PT-ore" having more than 1300 physical inputs/outputs proved its operability and reliability, provided the high level of supervision, protection, control and monitoring which assist to the increase of the readiness of the continuous transport to the operation and to the reduction of the time of operability restoration after the emergency shutdowns.

The practical implementation of the ASMODU with positive result can be regarded as the base one for the quarries and mining reductions of concentrating factories, has the sufficient flexibility of the software-hardware that will allow to increase, to enhance and to adapt it to the different technological schemes with continuous transport in which the screens and the crushers are integrated.

The cyclic-and-continuous transport on the base of steeply inclined conveyer SIC-270 with deep input into the quarry have next advantages.

SIC is installed practically without additional mining preparatory works on the narrow section of the pit wall that does not restrict the expansion of the mining along the entire contour of the quarry, does not require the laying of the special trenches or sinking operations (comparing to the traditional slightly inclined conveyers the steeply inclined conveyer takes up by 25-50% less area of the workspace).

Creation of the transport berms in the mountain part with the corresponding separation of the pit wall in this zone is not necessary for entrance of technological transport to CRP and that allows to retain its general angle of inclination. The combined parallel operation of two belts increase their resulting strength and tractive ability of the drives that increases the height of the lifting of the mountain mass with one conveyer without overloading.

Proposed technological chart allows the further depth increase of the continuous part sinking of CCT up to 450 m. When transporting the ore the necessity of its primary crushing in the process of processing at the plant. The loads losses due to the reduction of blowing and spilling of the transported cargo, the environmental situation of mining enterprise is improved.

It should be specially noted that with the average operation life of the dump trucks being 6-7 years, during the resource life of CCT operation during 20 years it would be necessary to renew three times the park of dump trucks which is released by the continuous transport;

For analyse effectiveness of proposed CCT in condition of "Muruntau" its parameters was compare with well known analogues. In the domestic mining industry and CIS countries there are no samples of CCT with steeply inclined conveyer. The nearest analogue in the world's practice of open pit mining (Majdanpek, Serbia) according to the height of the lifting is approximately one third of the created one (table 1).

Table 1 Comparative data of steeply inclined conveyers

Place of operation	Bulk loose density, t/m^3	Capacity, t/h	Inclination angle, degrees	Lifting height, m	Belts width, mm	Belts speed, m/s	Drives power, kW (rider / load)
Muruntau	**1,75**	**3500**	**37**	**270**	**2000**	**3,15**	**1260/3780**
Majdanpek	2,08	2000*	36	94	2000	2,80	450/900

* the achieved capacity is given (according to design 4000 t/h, but only one of two technological lines operating for one common assembly SIC was put into operation).

Industrial operation of steeply inclined conveyor with a pressure belt in the "Muruntau" pit was started from 2011. On this time were shipped more then 17 million tons of rock mass. The maximum shift capacity of 46.3 thousand tons for 12 hours shift or 3860 t/h was achieved. The maximum monthly capacity of 458000 m^3 or 1.19 million tons per month was achieved that will allow the capacity of the continuous transport being more than 13 million tons of the mountain mass per annum.

The same continuous transport has possibility to put out the overburden from the quarry which through the system of additional conveyers is delivered to the dump passing the ore warehouse (thus, the CCT system was given additional flexibility which allows to adapt it to the main cargo flow being formed in the quarry: ore or overburden).

At the transportation of mixed overburden and ore with application of the continuous transport on the base of SIC-270, the distance of the transportation with motor transport was decreased by 3.6 km, and the transport run was decreased by 40 %.

Developed technological decision allow to increase the coefficient of the readiness of the continuous transport from Cr=0.717 to Cr=0.842 of operation periods. The reliability of adopted and realized solutions will not restrain the increase of the transportation of the mountain mass in the cyclic elements of the quarry and annual capacity reach to 16 million tons.

According to the data of Navoi MMIC the economic effect for the mixed mountain mass (overburden and ore) which has been already obtained from the creation of the cyclic-and-continuous transport on the base of steeply inclined conveyer SIC-270 for year 2012 amount 1.53 million USD.

Development of an Expert System for the Prediction of the Performance of Bucket-Wheel Excavators Used for the Selective Mining of Multiple-layered Lignite Deposits

Michael J. Galetakis[1,*], Stylianos Papadopoulos[1], Anthoula Vasiliou[1], Christos P. Roumpos[2], and Theodoros Michalakopoulos[3]

[1] Technical University of Crete, Department of Mineral Resources Engineering, Chania, Greece
[2] Public Power Corporation SA, Athens, Greece
[3] National Technical University of Athens, Department of Mining and Metallurgical Engineering, Greece
E galetaki@mred.tuc.gr

Abstract. The performance of bucket-wheel excavators used for the selective mining of multiple-layered lignite deposits is mainly determined by the physical and mechanical properties of the excavated material (digability parameters), as well as by the thickness and the inclination of the excavated layer. The lack of sufficient information regarding the diggability of the excavated layers during mine design stage results in a poor estimation of the bucket-wheel excavator's performance. Such inaccurate estimation could lead to inappropriate equipment selection. The possibility of prediction of the bucket-wheel excavator performance by using an expert system was investigated in the present study. The development of the expert system was based on the existing boreholes' data related to physical and mechanical properties of waste and lignite layers, on their geological descriptions and on the existing experiential knowledge about selective mining by bucket-wheel excavators.

The developed expert system was applied for the prediction of the performance of the bucket-wheel excavators used for the exploitation of Mavropigi Lignite Mine in Ptolemais area (Northern Greece).

Keywords: Selective mining, bucket-wheel excavators (BWE), digability, expert systems.

1 Introduction

Lignite is the most important domestic fuel of Greece, used almost exclusively for power generation. The annual lignite production in Greece in 2013 was ~53 Mt,

* Corresponding author.

C. Niemann-Delius. (ed.), *Proc. of the 12th Int. Symp Continuous Surface Mining - Aachen 2014*,
Lecture Notes in Production Engineering, DOI: 10.1007/978-3-319-12301-1_6

while the electricity produced from the lignite-fired power plants comprised the 48% of the total power generation. Lignite is mined mainly by the Public Power Corporation S.A. (PPC) and is considered a strategic fuel for Greece due to secure supply and controllable cost; consequently it gives a competitive strength in Greece's fuel mix. The remaining lignite reserves that are suitable for electricity generation are 2.5 billion tones (2013) [1].

Most of the Greek lignite deposits have a multiple-layer structure and are located in the Ptolemais - Amynteon basin, in the area of Western Macedonia in northern Greece. For the exploitation of these deposits, the continuous surface mining method is used. High capacity bucket wheel excavators (BWE), conveyor belts and spreaders are used to achieve high output rate with low cost of mined lignite per ton [2,3]. Nowadays, the Western Macedonia Lignite Centre (WMLC) has four large scale open pit mines, which handle 315 Mm^3 of material and produce 43.6 Mt of lignite (2013). The lignite deposits under exploitation cover an area of 160 Km^2 including 1200 Mt of proven geological reserves and 1000 Mt of exploitable reserves under the current economic and technological criteria. It is estimated that the lignite of the Ptolemais-Amynteon basin is sufficient to supply the existing power plants for another 15-25 years.

The multiple-layered lignite deposits, which are characterized by the extreme splitting of lignite seams separated by non-lignite layers, make the selective mining a necessity (Figure 1). The alternated waste layers, mainly marls and clays, are of varying thickness and cohesion and also exhibit intense spatial fluctuation. In addition, the presence of hard formations in the overburden, consisting mostly of conglomerates and stiff sandstones, impose the use of discontinuous exploitation methods (hydraulic excavators and/or blasting and trucks) either independently or in combination with the existing continuous surface equipment. The implementation of selective mining procedures by terrace cutting and the occurrence of cohesive and hard formations affect decisively the performance of the used bucket wheel excavators. Hence the estimation of the effect of the geometrical, physical and mechanical properties of the excavated material, which determine decisively its digability, on the performance of the bucket wheel excavators is crucial for the appropriate long and short-term mine planning and design activities. However, the lack of sufficient information regarding the digability of the excavated layers during mine design stage in combination with the observed intense spatial fluctuation of the excavated material could result in poor estimation of the bucket-wheel excavator's performance. Such inaccurate estimation could lead to inappropriate equipment selection.

In the present study a new methodology for the prediction of the bucket-wheel excavator performance by using expert systems is proposed. The developed expert system is used to assess the diggability of the excavated blocks, while their thickness is estimated by the existing conventional practices used for the evaluation of multiple-layered lignite deposits. Expert system reasoning is based on the existing boreholes' data describing the lithology of the drilled geological formations, as well as on the acquired experiential knowledge about mining these formations by bucket-wheel excavators.

Fig. 1 Typical structure of multiple-layered lignite deposit in the area of Ptolemais (left) and selective mining by bucket wheel excavators (right).

2 The Effect of the Diggability and Thickness of Excavated Blocks on the Performance of BWE during Selective Mining

The correct selection of a BWE requires proper determination of the diggability of the excavated material. This diggability is defined as the ability with which a geological formation can be excavated, removed, transported and discharged to a conveyor belt by a BWE. The relationship between the physical and mechanical properties of different geological formations, as determined by geologists and engineers, and the diggability of these formations has been investigated by several researchers. The main parameters that affect the diggability by BWE are [4-9]:

- Mechanical and physical properties of the material (uniaxial compressive strength, cutting resistance, seismic waves velocity, stickiness, etc.)
- Geotechnical characteristics of the geological formation (joints, bedding, etc.)
- Thickness and inclination of the excavated layer.

From the literature review it is clear that there is not a single widely accepted and reliable classification of the diggability of geological formations. In addition, diggability cannot be assessed by a single factor but requires the determination of a set of geotechnical parameters. Most of the diggability classification systems use a five-level scale to quantify the ability of a geo-material to be excavated by a BWE as follows [4]: Class I: easily diggable, Class II: diggable, Class III: difficult to dig, Class IV: very difficult to dig and Class V: diggable after loosening. The effect of material diggability on the efficiency of the BWE has also been investigated [4,6,7] and indicative values are proposed for the rough estimation of the reduction of the BWE efficiency as the diggability decreases.

Except for diggability, the performance of a BWE, especially during selective mining, is also affected by the thickness of the excavated block. The determination of the mineable lignite and waste blocks is one of the most vital considerations in selective mining by BWE. For the exploitation of multiple-layered lignite deposits in the area of Ptolemais in Northern Greece, a sophisticated evaluation methodology, based on exploitation and quality criteria, has been developed [10-12]. It takes into account the minimum thickness for the selective mining of lignite and waste layers, as well as the maximum allowable ash content of mined lignite. These parameters affect significantly not only the recovery of the deposit and the quality of the mined lignite, but also the efficiency of the excavating equipment. According to this methodology thin non-lignite layers (with thickness less than the minimum thickness for selective mining by BWE) can be included in the mineable lignite blocks if the quality requirements for the excavated lignite are fulfilled. On the other hand, thin lignite layers which cannot be excavated selectively are included in the waste blocks. The thickness of the formed mineable lignite or waste blocks have a significant effect on the efficiency of the bucket-wheel excavators which are used in these selective mining operations. The productivity of the bucket-wheel excavators drops significantly as the ratio h/R decreases (where, R is the radius of bucket wheel and h is the thickness of the excavated block) as shown in figure 2 [12].

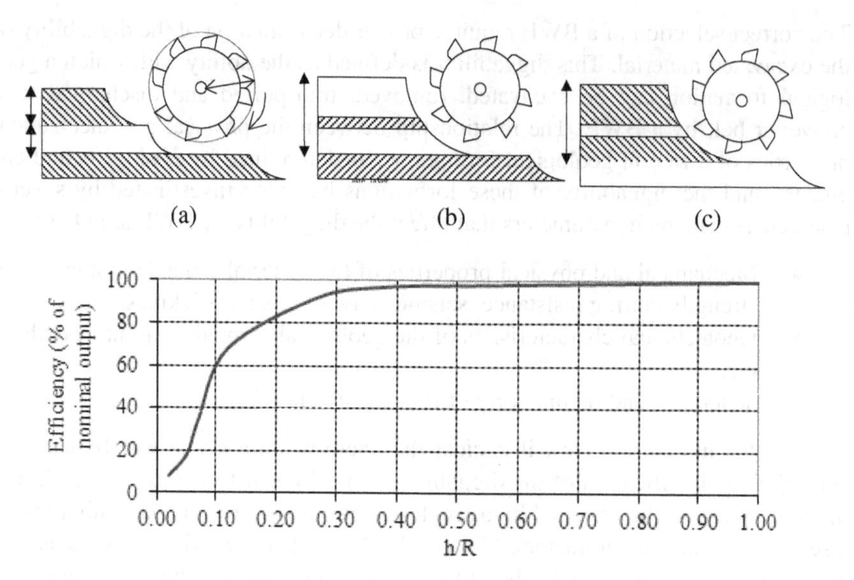

Fig. 2 (Above) a: Selective excavation of single lignite and waste layers, b: Selective excavation of waste block including a thin lignite layer and c: Selective excavation of a lignite block including a thin waste layer. (Below) Effect of h/R ratio to the efficiency of a BWE.

Apart from diggability and excavated block thickness, the performance of a BWE is also affected by BWE type and rated capacity, operation cycle, equipment breakdowns, external delay factors, management, personnel skills and working conditions. In this study we focus on the diggability and thickness since these factors are considered crucial for the BWE performance during the selective mining of multiple-layered lignite deposits [12].

3 Expert Systems (ES)

The Expert Systems (ES) derive from the scientific area of Artificial Intelligence and were developed to solve specific problems by using specialized knowledge. An expert system, also known as knowledge based system (KBS), is in effect a computer program that encompasses the knowledge and analytical skills of one or more human experts in the domain of a specific problem. The goal of the design of the expert system is to capture the knowledge of a human expert relative to some specific domain and code this in a computer in such a way that the knowledge of the expert is available to a less experienced user [13,14]

The basic components of an ES, shown schematically in figure 3, are: the knowledge acquisition module, the knowledge base, the inference engine, the explanation facility and the user interface [13].

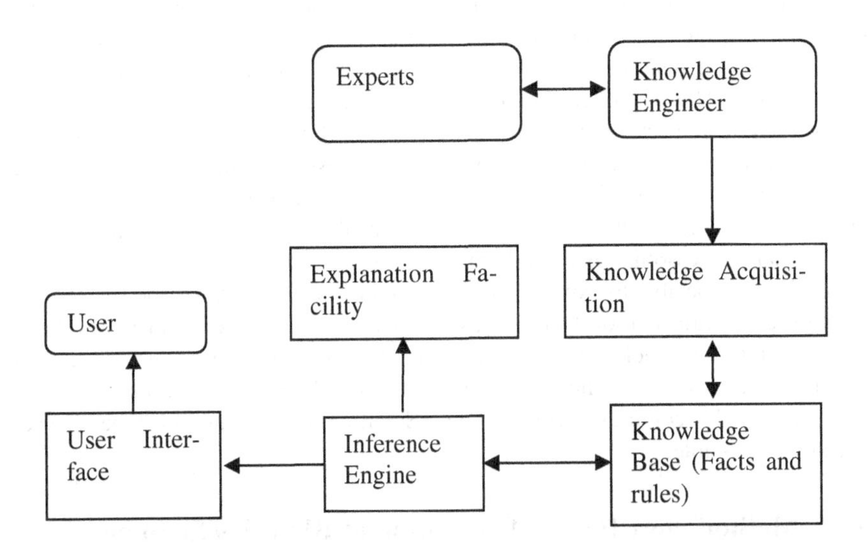

Fig. 3 Schematic representation of an expert system structure

Knowledge acquisition refers to acquiring, restoring the knowledge collected and organized in cooperation with the experts in the specific domain. It is a subsystem used to build knowledge bases. The techniques used for knowledge acquisition are protocol analysis, interviews, and observation. The process of acquiring knowledge is the most critical point in the development of an ES.

The knowledge base contains the knowledge necessary for understanding, formulating and solving problems. The acquired knowledge can be distinguished into factual and heuristic knowledge. Factual knowledge is that knowledge of the task domain that is widely shared, while heuristic knowledge is mainly experiential and largely individualistic. Factual knowledge is represented as facts, while heuristic is represented as rules. The facts are information that is generally known or published and generally accepted by the experts of the specific domain. The rules, unlike the facts, are a way of thinking and judgment characterizing the modus operandi and the skills of an expert. For the creation of rules the logical operators AND, OR and NOT in conjunction with the operator IF and THEN are used. The general form of these rules is:

IF F(1) AND F(2) AND F(3)……AND F(N)
THEN H
or
IF F(1) OR F(2) OR F(3)……OR F(N)
THEN H

where: F(1)……F(N) are the facts and H is the consequent

The inference engine must be independent of the acquired knowledge. The reason is that our knowledge in a specific field is changing but the way of reasoning and drawing conclusions should remain stable to ensure consistency, accuracy and completeness. The way of reasoning used by the ES inference engine is: forward, backward and sideways chain. The forward chain of reasoning takes into account the facts and is driven to the assumption or conclusion (synthetic thinking). The backward chain of reasoning takes into account the conclusion or the fact and tries to find data supporting this fact or conclusion (analytical thinking). Finally, the sideway chain of reasoning is a combination of the two above.

An ES must be able to explain how the conclusions were drawn and the need to obtain specific information from the user. For this purpose special programs have been developed to track the inference path. The user can at any time ask the ES why this rule was fired or how that conclusion was drawn. Finally, the user interface enables the communication of the ES with the user and must be as user friendly as possible.

4 Methodology for the Estimation of BWE Performance

The proposed methodology for the estimation of the effect of material diggability and excavated blocks thickness on the BWE performance includes the following stages:

- Development of an ES for the assessment of the diggability of excavated geological formations. This assessment is based on the lithology description of the related drillhole cores.
- Determination of the mineable lignite and waste blocks for selective mining as described above (section 2). This includes the estimation of the thickness of the blocks as well as the assessment of the diggability of the excavated block.
- Estimation of the impact of diggability and thickness of excavated block on BWE performance.

4.1 Development of an ES for the Assessment of Diggability

The assessment of the diggability was based on the existing information obtained during the detailed exploration of the lignite deposit. The coding of the boreholes' data is particularly useful, since it significantly reduces the size of the created files, makes data entry and correction very easy and, finally, significantly improves searching, matching and comparison efficiency. For each drilled geological layer there is a record describing its lithology, texture, inclusions, color, etc. After rigorous exanimation of these recorded parameters by experienced mining engineers and geologists of the Mine Division of the Power Public Corporation of Greece, three of them were selected to be used by ES for diggability assessment: the major lithology description (MM), the additional lithology description (S) and the texture (T). Typical coding cases for MM, S and T are shown in Table 1.

Regarding the classification of the excavated material according to its diggability a five-level scale is proposed, as indicated in Table 2. Furthermore a rough estimation of the expected performance of the BWE in each diggability class is also suggested. This performance is expressed as percentage of the BWE performance achieved when excavates easily diggable materials (diggability=1).

In the developed knowledge base facts are considered to be the material properties as described by the selected parameters (MM, S and T). These fact are connected with the logical operators AND, OR and NOT and in conjunction with IF – THEN operators create a set of rules. A typical form of such a rule is:

IF MM = "MR" AND AND S = "K" AND T = "M"
THEN Diggabilty = 2

Table 1 Coding of the most common cases regarding the geological parameters used for the assessment of diggability

Major lithology descriptor (MM)	Supplementary lithology descriptor (S)	Texture (T)
AL=Clay	A=Cherty	A=Loose
CO=Lignite	C=Carboniferous	C=Compacted
CP=Peat	H=Humic	D=Stiff
GR=Gravel	K=Calcitic	F=Brittle
KC=Conglomerate	M=With mica	G=Fractured
MR=Marl	S=Sandy	H=Hard
SD=Sandstone	T=Clayey	I=Bedded
SH=Shale	U=Silty	K=Crystalline
SI=Silt	X=Quartzitic	M=Soft
SL=Siltstone	Z=with sandstone inclusions	R=Porous
SN=Sand		S=Stratified
SO=Top soil	....	Y=Semi-soft
ZY=Xylite		....
....		

The above rule assigns the diggability value of 2 (diggable) to a soft (T="M") calcitic (S="K") marl layer (MM= "MR"). Moreover the BWE performance is expected to be 75% of the achieved in easily diggable material.

The developed knowledge base of the current ES version contains 104 rules and uses the forward chain of reasoning. The ES was developed in Visual Basic within the Excel environment.

Table 2 Diggability values of excavated material and BWE performance

Description of material	Diggability value	BWE performance*
Easily diggable	1	100%
Diggable	2	75%
Difficult to dig	3	50%
Very difficult to dig	4	30%
Diggable after loosening	5	12%

* Expressed as percentage of the BWE performance when excavates materials with diggability=1.

4.2 *Estimation of the Impact of Excavated Block Thickness to the BWE Performance*

The performance of a BWE during selective mining is also significantly affected by the thickness of the excavated block. The determination of the mineable lignite and waste blocks and their thickness was carried out by using the software METAL developed by PPC [10]. The impact of excavated block thickness on a BWE performance is calculated from the diagram shown in figure 2. This performance is also given as a percentage of the nominal BWE output rate.

Finally the overall performance of a BWE is estimated by multiplying the performance value considering the diggability and the performance value considering the thickness of the excavated block.

5 Application of the Developed Methodology in the Mavropigi Lignite Mine

The Mavropigi lignite mine is the newest of the four operational surface mines at the Lignite Center of Western Macedonia, covering an area of 11km^2. The remaining mineable reserves were estimated at ~ 140 Mt of lignite. The annual planned lignite production of the mine is ~ 8 Mt, while the majority of the lignite mined is fed to the nearby power plant of Ptolemais and Liptol with a total nominal output of 663 MW. The Mavropigi lignite deposit was explored by 230 drillholes, with an average spacing of ~230 m. The thickness of the lignite deposit (lignite and interbedded waste layers) varies from 50 to 250 m while the thickness of the overburden from 10 to 50m. At present there are two overburden benches and six lignite benches in operation. The installed equipment includes eight bucket wheel excavators, three spreaders and 40km of conveyor belts.

The most frequently occurring geological formations, as coded according to Table 1, are lignite (CO), marl (MR), clay (AL) and sand. Typical additional lithological characterizations of these formations include the terms: clayey (T), sandy (S), carbonaceous (C), calcitic (K) and sandstone inclusions (Z). Finally the most frequently observed textures were described as loose (A), compacted (C), hard (H), and semisoft (Y).

The application of the developed methodology for the estimation of BWE performance in Mavropigi mine was based on data selected from 211 drillholes' data. By using the developed ES a diggability value was assigned for each geological formation included in the drillhole database. After that, the mineable lignite and waste blocks for selective mining were determined and the thickness and the digability of each block were estimated. Taking into consideration the technical characteristics of the used BWEs (i.e. bucket wheel diameter) the performance of BWE during the excavation of each mineable block was estimated. Finally a three dimensional interpolation and representation of these estimated values was performed by using the software Voxler (Golden Software).

Figure 4 shows the spatial variation of the estimated diggability values along the major axis and along the one perpendicular to the major axis. Diggability varies mainly depending on depth. The majority of the formations indicate values between 1-3 and rarely 3-4. Only at the border of the deposit have formations been identified which cannot be excavated by a BWE without prior loosening by ripping or blasting (diggabilty value from 4 to 5). More specifically, the formations above the level of 670m (overburden) present low diggability values (1-2) and can be classified as easily excavated by BWE according to Table 2. This is due to the fact that these layers consist mainly of loose formations (gravel, sand and soft clays) which show small cutting resistance and low adhesiveness. The deposit formations (below the level of 670m) show diggability values varying from 2 to 3 and rarely 4 and are classified accordingly either as diggable or as difficult to dig by BWE. The difficulty of extraction of these formations by BWE is due to the relatively high cutting resistance values which cohesive layers like the compacted marls and stiff clays with hard sandstone inclusions exhibit.

Figure 5 represents the effect of the mineable blocks thickness on BWE performance. BWE performance is relatively high since it varies from 40% to 100% of the maximum expected value as defined in Table 2. More specifically, the performance during the excavation of relatively thin blocks, located below the level of 550m, is expected to fluctuate between 40 to 70%. From the level of 550m to 700m the performance of a BWE increases significantly since it varies from 60 to 100%. In the upper level (above 700m) BWE performance reach its maximum value since almost all the excavating blocks have considerable thickness.

Finally, Figure 6 shows the overall performance of a BWE (includes both diggability and thickness effect). The higher BWE performances observed in the upper part of the deposit in the overburden benches (above the level of 700m) are related to thick excavated blocks (selective mining is not required in these benches)

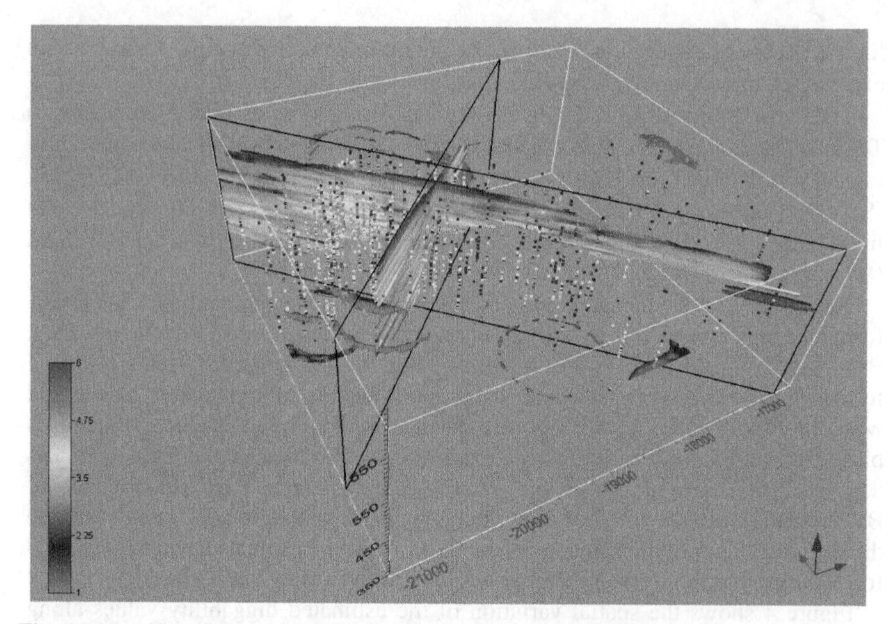

Fig. 4 Diggability of the geological formations of Mavropigi lignite mine

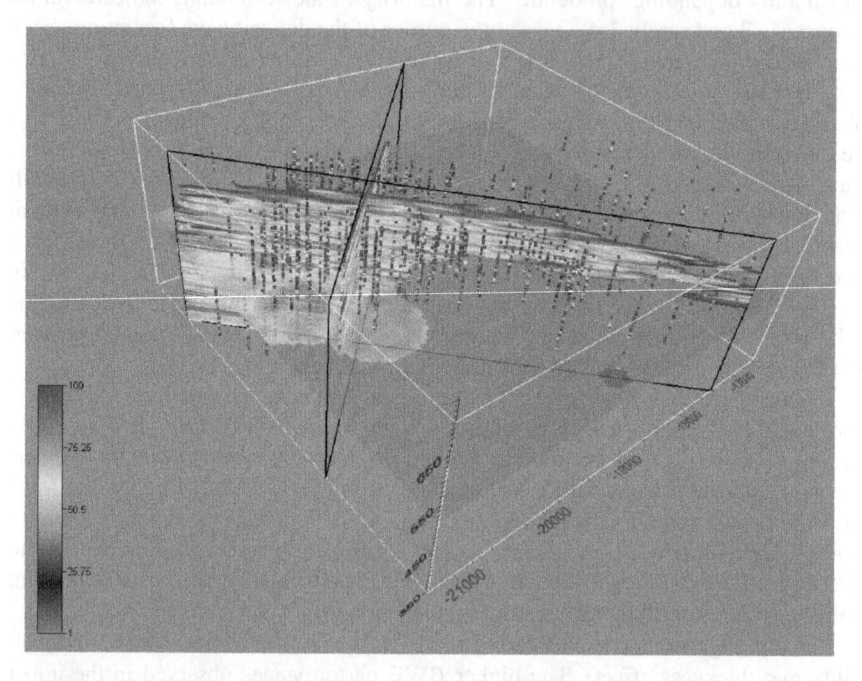

Fig. 5 Performance of a BWE with bucket-wheel diameter=8m when considering only the effect of the thickness of the excavated blocks

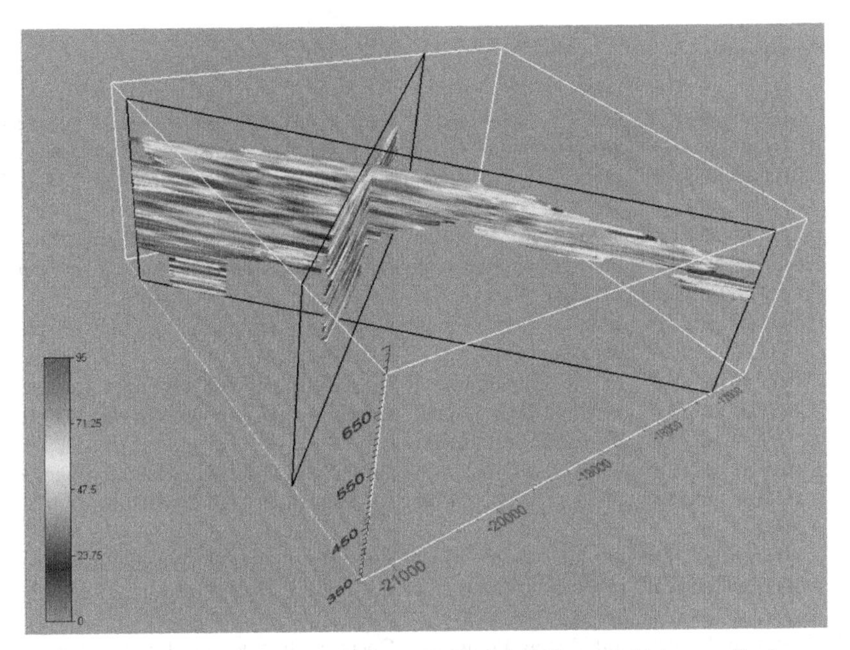

Fig. 6 Overall performance of BWE (includes both diggability and thickness effect).

on the one hand and to their relatively low diggability values on the other. From the level of 700m to 550m the performance decreases since it varies from 40-50%, while below the level of 550m the performance is further reduced to 20-25%. The drop in BWE performance is due to the decrease of the thickness of the formed mineable blocks (selective mining) as well as to the relatively lower diggability that the excavated material exhibits.

6 Conclusions

The applied methodology for the evaluation of multiple-layered lignite deposits is significantly improved when combined with an expert system for the assessment of the diggability of the excavated geological formations since it leads to the prediction of the performance of the used bucket-wheel excavators, which are considered to be critical elements in the overall continuous mining system.

Acknowledgment. Part of this study was carried out within the framework of DURECOBEL_11SYN_8_584. The authors affiliated with Technical University of Crete, would like to thank the General Secretary of Research and Technology of Greece for the financial support of the aforementioned project.

The authors also wish to thank Mr. Alexandros Kritikos, student of the Department of Mineral Resources Engineering-Technical University of Crete, for his valuable help in programming with Visual Basic.

References

1. Vamvuka, D., Galetakis, M., Roumpos, C.: Coal quality control techniques and selective grinding as means to reduce CO2 emissions. In: 6th International Conference on Sustainable Development in the Minerals Industry, Milos island, Greece, pp. 532–537 (2013)
2. Papanikolaou, C., Galetakis, M., Foscolos, A.: Quality characteristics of greek brown coals and their relation to the applied exploitation and utilization methods. Energy and Fuels 19(1), 230–239 (2004)
3. Galetakis, M., Vamvuka, D.: Lignite quality uncertainty estimation for the assessment of CO2 emissions. Energy and Fuels 23, 2103–2110 (2009)
4. Wade, N.H., Ogilvie, G.M., Krzanowski, R.M.: Assessment of BWE Diggability from Geotechnical Geological and Geophysical Parameters. In: Continuous Surface Mining, pp. 375–380. Trans. Tech. Publications (1987)
5. Ladányi, G., Sümegi, I.: Some issues of the technological design of bucketw-heel excavators. International Journal of Mining, Reclamation and Environment, 63–72 (2006)
6. O' Regan, G., Davies, A.L., Ellery, B.I.: Correlation of Bucket Wheel Performance with Geotechnical Properties of Overburden at Goonyella Mine. In: Trans. Tech. Publications, Australia, pp. 381–396 (1987)
7. Orenstein, Koppel: Soil Testing Equipment, Operating Instructions, Ref No. 834, 601-12 (1985)
8. Rasper, L., Rittner, H.: Der Aufschluss des Braunkohlentagebausder Neyveli Lignite Corporation und Erfahrungen mit Schaufelrädern in Hartem braum. Braunkohle, Heft 10, Okt. 390–400 (1961)
9. Kolovos, C.: Efficiency of a Bucket Wheel Excavator Lignite Mining System. International Journal of Surface Mining, Reclamation and Environment 18(1), 21–29 (2004)
10. Karamalikis, N.: Computer software for the evaluation of lignite deposits. Mineral Wealth, 76, 39–50 (1992) (in Greek)
11. Kavouridis, C., Leontidis, M., Roumpos, C., Liakoura, K.: The effect of dilution on lignite reserves estimation. Application in Ptolemais multi-seam deposits, Braunkohle 52(1), 37–45 (2000)
12. Galetakis, M., Vasiliou, A.: Selective mining of multiple-layer lignite deposits. A fuzzy approach. Expert Systems with Applications 37(6), 4266–4275 (2010)
13. Tripathi, K.A.: Review on Knowledge-based Expert System: Concept and Architecture. IJCA Special Issue on Artificial Intelligence Techniques - Novel Approaches & Practical Applications, 19–23 (2011)
14. Ritchi, D.: Artificial intelligence, Tata McGraw-Hill, New Delhi (1996)

Problems of Bucket-Wheel Excavators Body in Hardly-Workable Grounds in Polish Open Pit Mines

Weronika Huss

Mechanical Division, Wroclaw University of Technology, Poland

1 An Influence of Ground on Specific Operational Loadings of Bucket-Wheel Excavators

During Pleistocene the area of Poland has underwent an influence of several glaciers. This resulted in high density of ground and pushed in stones and erratic boulders of various size. Such a situation causes not only an increase of mining resistance, but also an increase of number of impact loadings and their value. According to the data, 13% of machine breakdowns in Polish open pit mines is a result of extreme geological conditions. As much again happens because of maladjustment of mining technology to a current circumstance [2]. Altogether – above a quarter of all breakdowns.

In difficult ground conditions (IV and V class of gettability) mining absorbs up to 50% of excavator's total power consumption. Costs of repairs and changes of buckets working in the same conditions are also about 50% of year expenses on excavators repairs. This situation worsens with the depth of mining (fig. 1). Additionally, majority of excavators working in Polish open pit mines exceeded their expected lifetime. Thus, residual durability of their structures is decreasing in quite high rate. In this situation no attempts of reducing the influence of excessive loadings is going to result in increase of excavators failure frequency and decrease of profitability of mines. Yet a cost of purchasing new machines is huge, so the exploiters aim at maintenance of existing fleet of machines in safe operation.

Mentioned heterogeneity of mined ground provides also significant difficulties in load model building. Due to mining technology, load vectors are random in every respect – their number, directions, values and points of application are changeable. This state enables a modelling only at a high degree of generalization. And what follows – the influence of operational loadings on durability of a structure's individual subassemblies is estimated with not sufficient precision. Also there is there is no certainty about assumption of linearity of a system, which is the structure of the excavator. Especially in the matter of extreme loads.

The article presents the most important achievements of Polish centres (research and industrial) connected with open pit mining branch. These solutions were developed with intention to solve problems resulting from more and more difficult mining conditions.

C. Niemann-Delius. (ed.), *Proc. of the 12th Int. Symp Continuous Surface Mining - Aachen 2014*,
Lecture Notes in Production Engineering, DOI: 10.1007/978-3-319-12301-1_7

Fig. 1 Boulders digged out during exploitation of Jozwin II pit in The Konin Mine

2 Resonance – The First Diagnosed Problem Resulting from Dynamic Overloads

At the beginning of 90's a worrying effect had been noticed – an excessive dynamics of body's structure of some bucket-wheel excavators [1]. The amplitudes of vibrations not only lowered working comfort, but also were so high that significantly disturbed any control of machine operation. In extreme cases it was impossible for staff to walk safely along the bucket wheel boom. Next, numerous and fast-propagating fatigue cracks started to appear (fig. 2) [2,8]. A reduction of structure's vibration can be achieved by lowering on intensity of operation. Yet it is not in mines' economic interest and cannot be accepted as a final solution.

Before taking any remedial actions, reasons of such state had to be identified. During the research, it was found that the cause of the increase of amplitudes was a formation of resonant vibration. For this purpose specially designed in Machine Designing and Operation Institute (Mechanical Division, Wroclaw University of Technology) strain gauge measuring circuit enabling measures of vibration in orthogonal directions (vertical bending, horizontal bending and compression) was used. Then the structure was excitated by impact loadings of various values (fig. 3). This way excavator's structure modal vibrations had been obtained. One of the frequencies present in those signals was same as the frequency resulting from the buckets and rotation of the bucket wheel. Operational vibration were also registered and they confirmed a resonance hypothesis [1].

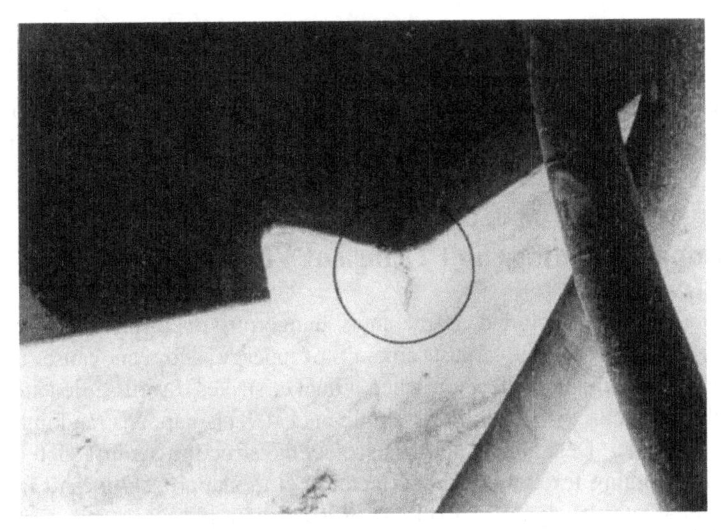

Fig. 2 A crack of upper shelf of lower box girder in bucket wheel boom – ca. 2,1 m from support joint on turn-table structure. SchRs-4000.37,5 excavator [2]

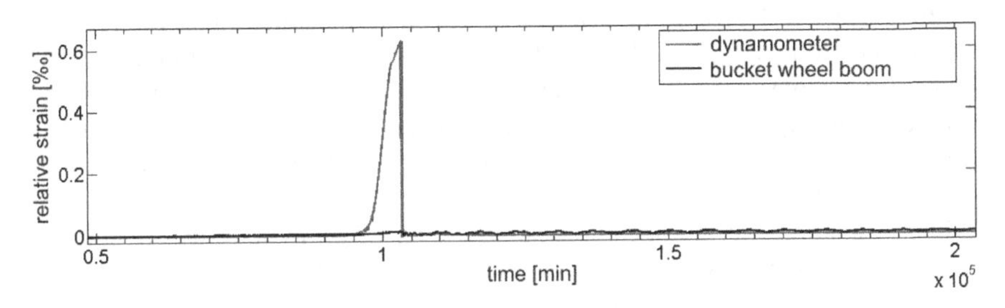

Fig. 3 An example of impact exciting an excavator's structure in order to obtain modal frequencies

Due to such situation, according to known formula:

$$m\ddot{x}(t) + c\dot{x}(t) + kx(t) = F(t), \quad (1)$$

where: m – mass, c – damping, k – stiffness, $F(t)$ – external force, to change machine's natural frequencies and move it out of resonant area, one can:

- change the frequency of pouring outs (from buckets to belt conveyor),
- change the body's mass,
- increase body's stiffness.

It's obvious that a change of excavator's body mass isn't preferred solution. Increasing of structure's stiffness is also troublesome. So the easiest and most preferred solution is a change in external force frequency. It is achieved by changing the final drive ratio (bucket wheel drive). It is also preferable, because it enables increasing of output.

Described procedure is successfully repeated on many other bucket-wheel excavators. Since the time of first such application, identification of modal frequencies of an individual machine has become a standard procedure carried out before placing it in service (new excavators, after major repairs and modernizations).

3 Impact Loadings and Transient Vibrations

Impact loadings are regarded as the most dangerous operational loadings. It's because they provide the largest amount of energy and can cause transient vibrations. Those loadings develop when a bucket strikes a well settled stone or a hard rock of dimensions larger than a single bucket. This happens randomly and is difficult to predict. Designing standards protect the structure against such loads by multiplying digging force by dynamic factor. Yet this approach doesn't include a different character of dynamic loadings. Results of rapid stroke are transmitted between subassemblies to the further parts of the excavator's structure. It also happened that it resulted in serious damages and crashes of bucket-wheel excavators [2]. Model research carried out at Machine Designing and Operation Institute, approved also numerically, showed that maximal impact effects in bucket wheel boom appear when the force is applied at angle 60° to the vertical. At that time compressed bars of main truss are exposed to the greatest effort and can easily undergo buckling. For example, sometimes bucket wheel booms in SRs-1200 excavators can reach 64 MPa during impact loading with mean dynamic stress of 24 MPa.

Impact loadings during operation can't be avoided. Only an effective solution for cutting off an energy excess against transmitting it through the structure can be proposed. It is implemented by various kinds of couplings:

- a mechanical friction clutch located on a beginning of gear shaft – a traditional solution, unreliable, requires frequent and burdensome regulation, isn't a good protection against impacts,
- a mechanical friction clutch together with hydrodynamic coupling located between an engine and a gear – proper for moderating a start-up, but not for as overload coupling, besides, it is also an additional rotational mass,
- an electrical braking by counter-current – instantly stops a bucket wheel, but much overloads elements of a gear.

So far, the most efficient solution was a strain gauge overload balance. It cuts off a drive inside main planetary gear. This led to significant decrease in number of break-downs of excavators SchRs-4000.

In the history of Polish open pit mining there were several cases of serious damages caused by impact loadings. The most interesting is the case of two excavators SRs-1200, among which one underwent a damage twice. In those three accidents consequences concentrated on bucket wheel boom. However, despite the same body structure, each time damaged elements were different (fig. 4). All cases of impact kind, but less harmful for SRs-1200 excavators were found 78. Those failures consisted of bar buckling, loosening and shearing of rivets and bolts as well as weld cracking. Standard overload friction coupling turned out to be inefficient. Other most serious damages caused by impact are presented in table 1.

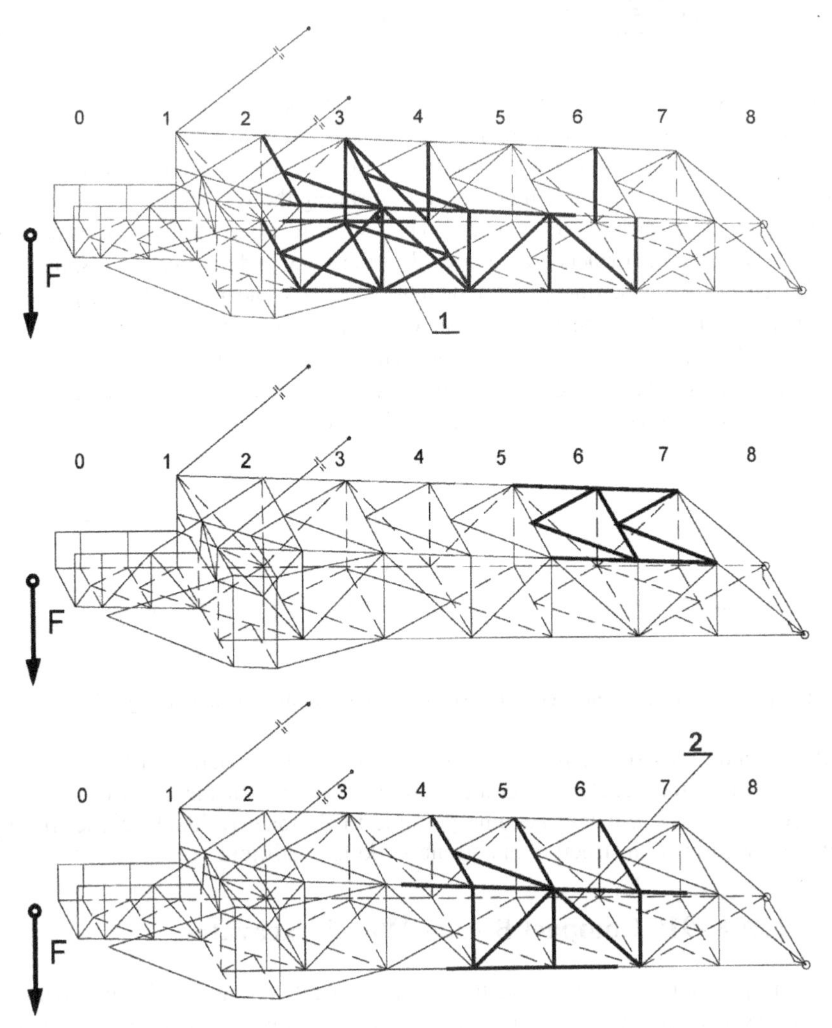

Fig. 4 Three cases of damages in bucket wheel boom caused by impact loadings in excavators SRs-1200

Table 1 The most important damages due to impact loadings bucket-wheel excavators in Polish open pit mines

excavator	damage
SchRs-4000.50	a breakage of moment beam breaking of wheels in main gear
SchRs-4600.30	a damage of bucket wheel drive shaft
SchRs-900	a distortion of whole body a deformation of moment beams of main drive
SchRs-4000.37,5	a cracking of bucket wheel boom and mast

In the result of further research it turned out that impact loadings may be dangerous not only due to their high energy. It was found that some of such events are accompanied by formation of additional frequencies in a structure [5]. This research is currently continued in a direction of stating the way of impact's dynamic influence on excavator's body structure. The reason for that is a hypothesis saying that at least some of impacts are connected to transients, which may be responsible for fatigue cracks appearing in unpredictable locations and progressing quite fast. Moreover it was noticed, that coincidence of impacts and untypical vibrations is the more probable, the highest is the excess of such impact above the rest of the values in its nearest neighbourhood (fig. 5 a). Not like it was believed earlier [4] – connected to absolute value of impacts (fig. 5 b). Other details can be found in [4].

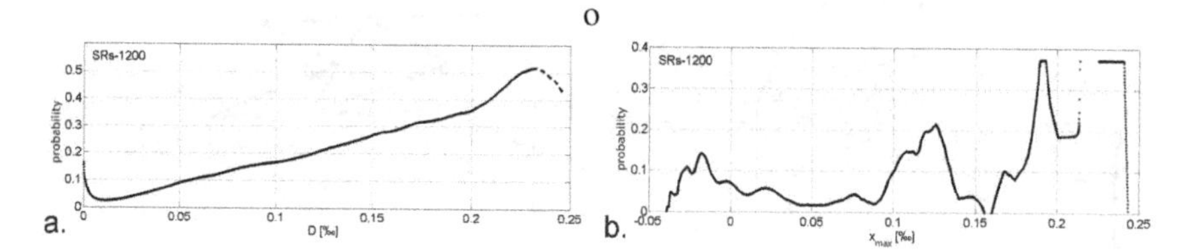

Fig. 5 A probability of appearance of transients accompanying impact loadings [5]

After conducting such a research, the mechanism of transients' influence on an effort of excavator structure is going to be uncovered. On the basis of a graph analogous to fig. 5 a. it will become possible to estimate residual fatigue life of neuralgic sections of a structure in current working conditions.

4 KWK-910 Compact Bucket-Wheel Excavator

Domestic production of basic machines for open pit mining has been extended with a new generation excavator. This is a compact bucket-wheel excavator for hardly workable geological formations. This KWK-910 excavator (fig.6) has been designed, constructed and assembled completely using polish companies only. The author of this project was SKW Design and Technical Office, which also supervised a building process of this excavator.

The machine has been equipped with a modern mining system with ability of cutting off impact excesses. Allows mining many types of rocks (with a linear resistance up to 200 kN/m) without a need of loosening them.

Fig. 6 KWK-910 excavator during a passage from assembly site to working level – clear signs of driving along the curves of with a small radius (25 m)

A double 500 kW bucket-wheel drive consists of a special bevel planetary gear, in which a special overload coupling was built in. The coupling allows instant (0,1 - 0,15 s) cutting off of dynamic excesses of certain values. One of the most important aims in choosing such drive construction is a protection of buckets and a steel structure of body from dynamics of mining process. For this purpose also a bucket-wheel with a large number (16 pcs) of buckets was chosen. This enables a simultaneous participation of four buckets in cutting process.

From December 2006 to December 2008 over 4000 cut-offs was noticed (impacts ≥460 kN). Although such large number of coupling cut-offs, no significant wear of break disk was found.

Parameters of mining system (fig. 7, other parameters of the excavator are presented in [7]):

- power: 2x 500 kW,
- bucket-wheel diameter: 10,2 m,
- number of buckets: 16 pcs,
- nominal bucket capacity: 910 l,
- number of pours: 66,5 1/min.

Fig. 7 Mining system with 2x 500 kW power unit

The body's structure has been designed for high fatigue resistance. Structural nodes have been designed to avoid stress concentration in welded joints areas. In the structure's main assemblies connections, movable joints have been introduced. This solution eliminates harmful influences from bending moments in places of rigidity step-change.

KWK-910 excavator's compact outline has been shaped in a way that allows advisable dynamic characteristics. This is to limit resonant vibrations due to input force frequency resulting from the mining process. Additionally, a continuous recording of stress spectrum is conducted in selected important locations at the body's structure. This enables an on-line estimation of fatigue life (details in [6]).

Special advantages of designed crawlers allow for transportation at slope 1:15 and mining at 1:20. The smallest turning radius is 25 m.

SKW Design and Technical Office – a leading Polish design office for mining machines [15], in cooperation with "Turów" open pit mine, managed to introduce into operation a modern, highly automated bucket-wheel excavator KWK-910. Equipped with visualization of control and diagnostic process, with cctv, alarm system, fire-fighting system etc. Widely applied user-friendly and nature-friendly solutions places this excavator among the most up-to-date machines for open pit mining.

5 New Buckets

On the basis of many years of experiences, SKW Design and Technical Office proposed two projects of buckets for cutting hard geological formations [14]. First of them – R12-N2 – has been designed for already existing KWK-1200 excavators (SchRs-1200 after modernization). The second one – CK910-N3 – has aroused for the newly designed compact KWK-910 excavator. The main idea behind both projects was to reduce repair costs and prolong the life of the whole bucket. Therefore a new shape of the cutting edges – corners and knife – have been proposed (Fig. 8).

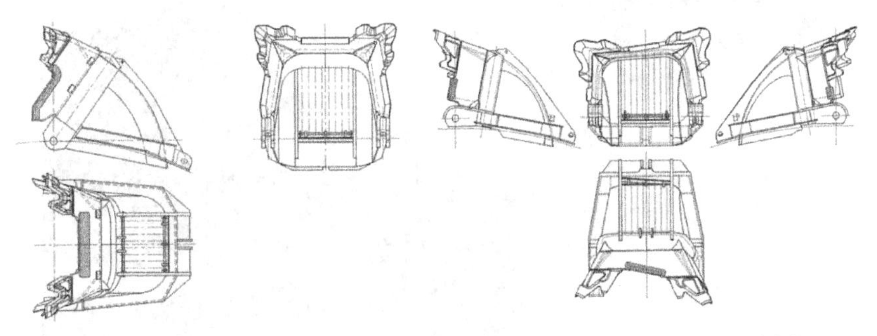

Fig. 8 Buckets for hard geological formations R12-N2 (at left) i CK910-N3 (at right)

Three pyramidal teeth in corners are located properly to the trajectory of cutting. In the initial stages of wear it ensures an optimal cutting angle 45°. Such a shape of cutting edges provides lower thrust force during whole wearing process. It also allows maintaining still advantageous cutting angle ≤60° in the final stages of wear with a requirement of self-sharpening (fig. 9). Special configuration of the lower tool flank (behind the teeth) ensures lowering influence of forces, which is particularly important while working in hard grounds.

In order to minimalize the volume of solids in the process of digging, each of the three corner teeth crushes independently and with a quite large advance in relation to knife. It results in lightening the abrasive wear of the knife. Therefore it doesn't wear as fast as it was for an old-type buckets. Hence main repairs of buckets are reduced to an easy exchange of corners. Additionally, such a structure concentrates stress in the corner part of the knife (the most stiff) and in the pre-knife.

Cutting corners for the buckets are made of a special alloy cast steel (chromium-nickel-molybdenum), which has high strength properties. It also enables deep hardening (ca. 150 mm) and multiple welding of corner and knife (used corners exchange). During intensive abrasive wear and impact loadings, padding welds on teeth undergo crumbling, distortion, cause high growth of digging resistance and finally premature destruction of knives and whole buckets. Therefore pad welding in R12-N2 and CK910-N3 buckets should be limited only to surfaces of knives. In the case of exchanging corners, annealing of the whole bucket is also not recommended. The design of both buckets allows for changing corners 10-15 times without a necessity of changing the knife. The shape of the corner has been designed so the stress in weld between corner and knife is quite low. During three months of using CK910-N3 buckets for mining very hard rocks, only a few number of weld breaks had been noticed.

Fig. 9 Wear of CK910-N3 bucket corners after ca. 136 hours of digging hard grounds

A geometric and strength durability of teeth, low failure frequency of buckets, self-sharpening and minimization of thrust force are the technical and economic advantages of these dwo proposed buckets.

6 Materials in Construction of Bucket-Wheel Excavators

Simultaneously with designing of mining machines in Poland, a research on the change of materials for newly built and modernized machines has been taken. This research was undertaken by Materials Science and Technical Mechanics Institute from Mechanical Division at Wroclaw University of Technology.

Research conducted within the theory of degradation showed that usage of unalloyed, low-carbon and effervescing steel is quite dangerous for structures of excavators bodies. However, an influence of degradation process on machines operation has been eliminated due to an improvement of welding technology. Another issue raising durability of structures was a recognition [11] and application [13] of modern corrosion protection methods (not only passive, but also through design solutions).

Other achievements was a deep recognition of structures and properties of L35GSM cast steel [10] as well as its multi-variant heat treatment (also austenitic cast steel L120G13). Despite that, some cases of fracture of track links eyes (in pre-operational state) or fracture of wheel rim teeth (after short time of operation, ex. 4 years) still happen.

Finally, a significant progress was noticed in the field of selection of chemical components of electrodes and technology of pad welding for surfaces exposed to friction wear during dynamic loading [9]. It turned out that it isn't true that the harder padding welt, the better it is.

Also some new materials were introduced. Never before low-alloyed, martensitic Hardox steel were used for open pit mining machines.

7 Hardox Group Steels

Hardox have been produced since 1970 (as Hardox 400) and up to now six grades of them are available. An evaluation of their operational behaviour wasn't carried or was negative. Beside a manufacturer's information, there was hardly any other research.

The manufacturer describes them as "high quality, resistant to abrasion steel". It is characterized by high wear resistance, ability to being machined by specialized tools, good weldability, high mechanical properties and impact resistance. Martensitic steel from Hardox group have a tensile strength from 1250 MPa (Hardox 400) to 2000 MPa (Hardox 600). Their microstructures are obtained from the normalized states through hardening in water and tempering in temperatures 200-700°C.

A comparative research has been carried out for Hardox 400, Hardox 500 and 18G2A steel. For this purpose plates made of those kinds were prepared and placed in the most abrasively loaded location at the fixed chute of bucket wheel (fig. 10). Then they were put into operation for 595 hours – an overburden of sanded clay was being mined out. Recorded values were: time of operation until a change of each plate, degree of abrasive wear and a place of attaching.

The comparative evaluation of chute condition was:

- 18G2A steel with padding welt – fractures of padding welt and attrition in some places throughout, rounded edges.
- Hardox stell – attrition to ca. 3-4 mm in the most intensively loaded places, but flat and smooth surface kept, no fractres or chipping, rounded edges.
- Lining plates made of Hardox 500 showed much lower abrasive wear than plates of Hardox 400.

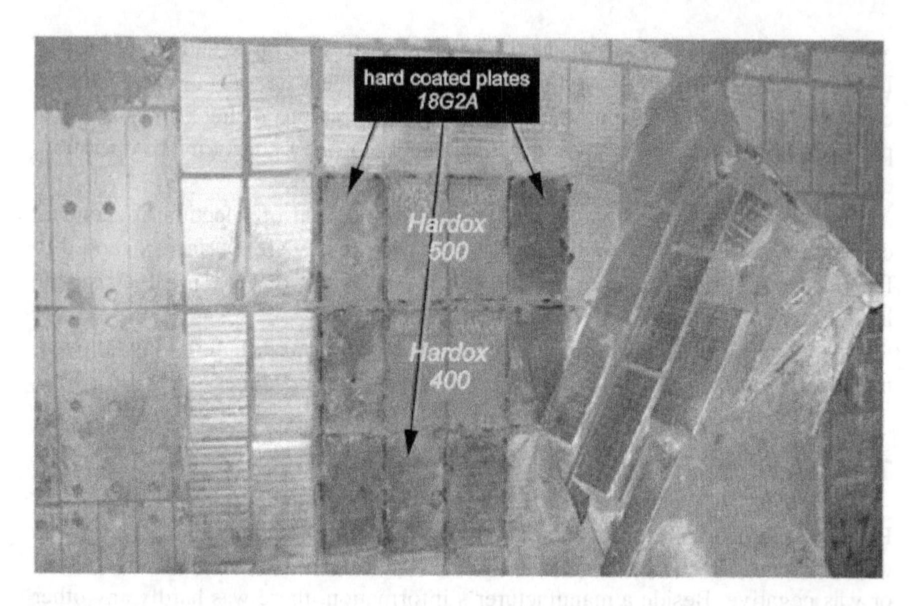

Fig. 10 A distribution of experimental plates on fixed chute surface [3]

Plates made of 18G2A with padding welt undergo replacement after about 50% of wear (a loss of operating properties). Plates made of Hardox steel may be used up to total wear and still they keep their mechanical properties. Because of an uniform wear, Hardox plates enable more effective usage. Welding of plates during their fixing causes increased wearing on edges. This disadvantage can be eliminated by fixing a plate by bolts and cutting them by a high-energetic water jet. Hardox plates visibly better receive and dampen dynamic loads coming from an output. This may result in decrease of loadings on excavator's body. The costs analysis of considered plates showed that the usage of Hardox may result in several dozen reduction of costs of fixed chute operation than 18G2A.

8 Summary

Undoubted achievement of research undertaken by Polish entrepreneurs and researchers is the "lifetime" extension of machines that have been in motion for more than 30 years. The adverse situation in which Polish brown coal open pit mines have found in due to geological conditions, has become an incentive to the development in designing of structures, their operating and diagnostics as well as in material research. These issues, of course, are still developed and implemented in practise.

References

[1] Augustynowicz, J., Dudek, K., Figiel, A.: Problemy drgań rezonansowych nadwozi koparek kołowych. In: XIV Konferencja Naukowa Problemy Rozwoju Maszyn Roboczych, Zakopane (2001)
[2] Babiarz, S., Dudek, D.: Kronika awarii i katastrof maszyn podstawowych w polskim górnictwie odkrywkowym, OWPWr, Wrocław (2007)
[3] Cegiel, L., Konat, Ł., Pawłowski, T., Pękalski, G.: Stale Hardox – nowe generacje materiałów konstrukcyjnych maszyn górnictwa odkrywkowego. Węgiel Brunatny 3(56) (2006)
[4] Dudek, D.: Elementy dynamiki maszyn górnictwa odkrywkowego: Akwizycja sygnałów, analiza układów, OWPWr, Wrocław (1994)
[5] Huss, W.: Metoda identyfikacji stanów nieustalonych ustroju nośnego koparki kołowej przy obciążeniach losowych, doktorat, Politechnika Wrocławska (2012)
[6] Kowalczyk, M.: Wymiarowanie spawanych konstrukcji nośnych maszyn podstawowych górnictwa odkrywkowego w zakresie trwałości zmęczeniowej, PhD thesis, Wroclaw (2010)
[7] Kowalczyk, M., Wocka, N.: Koparka KWK-910 – koparka specjalna do pracy w pokładach trudnourabialnych w PGE Kopalnia Węgla Brunatnego Turów S.A. Węgiel Brunatny 1(66) (2009)
[8] Muchaczow, J.: Doświadczalno numeryczna metoda estymacji obszarów rezonansowych w ustrojach nośnych maszyn roboczych, PhD thesis, Wroclaw (2009)
[9] Napiórkowski, J., Pękalski, G., Sochadel, U.: Optymalizacja doboru napoin w ujęciu materiałoznawczym, Mat. VIII Forum Energetyków, Politechnika Opolska (2002)
[10] Pękalski, G.: Kształtowanie struktur i własności odlewów ze staliwa manganowo-krzemowego poprzez obróbkę cieplną, Acta Metall. Slovaca (2001)
[11] Pękalski, G.: Przyczyny i skutki korozji koparek węgla brunatnego, Górnictwo Odkrywkowe (4) (2002)
[12] Waroch, M.: Wybrane zagadnienia odtwarzania potencjału produkcyjnego w kopalniach węgla brunatnego – budowa i modernizacja maszyn podstawowych. Węgiel Brunatny 1(50) (2005)
[13] Wocka, N.: Zwałowarka ZGOT-15400.120 - Maszyna nowej generacji przekazana do eksploatacji w BOT KWB Bełchatów SA o/Szczerców. Węgiel Brunatny 1(50) (2005)
[14] Wocka, N.: Buckets for extracton of very hard for mining formations with Bucket Wheel Excavators. Górnictwo Odkrywkowe 5-6 (2007)
[15] `http://www.skw.pl` (January 29, 2014)

Impact of the Bucket Wheel Support at Technical Parameters of the Block and Bucket Wheel Excavator Capacity

Dragan Ignjatović[1], Branko Petrović[2], Predrag Jovančić[2], and Saša Bošković[3]

[1] Mining Basin Kolubara, Lazarevac, Serbia
[2] Faculty of Mining and Geology Belgrade, Serbia
[3] Mine Gacko, Gacko, BiH-Republic of Srpska

Abstract. Bucket wheel excavators are machines that on the current level of surface mining development have gained the widest application. Its work consists of the main movements (slewing of the bucket wheel and superstructure) and auxiliary movements (moving of the excavator toward advance direction and modification of the height position for the bucket wheel in the vertical plane).

The support of the bucket wheel (bucket wheel boom) is the most loaded part of the excavator structure participating with 6 to 13% to the complete excavator weight. Depending on the soil-mechanic properties of the deposit, as well as mining technology, it is selected the optimal parameters of the excavator and therefore the required boom length, of course with the optimization of economic indicators.

This paper provides comparative analysis of the BW boom length influence on the technical parameters of the block as well as on the capacity of different types of excavators.

1 Introduction

Block parameters of the bucket wheel excavator depend on its size and vice versa - the required parameters of the site cause the size of the bucket wheel excavator. In other words, on the parameters of the site a large impact has bucket wheel boom structure, and bucket wheel boom structure to the overall design and weight of the bucket wheel excavator. Bucket wheel excavator without moving of the bucket wheel boom excavates face slope of the site through the operation of machine transport in the longitudinal direction, and side slope by reducing the slewing angle in the lower subbenches/cuts.

Together with the bucket wheel, the bucket wheel boom represents the operating element of the bucket wheel excavator. Heavy loads of the boom by dynamic and static forces, as well as its own mechanism and devices weight make it extremely loaded structure with the complex stress conditions of high value. It can be said that boom is the most loaded part of the bucket wheel excavator, and therefore during design and construction of the bucket wheel excavator it should been seek optimization for this part.

C. Niemann-Delius. (ed.), *Proc. of the 12th Int. Symp Continuous Surface Mining - Aachen 2014*, Lecture Notes in Production Engineering, DOI: 10.1007/978-3-319-12301-1_8

Supporting structure of the boom should transfer all forces occurring (during operation, transportation, and state of rest) to the central structure of excavators. The shape of the supporting structure is to aligned with the direction of load action and to the requirements of the technological parameters of the excavator, too. Great influence on the shape of the boom structure has disposition of the bucket wheel with drive aggregates and transfer point. The main boom structure must be designed to allow the installation of auxiliary structures taking into account the free space between parts being in the interrelated movements. In addition to these requirements, the structure must be suitable for carrying out maintenance operations often carried out on the drive mechanisms, components, and installations. Type of the designed boom structure depends primarily on the excavator sizes, and therefore the boom type. There are applied two types of structures:

- Boom structures in the form of solid walls (usually on larger sizes and weights of excavators).
- Boom structures in the form of spatial lattice structure (usually on larger sizes and weights of excavators).

Crossing part of the structure, boom bearing to the central pillar, is arranged by specially shaped sheet metal girders, box-section for both types of booms. In the box-girders are embedding boom bearings through which all loads are transferred to the central pillar. Part of the boom in the area of the bucket wheel is specially shaped depending on the bucket wheel location and its aggregates.

The material from which boom is made should have the following basic characteristics: high tensile strength, high yield strength, high resistance to brittle fracture, good weldability, and affordable price. The modern designed excavators have the following design features:

- The main structure material is steel S 355 J2 +N;
- The structure is carried out by welding technology;
- Only the main node points are carried out by screws, unregulated with a high-tensile force (HV);
- Application of other connecting ways is present to a much smaller extent.

1.1 Length of the Bucket Wheel Support (Bucket Wheel Excavator Boom)

Boom length is selected according to the technological requirements of mining deposit, and must be consistent with the structural possibilities for this type of machine. Anticipated mining technology is a starting point in the design of excavators, and thus determines the length of the boom. Only well-designed excavator with the optimal selection of mining technology, i.e. in compliance with the soil-mechanical and geological characteristics of the deposit, can provide full effects in the exploitation.

Physical-mechanical properties of the material to be excavated with the geological features of the deposit determine the optimal type of excavator for this deposit, i.e. its technical characteristics. Knowledge of the mentioned properties of materials to be excavated is essential due to the bench maximum height, where slopes are stable. Slope stability (face and side) is provided in two basic ways: with lower benches and smaller angles of slopes inclination. Hydro-geological characteristics of the deposit, the water bearing degree and working plane capacity are directly related to the size of the transport mechanism. Based on the mining technology requirements complying with the soil-mechanical properties of the deposit create conditions for the selection of optimal parameters for excavators, and thus determine the required length of the boom. Designing implementation of the excavator requires in most cases compromised solutions between the demands of mining technology and machine technical capabilities, of course, with the optimization of economic indicators.

Designing performance of the excavator, that is, implementation of single devices is closely related to the length of the boom. Boom length is in specific proportion with the following parameters: bucket wheel diameter, location of devices and aggregates installed on the boom, height of the suspension point of the boom from the working plane, the boom cross-section and size of travelling device.

Capacity of the bucket wheel, as already noted, is in a function of its diameter. Correlation ratio between the bucket wheel diameter and boom length is a function of the excavator type. Namely, by the intensive development of compact excavators in the last decades it is possible to divide it into two basic types of excavators, as follow: compact excavators, whose main characteristics are reflected in the small height range, short boom, the low drive weight, and high capacity compared to the so-called classical structures of excavators.

Important structural differences between these two types of excavators can be best noted in correlation between the bucket wheel diameter and boom length. Based on the statistical processing of 18 models of compact excavators (Krupp 6, O&K 6, Demag 1 and Takraf 5 pcs.), with capacity of Q_{min} = 1250 m^3/h to Q_{max} = 7500 m^3/h, it is evident the following ratios between bucket wheel diameter and the length of the boom (Table 1).

Table 1

Manufacturer	Type of excavator	The ratio of boom length L_s and bucket wheel diameter D
KRUPP	Compact excavators	$L_s = (1.7 - 2.1) \cdot D$
O&K		$L_s = (1.5 - 1.9) \cdot D$
TAKRAF		$L_s = (1.95 - 2.4) \cdot D$

Table 2

Manufacturer	Type of excavator	The ratio of boom length L_s and bucket wheel diameter D
KRUPP and O&K	Classical structures excavators	$L_s = (3.22 - 4.02) \cdot D$
TAKRAF		$L_s = (2.40 - 4.55) \cdot D$

At classical structures ofexcavators (statistically treated) larger number of excavators manufactured Krupp, O&K and Takraf), ratio of boom length and bucket wheel diameter is different (Table 2).

On the basis of this comparative review it may be conclude that ratio of boom lengths and bucket wheel diameter range in a wide scope from $L_s = (1.5 - 4.55) \cdot D$. The above ratio depends on the structural excavator type and level of bucket wheel and boom adjustment to the working environment.

Disposition of the bucket wheel drive, transferring point and location of conveyors in the boom structure are elements on which free cutting angles of excavator depend. The above angles are dependent on the boom length, where by the increase in the length are created more favourable values for these angles, being discussed with more details in the fifth Chapter.

Cross-section of the boom depends on its length and the installed drive power for the superstructure slewing movement. Expected loads in the horizontal and vertical planes are the starting points for designing the boom cross-section and are directly related to its length. Booms on excavators excavating materials of low strength and low density are with smaller cross-section, that is, have a greater slenderness than excavators provided for solid materials, which is particularly evident on the dumping bucket wheel excavators. Namely, these combined machines excavate, to be more precise, load already loose material from the stockyard, which provides a very low resistance to digging.

1.2 Bucket Wheel Influence to the Total Structure and Weight of the Bucket Wheel Excavator

Support length, i.e. bucket wheel boom depends on:

- Excavator designing scheme: structure of the bucket wheel boom with shifting or without shifting, the bucket wheel diameter and angle of its equipment arrangement, the height of the joint fixing for the bucket wheel boom suspension and its location relative to the axis of the excavator slewing, the size/length of the articulated joint feed (if in question is bucket wheel boom with shifting capability), cross-sectional radius of the boom and its location relative to the boom, weight of transported material, the height of transport device parts forward positions out of the turntable;
- The set of technology and physical and mechanical working conditions: type and composition of rocks/material, the digging height (during excavation in horizontal or vertical cuts), digging depth, angles of benches, face and subbenches site.

The main initial conditions are height and digging depth, position of the boom suspension joint, transport device size, bucket wheel diameter and soil conditions. All other conditions are in the some extent determined based on initial or can be corrected during design process. Stable factor, which practically is not modified due to the technology and machines structure, is the limit angle of the belt conveyor inclination on the bucket wheel boom. Boom length as well as the bucket wheel diameter significantly affect the weight of the excavator, and even more, if excavator is larger, that is, larger type (higher capacity) and higher capacity (Figure 1). Boom of normal size (for a given size of the excavator), the slope of the tangent for curve Qteor is, according to data by NKMZ, approx. 30-50° (more for larger capacity of excavators, that is, power, i.e. larger type).

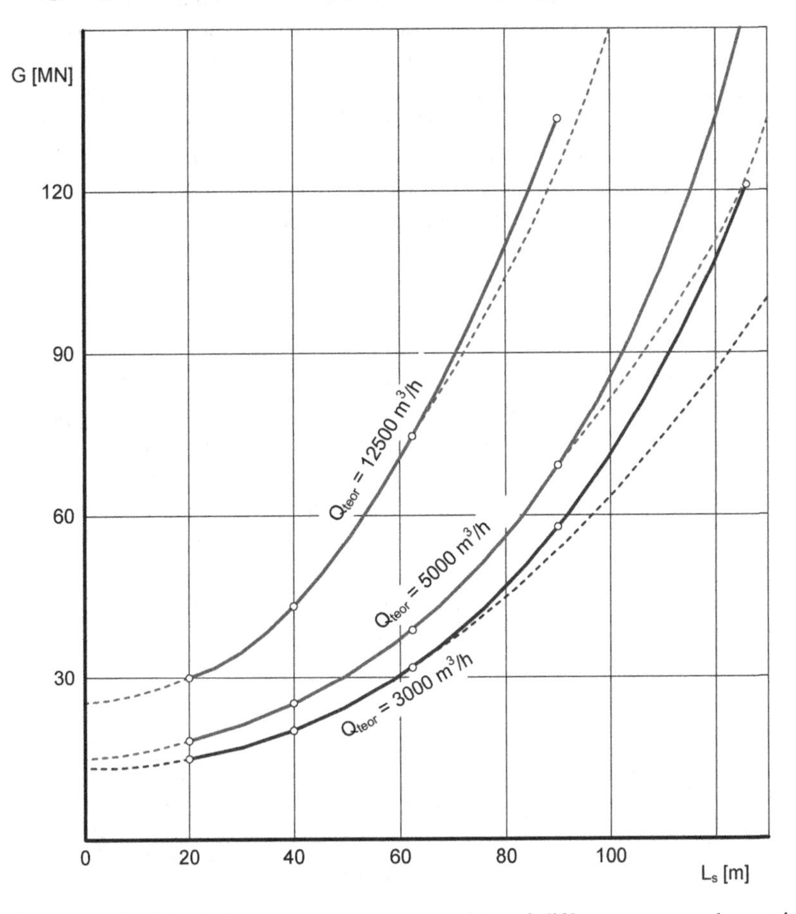

Fig. 1 Bucket wheel Ls influence on excavators weight of different types and capacities (acc. to data by NMKZ)

2 Influential Factors on the Bucket Wheel Excavator Capacity in the Function of the Boom Length

Factors that affect the excavator capacity can be divided into two basic groups:

- Factors that limit technical capabilities of the excavator by capacity,
- Factors that cause delays of the excavator and its utilisation by the time.

Capacity of the excavator can be significantly limited by technological excavation conditions. Maximum level of bucket wheel excavator capacities at the height bench excavation at a given height and angle of bench inclination is achieved during block excavation, determined by the bucket wheel boom slewing to the side slope and the highest subbench for 80^0, according to the excavated area within the lowest subbench for 30^0. During the block parameters reducing cause reduction of capacity, and further a reduction of operation efficiency within the block, as it reduces the effective time of excavation, and increase delays in the framework of the operation (changing of the subbench, block change, etc.). The reduction is particularly important during reducing of the block width and during the height and cuts thickness reducing.

Beside the main technological conditions that affect capacity (width of the block, block height) boom length directly affects the length of excavation within a single subbench thus reducing the non-productive operations in the block, the length of the boom directly affects the possibility of obtaining the angle of slope inclination being excavated, which is very important for the safe operation of excavators, as well as the concentration of mining operations.

Figure 2 shows dependence of the block width for three different types of excavators operating in the MB Kolubara - C 700 (compact excavator type A and boom length of 14.5 m), the standard excavator SchRs 630 type B and boom length of 35 meters) and excavator type C SRs 2000 with boom length of 44 meters. For the each excavator on the diagram is shown variation in the function of the boom length (increasing and decreasing of the boom length). The figure clearly indicates an increasing in the block width, and therefore the possibility of achieving greater efficiency in the block. Greater block width reduces the amount of auxiliary operations on the shifting of conveyors.

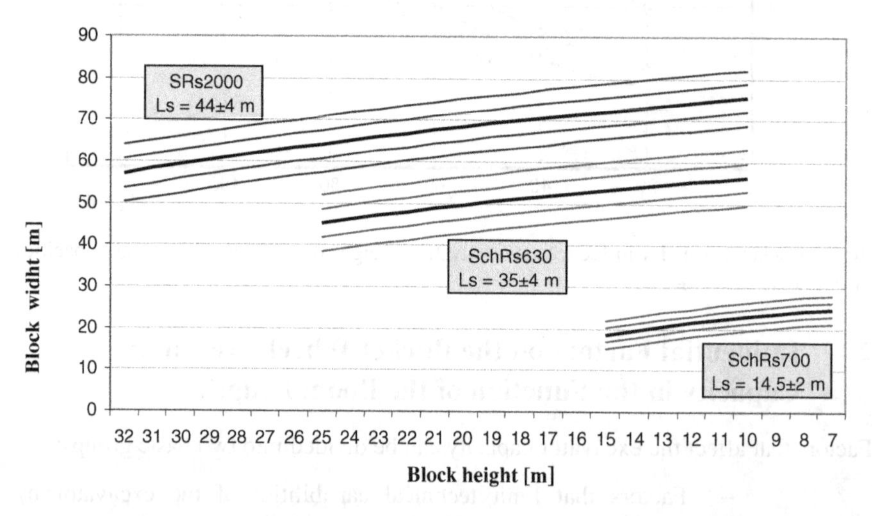

Fig. 2 Block width dependence of the boom length on the block height for different types of excavator

Figure 3 shows dependence of the advance direction within one subbench for different boom lengths and block heights. The figure also shows an increasing in the length of excavation within one subbench, which creates the possibility of achieving greater efficiency in the block. Larger advance direction within one subbench reduces scope of operation for auxiliary machinery on smoothing the plane and involvement with the bucket wheel excavators.

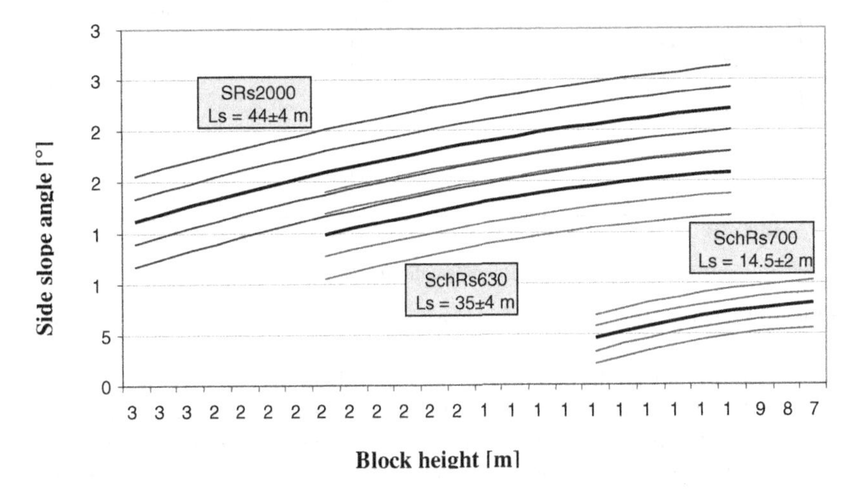

Fig. 3 Advance direction of one technology cycle dependence within the subbench on the boom length and block height for the various types of excavator

Figure 4 shows dependence of the possibilities in achieving inclination angle of the side slope for the different boom lengths and block heights. Possibility to create more smooth inclination angles of the side slopes is achieved by increasing the boom length, being very important for the stability of slopes.

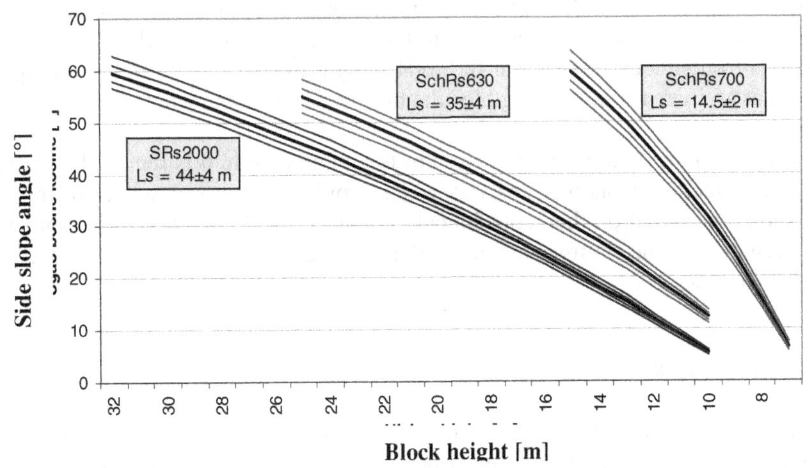

Fig. 4 Side slope angle dependence by the boom length and block height for the various types of excavators

The above-mentioned technological benefits of excavators with longer booms affect the efficiency operation increasing in the block (Figure 5), therefore depending on the excavator efficiency height with longer booms it is around 80%, while for compact excavators with short boom it is up to 70%.

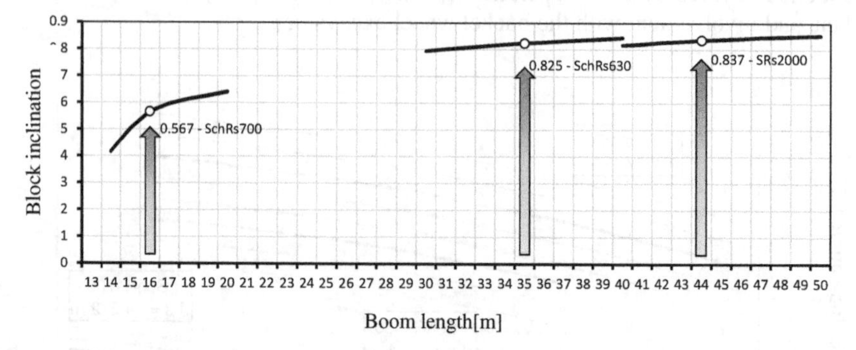

Fig. 5 Block efficiency dependence on the boom lengths for the various types of excavators

Undoubtedly, from all figures it can be seen advantage of excavator technological opportunities in function of boom length increasing. However, it should be considered the graphical illustration on Figure 1, which shows an increase in weight of excavators and therefore the purchase price of excavators in function of the boom length. Due to it, it is necessary to perform optimisation of the boom length in the function of the working conditions prevailing in actual opencast mine.

Researching the impact of the boom length on the efficiency [4] has prove that for the increasing of the block efficiency of +6%, it is required an extension of the bucket wheel boom from 35m to 40m based on the angle of slope of 45° and excavation height of 25m. This leads to an increase in device weight of about 15%. Further extension of the bucket wheel boom for an additional 5 meters (45m) causes only an increase of about 1.5% in the efficiency of the block. However, it requires a tremendous increase in the total weight of even +25%. This example shows that an increase in the length of the bucket wheel boom cannot be taken indefinitely.

In order to perform a proper selection of the boom length, it is required to carry out detailed techno-economic analysis which would for the each actual opencast mine should be based on the geotechnical parameters of the working environment for determination the optimal length of the boom and the concept of the bucket wheel excavator.

3 Conclusion

Length of bucket wheel boom has a crucial impact on the effectiveness of the excavator operations that is the length of the bucket wheel boom as the main factor affecting the efficiency of the excavator operations, i.e. its effective capacity. This relates to the standard operation and in particular to the selective mining.

The overall height and width of the block to be mined, and the block length depend on the length of the bucket wheel boom. The longer the bucket wheel boom is the better is operation efficiency in the block and actual capacity of the excavator. However, it should be noted that up to a certain value, efficiency is growing faster and then increasing the boom length does not significantly affect on the efficiency increasing, but significantly affects the width and therefore the purchase price of the excavator.

Further advantages of larger bucket wheel boom are as follows:

- Suitability for selective mining
- A small inclination of the belt on the bucket wheel boom in operation and in upper subbenches
- Reducing of interval for shifting of conveyors
- Larger depth of operation below the level of crawlers, provides greater flexibility during selective operations
- Optimization of ratio Q_{th}/Q_{eff}, due to better efficiency reduces the required width of the belt and the required speed of other the equipment in the system.

In order to perform a proper selection of the boom length, it is required to carry out detailed techno-economic analysis which would for the each actual opencast mine should be based on the geotechnical parameters of the working environment for determination the optimal length of the boom and the concept of the bucket wheel excavator.

References

[1] Mayer, T.: Mining anh hulage of coal in open pit mines particulary with regard to selective mining. In: IX International Symposium MAREN, Lazarevac, Serbia (2011)
[2] Pavlović, V., Ignjatovic, D.: Selective coal mining by continuous systems. In: Faculty of Mining and Geology, Belgrade, Serbia (2012)
[3] Petrović, B., Pavlović, V., Ignjatovic, D., Stepanović, S.: Bucket wheel excavator boom length optimization for selective operation on Radljevo open pit mine in Kolubara. In: IX International Symposium of Continuous Surface Mining, Dresden, Germany (2013)
[4] Veljković, N.: Development of continuous mining machines at EPS lignite mines and its application in view of selective mining. In: Coal 2013, Zlatibor, Serbia (2013)
[5] Study The Selection of Excavation-Transport-Dumping Equipment During Selec-tive Mining of Coal Series. In: Faculty of Mining and Geology, Belgrade, Serbia (2010)
[6] Technical documentation by TKF company

Typification of the Mining Technological Complexes at the Quarries

Bayan R. Rakishev and Serik K. Moldabaev

Kazakh National Technical University,
Almaty, Republic of Kazakhstan
{b.rakishev,moldabaev_s_k}@mail.ru

Abstract. The definition of technological complexes of opencast mining is given. Their systemization is done. It shows that in the open cast mining are used the following technological complexes (TC).

Simple systems complexes: excavator, bulldozer, scraper, drainage and hydro-mechanized technological complexes. At these TC the excavation and movement of rock mass is realized by a single machine.

Two-component complexes: excavator-rail, excavator-automobile, excavator-conveyor, excavator-console, excavator-dump-bridge technological complexes. At these TC the excavation and loading of mining mass is produced by one machine - an excavator (shovel, bucket wheel, chain), and its transportation is carried out by another machine (different type of transport).

Three-component complexes: excavator-automobile-rail, excavator-automobile-conveyor, excavator-automobile-skip, excavator-automobile-crate, excavator-automobile-of-different-types technological complexes. At these TC the excavation and loading of mining mass is produced by one machine (excavator), and its transportation is realized by two types of transport.Examples of their implementation in the quarries of Kazakhstan and Uzbekistan are given. The comptetive of the three-component technological complexes of deep pits is proved.

Keywords: technological complexes; single-component, two- and three-component complexes; cyclic, cyclic-stream and stream technology.

1 Statement of the Question

The following definitions of the basic concepts of geotechnology are used [1,2].

Mining technology - a set of methods and techniques of the mechanized implementation of mining processes to extract mineral resources of the required volume and given quality from the Earth crust.

Technology of stripping and mining operations - a combination of techniques and methods of excavation, loading and movement of overburden and minerals from the quarry to the destination.

C. Niemann-Delius. (ed.), *Proc. of the 12th Int. Symp Continuous Surface Mining - Aachen 2014*,
Lecture Notes in Production Engineering, DOI: 10.1007/978-3-319-12301-1_9

Technology of opencast mining - a collection of excavation and loading and transport operations technologies and technical means of their implementation.

In accordance with this definition, the structure of the technological complex (TC) of opencast mining can be represented as shown in Figure 1.

If technology excavation and loading and transport operations predetermined by geological conditions of occurrence of mineral deposits and displacement character of mining workings in quarry field, the mechanical equipments are intended to provide productive and quality execution. It requires a high-performance means of mechanization in each process, their reliability and durability in operation. Reasonable selection of means of mechanization of the main processes relevant to each other in performance, size and parameters of design subsystems ensures consistent, reliable operation of the technological complex the opencast mining [1-3].

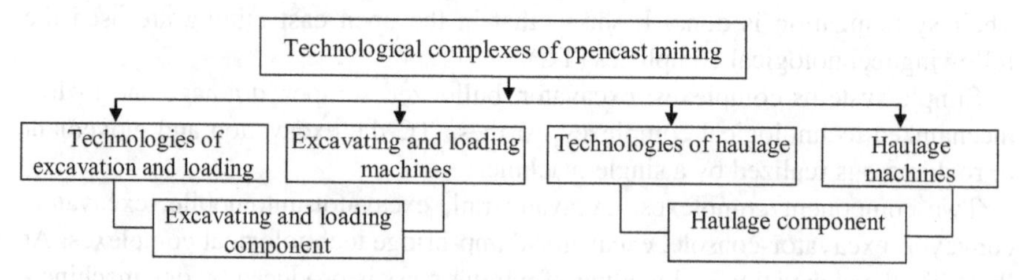

Fig. 1 The structure of the technological complex of the opencast mining

2 Typification of the Mining Technological Complexes of the Opencast Mining

Four elements of TC combining by the functions perform two components of complex: excavation and loading and transport (see Figure 1.).

In the excavation and the loading component as a means of mechanization used machines by cyclic: hydraulic excavators, mechanical shovels, draglines, scrapers, bulldozers, shovel loaders; and continuous machine: bucket-wheel and chain excavators, milling machines, hydromechanization tools - hydromonitors, floating suction dredger, dredge, etc.

In the transport component as a means of mechanization used cyclical transport: railway, automobile transport, diesel and trolley cars, lifting devices and continuous machines: conveyors, belt console; conveyor bridges, means of gravitational and hydraulic transport and cable-suspended road.

According to the number of the components, technological complexes of opencast mining can be divided into single-, two-and three-component. They may be referred to the name of the participating machines, such as excavating complex excavator-conveyor complex etc. Detailed systematic of the TC is shown in Fig. 2.

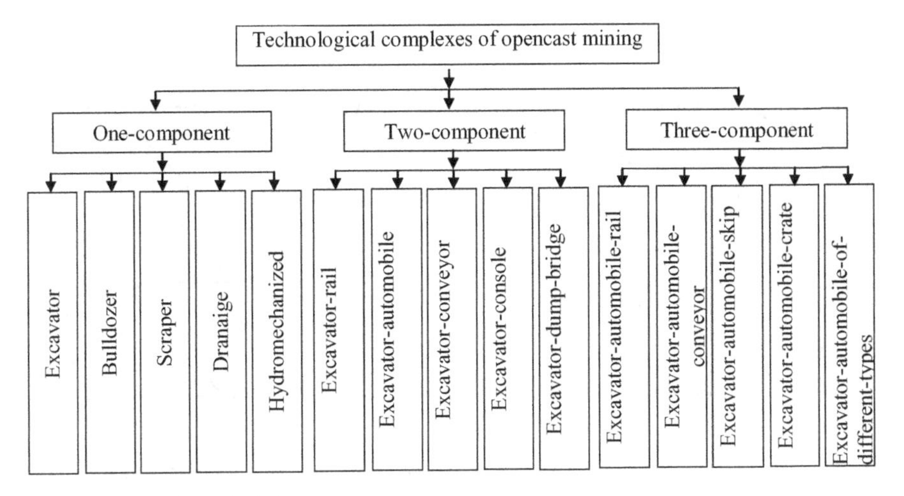

Fig. 2 Typification of TC of opencast mining

Simple systems complexes include: excavator, bulldozer, scraper, drainage and hydro-mechanized technological complexes. At these TC the excavation and movement of rock mass is realized by a single machine. Huge stripping excavators and draglines move the overburden into mined-out space, bulldozers and scrapers deliver to the destination. Dranaige and hydromechanized complexes also combined these processes.

Two-component complexes can be: excavator-rail, excavator-automobile, excavator-conveyor, excavator-console, excavator-dump-bridge technological complexes. At these TC the excavation and loading of mining mass is produced by one machine - an excavator (shovel, bucket wheel, chain), and its transportation is carried out by another machine (different type of transport).

Three-component complexes: excavator-automobile-rail, excavator-automobile-conveyor, excavator-automobile-skip, excavator-automobile-crate, excavator-automobile-of-different-types technological complexes. At these TC the excavation and loading of mining mass is produced by one machine (excavator), and its transportation is realized by two types of transport. It may be several types of TC in the pit. They are divided into TC of stripping and mining operations by type of rock mass produced.

Structure of the technological complexes are different depending on the content of the components. Transport component may be different at the same excavation and loading component.

3 Technological Complexes of Different Systems of Mining

Technological complexes of stripping and mining operations aren`t normal in coal mines and they don`t differ in mines of hard-rock. Also they can be same at different mining systems.

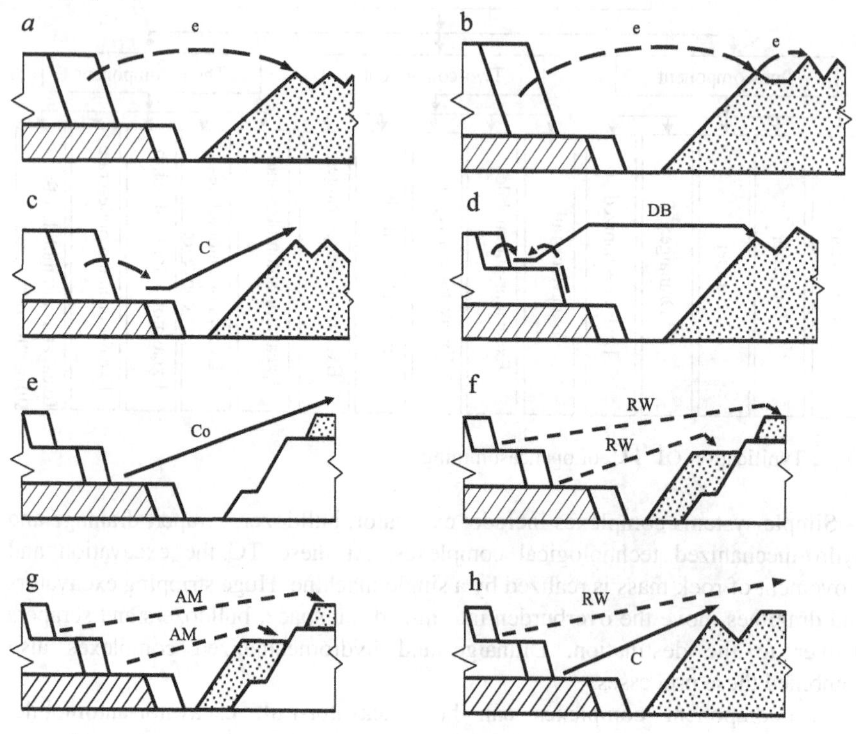

Fig. 3 Technological complexes of stripping works in continuous mining system

I. Following technological complexes are widely used in continuous mining system (Fig. 3):

1. Excavator technological complex of stripping works with transshipment of overburden into mined-out (fig. 3, *a, b*);
2. Excavator-console technological complex of stripping works (fig. 3, *c*);
3. Excavator-dump-bridge technological complex of stripping works (fig. 3, d);
4. Excavator-conveyor technological complex of stripping works (fig. 3, e);
5. Excavator-rail technological complex of stripping works (fig. 3, f);
6. Excavator-automobile technological complex of stripping works (fig. 3, g);
7. Combined technological complex of stripping works (fig. 3, h);
8. Dranaige technological complex of mining works;
9. Hydromechanized technological complex of mining works;
10. Scraper technological complex of mining works;
11. Buldozzer technological complex of mining works;
12. Technological complex of building rocks mining.

II. Following technological complexes are used in sinking mining system:

1. Excavator-rail technological complex of stripping works and mining works (fig. 4, *a*);

2. Excavator-automobile technological complex of stripping works and mining works (fig. 4, b);
3. Excavator-conveyor technological complex of stripping works and mining works (fig. 4, c);
4. Excavator-automobile-of-different complex of stripping works and mining works (fig. 4, d);
5. Excavator-automobile-skip complex of stripping works and mining works (fig. 4, e);
6. Technological complex of combined mining system (fig. 4, *f*).

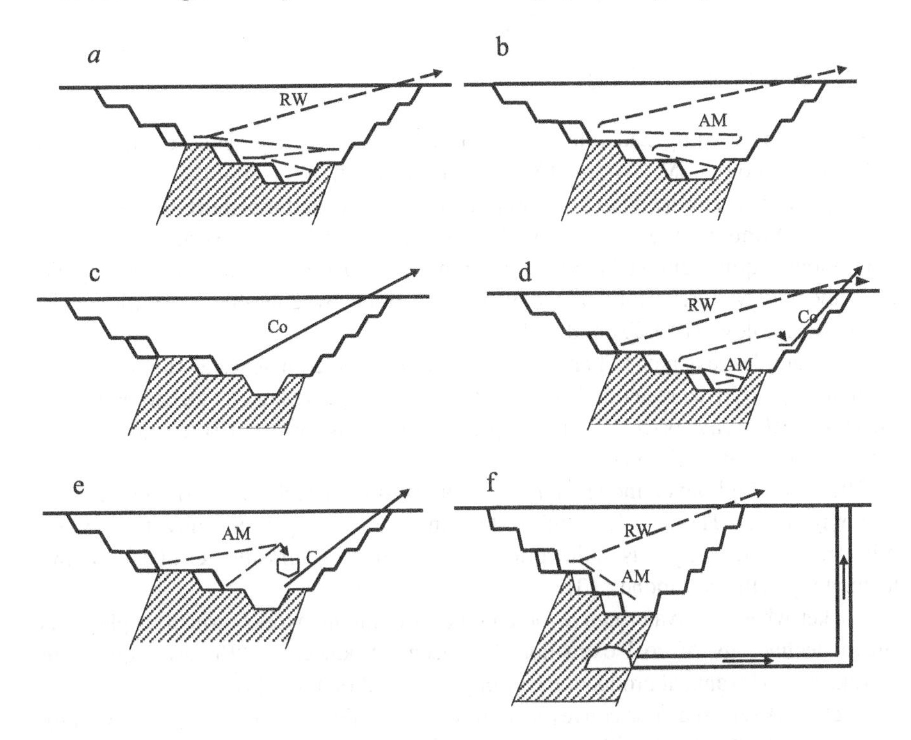

Fig. 4 Technological complexes of stripping and mining works in sinking mining system

4 Interconnection of Technological Systems with Mining Technology

Technological complexes are integral part of mining technology, they are instrument of its implementation. They serve technology of stripping and mining operations. They divide into cyclical, cyclic-stream and stream technology of mining (overburden and mining) activities by the nature of the recess and rocks moving (Fig. 5).

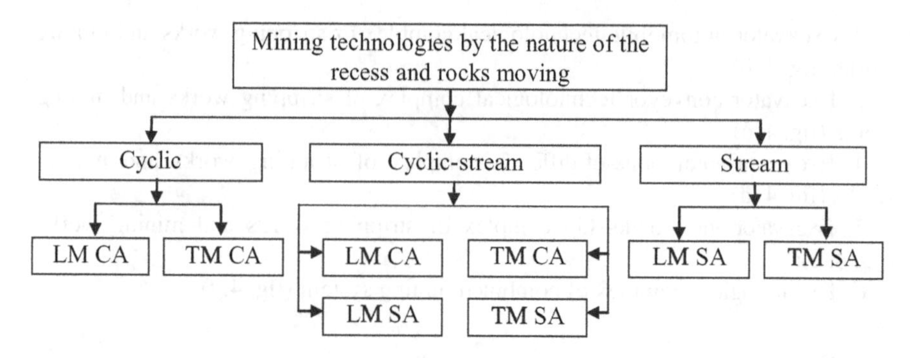

Fig. 5 Levels of opencast mining

Excavation-loading and transport operations are done by cyclic equipment in cyclic technology (LM CA and TM CA). In cyclic-stream technology excavation-loading works are done by cyclic equipment (LM CA) and transport operations are done by continuous equipment (TM SA) or excavation-loading works are done by continuous equipment (LM SA) and transport operations are done by cyclic equipment (TM CA). Both of the works are done by continuous equipment in stream technology (LM CA and TM CA).

For example excavator-rail and excavator-automobile complexes are used in cyclic technology, excavator-automobile-conveyor – in cyclic-stream technology, and bucket-wheal-excavator-conveyor complex, complexes with consoles and dump-bridges – in stream technology.

The most high-level mode of production is stream technology of mining and stripping works. The next level of production represents by cyclic-flow technology mining. This technique is quite effective and more promising. Cyclical mining technology is the commonest [3].

Bucket-wheel-excavator-conveyor complex of mining works, which implements stream technology of coal extraction, has been introduced in "Bogatyr" quarry in Kazakhstan. The annual production capacity is 50 million tones [4].

Bucket-wheel-excavator-conveyor complex of mining works are successfully operating in "Vostochnyi" quarry of Ekibastuz coal deposits. Excavation, transportation, averaging and movement of coal is made by complex, which includes bucket-wheel excavator SRs (k) -2000, face and bench cranes, connecting, lifting and trunk conveyors, averaging-loading machine and loading point [4].

Also excavator-automobile-conveyor technological complex is used in cyclic-stream technology in the quarry. Its reliability is ensured by two excavator-automobile complexes located on the flanks of the quarry with same crushing and transfer points.

Excavator-automobile-conveyor technological complex of mining works introduced in pit of JSC "Altyntau" [5]. Crushing-reloading setting CJ615 provides reception and crushing of ore mass, which is then transported by conveyor to gold factory.

Cyclic-stream technology also will be used in projects of opencast mining of large copper-molybdenum deposits in Kazakhstan, such as Aktogai, Bozshakol, and in Kacharsk iron deposits.

The unique project is project "Creation and implementation of cycle-stream transport (CLT) with steeply inclined conveyor KNK-270's in mine " Muruntau" of Navoi Mining and Metallurgical Combine of Uzbekistan, which has been performed by the Uzbek and Ukrainian scientists and specialists [6].

CLT of ore includes: CLP - crushing and reloading point; KNK-270 - steeply inclined conveyor; CWH - cargo warehouse; Asmodeus - automated system of monitoring and supervisory control (see Figure 6).

The developed design of steeply inclined conveyor with presser belt showed high performance and reliability, and provides stable reception, climbing at an angle of 37 degrees to a height of 270 m and unloading of crushed rock mass with a capacity of more than 3500 t/h.

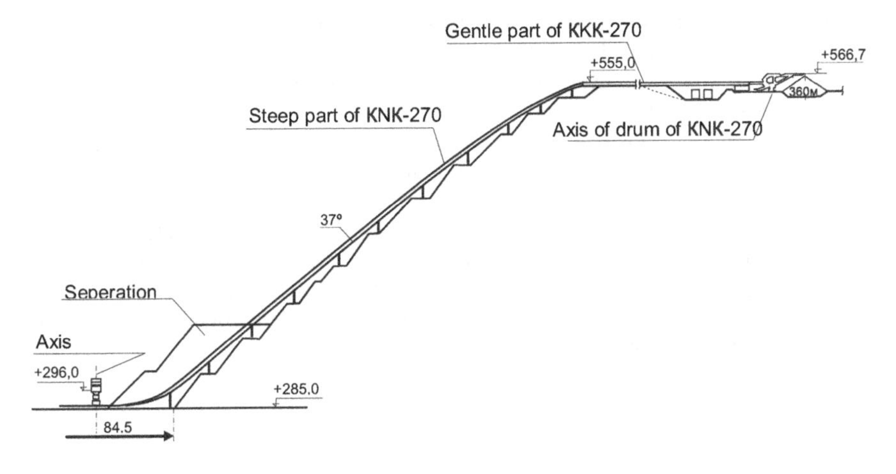

Fig. 6 Stream transport with KNK-270 in Muruntau pit

Trail of steeply inclined conveyor is situated perpendicular to general angle of the highwall.

According to the authors [6] complex KNK-270 will increase the depth of the mine development to 950 m.

Nowadays excavator-different-type-automobile technological systems successfully operates in medium and large quarries, which implement the cyclic technology. Their competitiveness may increase if depth of quarry is used by zones [5,7]. For example, it is advisable to share quarries space into two zones by height in deep quarries (H = 480 m) with annual volume more than 80 million tons. It needs to use excavators with bucket capacity of 10-20 m3 and dump trucks with payload 90-200 t in the upper zone and excavators with bucket capacity of 5-10 m3 and dump trucks with payload 45-90 t in the lower zone.

It is necessary to divide quarry space into three zones by height in ultra-deep pits (H = 600 m) with an annual traffic volume more than 100 million tons. It needs to use excavators with bucket capacity of 20-32 m3 and dump trucks with payload 200-320 t in the middle zone, excavators with bucket capacity of 10-20 m3 and dump trucks with payload 90-200 t in the middle zone and excavators with bucket capacity of 5-10 m3 and dump trucks with payload 45-90 t in the lower zone.

Thus, the three-technological complexes, accompanying as cyclic, and cyclic-stream technology development are quite effective.

References

1. Rakishev, B.R.: Systems and technologies of opet-cast mining. – Almaty: SRC «GYLYM», 328 p. (2003)
2. Rzhevski, V.V.: Open-cast mining. P.2 M.: Nedra, 549 p. (1985)
3. Directory «Open-cast mining». – M.: Gornoe buro, 590 p. (1994)
4. Rakishev, B.R., Moldabayev, S.K.: Resource saving technologies in coal mines:– Almaty: KazNTU, 348 p. (2012)
5. Mahambetov, D., Rakishev, B., Samenov, G.K., Sladkowski, A.: Efficient using of automobile transport for the deep open-pit mines. Transport Problems. Poland: Katowice 8(3), 25–33 (2013)
6. Sanakulov, K.S., Shemetov, P.A.: Development of cyclic-flow technology based on steeply inclined conveyors in deep pits. Mining Magazine (8), 34–37 (2011)
7. Yakovlev, V.L., Tarasova, P.I., Zhuravlev, A.G., Furin, B.O., Voroshilov, A.G., Tarasov, A.P., Fevelev, E.V.: A new look at the automobile transport. Proceedings of Higher Education. Mining Magazine 6, 97–107 (2011)

The Inden Residual Lake and the Federal Autobahn A44n – Geotechnical Requirements to Be Met by Large-Scale Structures in the Rhenish Lignite Mining Area

Gero Vinzelberg and Dieter Dahmen

RWE Power AG

Abstract. The geotechnical requirements of two major structures in the Rhenish lignite mining area will be discussed – the Inden residual lake and the federal autobahn A44n.

According to current planning, the Inden residual lake is going to be the first of three large residual lakes to be filled from about 2030 onwards when coal extraction at the Inden opencast mine comes to an end. However, the construction of the lakes' final slopes will start as soon as 2014. Compared to the existing man-made lakes in the Rhenish lignite mining area, with the largest being the 100-hectare large Blausteinsee, the area of the Inden residual lake will cover approximately 1,100 hectares and thus become the largest lake in North-Rhine Westphalia. Consequently, geotechnical aspects such as material distribution in the slopes have to be considered carefully and overall stability of single slopes as well as the slope system has to be ensured. Furthermore, the influence of waves, erosion and earthquakes on the overall stability will be discussed. A focus will be on additional loads caused by earthquakes that play an important role due to the fact that the Lower Rhine Embayment is one of the most active earthquake areas in Germany.

The second project presented here is the construction of the federal autobahn A44n. The A61 between the Jackerath and Wanlo interchanges will be decommissioned due to the progress made by the Garzweiler opencast mine. By then, the new autobahn A44n has to be commissioned in order to replace the above-mentioned A61 and the former autobahn A44 that had to make way for mining in 2006. More than 7 km of the new A44n including several bridges will be located on dumped soil with a thickness of up to 190 m. Geotechnical aspects including settlement forecasts and field measurements will be presented. Furthermore, this paper outlines measures to offset settlements including a so-called precautionary gradient as well as material distribution beneath the autobahn. Finally, soil improvement measures beneath a bridge with a clear span of over 100 m will be discussed.

C. Niemann-Delius. (ed.), *Proc. of the 12th Int. Symp Continuous Surface Mining - Aachen 2014*,
Lecture Notes in Production Engineering, DOI: 10.1007/978-3-319-12301-1_10

The Inden Residual Lake

1 Introduction

The Inden opencast mine is located in the western part of the Rhenish lignite mining area and – at a planned annual coal output of about 20 million tons – will be depleted around the year 2030. An opencast mine residual lake covering about 1,100 hectares, filled over a period of 20 to 25 years, is to be created in order to offset the deficit in volume resulting from many years of coal extraction. Apart from sump water, the lake will mainly be filled with water taken from the Rur River situated to the north-east. Although the Rhenish lignite mining area has already seen the creation of numerous residual and other man-made lakes that are now used for a variety of purposes, the Inden residual lake, owing to its surface area and volume, will be on an entirely different scale. As early as the planning stage, issues of stability and resistance to erosion were thus explored in depth with the aid of scientific evaluations and the lakes already in existence were examined.

2 Underlying Conditions for Dimensioning

Initial studies on the possible creation of a residual lake were conducted as early as 1984; the assumption at the time being that the lake would be flooded by means of the rising groundwater table. In addition to the disadvantage of a very long filling process, this procedure would have led to an uneconomic general inclination of the residual lake's slopes; consequently, the mining authority declined its approval on account of the objectives of the lignite mining plan. At the time it was instead decided to fill the Inden residual lake with overburden masses extracted from the neighbouring Hambach opencast mine. This plan was challenged by the Inden township in 2000, and the creation of a residual lake as an alternative reviewed. Following the revision of and amendments to the lignite mining plan, the master operating plan was approved on 20 December 2012 [1]. Concurrently with the procedure under approval law, basic investigations into the dimensioning of the planned residual lake were carried out. These investigations are summarised in the following.

In an attempt to both allow for the deposit to be exploited – which is an economic imperative – and ensure sufficient stability of the permanent residual-lake slope system, the initial plan is to deplete the deposit, setting up a border slope system inclined at 1:3 as the mine advances. Subsequently, a wedge-shaped fill with a general inclination of 1:5 will be banked up in front of the dump, resulting in a lake area of about 1,100 hectares. According to current plans, this slope system will consist of single slopes inclined at 1:2.5 and intermediate horizontal berms functioning as working levels for the spreaders. An underwater slope inclined at 1:5 and a wave attack zone border on the top berm. Depending on the level difference to the surrounding terrain, some areas above the wave attack zone are abutted by a surface slope with an inclination of 1:3. The general layout of the planned residual lake is shown in Figure 1.

To guarantee slope stability and controlled filling, the residual lake is to be flooded with outside water supplied by the Rur River to the north-east of the lake and by drainage operations at the Hambach opencast mine. The ongoing, selective operation of dewatering wells in the vicinity of the lake ensures that the lake's water level remains above the surrounding rock water level, enabling stabilising forces to act on the slopes.

The shape of the future wave attack zone is mainly determined by waves formed by wind. The wind velocities and fetch encountered in the surroundings as well as the contours of the lake shore have the greatest effect on the height of these waves. All of these factors were analysed by experts. Based on the maximum wave height to be expected (1 m) and the materials available in the opencast mine, a required "wave attack zone" inclination of 1:20 in the prevailing wind direction was established and, to be on the safe side, was transferred to the entire wave attack zone around the lake.

Since the Lower Rhine Embayment is one of the most active seismic areas in Germany [2], the stability of the residual lake slopes must also be guaranteed in the case of an earthquake. Both deterministic and probabilistic approaches can be taken to ascertain the ground acceleration caused by earthquakes; more information on how earthquakes are considered is given under the point "Stability analyses". In addition, we carried out investigations to determine whether the creation of a residual lake would affect the region's seismic activity within the scope of the approval procedure. As it turned out, the creation of the lake would lead to only a very minor relief of stress on the terrestrial crust that would not alter natural seismicity to such a degree as to put the shore or slopes at risk.

Triggered especially by the incident that occurred in Nachterstedt in 2009, when part of the dump of the Concordia lake failed, entailing a massive landslide, a renewed detailed analysis of the experience gathered in the Rhenish mining area up to now with the creation of residual and other man-made lakes as well as high dumps was performed as part of an assessment for monitoring purposes and described in [3]: "In stability terms, the examined lakes show no abnormalities. In two objects, there were minor slope deformations in the past. These concern the 13-ha Neurather See lake on whose shores a deformation surface measuring some 900 m^2 occurred in the wake of the 1992 Roermond earthquake. No remediation was necessary. Likewise, there was local deformation during the filling phase of the 85-ha Zülpicher See lake, which was remediated at short notice. At no point in time were material goods or people in danger from these two events. (…). In order to find out how the lake troughs profiled in dumping operations behave below the water surface, two exemplary sonar measurements were taken in the western mining area. Lakes Lucherberger See and Blausteinsee were selected as objects. With a surface measuring some 100 ha and a max depth of 41 m, Blausteinsee is the biggest and one of the youngest opencast-pit lakes in the Rhenish mining area. (…). A comparison of the sonar measurements with the underlying planning and the mapping system after completion of the lake troughs shows no abnormalities (Figure 2). Minor changes in the range of a few decimetres to the lake contour of Lucherberger See are mainly due to organic sedimentation at deep points, according to specialist interpretation of the measurement results. These phenomena are above all due to the lake's age and not unusual for standing waters."

3 Stability Analyses

The "Guideline for Stability Analyses" (RfS) governs the conducting of stability analyses in the Rhenish lignite mining area. In its 2003 version already, the guideline requests an adequate consideration of earthquakes. Planning bulletins issued in 2006 and 2008 consequently made allowance for earthquakes that might occur in the various sections around the future residual lake. The following sections were examined (Table, Figure 1):

Area	Examined sections
Schophoven	S 5/2, 65 C
Merken	C, D, S 96
Lucherberg	S 24, 65 ab, S 99
Inden/Altdorf	S 81
Autobahn A4	S 4, S 23

The planning bulletins were subsequently reviewed on behalf of the mining authority by experts of the Geological Service NRW who checked calculations, performed their own calculations and evaluated the bulletins within a report produced for the Arnsberg regional government. Taking account of the planning bulletins of the mining company and the opinion of the Geological Service, the Arnsberg regional government conducted a final check verified by an inspection report included in the documentation of the environmental impact assessment that followed. The report confirmed that geomechanical feasibility and the required stability will be assured at all stages: when depleted, during filling and in the final, filled state. The approach taken at the time pseudo-statically factored in earthquakes by assuming part of the seismic acceleration to be expected as permanent additional acceleration in the horizontal plane of up to 0.5 m/s^2.

In 2013, the RfS Guideline mentioned above was again updated and the consideration of earthquakes further defined by taking the advancement of geotechnical standards and bulletins into account. Specifically, this applies to DIN 19700 [4] and NRW bulletin No. 58 [5], which deal with the dimensioning of barrages. Accordingly, a probabilistic approach which, in the style of the mentioned standards and by way of simplification, assumes earthquakes with a recurrence period of 500 years for the filling phase and earthquakes with a recurrence period of 2,500 years for the final, filled state was inter alia specified for determining the ground acceleration to be expected. In this way, a safety level in the filled sate similar to retaining structures of major Class-1 reservoirs is taken as a basis although, unlike with reservoirs, no neighbours are at risk below the lake's water level. According to the RfS, the ground acceleration to be assumed can be ascertained inter alia by submitting a location-specific earthquake-risk data query to the German Research Centre for Geosciences (GFZ Potsdam). This ensures a transparent procedure since the relevant data is available to everyone [6]. For the Inden residual lake this approach pursuant to the current RfS yields a maximum ground acceleration of between 1.06 m/s² for the filling phase and 2.10 m/s² for the final, filled state. When this maximum acceleration is factored into pseudo-static stability calculations it is treated as a long-term parameter and converted using pseudo-static coefficients. The pseudo-static coefficients were determined by the Karlsruhe Institute of Technology's Soil and Rock Mechanics Department in connection with an evaluation commissioned by the Arnsberg regional government [7] by means of a dynamic comparative calculation on the basis of time-acceleration curves derived from recorded earthquakes and ascertained within the scope of an evaluation [8] prepared by Cologne University's Department of Earthquake Geology. Taking the values calculated in these evaluations as a basis and depending on the location of the examined rupture mechanisms, the result is horizontal acceleration of between 0.11 m/s² and 0.53 m/s², which is factored into the stability calculations as additional load acting on the slopes. Thus, the results from the above-mentioned stability calculations performed in accordance with the 2003 RfS were essentially verified.

4 Federal Autobahn A44n

4.1 Introduction

The Garzweiler II opencast mine interferes with the federal network of trunk roads encompassing the autobahn A44 and the autobahn A61, which are situated in the extraction area of the opencast mine. To replace the stretch of the A44 between the Jackerath and Holz interchanges that had to make way for mining in 2006, the A61 between the Jackerath and Wanlo interchanges was expanded to six lanes and the Wanlo junction was built as a replacement for the Otzenrath (A44) junction by way of a preliminary measure.

In the course of the mine's further development, the six-lane autobahn A61 will be reached by advancing mining operations and interrupted. By this time, the A44 will be re-constructed and widened to six lanes to replace the A61 over a distance of 10 km. In accordance with the specifications set forth in the requirement plan for federal trunk roads, this stretch of motorway will also be complemented by the six-lane expansion of the A46 between the Holz and Wanlo interchanges (Figure 3).

More than seven kilometres of the A44n will be located on dumped soil with a thickness of up to 190 m extracted from the Garzweiler opencast mine. Apart from the autobahn proper, which is a six-lane road, several bridge structures will be erected on dumped terrain. Although the construction of roads and motorways (A540) on dumped soil is state of the art in the Rhenish mining area, the building of the A44n is extensively supervised by engineers and scientists on account of the considerable dump thickness, the tight schedule and the autobahn length.

5 Geotechnical Specifics

Unlike the procedure that is usual followed when building a motorway on existing terrain, where geotechnical soil examinations are a crucial basis for planning, the construction of the A44 n is special in that an area measuring more than three kilometres has yet to be dumped, meaning that until now no soil examinations have been carried out. To deal with this situation, a pinpointed dumping scheme that predetermines the structure of the subgrade while simultaneously reducing settlements is applied (Figure 4). The fine-grained and loose in-situ soils are combined into mixed soil classes and there is a significant reduction in the structure's cohesive proportions from bottom to top. Specifically the top layer of 10-m thick gravel is well-graded, which on the one hand allows any compaction measures to be successfully implemented and on the other ensures sufficient percolation capacity to permit road drainage along the central reservation and hard shoulders. To enable the exact elevation to be reached, earthworking equipment places a sand/gravel layer on top of the unfinished dump produced in opencast mining operations with main mine equipment. This layer serves as formation for the motorway surfacing, i.e. it has to reach Ev2 values of 45 MN/m².

Apart from the materials used, the settlements occurring naturally in dump areas are influenced mainly by the thickness and life of the dump. In total, these settlements will be in the order of nearly two metres. Accordingly, when the autobahn is completed, the thickest areas dumped the latest will be subject to the greatest settlements yet to be expected (Figure 5). The fact that the settlements vary along the length of the autobahn and before and after commissioning is accounted for by a so-called precautionary gradient. This means that the different areas of the autobahn are built at a higher level than officially approved in order to ensure that the alignment in elevation complies with the relevant guidelines both when the autobahn is opened and after the subsequent settlements have subsided.

6 Mine Surveying Settlement Measurements and Soil-Mechanical Model

As mentioned above, settlement forecasts are of paramount importance in the construction of motorways. The stretches of autobahn that have already been built are measured by mine surveying methods for the purpose of validating and, if necessary, adjusting the settlement forecasts. This is done by means of a tight grid of geodetic points that are placed at least every 50 m along both carriageways and in the area of the axes and refined section by section. In this way, several hundred geodetic points, levelled according to the dump's life between every three months and every twelve months, have already been used up in building. These measuring points verify a semi-logarithmic/linear settlement progression and permit the settlement forecasts to be updated. All things considered, the targeted dump structure results in a reduction compared with the initial settlement forecasts, prepared on the basis of long-term measurement data collected in dump areas without the special soil structure described above. Munich Technical University provided scientific assistance in the way of a soil-mechanical finite-element model enabling settlement measurements to be simulated; if need be this model is spatially refined and used for settlement forecasts for bridge structures [9].

7 Hydrostatic Settlement Measurements

Hydrostatic line-measuring systems functioning on the principle of tube levels and permitting a 25-cm grid resolution were installed in various places so as to be able to monitor small-scale settlements occurring at different depths. Using these, settlements of the gravel layer occurring in the region of measuring field No. 1 were examined more closely. The findings showed that the settlements within the gravel layer resemble the settlement behaviour of the entire dump body below the autobahn bed. After a life of six months, settlements affecting only a very small area show differences of 1 to 3 cm for wavelengths between 2 and 5 m, occurring below the gravel layer after another six months.

The hydrostatic line-measuring system was also employed in the area of a bridge structure passing over a belt conveyor (measuring field No. 3) in order to record the spatial and time-dependent progression of settlements by means of a pre-load fill. The placing of an overload fill with a thickness of about 14 m allowed settlements amounting to about 30 m to be prevented in the space of four months.

8 Compaction Measures

To analyse the compactibility of the gravel layer, a test field where small-scale settlement differences within the gravel layer and the effects of different compacting methods can be studied by means of another hydrostatic measuring system (measuring field No. 2) was prepared next to the actual autobahn bed at the end of 2011. In addition, the planned percolation scheme and the effects of rainwater percolation on settlement behaviour were investigated in another part of the test field. Figure 6 gives a schematic overview of the test field. To prevent the tests from having an impact on the future autobahn bed, the gravel layer to the right and left of the bed was widened and the test field arranged next to the bed proper as described above. Compaction and percolation were studied over a period of 1.5 months by experimenting with the vibroflotation, pulse, and roller compaction methods. To evaluate compaction, both cone-penetration and dynamic-probing structural test holes were drilled all over the test field (in total 43 for cone penetration and 38 for dynamic probing). In uncompacted condition, packing density in the area close to the surface, down to a depth of about 1 m, was medium, followed by largely loose packing down to about 6 m. Farther down, packing density again increases to medium, rising further as depth increases. The core penetration and dynamic probing tests conducted after compaction were assessed to evaluate the various methods used.

In the area close to the surface, roller compaction with a heavy-duty polygonal roller increased packing to medium and dense. Down to a depth of 6 m, packing density was largely medium, with no change in compactness being noted at even greater depths. Pulse compaction in a 3-m grid, with an additional compaction point at the grid's centre in certain areas, led to an increase in compactness to medium or dense down to a depth of 6 m; and while a compaction effect was still detectable at a depth of up to 12 m, packing density could not be increased beyond medium. Vibroflotation compaction was applied in 3 and 4-m grids down to a depth of 17 m. Down to about 6 m, the result was a heterogeneous change in density so that both loosening (very loose packing) and compaction (dense packing) of the soil were locally noted. Beyond this point down to the final depth, however, compactness significantly increased to medium and dense packing.

The evaluation of the settlement measurements taken in the test field so far shows that the compaction methods do not affect the settlement behaviour of the dump, which is considerably thicker. Close to the surface, immediate settlements in the range of a few centimetres or decimetres were observed. To what extent compaction of the gravel layer can contribute to a reduction in small-scale settlement differences is a question that can only be answered once future line measurements have been appraised. On the whole, the test results attest to the good compactibility of the gravel layer.

Proceeding from the aforementioned findings, we intend to compact the autobahn bed by means of a heavy-duty polygonal roller whereas vibrodisplacement columns, in addition to the preliminary fill already in place, are to be erected for the mentioned bridge structure situated in the area of the belt conveyor due to the greater depth of penetration of the structural loads. Both measures are supported by extensive investigations and measurements.

Apart from the compaction tests, multi-stage percolation trials were performed on two of the test fields (previously compacted by polygonal roller and uncompacted) in surface areas measuring 252 m² and 230 m². The trials ran for two to six hours, with an inflow of between 10 and 70 m³/h. Relative to the percolation area, rainfall of approx. 280 mm/h was simulated in this way, which also resulted in the flooding of the basins up to a water depth of 0.5 m. With permeability coefficients of kf $\geq$ 1•10-5m/s, the requirements imposed on the percolation capacity of the soil have thus been met both in uncompacted and compacted condition. In addition, the settlement measurements taken concurrently show that even the highest inflow rates do not lead to further subsidence. The percolation scheme via the gravel layer has thus proven suitable.

9 Summary

Planning for the Inden residual lake dates back to the year 1984 and is regularly updated on the basis of new developments for instance in geotechnics and seismology. Since the Inden residual lake, once created, will have a surface area ten times that of the largest man-made lake in the Rhenish mining area to date (Blausteinsee), comprehensive dimensioning and stability studies were carried out. Thus, the special requirements to be fulfilled by a geotechnical structure that is erected over several decades and which, as a permanent part of the scenery, will leave its mark on the region are addressed in an adequate manner. To take account of the special geotechnical requirements needing to be met when constructing a motorway on an opencast mine dump with a thickness of up to 190 m, settlement forecasts underpinned by extensive mine surveying settlement measurements are produced to ensure that the alignment in elevation complies with the relevant guidelines both when the autobahn is opened and after the subsequent settlements have subsided. Test fields prepared to conduct compaction and percolation trials enable specific technical and economic soil improvement measures geared to the different load situations in the area of the autobahn bed and bridge structures to be identified.

Both major projects presented here highlight the wide range of tasks involved in ensuring the organised post-mine use of surfaces that had to make way for mining and underline the significance of geotechnics as a key process for safe and economic opencast mining operations.

References

[1] Petri, R., Buschhüter, K., Dahmen, D.: Standsicherheitsuntersuchungen für den geplanten Restsee Inden unter Berücksichtigung von Erdbeben. WOM (February 2014)

[2] Pelzing, R.: Erdbeben in Nordrhein-Westfalen, Geologischer Dienst Nordrhein-Westfalen (2008)

[3] Petri, R., Stein, W., Dahmen, D., Buschhüter, K.: Nachhaltige Folgenutzung rekultivierter Flächen - Evaluierung von Restseen und Hochkippen mit beendeter Bergaufsicht im Rheinischen Braunkohlenrevier. WOM (February 2013)

[4] DIN 19700: Stauanlagen (Juli 2004)

[5] Merkblatt 58: Berücksichtigung von Erdbebenbelastungen nach DIN 19700, Landesumweltamt Nordrhein-Westfalen (2006)

[6] Deutsches GeoForschungsZentrum (GFZ Potsdam): Interaktive Abfrage von Karten der Erdbebengefährdung und Beschleunigungs-Antwortspektren für die Gefährdungsniveaus gemäß DIN 197005, http://dx.doi.org/10.5880/GFZ.2.6.2012.001

[7] Triantafyllidis, T.: Institut für Bodenmechanik und Felsmechanik Universität Karlsruhe (TH): Gutachterliche Stellungnahme zu Standsicherheitsberechnungen mit Ansatz von "Erdbebenbeschleunigungen für Böschungen im Rheinischen Braunkohlenbergbau" (Mai 2012 mit Ergänzungen im Juni 2013)

[8] Hinzen, K.-G.: Abt. Erdbebengeologie, Institut für Geologie und Mineralogie, Universität Köln: Seismische Lasten für die Ermittlung von Böschungsstandsicherheiten, Bericht im Auftrag der RWE Power AG (Oktober 2006)
[9] Zentrum Geotechnik, Technische Universität München: Wiederherstellung der Autobahn A 44 im Bereich des Tagebaus Garzweiler - Kurzbericht zu den geotechnischen Untersuchungen der Phasen I und II, Bericht im Auftrag der RWE Power AG (December 11, 2012)

Design of the Opencast Coal Mine Drmno Dewatering System

Vladimir Pavlovic[1], Dušan Polomčić[1], and Tomislav Šubaranovič[2]

[1] Faculty of Mining and Geology, University of Belgrade, Belgrade
[2] Republic of Serbia State Secretary, Belgrade

Abstract. Development of mining in the opencast mine Drmno from Kostolac basin in Serbia is carried out in the increasingly complex geological and hydrogeological conditions with the decline of the coal seams to the alluvial part of Europe's largest river Danube. Greater depth and inflow of groundwater significantly increase costs for dewatering facilities construction with the requirement for optimizing of this more and more important surface mining process. Making decision on screen construction in the opencast mine with redesign of line wells parameters for the defence against groundwater, in addition to geology and hydrogeology exploration activities, analyzing the elements of the opencast mining systems and a number of working environment parameters with the development of the hydrodynamic model requires necessary review of economic and environmental effects. A detailed techno-economic analysis has been done in order to demonstrate technological justification and economy of screen construction as the system for protection against groundwater in the opencast mine. The defence of the opencast mine against the groundwater without the construction of screen causes construction of several dewatering wells and lines with increased pumping capacities by the end of deposit mining. On the other hand, screen construction means big initial investments but with a smaller number of dewatering wells with reduced capacities, lower cost of electricity and larger reliability of the system operation in the long period until the closure of the opencast mine. Estimates of environmental impact and efficiency comparison between these systems options for protection against of groundwater have been done. DCF analysis has shown that the long-term averaged costs for opencast mine Drmno dewatering against groundwater per ton of coal are approximate for the both system options. Final decision on the approval of the combined dewatering system with a reduced number of dewatering wells and screen construction has been caused by the system reliability and environmental effects related to the impact on the groundwater level outside of the opencast mine boundaries and safer operation of the hydro power plant Djerdap system on the Danube River.

Keywords: dewatering system, groundwater, hydrodynamic model, dewatering wells, screen, reliability, environmental effects.

C. Niemann-Delius. (ed.), *Proc. of the 12th Int. Symp Continuous Surface Mining - Aachen 2014*,
Lecture Notes in Production Engineering, DOI: 10.1007/978-3-319-12301-1_11

1 Introduction

Opencast mining, especially today when mining is performed at greater depths in complex hydrogeological conditions, requires special attention of the scientific and experts community in terms of dewatering in all phases of the opencast mine development. Therefore, in recent decades, significant progress has been made primarily by developing methods, models and software that significantly facilitates the design of optimized groundwater dewatering system.

From the standpoint of dewatering, opencast mines are typically very dynamic systems influenced by a large number of natural, technical-technological, economic, environmental and safety factors and constraints in all periods of the life cycle. The entire life cycle is defined by periods of surface mining, realized through a set of business processes in each of the periods that allow the efficient, effective and reliable mineral resources mining in the opencast mine. One of the key process is the process of opencast mine dewatering, which belongs to the group of the *technical support processes* for the implementation of the opening process, production and closing of the opencast mine, as well the reclamation process.

Established dewatering process flow is at the general content level and is applicable regardless of the opencast mine size, the type of mineral resources being mined and complexity of deposit conditions. In addition, it is applicable during a new opencast mine opening, but also when it is required to design or redesign dewatering process in the opencast mine where is performed mining.

Groundwater dewatering system is a complex of more than one dewatering facilities and it represents the most complex sub-process of dewatering process, to which is necessary to pay the greatest attention [6]. In practice, as a base, are the most commonly used systems of dewatering wells, dewatering channels system and the water collector and waterproof screens.

2 Dewatering Planning Process of Opencast Mines

Successful implementation of the opencast mine protection against ground and surface water depends on the knowledge reliability degree of hydrological, hydrogeological and geotechnical characteristics of the near and distant deposit areas and their correct interpretation. The initial parameters for the dewatering requirements, in addition to these mentioned, are hydraulic and hydrodynamic parameters and dynamic parameters of groundwater.

Through a detailed analysis determine data on the possible methods, systems, facilities and equipment for dewatering in the opencast mine. Based on these data and data from the deposit model, too, (hydrological, hydrogeological, hydraulic, geological, geophysical and geotechnical), as well as data on applied mining technology and the conditions of the opencast mine stability (opencast mine model data), is carried out dewatering system selection for the opencast mine (fig. 1). For sure, it has to be emphasized that for the dewatering system selection is required for technical, economic and environmental risks to be analyzed, including the reliability of the system and the inputs data of the working environment.

Selection of dewatering system is an *iterative process* that includes risk, reliability, effectiveness and efficiency analysis for the selected system. The process model for *Opencast Mines Dewatering System Selection* is shown at figure 1. The presented process model is the part of the overall planning procedure and implementation of dewatering system. It provides high quality and reliable performance of the dewatering process in all phases, from preliminary plans, through the development of investment-technical documentation up to the construction and operation of the dewatering system since it has flow determined by the process and clear procedure.

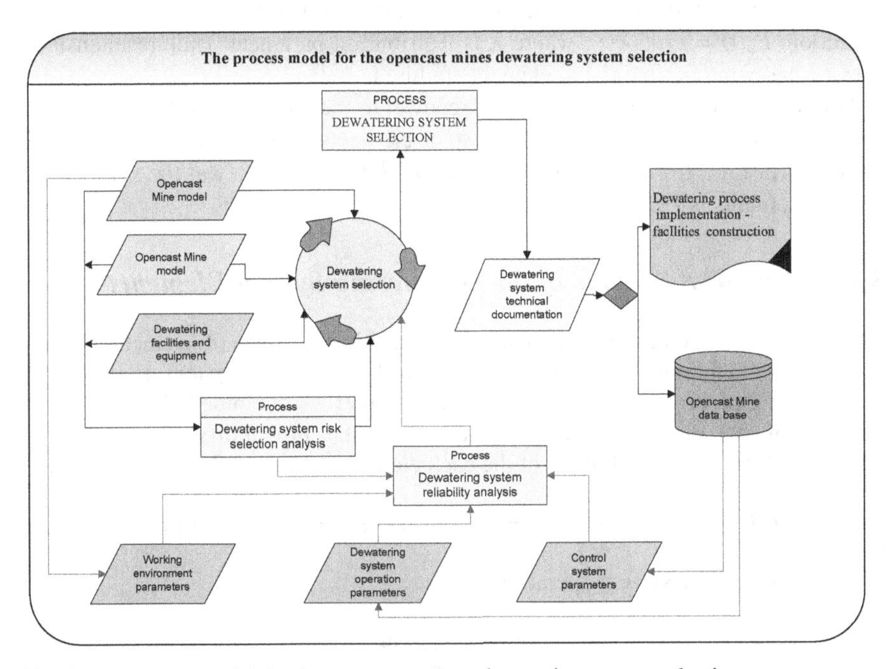

Fig. 1 The process model for the opencast mines dewatering system selection

3 Opencast Mines Dewatering System Reliability

Dewatering systems in opencast mines, as well as other technical systems have to perform its function successfully and reliably and with all its elements to serve all anticipated stresses, and at the same time to be simple and economical. The parameters of the system reliability are obtained on the basis of an analysis of possible system conditions related not only to the operation time and equipment failure but also caused by delays due to the characteristics of the working environment in real space, manpower and other unscheduled and planned outages in real time [1, 2, 7].

Functioning of dewatering system lines of wells and its elements is defined as a random process with exponentially distributed time of the system operational conditions implementation in the function of groundwater inflow in the opencast mine. By analyzing information on the status of dewatering systems is determined the strategy by which in the case of full or partial failure is performed renewal and scheduled servicing. Designed, implemented and controlled water screen permeability defines its functional reliability [4].

Operation to failure of dewatering system elements, as continuous random size, can be described by different laws of distribution, depending on the characteristics of the system and its elements, working conditions, characters of failure, and so on. The simplest tested and most widely used is the exponential distribution with the following set operating time (t) distribution function of system elements operation: $P_o(t) = exp(-\lambda*t)$, where λ is distribution parameter (failure intensity). The mean operation time (t_o) to elements failure is $1/\lambda$..

On the other hand, the probability of renewal to the moment t at the exponential law of renewal time distribution of system elements with β parameter (renewal intensity) in time from 0 to t is: $P_r(t) = exp(-\beta*t)$. Mean renewal time (t_r) of the system elements is $1/\beta$.

3.1 Structural Scheme of the Dewatering System Elements

The structural schemas that represent the graphical display of elements in the system can unambiguously define operation or failure of the system. Elements of the system can be connected in series, in parallel or combined. If the system consists of (n) elements connected in series, the probability of the system operation $P_{os}(t)$, for probabilities of the each element operation $P_{oi}(t)$, is:

$$P_{os}(t) = P_1(t)*P_2(t)*,...,*P_n(t) = \prod_{i=1}^{n} P_{oi}(t) = exp(-\lambda_1*t)\, exp(-\lambda_2*t),..., exp(-\lambda_n*t) = exp(-t*\sum_{i=1}^{n} \lambda_i) \quad (1)$$

Mean time of the system operation is:

$$t_{os} = 1/\sum_{i=1}^{n} \lambda_I \quad (2)$$

If the system consists of (m) elements connected in parallel, where the failure probability of each $P_{rj}(t) = 1\text{-}P_{oj}(t)$, the probability of system failure is:

$$P_{rs}(t) = P_{r1}(t)*P_{r2}(t)*,...,*P_{rm}(t) = \prod_{j=1}^{m} P_{rj}(t) = (1\text{-}exp(-\lambda_1*t)),..., (1\text{-}exp(-\lambda_m*t)) = \prod_{j=1}^{m} (1\text{-}exp(-\lambda_j*t)) \quad (3)$$

If the structure of the system is combined, the operation probability is calculated with the participation of both the above-mentioned formulas. The combined parallel-serial systems are present in the dewatering of groundwater in the larger opencast mines with the complex hydrogeological conditions, when the single-well pumping facilities are connect as a parallel subsystem, and the water is re-pumped to a common serial collectors and dewatering lines.

3.2 *The Functioning of the Dewatering System*

The function of groundwater dewatering system is that at a certain time it protects opencast mine against a model or practically prescribed amounts of water inflow. During system operation it the can lead to distortion of the normal subsystems and elements operation, which leads to increased flow of groundwater in the opencast mine, and when it exceeds the permissible level to the conditional system function failure and mining endanger.

A complete system failure occurs during the failure of any system connected elements in series, or other unforeseen failures such as, for example, electricity outage. Possibility of reliable dewatering system against groundwater operation, installed as a stationary process can be analyzed on the basis of the created graph with three conditions. Installed dewatering system in the function of groundwater inflow can be in three conditions (S_0, S_1, S_2).

The normal condition of the system is denoted by S_0, condition S_1 represents endanger within the permissible limits, while the condition S_2 is extreme condition, treated as a conditional failure, when there is a threat to the functioning of the dewatering system in the opencast mine over the allowed limits. Then the system correction and active protection against groundwater leads to reduction in the system distortion and the possible transition from condition S_2 to condition S_1 or S_0 condition.

To calculate the probability of condition P_0, P_1 and P_2 is necessary to solve the corresponding system of equations with the different number of transition intensity probabilities between conditions S_0, S_1 and S_2, according to the graph condition shown at figure 2.

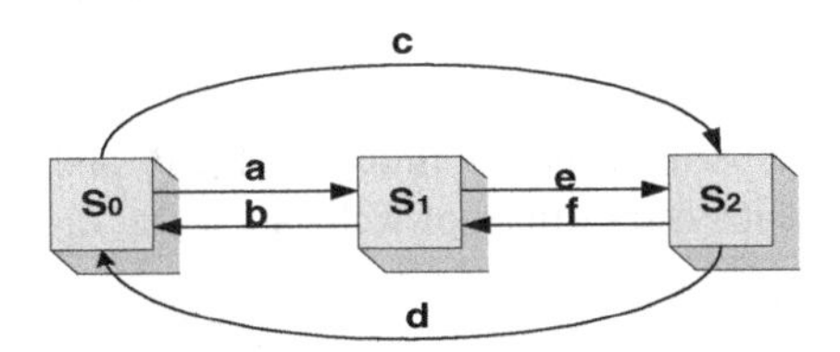

Fig. 2 Dewatering system graph sheet

It is accepted simplification to solve equations by using dewatering system graph sheet with the rule that the each condition probability is equal to the sum of all flows probability being transferred from any condition in the given condition, reduced for the sum of all flows that from a given condition go to other conditions (fig. 2).

Provided system can be in the operation condition (S_0) with the transition intensities (a) to (S_1), and (c) to (S_2), the condition of operation within the acceptable limits (S_1) with the transition intensities (e) in (S_2) and (b) in the (S_0) and the failure condition (S_2) to the transition intensities (d) to (S_0), and (f) to (S_1). For dewatering system is created graph with three conditions in the form of a chain with limit (stationary) condition probabilities by the provided formulas as follow:

$$P_0 = (1+ (a*b+f*a+f*d)/(c*f+b*c+b*e)+(a*e+d*c+d*e)/(c*f+b*c+b*e))^{-1}$$

$$P_1 = (1+(a*e+d*c+d*e)/(a*b+f*a+f*d)+(c*f+b*c+b*e)/(a*b+f*a+f*d))^{-1}$$

$$P_2 = (1+(c*f+b*c+b*e)/(a*e+d*c+d*e)+(a*b+f*a+f*d)/(a*e+d*c+d*e))^{-1} \quad (4)$$

The process of the dewatering wells system operation is random, and transitions from condition to condition does not depend only on the elements failure but also on deviations from the scheduled values. Analysis of the system operation probability is carried out based on the random sizes exponential distribution law.

3.3 Reliability of Dewatering System Combined Lines

Operation in increasingly more and more complex hydrogeological deposit conditions and the need to reduce costs has led to the requirement for the significant use of optimization for new and rationalization of the existing combined dewatering systems in surface mining. Analysis of the dewatering system reliability in opencast mines on the basis of well-established information system can significantly ease the process of control, increase the effects of maintenance and to enable more cost-effective operation.

The combined system of dewatering lines, dewatering wells with consolidating the flow of water can be represented by three parallel lines of wells that by pumps transport water to the serial subsystem consisting of a common first collecting pipeline, transferring point, other common gravity or pressure pipeline and refilling facility with operation probabilities P_c, P_v, P_p and P_r (fig. 3). Reliability indicators of parallel subsystems lines of wells are a_1, a_2, a_3 and b_1, b_2, b_3.

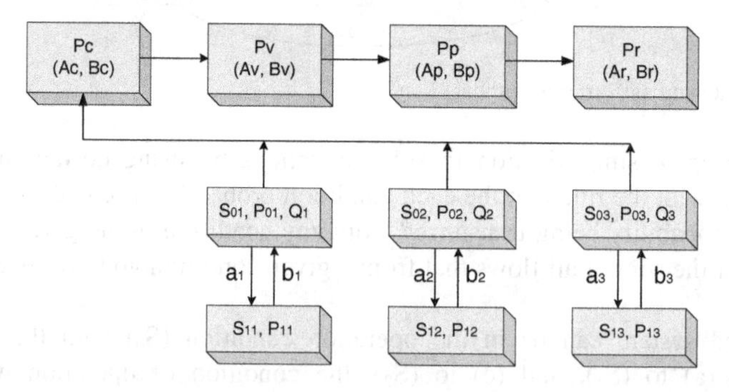

Fig. 3 Graph of combined dewatering system condition

The system operates if all three parallel lines of wells and serial subsystems are in order but, also, when one of the three lines of wells is in failure. In case of serial subsystems failure the system is out of order, but at this point, since conditionally they are out of operation, lines of parallel wells subsystems cannot be in failure. Conditional failure for parallel subsystems lines of wells as a whole resulting from malfunctioning or upon failure of two lines.

The functioning of such installed parallel dewatering subsystem, for each well system can be represented by a stationary process with two conditions. Elementary failure flow of each well plant has intensity (a). Recovery rate has an exponential distribution with parameter (b). Conditions of each parallel line are (S_0) if it is in operation and (S_1) if it is in failure, that is, if the system is in renewal (fig. 3).

Stationary probabilities can be obtained easily by solving general linear algebraic equations for the limit subsystems condition probabilities: $a*P_0 = b*P_1$, $P_0+P_1 = 1$.

So for the first line it is for example:

$$P_{01} = 1/(1+a_1/b_1),\ P_{11} = (a_1/b_1)(1/(1+a_1/b_1)) \tag{5}$$

Stationary probability of the parallel subsystem operation with two lines of wells, which is out of operation in the event of failure of one of the wells or lines is:

$$P_{oII} = P_{01}*P_{02} = (1/(1+ a_1/b_1))(1/(1+a_2/b_2)) \tag{6}$$

Operation probability for the parallel sub-systems installed with three lines of wells is its whole:

$$P_{oIII} = P_{01}*P_{02}*P_{03} = (1/(1+ a_1/b_1))(1/(1+a_2/b_2))(1/(1+a_3/b_3)) \tag{7}$$

Operation probability of the subsystem when in failure is the first line of wells is:

$$P_{o1} = (a_1/b_1)(1/(1+a_1/b_1))(1/(1+a_2/b_2))(1/(1+a_3/b_3)) \tag{8}$$

Operation probability of the subsystem when in failure is the second line of wells is:

$$P_{o2} = (a_2/b_2)(1/(1+ a_1/b_1))(1/(1+a_2/b_2))(1/(1+a_3/b_3)) \tag{9}$$

Operation probability of the subsystem when in failure is the third line of wells is:

$$P_{o3} = (a_3/b_3)(1/(1+a_1/b_1))(1/(1+a_2/b_2))(1/(1+a_3/b_3)) \tag{10}$$

Stationary probability of the parallel system conditional failure when in failure is two lines of wells are provided by equations:

$$P_{12} = (a_1a_2/b_1b_2)(1/(1+a_1/b_1))(1/(1+a_2/b_2))(1/(1+a_3/b_3))$$

$$P_{13} = (a_1a_3/b_1b_3)(1/(1+a_1/b_1))(1/(1+a_2/b_2))(1/(1+a_3/b_3))$$

$$P_{23} = (a_2a_3/b_2b_3)(1/(1+a_1/b_1))(1/(1+a_2/b_2))(1/(1+a_3/b_3)) \tag{11}$$

Stationary probability of system operation with three lines of wells of the same reliability characteristics is:

$$P_{rIII} = 1-(1-P_{01})^3 \tag{12}$$

Stationary operation probability for combined system lines of wells is (fig. 3):

$$P_{rk} = (P_{rIII}+P_{r1}+P_{r2}+P_{r3})P_c{}^*P_v{}^*P_p{}^*P_r \quad (13)$$

3.4 Process Reliability Dewatering System Model

Reliability of dewatering systems operation in real time is the basis for analyzing the Model effects of the system as a whole and its individual elements. Within the established process model for the reliability dewatering system analysis (fig. 4), is performed establishing of lines of wells combined system reliability and combined flow by calculations of the stationary probabilities for subsystems and systems operation, including analysis of environmental impacts and verification and optimization of economic analysis reliability.

Presented model of the process analysis and reliability optimization is the part of the overall planning procedure and dewatering systems implementation and can be used as an element for development of an opencast mine integrated information dewatering system [3].

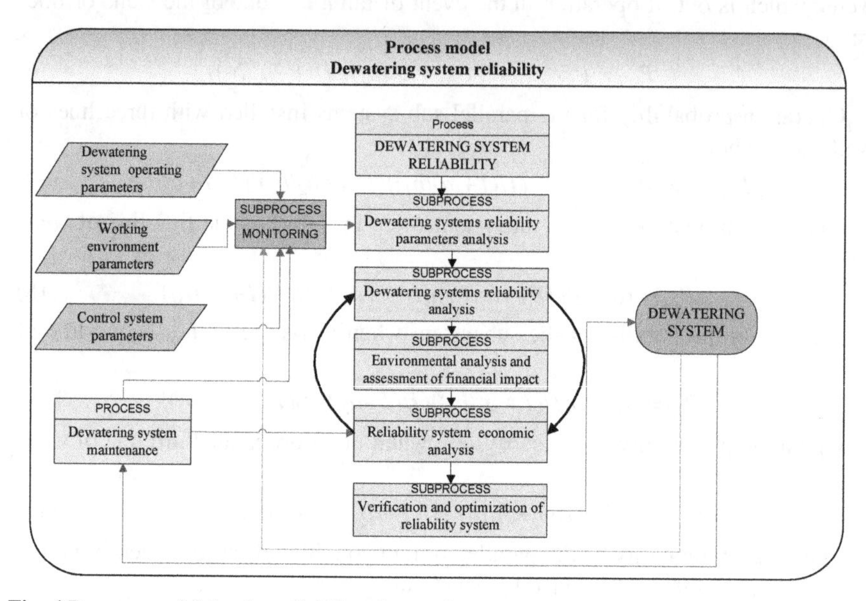

Fig. 4 Process model for the reliability dewatering system analysis

4 Selection of Groundwater Dewatering System for the Opencast Mine Drmno

Opencast mine Drmno in Kostolac coal basin is an ideal example for the selection of an optimal groundwater dewatering system. It is about 100 km east from

Belgrade. The Main Mining Design envisages excavation of $43.5*10^6$ m^3 of overburden and 9 Mio tons of coal annually.

Considering the very large inflow of groundwater in the alluvia of rivers Danube and Mlava, a detailed techno-economic analysis of capabilities to protect the opencast mine Drmno has been carried out, only by line of wells or by a combination of waterproof screen and line of wells with reduced number of wells. It has provided the bases of hydrodynamic calculation for reductions of groundwater levels and hydrodynamic model of coal deposit Drmno has been presented, and methodology includes hydrodynamic model of coal deposit Drmno recalibration, hydrodynamic calculation for groundwater levels reducing and the results of forecasting calculations, reliability analysis and techno-economic analysis.

Dewatering system for the surface water is not included since it is the same in both cases, and consists of a perimeter, bench and dewatering channels, water collector and pump stations.

4.1 Hydrodynamic Model of Deposit Drmno

Hydrodynamic calculations for requirements of sizing wells number, their common distances and individual capacities, as well as to forecast the effects for the defence system against groundwater have been implemented in a hydrodynamic model of the groundwater regime for the wider area of coal deposit Drmno.

Hydrodynamic model of the opencast mine Drmno was conceived and designed as multi seam model, with a total of six seams, as seen in the vertical profile. Each of these seams corresponds to a particular real seam, schematically presented and separated based on knowledge of the field and the results of analysis conducted via extensive field exploration works. The floor of the fifth or sixth seam (to and fro), makes the 3rd coal seam, which by its hydrogeological and hydraulic mechanisms represents an insulator, i.e. the current limit surface. The movement of the groundwater on the model has been calculated and simulated as the real movement under the pressure, or with free level, in the each discretization field separately, wherein aquifer flow conditions over time were changed in the model in accordance with the real conditions [5].

4.2 Estimated Calculation – Setting

In accordance with the accepted development schedule for the opencast mine Drmno, has been defined contours of the advance mine direction in general, the specific time sections. In the option calculations have been analyzed two options of the mine defence system against groundwater:

- Option 1 - opencast mine defence by dewatering wells, and
- Option 2 - combined opencast mine defence system by waterproof screen and dewatering wells.

In both cases, as the current situation is taken condition of mining operations at end of the year 2012, and the results of foreseen calculations are provided for the three time sections.

Option 1 - Mine Defence by Dewatering Wells
In Option 1 dewatering wells were set along the line of wells in parallel to the mine advance direction. Drained by wells are gravel and sand in the roof of the main coal seam.

Option 2 - Mine Defence by Combined System
This option for mine protection against groundwater implies the existence of the waterproof screen and dewatering wells. In Option 2, the western, northern and eastern borders of the mine (till the mining end) are contoured by waterproof screen with the depth to the second coal seam, that is, waterproof clays. Construction of waterproof screen begins in the year 2013, and ends at the end of the year 2017 (fig. 5).

As an example, is shown a condition of works, in both cases at the end of the year 2027 as at figure 5, when are fully express the effects of the constructed screen.

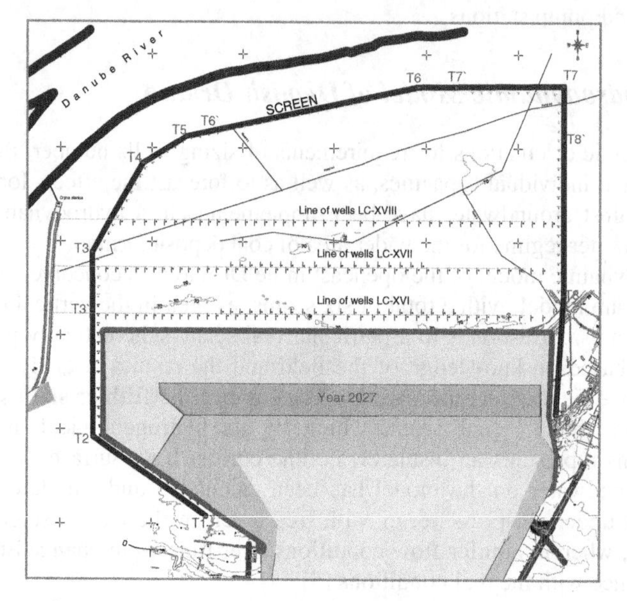

Fig. 5 Conditions of works of the opencast mine Drmno at the end of the year 2027

4.3 Results of the Forecasting Calculations

Calculation forecasted results for dewatering of the opencast mine Drmno, as per conducted hydrodynamic analysis are interpreted and presented as follows:

- To display the calculation results it has been accepted three characteristic time sections, namely: the end of the 2020, the end of the 2027 and the end of the year 2038 (for each option);
- Monitoring wells level in sandy sediments in the roof of the 3rd coal seam obtained by calculations are presented in the form of hidro contour lines map provided with the equidistance of 5 m, for each option and each time section;

- Balance of groundwater is provided by the balance of the lines of wells ahead of the advance mine direction, and through the balance of individual wells within these lines.

Figures 6 and 7 provide examples of layout maps for monitoring wells level in the sandy water bearing seam of the 3rd roof coal seam, for the selected time-section at the end of the year 2027.

As a result of wells system intensive operation, within the contours of the dewatering system occurs significant reduction of monitoring wells groundwater levels in roof sands. In the example for the second time-section (end of the year 2027) was achieved almost identical groundwater level overthrow in front of the opencast mine in both options, but in Option 2 within the area covered by the screen at the wider area reduced level of groundwater table is up to 10 m more than within the option 1. Table 1 shows the comparative values of the number of wells, the total capacity of new wells and their average values for both options and for all three time sections.

Table 1 Comparative review per options for mine defence against groundwater

Time section	Option 1 Number of wells	Option 1 Total capacity (l/s)	Option 1 Water inflow (l/s)	Option 2 Number of wells	Option 2 Total capacity (l/s)	Option 2 Water inflow (l/s)
2020	350	2610	105	350	2758	92
2027	237	1684	101	104	832	86
2038	391	2453	97	64	440	78
Total	978			518		

From table 1 it can be concluded as follow:

- In option 1, 460 wells more than in the option 2 is made where, in addition to a small number of dewatering wells, there is a waterproof screen.
- At the end of the year 2020 (the first analyzed of the dewatering system operation section) number of new constructed wells is identical with approximately the same total capacity of these wells.
- In the second time section, in the Option 2, 50% less groundwater is captured with locally higher reduction of levels.
- In the third time section, something bigger overthrow of the groundwater level has been achieved by defence system simulated in Option 2, with the participation of six times smaller number of dewatering wells than in Option 1 and by five times less capture of groundwater.
- Inflow of groundwater in the opencast mine, besides fro dewatering wells declines over a time, slightly lower values (from 13-20%) were obtained in Option 2.

At the end of the hydrodynamic analysis it can be concluded that the increased reduction of groundwater levels is achieved in Option 2 regarding mine protection, which includes dewatering wells and waterproof screen.

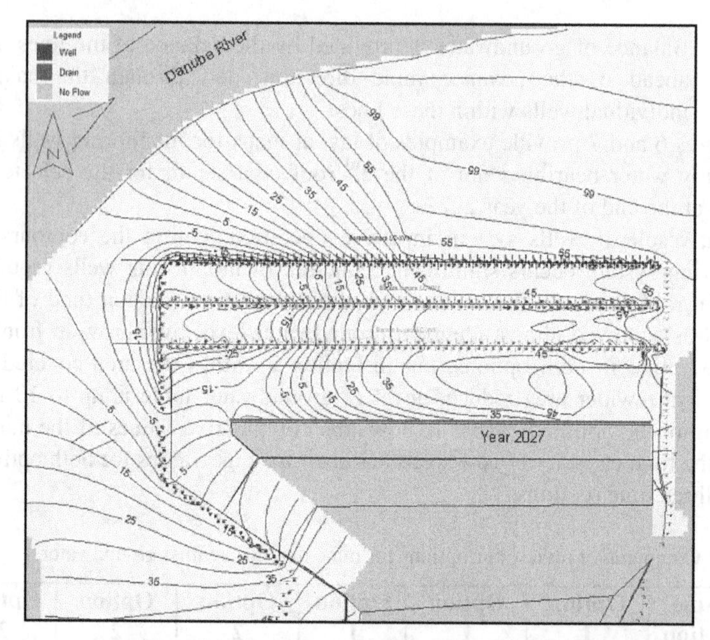

Fig. 6 Types of monitoring wells level in the sands above the 3rd coal seam (end of the year 2027) - Option 1

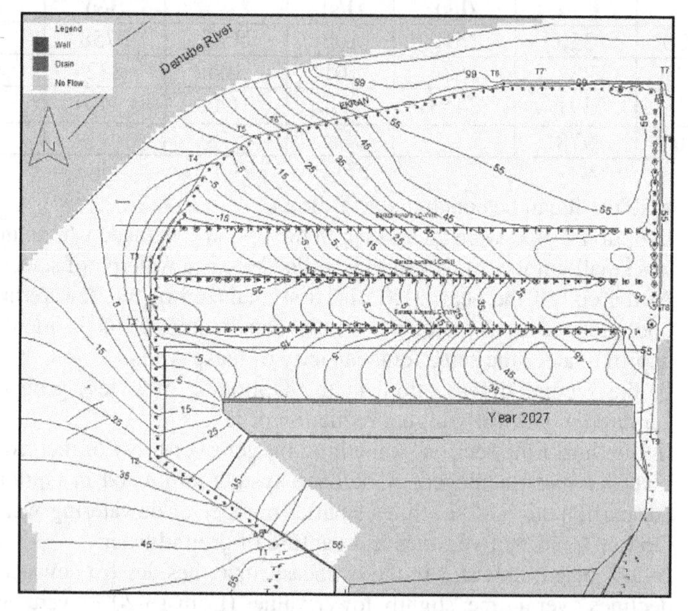

Fig. 7 Types of monitoring wells level in the sands above the 3rd coal seam (end of the year 2027) - Option 2

4.4 Reliability of Dewatering System for the Opencast Mine Drmno

The functioning of the dewatering system is defined as a random process with system condition realization exponentially distributed time. Installed dewatering system can have three conditions (S_0, S_1, S_2) according to the graph at figure 2.

An example is done for operation conditions in the year 2027 when according to the Option 2 waterproof screen is fully in operational with 104 wells, while in Option 1 in the operation are 237 wells. With dewatering by wells is permitted limited partial failure in the condition of S_1 to 75% from the normal condition S_0, so that in the condition S_1 in operation is 178, and in failure 59 wells for Option 1, and in operation are 78 and in failure 26 wells in Option 2. At transition from operation condition S_0 in the failure condition S_2 due to unforeseen extreme reasons, failure intensity is 0.00012, and the intensity of recovery 0.002 for both options. Average recovery time per well is 15 hours. From the condition S_1 occur conditional system failure with the condition S_2 during deterioration of operation of 75% compared to the condition S_1, when in the Option 1 in operation are 134 wells, and in failure additional 44 wells, while in the Option 2 in operation are 59 wells, and in failure additional 19 wells. Reliability parameters for the provided conditions in the calendar time, according to the equation (4), are shown in table 2.

Clearly, it is evident that combined system is for 22% more reliable compared to the dewatering system only by wells in the condition S_0 and with 14% lower probability of staying in an unfavourable condition with difficult operation of the S_1, while in the conditional condition failure S_2, Option 2 is reliable four times if compared with Option 1.

Table 2 Parameters of reliability by options

Parameter	**Option 1**	**Option 2**
a	0.009	0.004
b	0.017	0.067
c	0.00012	0.00012
d	0.002	0.002
e	0.009	0.004
f	0.0015	0.0035
P_0	0.385	0.469
P_1	0.424	0.484
P_2	0.191	0.047

Line of wells with parallel-series connection is provided in both cases as a system must have a reliability rate of operation in the condition P_{rk} larger than 90%. Required reliability is obtained by a model, in an iterative procedure. Condition graph for combined and connected dewatering system is provided in figure 3. Stationary probability of operation with three parallel lines of wells with the approximately same characteristics of reliability is ($P_{01} = P_{02} = P_{03} = 0.9$) (12):

$$P_{rIII} = 1\text{-}(1\text{-}P_{01})^3 = 1\text{-}(1\text{-}0.90)^3 = 0.999$$

Based on the statistical data of previous work for the installed lines of wells in the opencast mine Drmno, the operation probability for the combined system is (13):

$$P_{rk} = P_{03}{*}P_c{*}P_v{*}P_p{*}P_r = 0.999{*}0.98{*}0.97{*}0.98{*}0.98 = 0.91$$

Minimum are required three parallel lines of wells in operation for the implementation of the scheduled system dewatering reliability at Drmno mine.

4.5 Techno-economic Analysis

Option 1, provides the opencast mine Drmno protection against groundwater only by dewatering wells system, while Option 2, foreseen protection is by the combined method, that is, by system of dewatering wells and waterproof screen. Construction of wells, as well as waterproof screen construction is handed over to a third party.

By the end of the year 2038 for dewatering of groundwater in Option is required to construct less than 460 wells or to drill less than 54,335 m of boreholes in relation to Option 1 (tab. 1). In addition, it is necessary to procure and construct less than 10,530 m of gravity drain pipelines. Applying of Option 2 by the end of the year 2038 is to be used less than 127,125,558 kWh of electricity.

Since that by dewatering only with dewatering wells (Option 1) is disturbed the regime and level of groundwater, it is required to provide for village Kličevac drinking water and to invest €4 m for the construction of water and water supply factory. In addition it is to be disturbed humidity of the surrounding agricultural land, and due to it is required to provide water for the irrigation of 650 hectares of land.

Total investment in the system for protection of the opencast mine Drmno against groundwater by Option 1 amounts to €146 m, and by Option 2 it is €137 m. The implementation of Option 2 for the protection of the opencast mine Drmno against groundwater till the end of the year 2038 it is required to invest €9 m less than in Option 1. Based on the DCF analysis it was determined that in Option 1, investment in dewatering system amounts to 0.47 €/t of coal, while by investment implementation for Option 2 is 0.44 €/t of coal.

To protect the opencast mine Drmno against groundwater in Option 1 in the next 26 years it is required to be invested annually on average about €5.6 m, while for the protection by Option 2 is required to invested annually on average about €5.3 m. An average cost for dewatering of the opencast mine Drmno against groundwater per tonne of coal for a period of 26 years are more favourable in Option 2 (0.59 €/t compared to 0.62 €/t in Option 1). The dynamic growth of the total annual cost is provided in the chart at figure 8.

From the present chart it can be concluded that the total annual cost are to some extent higher in the initial phase of the protection system against groundwater implementation in Option 2, which is primarily due to investments in the construction of waterproof screen, but during later phases of implementation are significantly less than the total annual cost for Options 1. In addition the total annual cost of Option 1 shows a distinct cyclist in the occurrence of extreme high values as a result of constant investment in new lines of wells.

Considering the relatively small difference in the total and average costs it can be concluded that the final decision on the acceptance of the system for protection against groundwater is conditioned by other system elements, too, such as dewatering system reliability and environmental effects associated with the impact on the groundwater level outside the contour of the opencast mine and safer system operation of the HPP Iron Gate. The implementation of Option 2 reduces disturbance of groundwater regimes, and thus harm to the environment. A small number of wells significantly reduced the amount of water that is discharged into the Mlava and Danube rivers, which results in safer operation of the HPP Djerdap.

From the above is derived conclusion that the system of wells and the screen provided in Option 2, is economically, environmentally and in terms of reliability better if compared to Option 1.

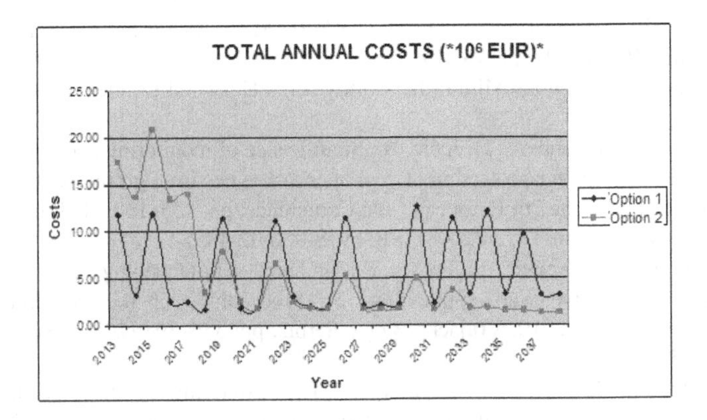

Fig. 8 Diagram of the total cost per year for Option 1 and Option 2

5 Concluding Remarks

In addition to a range of natural, technical, economic, environmental and safety factors and limitations, one of the most influential natural factors on the surface mining is the influence of surface and groundwater on the production processes and the stability of the site and the final slope system for the both, opencast mine and dump site. Due to the complexity and multidisciplinary nature, system selection and construction of opencast mine dewatering facilities cannot be implemented without geological and hydrodynamic modeling of the working environment, by the application of the modern approaches to determine the reliability of the dewatering system and optimization of dewatering facilities. For effective and efficient dewatering of opencast mines is not sufficient optimal selection of systems and facilities, but also well organized and manageable process of dewatering.

When planning dewatering until the end of mining period it is necessary to achieve, in addition to the requirements for safe and reliable mining, environmental and social conditions, that is to eliminate negative environmental impacts, effective use of land and water resources, and social and economic benefits in the course of sustainable development and operation of the opencast mine.

The present model is a synthesis of the theoretical basis and practical aspects for the dewatering system implementation and by the established procedures it can significantly support experts which in practice are dealing with dewatering projects. Defined model is fully verified on the example of dewatering facilities selection and optimization of the groundwater dewatering system in the opencast mine Drmno.

References

[1] Barlow, R.E.: Engineering Reliability. ASA-SIAM Series on Statistics and Applied Probability, Philadelphia, USA (1998)

[2] Pavlovic, V.: Continuous Mining Reliability. Ellis Horwood Limited, Chichester, UK (1989)

[3] Pavlovic, V., Subaranovic, T., Jocic, B.: Significance of monitoring and administration of dewatering wells for protection of coal opencast mines from ground water intrusion. In: Proceedings of the 7th European Coal Conference, pp. 136–140. National Academy of Sciences of Ukrain, Lvov (2008) ISBN 978-966-02-4855-7

[4] Pavlovic, V., Subaranovic, T., Prokic, S.: Technical and technology construction and possibility of slurry wall application. In: Proceedings of the 9th International Conference - OMC 2010, Vrnjacka Banja, Serbia, pp. 376–394 (2010) ISBN: 978-86-83497-15-7

[5] Polomcic, D., Pavlovic, V., Subaranovic, T.: Dewatering system selection at the opencast mine Drmno by using hydrodynamic forecasting calculations. In: Proceedings of the 10th International Conference - OMC 2012, Zlatibor, Serbia, pp. 376–394 (2012) ISBN 978-86-83497-15-7

[6] Stoll, R., Neiman-Delius, C., Drebenstedt, C., Mullenseifen, K.: Der Braunkohlen Tagebau. Springer, Berlin (2009) ISBN 978-3-540-78400-5

[7] Wolstenholme, L.C.: Realibility Modelling. Chapman and Hall, London (1999)

New Technological Developments in the Field of Vibro-Compaction in the Lusatian Lignite Mining Area

Charles-Andre Uhlig

GMB GmbH, Senftenberg, Germany

1 Introduction / Background

In order to create a safe and sustainable postmining landscape in the Lusatian lignite mining area dynamic compaction methods at great depths of about 65 m are required. This involves extremely high technical demands on the applied vibro-compaction (VC) methods. These include the vibro-flotation (VF) and a vibro-displacement method (VD).

The choice of vibrocompaction method applied in the Lusatian lignite mining depends on the local geologic conditions, such as particle size distribution and local ground water level. (Figure 1)

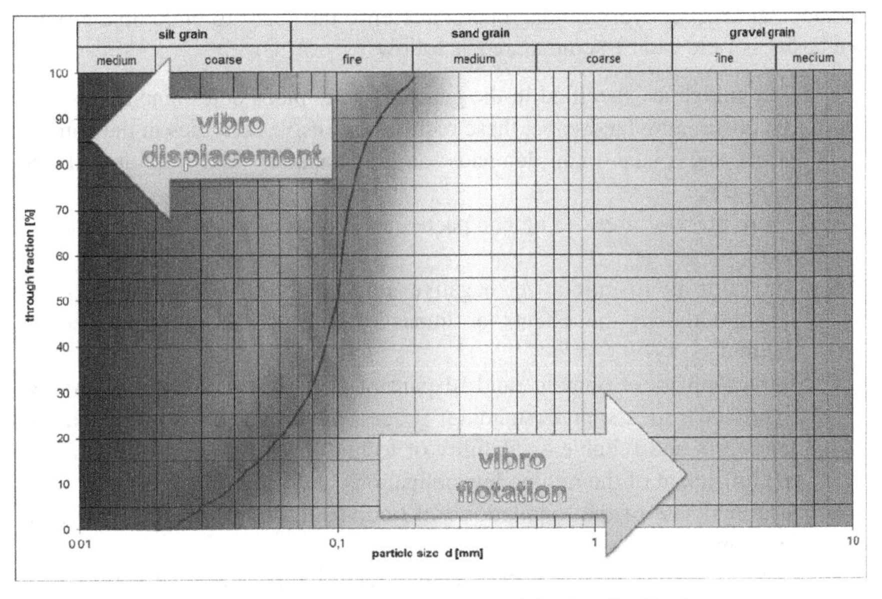

Fig. 1 Application of VF or VD depending on the particle size distribution

Advantages of the use of vibrators for ground improvement compared to the use of large-scale and cost-effective explosive compaction can be found in particular in the degree of compaction that can be reached subsurface. Thus, the

C. Niemann-Delius. (ed.), *Proc. of the 12th Int. Symp Continuous Surface Mining - Aachen 2014*,
Lecture Notes in Production Engineering, DOI: 10.1007/978-3-319-12301-1_12

compaction effect is nearly complete and the available pore space can be reduced to a maximum when using the VF. Other benefits of the VF are significant improvements to the soil parameters, such as increased shear strength and decreased consolidation parameters. The primary target for using the VF-procedure is to increase the density of the soil on site. Only this method of soil improvement can ensure that the geotechnical safety of the postmining landscape sustainably. The state of the art technology already in use, allows vibro-compaction operations at depths of up to about 70 m. With the development of construction projects on dumps and for the safe landscaping of the postmining areas vibro-compaction methods are of great importance and value. In particular, the dumps in the Lusatian mining area, with their loosely deposited, non-cohesive, narrow-classified sands tend with water saturation due to the gradual rising of the groundwater level to liquefaction and flow-slide. It is therefore sometimes necessary to vibro - compact open pit dumps, also those with a higher thickness of the F60-mining dumps, down to the footwall.

Due to the strict requirements mentioned, the mining operator Vattenfall Europe Mining AG (VE-M) decided to establish a new business area for vibro-compaction. During the financial year 2010, the GMB GmbH (GMB), a subsidiary of Vattenfall Europe Mining AG, began after extensive preparatory work with vibro compaction operations. The company has been operating successfully for about three years, currently with 2 vibro-compaction units. The establishment of a safe postmining landscape with a high geotechnical quality is one of the fundamental tasks of every mining operator. Hence the needs to build up a group's own business field can be summarized as follows:

- The guidelines specified in the general lignite plans determine the uses of the postmining landscape; these requires extensive activities in the coming years and place high demands on the capacity and stability of the subsurface
- There are vast areas to be compacted and thus a considerable volumes of soil
- The aim is to minimize negative impacts on the environment and neighbourhood; the effects or impacts by using the ideal compaction method
- Intragroup development and expansion of technical know-how for improved process understanding, process control and cost control; the long-term and reliable availability of technical know-how is essential for the fulfilment of the mining law obligations
- The thickness of the foundation soil / dumps to be improved is very high (> 45 m)
- Only when using a reliable method and by providing the necessary documentations accessibility also for vehicles of the postmining landscape depending on the type of use can be guaranteed in the long term by the mining operator

2 Construction Equipment and Complex State of the Art Technology

In preparation for the commencement of regular operation of the VF intensive preparation work was done by the departments mining services and engineering of GMB. As a result of these preparations relevant technological and geotechnical specifications anticipated, have a significant impact on the device parameters analysed. The determined project process parameters such as working depth and the necessary foreland between carrier and working point were the basis for the technical dimensioning of the VF complex. The experience gained in remediation work since the early 1990's about the use of technology and infrastructure of the VF were of course taken into consideration during the preparation of the project. With the practical experience already gained and as a result of the process analysis carried out by GMB the planned technology use was modified. In addition not purely technical aspects, but also extensive changes in the labour safety and process reliability were introduced by GMB.

Specifically, the first project required (1st section of tank transport road Nochten) compaction of depths up to 45 m with a required foreland > 20 m. These specifications were the basis for the sizing and selection of the lance set and the carrier. The real working tool the vibrator is a steel body and has the form of a torpedo. The dimensions and thus the performance of the working tool depend very much on the intended use. Common dimensions are between 3 m to 5 m in length, a diameter of 30cm to 40 cm with an impact force of up to 500 kN and a power consumption of 50 to 200 kW. The diameter of the circular arc described by the movement of the vibrator is device-specific between approximately 6 mm and 4 cm. The resting point of the working tool is below the clutch. By means of the use of a vibration-damping elastic coupling between vibrator and the lance the loss of vibration of the vibrator is minimized. In the working process the vibrator impacts directly on the surrounding soil. After extensive technical discussions with potential providers a vibrator of the performance class V 48 with an impact force of 450 kN, and a tip and side water supply and a corresponding lance probe of the company VIBRO Services were rented.

In addition to the work unit vibrator there is another primary component the carrier crawlercrane. Only on the basis of correct economical and technical consideration the correct dimensioning of the carrier crane for an efficient and safe VC-process is possible. The requirements of each single project vary depending on cost, safety and performance objectives. Primarily the carrier unit should always have the necessary lifting force and lifting height for each specific application.

The process of VF requires a reliable supply of water and compressed air. The media air and water are fed to the process through the so-called media-mixing-equipment in variable proportions.

Given the importance of the provision of air and water, the first activities for technical process improvement focused among others on the insulation of the media-mixing-device. By means of these technical improvements, stable winter operation down to very low temperatures (max. -25°C) is guaranteed.

Furthermore, the improvement of penetration process was optimized by a new arrangement of nozzles for the top and side water supply of the vibrator. The installation of a frequency converter, which controls the start-up process of the vibrator as well as the speed and direction of rotation, was a further important technical improvement.

The compressed air is supplied by an electrical compressor. The compressor installation is housed in a fully enclosed and thus noise-reduced container. The negative environmental forces acting directly on the technical components were also significantly reduced with this technical measure. Using an electrical compressor, there was significant improvement in the energy efficiency. The components compressed air supply, frequency converter and control cabinets were installed in a backpack design directly on the carrier unit. With the permanent installation of the components on the carrier the advance and efficiency of the VF-equipment complex increases. The winter safe water supply is accommodated in a compact designed container module that is carried at the working level behind the carrier. In the heated water supply module a water storage tanks and an electrically operated booster pump were installed.

A heavier lance with a diameter of 400 mm was used and it was assumed to have resulted in a reduced penetration period. A positive effect has not been established yet.

Until now GMB has successfully carried out compaction operations with the described methods for vibro-compaction construction sites at depths of up to at least +45 m up to +65 m in the standard process. The near-surface areas (depth < 10 m) are processed project specific with other compaction methods, such as the falling weight compaction or the use of a vibratory roller.

3 Experiences

3.1 Dimensioning Carrier Unit

With the experience gathered in the production process when using the Dragline Liebherr HS 895 HD it was concluded, that this carrier is not optimal for the intended purpose of utilization. The equipment was operated in VF-project at a depth of up to 45 m. Besides technical disadvantages especially the geotechnical conditions on site required a frequent operation at the absolute limit of the performance of the equipment. Furthermore the high installed motor power made the machine inefficient. The limitation in available lifting forces led to the decision to use the carrier crane type Liebherr LR 1300 for the production process (Figure 2).

For the case with compaction depths greater than 45 m it's from safety and process engineering point of view worth considering only cranes of the order of a LR 1600 and LR 1750. During the decision making process also the dimensioning and the planned applications for the selected unit by the equipment manufacturer had to be taken into account. The anticipated operating environment and thus the dimensioning of the LR 1600 carrier are found in the technical design of the manufacturer Liebherr. The carrier is normally used on construction sites and not in opencast mines. The manufacturer therefore proposed the acquisition of a LR1750 for the planned vibro-compaction operations. According to the manufacturer the underlying statics of a LR 1600 crane is not designed for long-term use in surface mining operations with difficult subterranean conditions. In particular, the undercarriage of the LR 1600 does not meet the long-term needs and requirements in mining operations.

Other factors to be considered are derived from the technological requirements. The lance including accessories may have at some constructions site, a length of 70 m and more. This means a total weight of about 40 tons. This requires, to ensure a satisfactory productivity, a significantly higher lifting capacity of the carrier. In ordinary forelands a crane similar to a LR1750 is required. Already today, at depths of 65m + +; lifting forces of up to 80 tonnes in the standard process are required in order to ensure a continuous production process. As a result of experiences gathered by GMB in 2013 a Liebherr LR 1750 was purchased. For example, a two- to three-times reserve in lifting capacity is beneficial and meaningful in difficult ground conditions. The lifting forces of up to 120 tonnes in the standard process should be applied.

Carrier class / type application criteria/ technical parameters	Liebherr HS 895	Liebherr LR 1300	Liebherr LR 1600 SL	Liebherr LR 1750 SL
winches	2 winches 350 kN	2 winches 150 kN	3 winches 180 kN	3 winches 160 kN
boom	optimized for dynamic loads	optimized for static loads	optimized for static loads	optimized for static loads
installed capacity	engine 670 kW	engine 450 kW	engine 370 kW	engine 400 kW
undercarriage	crawler pad 1,5 m (25 to)	crawler pad 1,5 m (22,2 to)	crawler pads 2 m (38 to)	crawler pads 2 m (55 to)
slewing speed	0 - 3,6 rpm	0 - 1,8 rpm	0 - 0,95 rpm	0 - 1,5 rpm
total weight	229,2 t (boom length 60,8m)	280,3 t (boom length 62,0m)	511 t (boom length 84 m)	657 t (boom length 84 m)
rope diameter	Ø 24 mm	Ø 28 mm	Ø 28 mm	Ø 28 mm
foreland [1]	25 m - 35 m	25 m - 35 m	25 m - 35 m	25 m - 35 m
lifting capacity [2]	31 m - 19 m	53 m - 33 m	88 m - 51 m	113 m - 73 m
field of application	crane	crane	crane	crane
	simple	simple	simple	simple
	-----	intermediate	intermediate	intermediate
	-----	-----	-----	**complex**
	dynamic compaction	-----	-----	-----
	drag line / clamshell / etc.	-----	-----	-----

1) *vibro compacation projects: simple* - foreland <= 25 m, depth <= 35 m / *intermediate* - foreland <= 25 m, depth <= 50 m / *complex* - foreland > 25 m, depth > 50 m
2) *total weight of vibrocompactor and lance tubes (65m) = 34 to*

Fig. 2 Technical comparison of carrier crawler cranes and suitability for vibro-compaction projects

3.2 *Labour and Process Safety*

The state of the art in the business field vibro-compaction of GMB is considered as reached. A variety of technical improvements were implemented in different process segments: So the technical equipment for example, was adapted such that reliability down to minus 20°C is guaranteed. In practice the new technical solutions have proven their suitability down to minus 26°C. Therefore year-round unit use (winter) provides an optimised equipment utilization. Furthermore there is now a hydraulic press being designed, which will allow in future a safe recovery of lance and vibrator when necessary. In addition the revised process data collection ensures a better process control through extensive and more precisely processed data. The highly important objective of establishing a long-term holistic and safe data processing (see 4.2) as an important part of the dump registry is realized in the process on site. A variety of organisational measures and definitions ensures a smooth and efficient process realization.

From an extensive risk assessment of the compaction process, in which the geotechnical, technical and behavioural hazards have been assessed in detail, an overall operational health and safety process documentation was compiled. Besides the best practice design of the production process, it is vital to maintain employee safety on site at all times. In this context, together with the fire brigade of VEM a new rescue technology has been developed. (Figure 3) This includes behavioural rules and new technical solutions which allow a quick and simple assistance in case of an emergency. Of course prevention is of highest priority.

Fig. 3 Rescue technology in the working area

4 Application of New Software Solutions

4.1 Planning, Implementation, Evaluation

Of particular importance for the improvement and optimization of the vibro-compaction process are the process data acquisition and control, as well as the process data analysis. The application of a software solution for the planning and evaluation of ground compaction measures of GMB is based on the software SCMS from the company Gicon. The software Soil Compaction Management System (SCMS) from Gicon existed as software solution for the control and preservation of process data as version 1.X. The program has been used to date for the analysis of extensive data in practice, and has proven its functionality. Despite this tests were carried out by GMB with the SCMS software version 1 to check suitability. Before that a pre-selection based on a market research and a cost evaluation for other solutions or an own software development was done. Among others software, especially the solution of remediation businesses used were considered and rated. The result of the assessment was the necessity of expanding the functionality for all the different software solutions investigated. As a result of the specifications listed by GMB the Gicon company's interest was stimulated to launch a new improved version of SCMS. The functionality of the software solution has been enhanced with a variety of new features. In addition to the existing exclusive use as evaluation tool for realized vibro-compaction points important innovations have been integrated into SCMS version 2. Examples thereof are:

- Inclusion of additional soil compaction methods (vibro-compaction with step-back technology, falling weight compaction)
- Im-/export planning data (import of MicroStation file into SCMS, export of SCMS for process data control (PDC) of vibro-compaction unit)
- Basic expansion of import functionality of SCMS (supplied data from PDC are accurately assigned to planning data)
- Significant enhancements in geotechnical monitoring (monitoring via preliminary assessment, construction site management and follow-up assessment including data comparisons between the different exploration phases)
- Functional extension for the following field exploration: drilling, geophysics, combined penetration tests, sampling, soil boring, trenches)
- Inclusion of cross-links to other information systems (LIMS) of GMB
- Free export options to Excel for detailed evaluations in detail
- AddIn MicroStation
- further extension of GIS functionality
    - display of packing-compaction-points
    - setting up an interface to GIS solution of VE-M
- Introducing the function of construction and planning of disintegrated compaction-grids

In Figure 4 the basic sequence of VC-project is presented. In phases 1 and 3 Office software solutions are used. These include the programs MicroStation, ArcGIS, SCMS.

The planning specifications for the compaction process are now created through the software solutions SCMS / MicroStation. The planning data is handed over to the production process through a pre-defined interface. Planning data is therefore directly available without further efforts for the production process. In the production process a separate software solution is used. Smooth operations are ensured by a clear rule on the interfaces between Phase 1 and 2 or 2 and 3. Upon completion of Phase 2 the process data is transferred, stored and displayed according to the defined parameters in the systems mentioned above. The preparation of the immediate success control as basis for the final documentation is efficient and reliable generated.

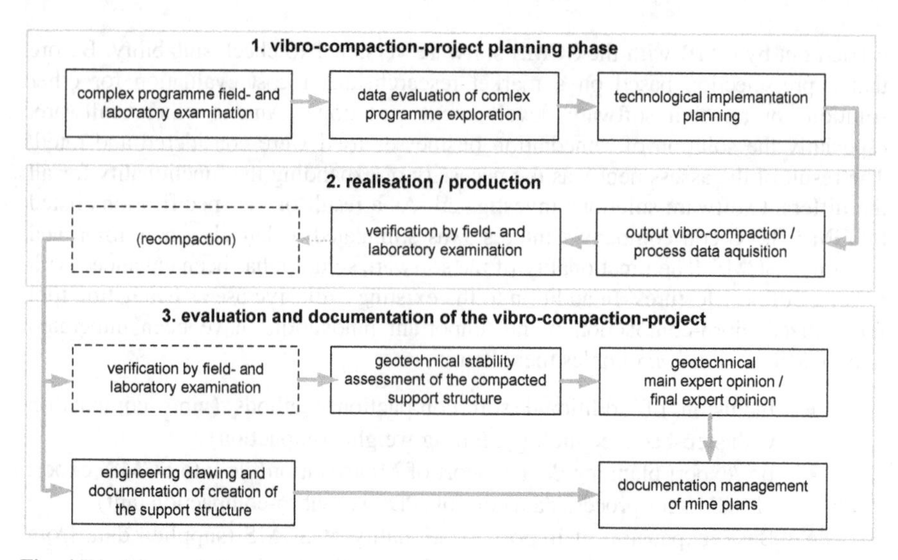

Fig. 4 Workflow of vibro-compaction projects

4.2 Process Control and Process Data Acquisition

The complete control of the compaction process, which includes the vibrator, the compressor, the media-mixing-device (control unit for water/air), the water pump and the frequency converter, is possible from the cockpit via touch panel in the cab of the carrier. The data recording of compaction process is done by a programmable logic control. All data required for the evaluation will be saved on

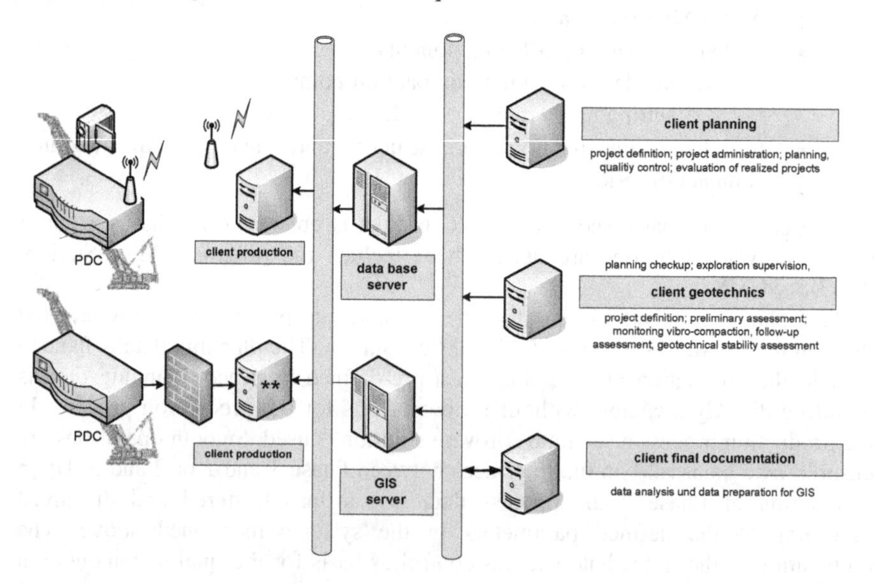

Fig. 5 Software architecture principle

a suitable storage medium over a period of 8 weeks. The reading of the data is done using USB-ports and Ethernet-interfaces at present. In perspective the data transfer is provided by radio transmission (Figure 5). In case of unplanned occurrences an evaluation afterwards is always possible using video cameras installed which are directed on the work point.

5 Optimization Potentials and Objectives

Basically a further reduction of the average penetration period has to be achieved. Sensors which monitor precise information about the location and position of vibrator and lance will be installed. Based on this information the overall availability of the equipment complex will be increased (e.g. reduced sticking of the lance underground, minimize number of total loss of equipment). In the penetration process operators of the complex can maintain a safe vertical penetration of the lance, thereby optimizing the entire penetration process. Aberration during the penetration process may be detected early. In future additional information and data will be stored in a complex database (GPS, power consumption, temperature, amount of water, etc.). This data will be compared with the knowledge about the dump geology (preliminary and follow-up field exploration) and is combined with the goals of process and cost optimization. The future aim is to develop a reliable online success control. This tool enables the persons responsible to carry out the process more accurately and efficiently during the process execution. An equally important element is a continuous quality control and documentation of the entire production process.

In the short term there is a continuous increase in the overall availability of the entire equipment complex at the center of all operations. This is predominantly an organisational advantage. The benchmark is an overall availability of 90 %.

The substantial expenses involved with stuck lance underground have to be reduced by using alternative recovery technologies. Up to now one solution to accelerate the recovery process was tested in practice. The support of the lifting forces of the carrier with a hydraulic recovery device is right in test mode at present. The main challenge is ensuring an optimum grip between the lance probe and hydraulic driven recovery device. Forces to be handled are up to 400 tonnes. Currently an supportive additional method is being prepared and planned to be tested soon. The accelerated recovery of the vibrator and lance was attempted by means of vibrations.

6 Summary

In order to create a safe and sustainable postmining landscape in the Lusatian lignite mining area dynamic compaction methods at great depths of about 65 m are required. This includes extremely high technical demands on the applied vibro-compaction methods. Due to the existing requirements GMB began in 2010 with vibro-compaction operations. GMB has been able to demonstrate a very positive development for the first 3 years since the commissioning of the first complex of vibro compaction. Newly introduced technical solutions and the use of improved or new software solutions led to a sustainable process improvement. In practice the suitability of all new introduced features could be verified to external temperatures of minus 26 degrees centigrade. Occupational safety has been improved by a new fuse technology. Especially significant for the further improvement and optimization of the vibro compaction processes are the recently launched process data acquisition and process control, as well as the subsequent process data analysis. Tasks and objectives for further process development are described and milestones for their implementation in the coming months and years are defined. Apart from the exact determination of the position of vibrator and lance during the production process the 'online-process control', which is pooling a variety of process parameters, is of the utmost importance.

References

[1] Kirsch, F.: Experimentelle und numerische Untersuchungen zum Tragverhalten von Rüttelstopfsäulen. TU Braunschweig, Fachbereich Bauingenieurswesen, Dissertation (2004)
[2] Kirsch, K., Sondermann, W.: Baugrundverbesserung. In: Grundbau Taschenbuch, Teil 2: Geotechnische Verfahren, sechste Auflage. Ernst & Sohn, Berlin (2001)
[3] Möller, G.: Geotechnik Grundbau. Ernst & Sohn, Berlin (2006)
[4] Schnell, W., Vahland, R.: Verfahrenstechnik der Baugrundverbesserung. B. G. Teubner, Stuttgart (1997)

Estimating Stability of Internal Overburden Dumps on the Inclined Foundation by Simplified Bishop Criterion

Bayan R. Rakishev[1], Oleksandr M. Shashenko[1], Oleksandr S. Kovrov[2], and G.K. Samenov[2]

[1] Kazakh National Technical University, Almaty, Republic of Kazakhstan
b.rakishev@mail.ru
[2] National Mining University
Dnipropetrovs'k, Ukraine

Abstract. Current development of open-cast mining is characterized by complication of geological conditions and increased requirements to reduce the negative environmental impact. A significant role in solving problems concerning management of mineral resources allocated to internal dumping and justification of such dump geometric dimensions which provide simultaneously stability and minimum spacing.

Stability of internal overburden dumps while surface mining depends on physical and mechanical properties of the rock mass, climatic factors, groundwater levels, shape and geometry of the foundation, and external loads. Complex influence of these factors leads to the emergence and outspreading geomechanical deformations in the dump core with the formation of rockslides, which complicate the mining operations and result in an increase of the specific mining capital expenditures. Therefore, effective management of open-cast mining technology, internal overburden dumping on inclined foundations is an important engineering issue.

Reliable estimation of high internal dumps stability in the conditions of gently sloping and inclined foundations by modern numerical geomechanical models allows optimize surface mining operations as a whole and decrease harmful impact on environment.

The paper deals with the estimating stability of internal dumps at open pit "Maikubenskiy" in Kazakhstan using GGU-Stability software. The safety factors for open-cast slopes depending on geomechanical properties of overburden mass, inclination of foundation and mining equipment loads by Bishop failure criterion are calculated. The most stable geometric parameters for internal dumps on the inclined foundation are determined.

This research was carried out at the Department of Hydrogeology and Engineering Geology of the RWTH Aachen University (Germany).

Keywords: stability of internal dumps, safety factor, inclined foundation, geomechanical evaluation of rock massif, sliding surface, Bishop failure criterion.

C. Niemann-Delius. (ed.), *Proc. of the 12th Int. Symp Continuous Surface Mining - Aachen 2014*, Lecture Notes in Production Engineering, DOI: 10.1007/978-3-319-12301-1_13

1 Introduction

Open-cast "Maikubenskiy" is a coal mine company specializing in mining brown coal of Shoptykolskoye coalfield with an annual project capacity of 25 million tons of coal per year. At present, the actual output is about 4 million tons per year and it is planned to increase an annual coal output up to 8 million tons. Brown coal seams are mined by the advanced open-cast technology with internal dumping.

Shoptykolskoye coalfield is characterized by flat and slightly inclined coal seams in the range of 4…10°. The overburden is deposited both inside and outside of the open-cast area by using traffic (Eastern tract) and non-traffic mining systems (Central tract). Considerable volumes of overburden reduce the technological and economic performance of the enterprise and negatively influence to the territory adjacent to the national park "Jasybay".

Flat bedding of brown coal seams (4°-12°) allows optimize mining operations by using internal space of the open-cast for disposal of overburden mass. Internal dumping reduces outlay on transportation of overburden into external dumps and minimize negative environmental load on the territory.

The oblective of this paper is estimating stability of internal overburden dumps on the flat and inclined foundations in application to geological and mining conditions of open-cast "Maikubenskiy" (Kazakhstan) by geomechanical calculations in the GGU-Stability software.

2 Methodology and Theoretical Background for Analysis in GGU-Stability Program

The GGU-Stability program system allows slope failure investigations according to German Standard DIN 4084, DIN 4084:1996 and DIN 4084:2009, using circular slip surfaces (Bishop) and polygonal slip surfaces (Janbu, General Wedge method and Vertical slice method) [1]. Furthermore it is possible to investigate soil nailing and reinforced earth walls.

Analysis uses the design variables φ_d and c_d:

$$\tan \varphi_d = \tan \varphi_k / \gamma_\varphi, \tag{1}$$

$$c_d = c_k / \gamma_c, \tag{2}$$

where φ_k and c_k are characteristic shear strength values; φ_d and c_d are the shear strength design values.

Stability is then analysed using the modified values. This gives the safety factor η in accordance with the global safety factor concept, but relative to the design values. The reciprocal of this "global safety factor η_d" gives the utilisation factor μ in accordance with the partial safety factor concept:

$$\mu = 1/\eta_d. \tag{3}$$

The GGU-Stability program integrated the Bishop (circular slip surfaces as presented on the Figure 1) and Janbu (polygonal slip surfaces) calculation methods based on DIN 4084. The basic equations related to these methods are the following for Bishop (circular slip surfaces)

$$\eta = \frac{r \cdot \sum T_i + \sum M_S}{r \cdot \sum G_i \cdot \sin \vartheta_i + \sum M} \tag{4}$$

with

$$T_i = \frac{\left[G_i - (u_i + \Delta u_i) \cdot b_i\right] \cdot \tan \varphi + c_i \cdot b_i}{\cos \vartheta_i + \frac{1}{\eta} \cdot \tan \varphi_i \cdot \sin \vartheta_i}, \tag{5}$$

and for Janbu (polygonal slip surfaces)

$$\eta = \frac{\sum T_i + \sum H_S}{\sum G_i \cdot \tan \vartheta_i + \sum H} \tag{6}$$

with

$$T_i = \frac{\left[G_i - (u_i + \Delta u_i) \cdot b_i\right] \cdot \tan \varphi_i + c_i \cdot b_i}{\cos^2 \vartheta_i \left(1 + \frac{1}{\eta} \cdot \tan \varphi_i \cdot \tan \vartheta_i\right)}, \tag{7}$$

taken from DIN 4084, where η – terrain or slope failure safety factor, dimensionless; G_i – self weight of an individual slice with consideration of the soil unit weight estimates including surcharges, kN/m; M – moments of loads and forces not included in G_i around the centre-point of the slip circle, positive when acting excitingly (H for Janbu analogous), kNm/m; M_S – moments around the centre-point of the slip circle from forces after Section 6e (DIN 4084), which are not considered in T_i (HS for Janbu analogous), kNm/m; T_i – the resisting tangential force of the soil at the slip surface for each slice (for polygonal slip surfaces the horizontal component), kN/m; ϑ_i – tangential angle of the slice to the horizontal, which for circles is equal to the polar coordinates, degrees; r – radius of slip circle, m; b_i – width of slice, m; φ_i – the decisive friction angle for the individual slice after Section 8 (DIN 4084), degrees; c_i – the decisive cohesion for the slice after Section 8 (DIN 4084), kN/m²; u_i – the decisive pore water pressure for the individual slice, kN/m²; Δu_i – the decisive pore water pressure for the slice as a result of soil consolidation, kN/m².

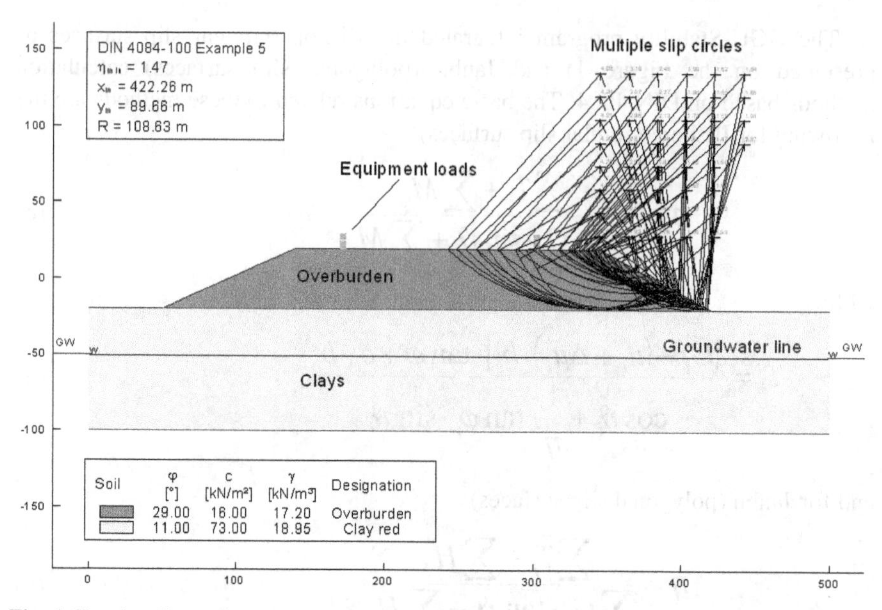

Fig. 1 Circular slip surfaces by the Bishop failure criterion

In the GGU-Stability program, Δu_i is calculated by multiplying the pore water pressure coefficient and the effective vertical stresses. Alternatively, you have the possibility of defining so-called consolidation layers. Using the required input data, the program carries out a one-dimensional consolidation calculation.

The above mentioned relationships are described in detail in DIN 4084 and DIN 4084:2009. Partial factors are used in DIN 4084:2009. The safety factors are thus already incorporated in the soil properties, loads, etc. The term "*safety factor*" is thus already allocated. Instead of "η", then, "μ" must be adopted in the above relationships, known as the "*utilisation factor*".

In accordance with DIN 4084 (old), the program uses the safety definition after Fellenius:

$$\eta = \tan \varphi_{\text{work.}} / \tan \varphi_{\text{req.}}, \tag{8}$$

where $\varphi_{\text{work.}}$ and $\varphi_{\text{req.}}$ are friction angles under working load and required [2].

3 Estimating Stability of Internal Overburden Dumps and Input Data

For the purpose of estimating stability of overburden dumps on the inclined surface, the standard Bishop (circles/slices) calculation method integrated in the GGU-Stability program is chosen. This method is based on determination of multiple circles as possible slip surfaces in the slope that will be described below. The further step is associated with building model geometry, selecting external surfaces and soil layers, assigning soil properties and entering water levels and loads. After assigning the centre-points of slip circles the system can be analysed. The minimum safety factors corresponding to each centre-point will be displayed. The slip circle with the lowest safety will also be graphically displayed. If the most unfavourable slip circle centre-point is at the edge of all defined centre-points, its position can be determined more precisely by assigning new centre-point array.

There are some important notes concerning calculation procedure. The first one refers to the water levels because the program calculates the water load in the area of a slice and the horizontal loading of the slope due to water pressure. If the water level is below surface level it has no meaning for calculations. The second note concerns the number of slices because small numbers of slices mean low precision and shorter calculation times. Large slice numbers mean a correspondingly longer calculation time and higher precision. The minimum number of slices is also dependent upon the complexity of the slope. A slope which is heavily layered will require a larger number of slices than one which is homogenous. For general approach using at least 50 slices is recommended.

According to the technology of internal dumping at the open-cast "Maikubenskiy" the overburden is expected to dispose in two layers. The lower layer with the height of 40 m is disposed by the walking dragline ESh–15/100 with average load on the foundation P_{drag} = 100 kPa, and the upper layer with the height up to 40 m by bulldozer D-394 with average load on the foundation P_b = 43 kPa. The slope angle for disposed overburden rocks is 35°. The safety bench between lower and upper layers is recommended as 75 m. As the inclination of the internal foundation varies in the range of 0…14°, estimation of dump stability and determination of their reasonable geometrical parameters is essential for open-pit operations. For these reasons the GGU-Stability program was applied.

Table 1 Geomechanical characteristics (calculated) of the dump foundation and overburden mass

Parameters	Units	Layers	
		Dump foundation	Overburden mass
Unit weight	kN/m^3	18.95	17.2
Poison's ratio	dimensionless	0.43	0.3
Young's modulus	kN/m^3	20000.0	5000.0
Tensile strength	kN/m^3	121.0	12.0
Friction angle	degrees	10.5	29.0
Cohesion	kN/m^3	73.0	16.0

According to the national standards for slope stability of open-cast benches the safety factor depends on the soil properties and their changes in time, mining technology, static and dynamic loads etc. Taking into consideration just the most essential parameters impacting slope stability and limit equilibrium in the soil or rock mass the recommended value of the safety factor for overburden dumps slope stability during surface mining is 1.20 [3].

Dump parameters depend mostly on bearing capacity of dump foundations, geomechanical properties of the overburden and mining equipment. For the conditions of the open-cast "Maikubenskiy" the calculated characteristics of the dump foundation and overburden mass are presented in the Table 1.

The main objective of geomechanical modelling is an estimation of the limit geometrical parameters for internal dumps, evaluation of their stability due to physical and mechanical characteristics of the overburden and rock mass, and loads from mining equipment.

According to the problem described above the overburden from mining operations is piled within the internal space of the quarry on the surface with inclination angles up to 15°. The external boundaries of the geomechanical model are: 500 m in width, 100 m below the ground surface as foundation layer consisted of clays and overburden mass with the height of 10...40 m above the ground surface (Figure 1). The angles of the overburden piles disposed on inclined foundation vary in the range 20…50 °. The groundwater level is at the 50 m below the surface but not considered at this research. The geomechanical model has two different layers and their properties are presented above in the Table 1. A permanent equipment load of 43 kN/m² on the slope is also considered.

The dump stability was estimated by stage-by-stage building-up of its height *H* with the step of 10 m. The position of the bulldozer D-394 towards the crest of the bench was also varied in the range of 2…10 m.

4 Results

The Figure 1 presents general approach to evaluating internal dump stability based on Bishop (circles-slices) method. The calculation procedure allows to determine the most appropriate slip circle for the object according to Bishop failure criterion. In this case the flat surface of the dump foundation was considered with the total height of the overburden layer 40 m and the slope angle 35°. The safety factor is 1.47 for the utmost justified height as for the first overburden layer.

The Figure 2 presents main graphical results of the calculations of the overburden dump stability on the inclined foundation with the angles of 0…15°.

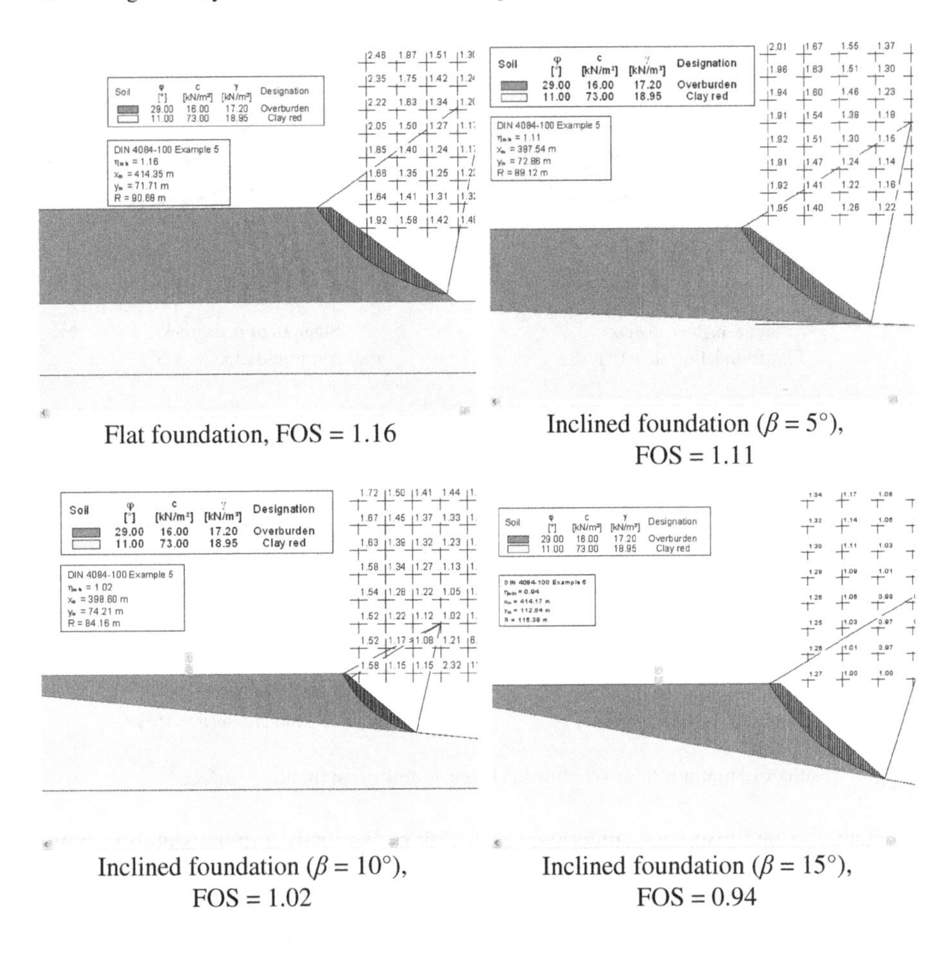

Fig. 2 FOS values for limit geometrical parameters of the overburden layer with the height of $H = 40$ m and the slope angle $\alpha = 40$

The approximate limit geometrical parameters of the overburden layer are the following: height $H = 40$ m and the slope angle $\alpha = 40°$. Depending on the inclination of the foundation the safety factor is decreasing steadily from 1.16 to 0.94. These conditions under assigned geometry and soil (rock) properties could be estimated as extreme stable for the dump. But this approach is not well appropriate for design because any external changes related to climatic variations, equipment loads etc. can reduce dump stability. So, the most desirable safety factor which should be oriented is above mentioned ≥ 1.2 to ensure long-term stability of the dump.

All results of simulation for overburden dump stability on inclined surface are presented in the Figure 3. A set of curves describes calculated values of FOS for specific geometry of the first layer of overburden on the foundation. A dotted line with the FOS = 1.0 just shows the state of failure.

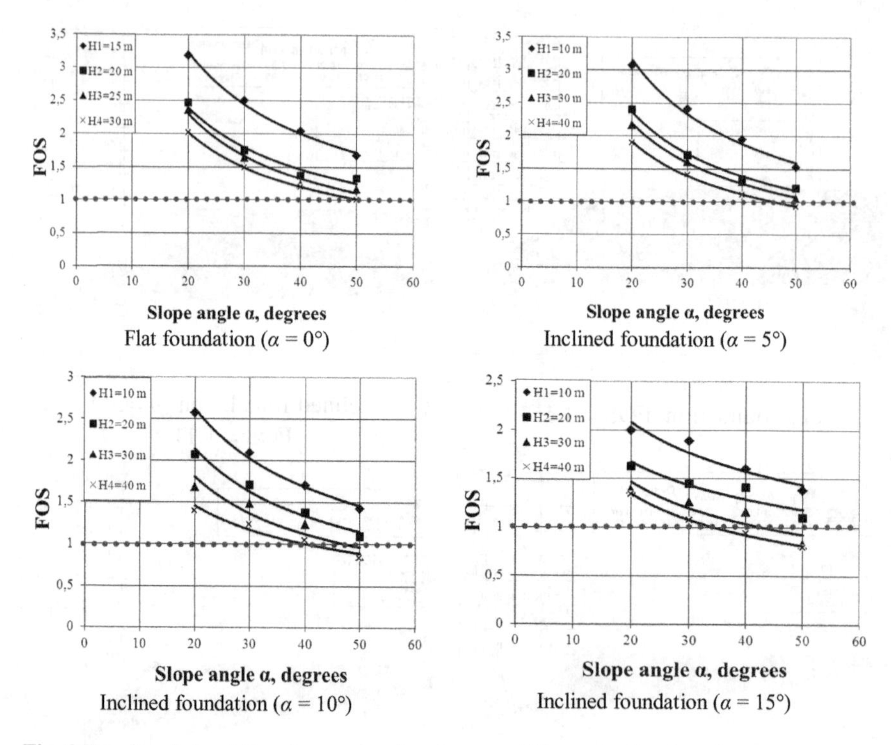

Fig. 3 Results of simulation for overburden dump stability on inclined surface

The obtained results of numerical simulation of the internal dump stability show that the height $H \leq 40$ м as project proposed value for the first layer of overburden should be accepted with a big caution. For flat or inclined surface with the $\beta \leq 10°$ the slope angle values should not exceed $\alpha = 30\ldots35°$ and the H is better to decrease to sufficiently safe value 30 m. If the foundation inclination angle increases up to $\beta \geq 10°$ the slope angle values should not exceed $\alpha = 30°$ and the $H \leq 30$ m as well. Establishment of the dumps with critical geometric parameters will result in large-scale rockslides on the designated areas within the open-cast and destabilize operations on overburden transportation and disposal [4].

According to the technological project of the open-cast "Maikubenskiy" the second layer of the overburden is recommended as an advisable environmental measure. This paper do not deal with estimation of double-benched dumps stability but some essential recommendations based on the obtained simulation results could be taken into consideration. The first one refers to the surface characteristics of the foundation and its interaction with the overburden layer that should be analyzed thoroughly regarding certain disposal site. The second recommendation concerns variable values of overburden properties depending on moisture content, climatic factors, consolidation etc. Reliable lab analysis of physical and mechanical properties allows obtain as accurate stability results as possible.

5 Conclusions

The paper presents a simulation results for stability of internal overburden dumps at the open-cast "Maikubenskiy" for which strength is modelled with the Bishop failure criterion in the GGU-Stability program.

For geological and technological conditions of the open-cast "Maikubenskiy" the application of above mentioned technology of internal dumping is rational engineering decision. But the most problematic issue while surface mining operations is inclined bedding of brown coal deposits. As a result, all the mined out area inside the open-cast has tilting surface. Disposal of the overburden on such foundation can cause to instable condition of the internal dumps. Therefore justification of rational geometry for internal dumps and estimation their stable conditions under assigned physical and mechanical properties is an important stage in engineering design.

The FOS values for different geometric profiles of foundation ($\beta = 0\ldots10°$) and overburden mass ($H = 10\ldots40$ m, $\alpha = 20\ldots50°$) were estimated. It is established that for flat or inclined foundations with the $\beta \leq 10°$ the slope angle values should not exceed $\alpha = 30\ldots35°$ and the value of the most stable height for the first overburden layer $H = 30$ m. If the foundation inclination angle increases up to $\beta \geq 10°$ the slope angle values should not exceed $\alpha = 30°$ and the $H \leq 30$ m.

The results obtained by numerical simulation serve as a basic justification for application of internal dumping that facilitate technological management of disposal operations at the open-cast and prognosis of potential rockslides.

References

1. DIN 4084 Gelande- und Boschungsbruchberechnungen. Deutsche Normen DK 624.137.2.001.24:624.131.537. Beuth Verlag GmbH, Berlin 30, 36 p.
2. Buß, J.: GGU-Stability. Slope stability analysis and analysis of soil nailing and reinforced earth walls to DIN 4084 and EC 7. – Version 10. – Civilserve GmbH, Steinfeld, 164 p. (2013)
3. Methodical instructions for determining inclination angles of pit edges, slopes and dumps of the open-casts in operation and to be developed. – Leningrad, VNIMI, 162 p. (1972)
4. Shashenko, O.M., Rakishev, B.R., Kovrov, O.S.: Slope stability analysis by hoek-brown failure criterion. In: Proceedings of the 22nd International Conference Mine Planning and Equipment Selection (MPES), pp. 541–550. Springer International Publishing, Switzerland (2013)

Rim Slopes Failure Mechanism and Kinematics in the Greek Deep Lignite Mines

Marios Leonardos

Public Power Corporation,Mines Planning & Performance Department,
Athens, Greece
M.Leonardos@dei.com.gr

Abstract. This paper presents the studies and the findings of various cases encountered during the development of slope failures detection methods at the Lignite mines of Ptolemais and Megalopolis, Greece. The type of failure is a compound one consisting of a nearly horizontal surface and a curved one at the back. The most important factor for the stability is the shear strength available in the planar part of the failure surface, which shows that a progressive failure is taking place. The investigations revealed that the development of a failure surface was from the toe to the crest and therefore impending slope failures can be detected and analysed long before any crack formation at the slope crest becomes visible. In addition, there are simple tools for failure monitoring that can be easily incorporated in the mining activities. The displacement velocity of a failure follows an exponential low with different parameters depending on failure condition.

Keywords: Lignite mines, slope stability, progressive failure, slope kinematics.

1 Introduction

Lignite is an essential component of the Greek energy policy, as the lignite's share in power generation was 45% in the year 2013 with installed power plant capacity of 4.8 GW. The total lignite production for the same year was 52.5 Mt corresponding to 315 Mm^3 of excavations (down from the past peak figures of 72 Mt and 370 Mm^3).

The lignite is mined and transported by bucket-wheel excavators (BWE), spreaders, tripper cars and conveyor belts. To achieve this production, new deeper pits started operation in the last decades. The maximum pit depth increased from 70m in 1980 to 170m in 2000, 240m in 2010 and it is expected to reach 300m by 2030.

This depth increase, combined with the non-flexible mining methods applied, created pressure to the geomechanical engineers involved in these mining activities. They had to develop methods for an early detection of slope failures, as mine pit design modifications were not easy to implement in a short period of

C. Niemann-Delius. (ed.), *Proc. of the 12th Int. Symp Continuous Surface Mining - Aachen 2014*,
Lecture Notes in Production Engineering, DOI: 10.1007/978-3-319-12301-1_14

time. Besides, the stabilization measures had to fit to mining machines capabilities and available personnel skills with the minimum production disturbance and production cost increase.

This paper presents the studies and the findings of various cases encountered during the development of slope failures detection methods at the Lignite mines of Ptolemais and Megalopolis.

The sediments of the Ptolemais basin, where the main lignite deposits are located, belong to Neogene (Pliocene) and are covered by those of Quaternary (Pleistocene and Recent). A typical stratigraphic column is given in Fig. 1 (Anastopoulos et al. 1972).

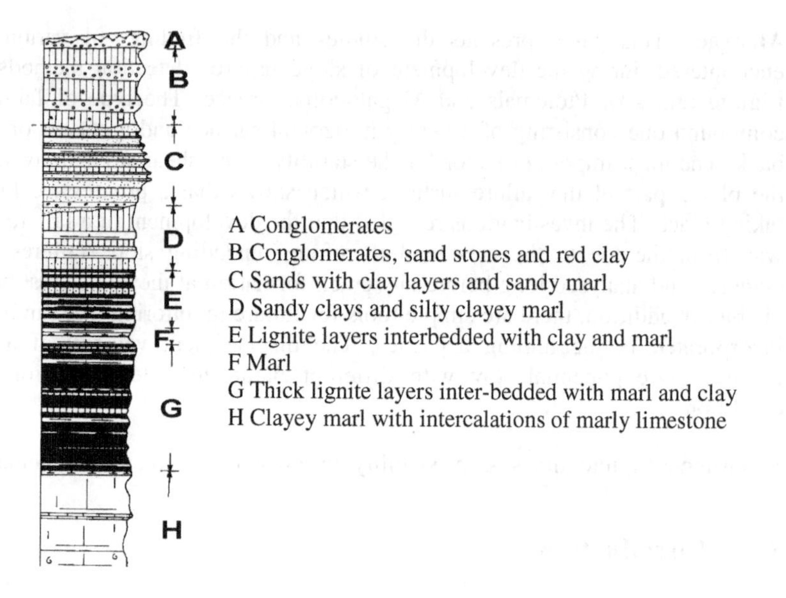

Fig. 1 Typical stratigraphic column

2 Type of Slope Failures

In the past, the most common among the failure types anticipated for the lignite pits rim slopes in Greece have been the rotational slips, either circular or non-circular. Numerous stability assessments for lignite mine slopes were based on rotational slips. Some of these slopes, with a high calculated safety factor, failed even in cases with excellent sampling, laboratory work and stability calculations. This tradition of using circular failure surface for lignite mines is still active (Singh at al., 2011).

These incidents have been examined and the following characteristics have been found common in all failures:

- The failures occurred at the time of the lowest lignite recovery
- The lower and central part of the slope moved as a block towards the excavation void, parallel to the lignite foot wall. This block had the shape of a truncated cone and it was slightly disturbed by few cracks.
- The surface, left at the slope after the slip, was a curved one with a high inclination in the crest area.
- A deep trench with crushed material was formed between the front block and the curved surface of the intact slope.

The failure mechanism can be further understood by examining the photo in Fig. 2, which covers part of the slope shown in Fig. 3. In this case, the slope moved due to the failure about 8m before the successful application of stabilisation measures. Afterwards, a bucket wheel excavator removed material up to a depth of 30m from the surface (total slope height 80m), creating a nice cut that depicted a well developed but not collapsed failure. The photo shows the curved slip surface (B) formed in this area. Region A is the intact material with the strata dipping to the pit while in region C the inclination had reversed due to the rotation of the upper part of the failure. Further to the right, the photo shows the gradual recovery of strata inclination, from the reversed condition (C) and (D) to normal (E). As the strata inclination did not change from region E up to the slope toe, it is reasonable to assume that this part of the slope moved as a block on a certain layer.

As can be seen, the type of this failure is not circular, as suggested by the computational models, but a compound one consisting of a nearly horizontal surface and a curved one (see Fig. 3).

Fig. 2 A cut through a failure. (A): The intact material. (B): The curved slip surface. (C): Strata inclination had reversed due to the rotation of the upper part of the failure. Further to te right, the gradual recovery of strata inclination, from the reversed condition (C) and (D) to normal (E)

The answer to the question of the failure type is not only of scientific significance but it has practical implications concerning the suitability and the effectiveness of the stabilizing measures to be taken and also the proper interpretation of the slope monitoring data. Therefore, it was decided to proceed with further investigations using geomechanical instruments, mainly inclinometers. The inclinometer measurements proved the compound type of the failures.

From our studies and investigations, it was found that compound failures can develop in slopes with low inclination. In a such a case a 140m high slope with inclination 1:5 (V:H), started to slip on a clay layer just below the lowest lignite seam. The reason for such a behaviour is that a compound failure is not affected so much by the slope inclination as in a circular one. The most important factor for the stability in a compound slip is the shear strength available in the planar subhorizontal part of the failure surface.

The safety factor for the compound slip can be lower than for the circular slip. In other words, in the case of a multilayer slope with one layer of lower shear strength compared to the other layers, the most probable failure type is the compound one with the planar part of the slip passing through the low strength layer. This theoretical conclusion is in agreement with the observations described above for the rim slopes failure in the Greek deep lignite mines, where more than one hundred layers can be identified.

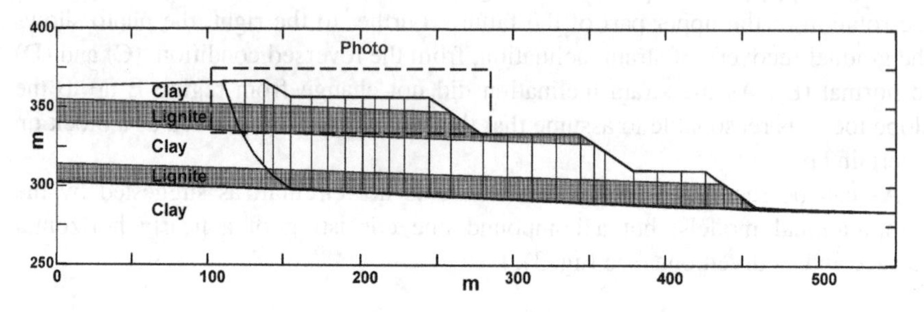

Fig. 3 Schematic cross section of the slope, where the photo of Fig. 2 was taken

3 Shear Strength and Stability Calculations

Back analysis of failures in the lignite mines revealed very low shear strength (6 - 11 deg) in clayey layers, where the planar part of the compound slip was passing through. These low values were associated with very large displacements and, therefore, they were considered to represent residual strength. As expected, these values where notably lower from those obtained in conventional triaxial testing. Direct shear tests, conventional and in precut samples, produced better but not totally acceptable results. As this test suffers from several disadvantages, the ring shear apparatus was implemented afterwards. The results that have been produced from slow shearing, using this testing technique (Bromhead ring shear apparatus), were in good accord with the figures from back analysis. As the results from this test had been used in stability calculations with notable success in slope behaviour prediction, the ring shear became the main apparatus for shear testing. Although shear box testing is still in use for lignite slope stability estimation (Ural et al., 2004, Kayabasi et al., 2012), in Greece it is not used anymore.

The use of residual strength in stability analysis implies that a progressive failure is taking place. As the materials in the lignite mines are over consolidated, the progressive failure is attributed to the initial stress conditions (differential soil strata relaxation during the stepped, bench by bench, excavation process). The same can happen during a dump construction, when a soft clay is present a the dump base (see Fig. 4).

The condition of progressive failure does not satisfy the requirements of the limit equilibrium concept that is extensively used in the conventional stability analysis of slopes because a continuous slip surface, along which the soil behaves as a rigid body, does not exist. Numerical and other methods have been used for progressive failure analysis (Tutluoglu et al., 2011). Although such complex techniques are based on more rigorous analysis, according to the author's experience, the simple methods based on the limit equilibrium analysis can be used with confidence for safety factor calculation. The simplicity combined with the averred prediction accuracy are very important especially when considering the number of sections to be examined with the alternative stabilisation measures. The limitations and the accuracy of a method must be taken into account, but the most critical parameters are the type failure, the prediction of the layer, that accommodates the planar part of the failure, and the corresponding shear strength.

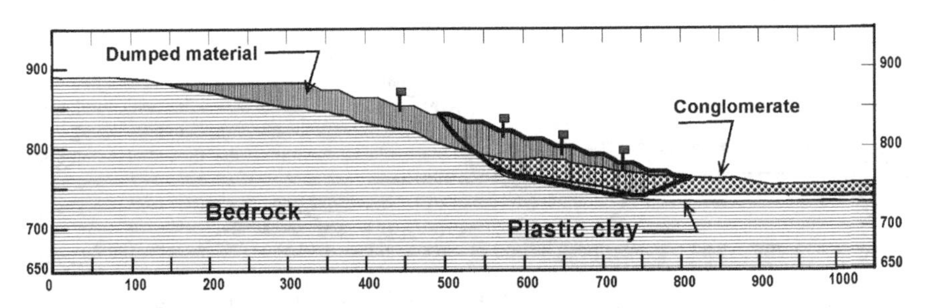

Fig. 4 Progressive failure in a plastic clay layer at the dump base

4 Direction of Failure Surface Development

In most cases, the mining personnel consider the formation of a crack, close to the slope crest, as the first sign of an impending slip. Even in most geo-mechanical textbooks, it is not clear where a slope failure starts from. It is implied that the development of the failure surface starts from the tension crack in the slope crest with gradual propagation towards the slope toe area. The investigation of the failure type described earlier, revealed that the development of a failure surface from the crest to the toe was not the case for the rim slopes of the lignite mines. On the contrary, it was found that the direction was the opposite, from the toe to the crest. The case is further explained from the measurements of inclinometer No16 shown in Fig. 5. This inclinometer was installed in the middle of a 120 m high slope, well before any displacement took place. The excavator mined a 20m high bench in the front of the inclinometer during the period June - July, leaving another 23m to the lignite floor. At that time, a failure surface started at the depth of 92m from the casing top, which coincided with the lignite floor. From then onwards, the displacement at the failure surface had a constant velocity of about 12mm/month, a figure which has been also confirmed by surveying methods.

The top of the slope was cleaned by dozers for better crack detection. After careful and regular inspection of the area, the first thin crack was detected in June of next year, ten months after the first displacement has taken place in the inclinometer.

This early appearance of displacements in inclinometers was confirmed afterwards, in other slopes with different heights and geology.

The above model of failure surface development is similar to the model proposed by Fleming & Johnson (1989) for translational landslides. The observations presented earlier, suggested that the slipping at the planar surface causes an increase in the tensile stresses in the masses above, turning the failure surface tip towards the slope crest. Muller & Martel (2000) came up with the same conclusion analysing translational landslides by numerical methods. They also anticipated that their conclusions apply also to compound slides exploiting pre-existing planes of weakness. The relevant findings in the lignite mines strongly support this claim.

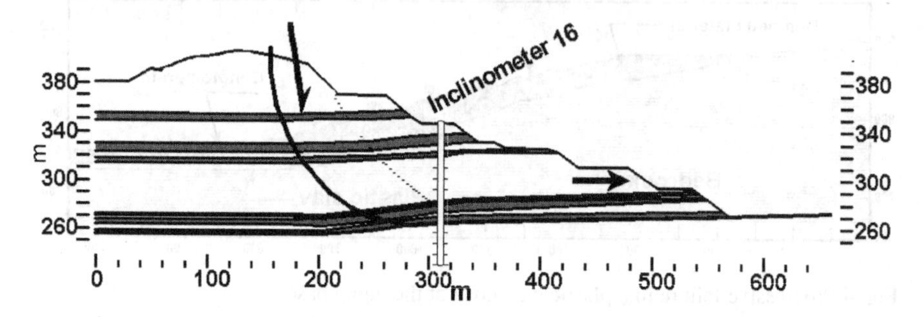

Fig. 5 Final slope profile, where inclinometer 16 was installed

5 Slope Kinematics

5.1 Monitoring Method

The type of failure, which is compound consisting of a nearly horizontal surface and a curved one at the back, permits the proper monitoring by simple surveying methods, that can be easily incorporated in the mining activities. The most common method is the distance measurement (using EDM instruments) between a distant, fixed point and various target (reflectors) placed on the slope. Afterwards, three diagrams can be drawn for each point vers time(see Figure 6):

- cumulative displacement (mm),
- displacement rate (velocity, mm/day) and
- acceleration (mm/day^2)

While the cumulative displacement gives a good indication, the acceleration is useful only for the last days before slope collapse. So, the displacement velocity was found as the best criterion for failure condition estimation.

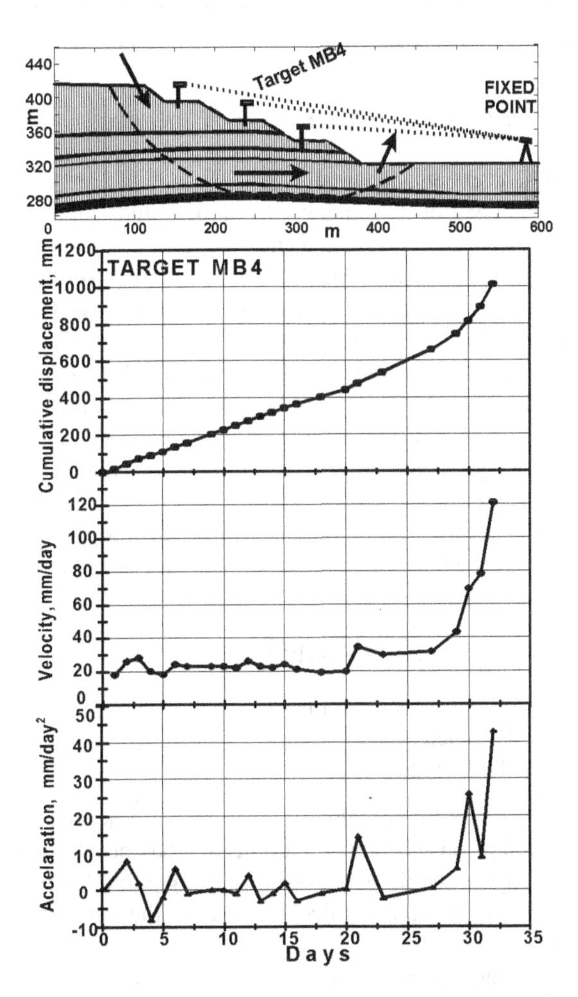

Fig. 6 Typical cumulative displacement, velocity and acceleration vers time diagrams for a failure up to the previous day of the slope collapse.

Table 1 Values of parameters a and c and number of cases analysed

Parameter	Failure to collapse	Retrogressive failures	
		Acceleration phase	Deceleration phase
a	22 ± 3.0	4 ± 0.7	7 ± 4.7
c	2.44 ± 0.29	1.87 ± 0.80	-3.03 ± 1.66
Cases	5	4	5

5.2 *Time Estimation of Slope Failure Collapse*

After analysis of five events, it was found that, in case the velocity is steadily increasing above the level of 20 mm/day, the slope failure collapse is expected to occur 6 to 12 days after the velocity exceeded the level of 20 mm/day. Most of the slopes collapsed hours after reaching the velocity of about 100 mm/day but other «survived» under velocities up to 600 mm/day without severe damage. This difference in slope behaviour is attributed to the variation of the residual shear strength of the failure surface clays with the rate of displacement (Tika et al. 1996). There are clays which exhibit a reduction of the residual shear strength for shearing rates above 100 mm/day. Furthermore, the displacement velocity can be used as a guide for stabilization measures timing and improvements in safety environment.

Statistical treatment of the measurements shows that the slope velocity, for the last 15 - 20 days before slope collapse, complies with the following exponential low:

$$V=a+be^{t/c} \tag{1}$$

Where V : Slope velocity in mm/day
a, b, c: Parameters
t : Time in days

The parameter a represents a velocity level, above it accelerating movements are expected («a failure onset point»). Parameter b has no physical meaning and its value depends only on the time measuring starting point. Parameter c is related to the acceleration, the lower the value the steeper the velocity curve after the failure onset point.

The values for a and c parameters are given in Table I. As the cases analysed are from different pits, the noticeable small values spread can be attributed to the common failure mechanism.

5.3 *Retrogressive Failures*

Not all the slope movements, even with high velocity, indicate an impending collapse. In many cases, where the slope after a bench excavation has a sufficient safety factor, the slope velocity, after a peak, returns to a low value. These retrogressive failures exhibit two phases:

a. Acceleration phase (the velocity increases), attributed to the fast lowering of the safety factor caused usually by the pit deepening in the front of the failure or by the water level rise in the failure cracks during heavy rainfalls.
b. Deceleration phase (the velocity decreases), which follows the withdrawal of the displacement cause.

The retrogressive failures follow the same exponential low and the values for a and c parameters are also given in Table I.

The values presented in Table I reveal some interesting points:

- The acceleration phase of a retrogressive failure start at significant lower velocity compared to the failure (collapse) onset point.
- The deceleration phase turns out with a higher velocity than at the starting point of a retrogressive failure acceleration phase. This shows the detrimental effect of the movements to slope condition.
- The acceleration phase of a retrogressive failure is faster than the deceleration phase.

Some other interesting observation data concerning the slope failure kinematics are the following:

Total displacement up to crack formation at the slope crest:	200 - 300 mm
Minimum total displacement up to slope collapse:	700 mm
Maximum total displacement without slope collapse:	9000 mm
Maximum velocity for a retrogressive failure:	205 mm/day

6 Practical Implications

The above findings can be easily exploited for practical purposes such as for the early failure detection, design of stabilisation measures, slope monitoring etc.

The first step is to take advantage of the early formation of the planar part of the failure surface. This is accomplished by constructing at least two inclinometers per section to be checked, when 30 - 50% of the total slope height has been excavated. In the case of a failure surface at the lignite floor, this method gives a clear picture months, or even a year, before the crack formation at the slope crest.

In the case of a compound slip, controlling the stability by modification of the slope inclination is less effective than in the case of a circular slip. Slope flattening, where compound slips occur, it is not the best measure to take considering effectiveness and money to be spent or income to be lost.

Unfortunately, there are cases where slope inclination is the only parameter which can be changed in order to improve stability.

Some of the failures developed in the rim slopes of the deep lignite mines are deliberately left to collapse because stabilisation is not feasible, either technically or economically. The time estimation of a mine slope failure collapse can be used to maximize the operation of the mine in the failure area without decreasing the safety level.

The slope kinematics together with the proper stability calculations can be used to distinguish a failure being in the collapse or retrogressive stage.

When mining in the front of a retrogressive failure, the slope maximum displacement velocity must be determined to avoid the transition from retrogressive to progressive failure. Mining is suspended when the velocity exceeds this level. The modelling of the retrogressive failure can be used for optimization of mining / suspension periods. It can be proved that keeping this maximum displacement velocity to a lower level, more time is given to excavators to mine in the front of the failure.

7 Conclusions

Miners, throughout the mining industry, recognize some signs of geomechanical phenomena as precursors of impending rock or soil mass failure and collapse. In surface mining, it is the crack formation, while for underground mining it is the convergence of the excavation. Every experienced miner would agree with the following statements:

- Slope failures do not occur spontaneously and
- A slope failure does not occur without warning.

This paper goes a step further for the rim slopes of the deep lignite pits: Impending slope failures can be detected and analysed long before any crack formation at the slope crest becomes visible. In addition, there are simple tools for failure monitoring, making the personal opinion and guesswork worthless. It is very important that the mining people skills cover the requirements of a slope monitoring system, thus the new task can be easily incorporated in mining.

Obviously, the first benefit from the early failure detection and the slope monitoring is an improved safety environment. The second is the deeper understanding and better control of the slope stability in the mine. The latter is often materialized with the application of steeper final slopes without decreasing the safety level. Both benefits have a positive effect in the mine economics, thus the money paid for an improved slope stability can be considered as good investment.

Acknowledgement. The author is graceful to Public Power Corporation, Greece for the permission to publish data from internal reports. However, Public Power Corporation does not necessarily endorse the views presented in this paper.

References

Anastopoulos, J.C., Koukouzas, C.N.: Economic Geology of the Southern part Ptolemais lignite basin (Macedonia - Greece) Institute of Geology and Mineral Exploration, Athens, pp.101–136 (1972)

Fleming, R.W., Johnson, A.M.: Structures Associated with Strike-slip; Faults that Bound Landslide Elements. Eng. Geol. 27, 39–114 (1989)

Kayabasi, A., Gokceoglu, C.: Coal mining under difficult geological conditions: The Can lignite open pit (Canakkale, Turkey). Engineering Geology 135-136, 66–82 (2012)

Leonardos, M., Terezopoulos, N.: Time estimation of slope failure collapse in the rim slopes of the deep lignite mines. Mineral Wealth, 124, 7–18 (2002) (in Greek)

Muller, J.R., Martel, S.J.: Numerical Models of Translational Landslide Rupture Surface Growth. Pure Appl. Geophys. 157, 1009–1038 (2000)

Singh, R.N., Pathan, A.G., Reddish, D.D.J., Atkins, A.S.: Geotechnical Appraisal of the Thar Open Cut Mining Project. In: 11th Underground Coal Operators' Conference, University of Wollongong & the Australasian Institute of Mining and Metallurgy, pp. 105–114 (2011)

Tika, T.E., Vaughan, P.R., Lemos, L.J.: Fast shearing of pre-existing shear zones in soil. Geotechnique 46(2), 197–233 (1996)

Tutluoglu, L., Oge, I.F., Karpuz, C.: Two and three dimensional analysis of a slope failure in a lignite mine. Computers & Geosciences 37, 232–240 (2011)

Ural, S., Yuksel, F.: Geotechnical characterization of lignite-bearing horizons in the Afsin–Elbistan lignite basin. SE Turkey, Engineering Geology 75, 129–146 (2004)

Innovative Methods to Improve Stability in Dumps and to Enlarge of Their Capacity

Lilian Draganov[1], Carsten Drebenstedt[2], and Georgi Konstantinov[1]

[1] UBG "St. Ivan Rilski"
ldraganov@yahoo.com,
gk@abv.bg
[2] TU Bergakademie, Freiberg

1 Problem Information on the Innovative Method HDD

An important reason for the existence of bank slides in natural terrains or in engineered dam excavations is that the water banks up and destroys engineered installations and soil by excessive moisturisation. The problem is usually solved by building drainage facilities – vertical filtration wells, horizontal segments and drill holes. The wells and segments have several major disadvantages. The vertical drill holes have a limited effective radius. This requires small spacings in-between, a large volume and expensive drilling work. Horizontal drillings require an initial shaft or a protective ditch for the drilling equipment, which is difficult to lay and is limited to drainage pipes of a maximum length of 100–150 m. To avoid the shortfalls of the filtration and drainage installations mentioned above the method HDD has been devised and introduced. In Germany, more than 15%-18% of the cable, pipe and filtration networks that may vary in type and purpose are installed without excavations (1). This is the so-called "Subsoil Revolution".

The essence is as follows: set-up in a ramp at a starting point a straight or curvilinear drilling is carried out by a controllable cutting drill head that uses a water jet with a pressure of 30 - 1.500 bar which penetrates the soil. When the pilot drilling has reached the final location the drill head is replaced by an extension head. There a pipe run is attached consisting of a protective, filtration and drainage pipe. On the return path the bore is expanded and the combined pipe run is inserted, which is pulled out all the way back to the pilot drilling. The last stage consists of the pulling out of the protective pipes in reverse from the final location which keeps the drainage filtration run in the bore to fulfil its function.

In order to realize the HDD method different installations with the following specifications are used: maximum hoist power 50-1000kN, maximum length of the drill pipe 250-2000m, maximum drilling depth 12-300m.

A significant positive feature of this method is the option to effect curvilinear clearing drillings. In this, cable and pipe runs are inserted according to the defined goal adapted to the concrete characteristics of the soil environment. This method

C. Niemann-Delius. (ed.), *Proc. of the 12th Int. Symp Continuous Surface Mining - Aachen 2014*,
Lecture Notes in Production Engineering, DOI: 10.1007/978-3-319-12301-1_15

is suited for the application in large-scale draining of saturated and non-saturated rocks in slide-endangered banks.

2 Analysis of the Introduction of the HDD Method in Germany Open-Pit Mines. (3,4)

The theoretical and experimental examinations and the introduction of horizontal drillings were started at the end of the last century, mainly for the laying of subsoil networks without ditch and to rehabilitate waste dumps.

Drainage is an important process in pits especially when the excavation is done in bedrock where water amounts to over $5m^3$ /t. Expenses and repercussions on the environment are significant. At the moment, vertical drillings are the preferred drainage method (preliminary drainage) in open-pit mines. In thin, aquifers the filter location is of limited effect. The maximum pump capacity from a drilling is $1\text{-}2m^3$ /min, where each one is equipped with a pump and an electrical installation. To achieve this goal major investments for the setting-up and expenditure for energy and labour are required. Instead of using vertical filters horizontal filter well drillings were used. The results of the laboratory and field tests have been described (2). It is also the technical basis for investments in technological variants of the HDD used for drilling operations in open-pit mines. The comparative technological configuration for the drainage of an aquifer is illustrated in figure 1.

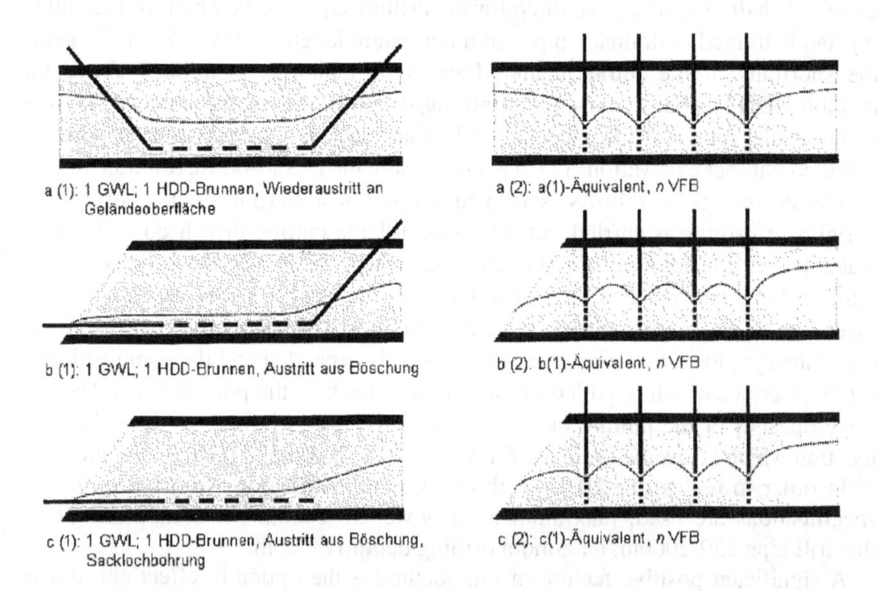

Fig. 1 Technological schemes for the drainage of bedrock by the HDD method and vertical drillings (3)

Obvious is the uniformity of the lowered water level in the case of the horizontal drillings. The water infiltrated in them drains off in a gravitating way. What is more, a larger zone of developed deposits is created. One HDD drilling can replace 30 or more vertical drillings. The result is a substantial reduction in materials, energy and a concentration on the drainage process without an interruption of the mining operations in the open-pit mine.

It has become evident that existing calculation methods are not suitable to evaluate the reliability of the drainage effect.

For an ore mine 10 drillings in rows are usually required.

To this end a reverse modeling based on a laboratory test (at the Institute for Geotechnical-engineering at T.U. Bergakademie Freiberg) has been carried out, and a field test in the lignite mine "Vereignigtes Schleenhein" based on single horizontal drillings. The results of this modelling went to confirm the methods to calculate the effects of draining of the calculation model of PCGEOFM® (4). Here the current in the HDD drillings was regarded as a stream whose capacity depends on a number of parameters: basic parameters are the length of the drilling filter and the strength of the colmated layer.

3 Rehabilitating Waste Dumps to Protect Them against Disintegration Including the Possibility to Enlarge Their Capacity

In Bulgaria, in mining and processing raw materials billions of sterile rocks are piled up. A large number of the dumps were built 40-50 years ago. There is probability that they are destroyed in their elements by destructive processes. The dry dumps are often set up on unlevelled, undrained surfaces that have no appropriate slope. The hydro dumps (tailing pond) have scooping facilities, a poor drainage system and colmated drainage. In the case of these two methods there is or we expect an increase of the piezometric level in the slopes in case of the first methods and the dams in case of the second method.

In the case of dry damps slow water saturation and subsequent, sliding of the banks leads to significant economic and ecological damage. The rehabilitation of the colmated drainage system in the hydro damps is virtually impossible. As a result, the piezometric level increases; the capacity of the drainage system is reduced. There is a risk for the dams breaking and of a discharge of gigantic amounts of dump material with unforeseeable economic and ecological consequences. The undrained state, i.e. the water saturation of the banks and dams, excludes the possibility to enlarge their capacity. The enlargement of the existing dumps saves resources, land and social efforts.

The examples given above illustrated in figures 2a and 2b demonstrate the advantages of the horizontal drainage drillings. They allow a periodic cleansing of the colmated material.

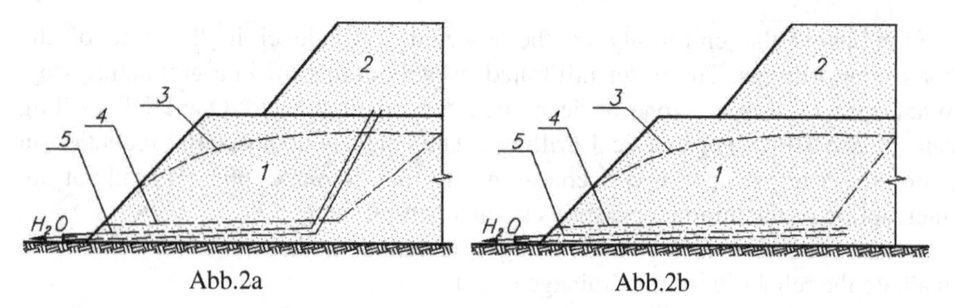

Fig. 2 Example scheme for the lowering of the piezometric level in the bank of the dry water-saturated dump

1. spread material; 2. newly spread material;
3.increased piezometric level 4. lowered piezometric
level; 5.horizontal draining drilling of gravitational drainage

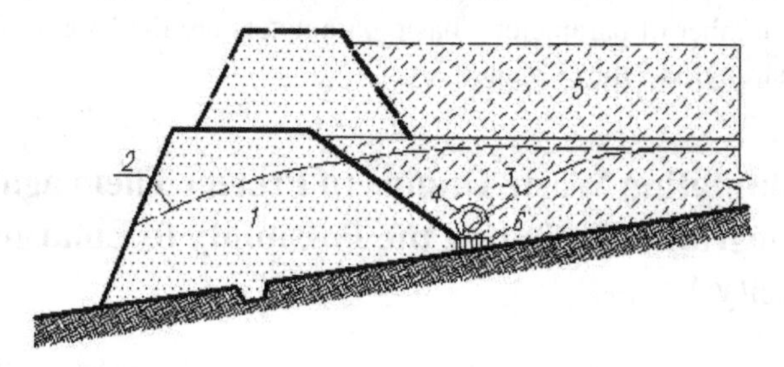

Fig. 3 Example scheme for lowering the piezometric level

in the dam of a tailing pond
1.dam, 2. increased piezometric level; 3. lowered
piezometric level, 4. horizontal drainage drilling
5. additional washed-in material (waste); 6. colmated
drainage.

For any concrete case it is necessary to devise ideas and elaborate a working project taking into consideration the elements of the draining systems. The capacities of the discharge and the effect of the drying drainage process are determined by the diameter and the length of the filters and the spacing between drillings.

The horizontal drainage drillings in dry and hydro dumps as well as the collectors of the latters are colmated in the course of time.

The result is a reduction of the draining or the complete colmatage of the drillings. The current innovative methods are first used for the observations registered and second for the cleansing of the colmated areas in horizontal and sloped drillings.

The first method is the "Supervision®" system. It allows an internal inspection of the channels of all kinds by using optical chambers fitted on a trolley. They can move in channels and drillings with a diameter of 100mm to 1500mm. To enable optimal inspection of the inside of the drilling the chamber head can be rotated using a small lever. The picture is transmitted from the inside onto a computer screen by cable. The trolleys are SVR9S, SVR140, SVR150 und SVR SA140 with a weight of 8kg to 25kg. Figure 4 shows the trolley SVR250 with the optical camera Supervision®.

Fig. 4 Trolley SVR20 with optical camera Supervision® (5)

Unless regularly cleaned small crystalline inner deposits are found on the walls of the channels (figure 5).

The second method to clean the colmated inner surfaces is to clear by water jets with a pressure of 150bar. Market leader is Rohrtechnologie KMG. In the process, the highly-effective turbine jet with a combined pressuring and sucking effect with a maximum length of the jetting hose of 1000m and an output of 320l/min. It has proved to clean channels with a length of 620m, diameter 250mm and a slope of up to 30° (7)

Fig. 5 a piece of incrustation on the inner wall of a drainage pipe (6)

4 Conclusion

1. Radical drying of the burden rock is realized by horizontal, inclined clearing drillings which are technically and economically more effective than vertical drainage drillings, filtration wells and tunnels
2. Slow, unnoticed increase of the piezometric level in the banks of the dry waste dumps needs to controlled and decreased by horizontal/inclined drainage drillings plus observation and cleaning systems. The dried burden material in the banks reduces its volume weight, cohesion is stepped up and also its angle of internal friction. As a result the coefficient of the resisting force is increased. The final result is the possibility of additional amounts of overburden so that the capacity of the dump can be enlarged.
3. Currently, the HDD method is the only possibility to rehabilitate the colmated drainage systems in tailing ponds. In doing so, one can attain the decrease of the piezometric level and the risk of the breaking of the dams.

References

1. Prospektmaterialien der Firma FlowTex – Grossbohr und Umwelttechnik – Deutschland
2. Struzina, M.C.: Drebenstedt Integration of HDD – wells in open pit mines with multiple acquifers. Continuous Surface Mining, TU BAF (September 13-15, 2010)
3. Struzina, M.: Beitrag zur Vorrausberechnung der Wirkung verlaufsgesteuerter Horizontalfilterbrunnen (HDD-Brunnen) bei der Entwässerung von Lockergstein, Freiberg, S.152 (2012)
4. Mueller, M.: Modellabbildung von Horizontalfiterbrunnen im Programmsystem PCGEOFIM Stand und Ziele. TU BAF, Freiberg, pp. 44–58 (2007)

5. Kanal – TV aus neuer Sicht vom Grund auf! Supervision® Firma SPR TEC Europe GmbH, Linz Austria
6. Edenberger, W., Kaessinger, J.: Adaequate Reinigung und Ueberwachung von Deponieentwaesserungssystem Tagungsband Depotech, Montanuniversitaet, Loeben, Austria (2012)
7. KMG Pipe Technologies Kein Respekt von langen Leitungen DepoTech, Montanuniversitaet, Loeben, Austria (2012)

Research on the Optimal Technology for Exploitation of the Thin Lignite Layers in the Open Pits from Oltenia Coalfield

Maria Lazar and Andras Iosif

University of Petrosani, Romania

Abstract. The lignite deposits from Oltenia's mining area consist of several lignite layers, separated by sterile formations. The thickness of the coal layers, as well the sterile intercalations varies in a wide enough range, not only among the entire coalfield, but even in the same exploitation field. Taking into account the geological specificities of the deposit and the large number of open pits in which is extracted, the analysis of the opportunity to introduce highly selective extraction of thin layers in each open pit is difficult because of the large amount of calculations to be performed. For this reason, in this paper an open pit model is discussed, in which the selective extraction of a thin lignite layer is analyzed, in terms of different layer thicknesses and three different lengths of the exploitation front. In order to determine the influence of layer thickness and front length on the main economic indicators of the exploitation activity, not only quantitatively but also qualitatively, on the proposed open pit model the sensitivity analysis realized, which takes into consideration three different extraction technologies: nonselective extraction, limited selective extraction and highly selective extraction. In order to establish the optimal technological variant for extracting of the thin lignite layers in various concrete conditions (length of the exploitation front, layer thicknesses) a flexible multi-criteria multi-attribute type optimization method has been used, namely the weighted average maximization method. In order to generalize the conducted study for other concrete exploitation conditions (front length and coal layer thickness), a nomograph has been drawn-out based on the extrapolation of the calculated values and on the regression equations of the curves representing the separation limit between optimal zones.

1 Problem Statement

The goal of a high as possible selective extraction is to improve the quality of the throughput coal while reducing at the minimum possible the losses of coal [5]. In the case of lignite deposits being mined out in Oltenia's open pits, two situations occur which requires a high backhoe-hydraulic shovel selective extraction to fulfill the above-mentioned goals.

The first case refers to the extraction of thin seams, which occurs relatively seldom in the overburden rocks, when the outcome of such an extraction is the

C. Niemann-Delius. (ed.), *Proc. of the 12th Int. Symp Continuous Surface Mining - Aachen 2014*,
Lecture Notes in Production Engineering, DOI: 10.1007/978-3-319-12301-1_16

avoidance of losses and reduction of the dilution. The second case is representative for the fascicular type seams, when many coal banks are separated by sterile intercalations.

In order to illustrate the effect of the selective extraction in the second case, we present below an example based on a case study. In this view, we consider a sector from a bench in which a lignite seam containing 5 coal layers separated by sterile intercalations (as in table 1) is to be mined out. We consider a length of a block of 1000 m, and its width equal to annual advance of the face, i.e. 200 m.

Table 1 Configuration of the lignite seam

Type of rock	Seam thickness [m]	Volumic weight , γ [t/m3]	Ash content [%]
Coal	1.2	1.25	25.5
Sterile	0.5	2.10	68.0
Coal	1.0	1.25	30.2
Sterile	0.9	2.10	65.0
Coal	0.7	1.25	20.8
Sterile	1.1	2.10	65.0
Coal	2.0	1.25	32.8
Sterile	1.0	2.10	67.0
Coal	1.1	1.25	27.5

The extraction of such a seam can be realized in non-selective, limited-selective and high-selective manner, depending on the equipment utilized (Table 2).

Table 2 Different variants of mining

<table>
<tr><th colspan="3">Non-selective extraction</th><th colspan="3">Limited -selective extraction</th><th colspan="3">High-selective extraction</th></tr>
<tr><th>Extraction</th><th>Width [m]</th><th>Ash [%]</th><th>Extraction</th><th>Width [m]</th><th>Ash [%]</th><th>Extraction</th><th>Width [m]</th><th>Ash [%]</th></tr>
<tr><td rowspan="9">Coal</td><td rowspan="9">9.5</td><td rowspan="9">47.07</td><td rowspan="3">Coal</td><td rowspan="3">2.7</td><td rowspan="3">38.78</td><td>Coal</td><td>1.25</td><td>25.5</td></tr>
<tr><td>Sterile</td><td>2.10</td><td>68.0</td></tr>
<tr><td>Coal</td><td>1.25</td><td>30.2</td></tr>
<tr><td rowspan="3">Sterile</td><td rowspan="3">2.7</td><td rowspan="3">57.38</td><td>Sterile</td><td>2.10</td><td>65.0</td></tr>
<tr><td>Coal</td><td>1.25</td><td>20.8</td></tr>
<tr><td>Sterile</td><td>2.10</td><td>65.0</td></tr>
<tr><td rowspan="3">Coal</td><td rowspan="3">4.1</td><td rowspan="3">43.60</td><td>Coal</td><td>1.25</td><td>32.8</td></tr>
<tr><td>Sterile</td><td>2.10</td><td>67.0</td></tr>
<tr><td>Coal</td><td>1.25</td><td>27.5</td></tr>
</table>

Variant 1 – Non-selective extraction

In this case, the equipment utilized is a Bucket Wheel Excavator (BWE), which cannot extract separately the sterile and the coal. As a result, from the entire block 1.9 mil. m^3 coal-sterile mixture will be extracted , with an average ash content of 47.07%. The average ash content A_{av} can be calculated as a weighted average using the formula:

$$A_{av} = \frac{\Sigma(m_i \gamma_i A_i)}{\Sigma(m_i \gamma_i)}, \quad [\%] \tag{1}$$

We can notice a high ash content of the throughput, because of mixing the coal with sterile rock, while the losses in this case are zero.

Variant 2 – Limited -selective extraction
In this case, we can utilize a smaller BWE, a hydraulic shovel or a Krupp Surface Miner. The throughput obtained from the entire block, in this case will be 540000 m^3 of lignite with an ash content of 38.78 % and 820000 m^3 of lignite with an ash content of 43.60 % while the loss of coal with reduced ash content will be 140000 m^3.

Variant 3 – High -selective extraction
This variant requires the utilization of Continuous Surface Miners, which are able to extract thin layers; in this case, the coal will be extracted without dilution, obtaining 1.2 mil. m^3 of lignite with an ash content of 28.53 %, while the losses are zero. Another advantage of high-selective extraction is that the ash content can be controlled, and the mixing in the Power plant stockpile in order to provide a constant quality coal is easy.

2 Sensitivity Analysis Based on an Open Pit Mine Model

2.1 *The Model Construction*

The Oltenia's coalfield lignite is mainly composed by many coal seams separated by sterile rock layers. Both the coal seams and sterile layers have a large variance of the thickness, not only inside the entire basin, but also inside the same perimeter.

Taking into account the characteristics of the coalfield and the large number of operating open pits, introducing the high selective extraction in each mine is difficult, because of their mentioned variability. For this reason, we will demonstrate the advantages of the high-selective extraction method we propose a surrogate model, in which the goal is to extract a thin coal seam, with different thicknesses, for three different values of the face length.

Related to the lignite properties, we consider a medium hardness of coal, the allowable wall slope is considered to be 45°. The specific weight is considered to be $\gamma_v = 1.3$ t/m^3 and the bulk factor $k_a = 1.35$.

The main "in situ" content parameters of the lignite are:

- Heating value, Qi= 1900 kcal/kg;
- Ash content, $A_{anh} = 22.5\%$;
- Total moisture $W_t = 35\%$.

The seam thickness “m” and face length “L” varies in the following ranges:

- Thickness : 0.2 m; 0.4 m; 0.6 m; 0.8 m; 1.0 m; 1.2 m; 1.4 m; 1.6 m; 1.8 m.
- Face length: 700 m; 1000 m; 1500 m.

The working bench is considered as: height h=25 m, slope angle α=45°, the coal seam is located in the middle of the bench height, regardless its thickness. Also we consider that the overburden of the block has been removed in due time, without influence on extraction activity.

The working time is two shifts per day, with 260 days/year, which means an amount of available working hours of t_{plan}= 4160 hours /year. If we consider a coefficient of charge η_e=0.9, the operating time is t_{ef}= 3744 hours/year.

2.2 *Extraction Using Surface Miner*

The Wirtgen 2100 SM surface miner has been chosen for high –selective extraction modeling [8, 9]. Having in mind the open pit mine model described, using the productivity calculation algorithm for this technology, we plotted the dependence of annual production as a function of seam thickness, m and different face lengths, as in figure 1.

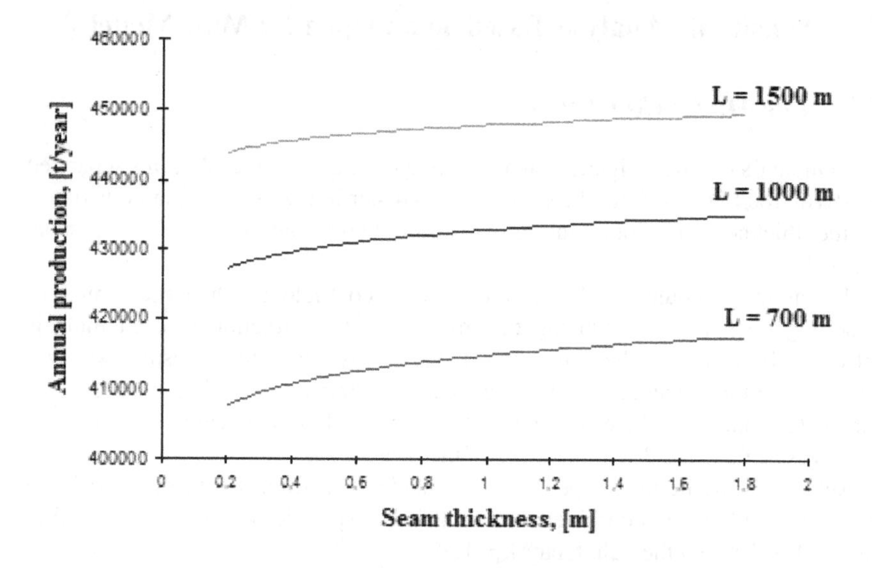

Fig. 1 Dependence of annual production on seam thickness and face length

From the above presented diagram, we can conclude that in case of SM, the annual production is mainly influenced by the face length and less by the thickness of the seam [3]. That is due to the fact that as the face is longer, as the unproductive time consumption for slice-end maneuvers. The influence of seam thickness on the annual output is not very significant, except the fact that in the small thickness range the decrease of output is more evident, while in greater thickness range the output increase trends towards zero, asymptotically, or can take even negative values, also due to longer time for face-end instrumentations. The dependences shown in fig. 1, being determined starting from discrete points, so we can derive by regression a general equation as:

$$Q_{an} = 214966.9\, L^{0.007}\, m^{0.008}\ [t/an] \tag{2}$$

Using this equation, we are able to calculate the maximal output that can be obtained with this technology for any value of the face length and seam thickness.

In order to derive the economic efficiency of the CSM technology, we estimated the specific extraction costs, for the first year of operation, and for the next five years, ignoring only the eventually maintenance (repair) costs, taking into account that the manufacturer guarantees a good functioning of 20000 hours, which means six years. In figure 2.a, we plotted the specific cost dependence on face length and seam thickness for the first year, while in figure 2.b the same, for the next five years.

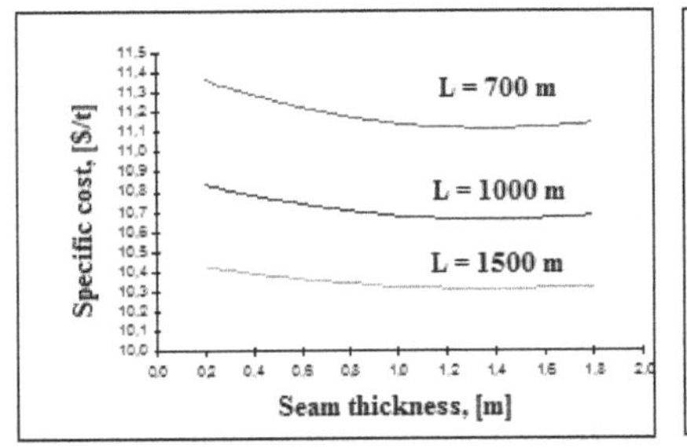

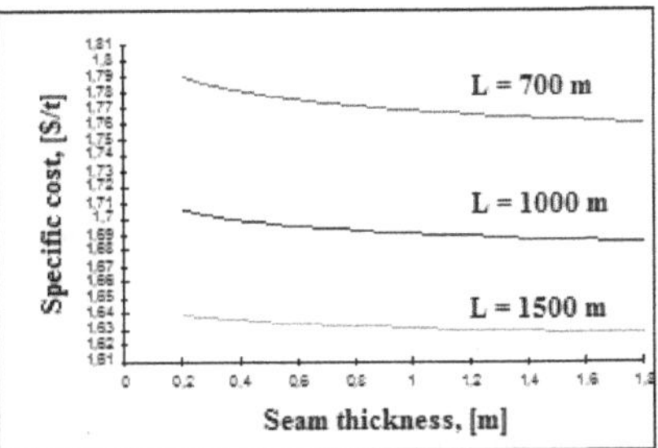

Fig. 2 Specific cost dependence on seam thickness and face length: a) first year; b) next five years

The analysis highlights that the highest specific costs are obtained at reduced thickness and reduced face lengths. The face length influence is more important than the thickness one. In order to have a calculable formula, we used same as above the regression, obtaining the formula below:

$$C_u = 3.564\, L^{-0.1073}\, m^{-0.0057},\ [\$/t] \tag{3}$$

In the conditions of the used model, we notice that in the first year, the specific costs are in the range from 11.408$/t (for L = 700 m and m=0.2 m) to 10.319 $/t (for = 1500 m and m = 1,8 m), and in the next years, after recovery of the initial investment they decrease to 1.797 $/t (for L = 700 m and m = 0,2 m) and to 1,629 $/t (for L = 1500 m and m = 1,8 m).

2.3 *Extraction Using Backhoe-Hydraulic Shovel*

For the considered open pit mine model, a backhoe-hydraulic shovel is proposed, (PROMEX S/SC/ 3602), with bucket capacity of 1,8 m^3, which will work in parallel with the SchRs 1400 BWE, without interfering with this main digging equipment. The BWE will extract the overburden and sterile intercalations, while the shovel the coal layers [2]. The option for a backhoe is due to its advantage in given conditions, in comparison with the face type ones. In figure 3 the annual production as a function of seam thickness and face lengths is plotted.

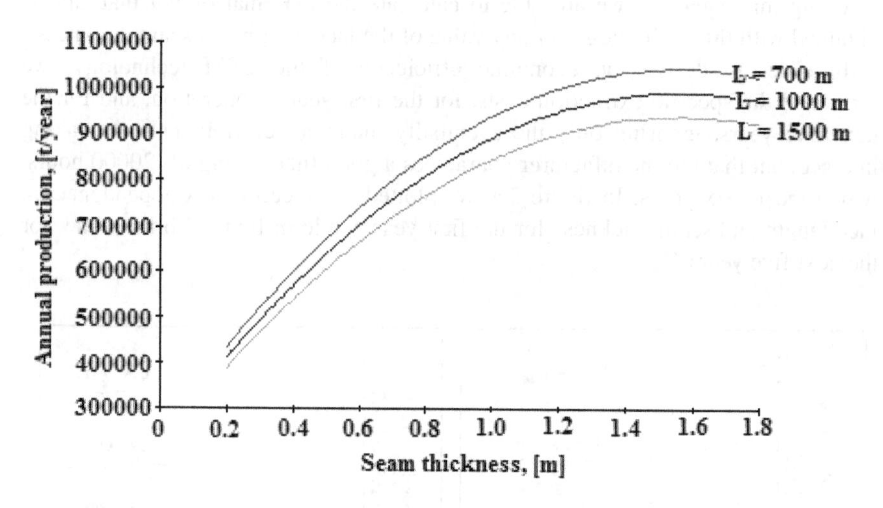

Fig. 3 The annual production as a function of seam thickness and face length

From the above, we can notice that, contrarily with the CSM technology, the productivity, i.e. the annual output in the case of hydraulic shovel are less influenced by the face length and more by the seam thickness. Also, the output decrease with the increase of the face length, which is due to the increased hauling distances with increased face lengths, the haulage of the coal extracted with shovels being realized with trucks. Another difference is that beyond a certain value of the seam thickness, the output decreases, after reaching a maximal value. The regression equation of the output dependence on seam thickness and face length is given by (4).

$$Q_{an} = 2198201\,L^{-0,1379}\,m^{0,3788}, [t/an] \tag{4}$$

Regarding the economic efficiency issue, we proceeded in the same way as in case of CSM adapted to some specific elements of this technology, as follows:

- no need to investment expenses, because all the open pits in the Oltenia's coalfield are endowed with hydraulic shovels used in auxiliary operations;

- the incomes were calculated taking into account a dilution of ca. 10% because of the weaker selectivity allowed in the case of this technology, so we considered a heating value of 1710 kcal/kg instead of 1900 kcal/kg, as in case of pure coal.

As a result, we obtained the total costs (figure 4) and the specific costs (figure 5). The regression equation of the dependence of specific costs on face length and seam thickness is given in (5):

$$C_u = 0.4367\, L^{0.166}\, m^{-0.4368}, [\$/t] \quad (5)$$

We can notice that the specific costs are relatively reduced, being in the range of 3,041 $/t for thickness of 0.2 m and face length of 1500 m to 1.127 $/t for thickness of 1.8 m and face length of 700 m.

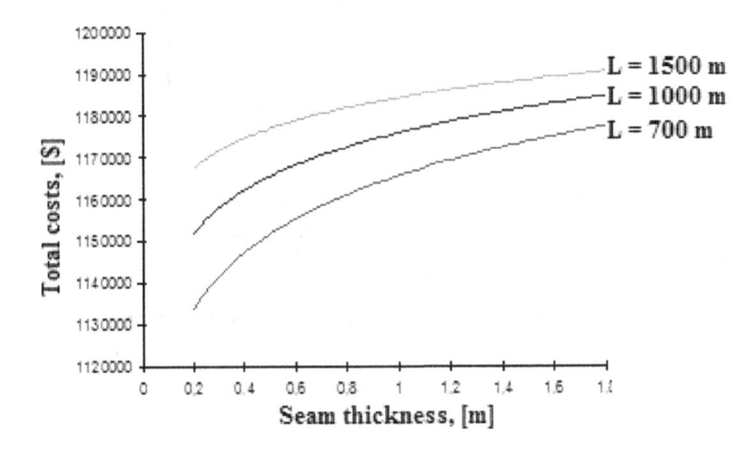

Fig. 4 Total costs dependence on seam thickness and face length

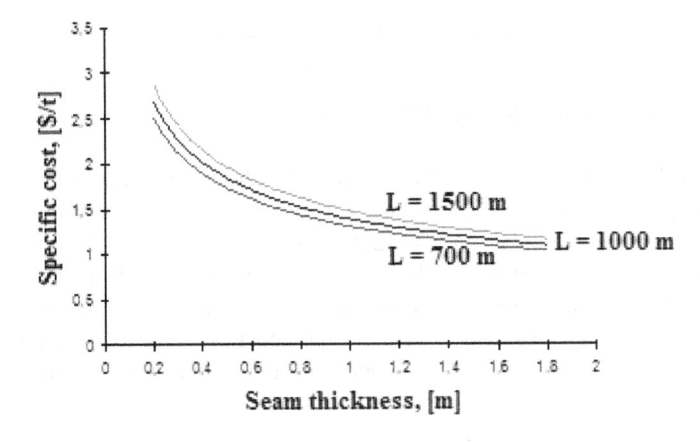

Fig. 5 Specific costs dependence on seam thickness and face length

2.4 Extraction with Small BWE

For this technology, given the strata conditions, we considered a SRs 130-9/0.5 Bucket Wheel Excavator. The possible obtainable outputs were determined, taking into account the inherent productivity decrease of this type of equipment with the decrease of the seam thickness [4].

The dependence of the BWE's productivity taking into account the excavated face height and the geometric elements of the obtained cut shape is shown in figure 6, which is influenced by the amount of uptimes and downtimes imposed by the above mentioned parameters.

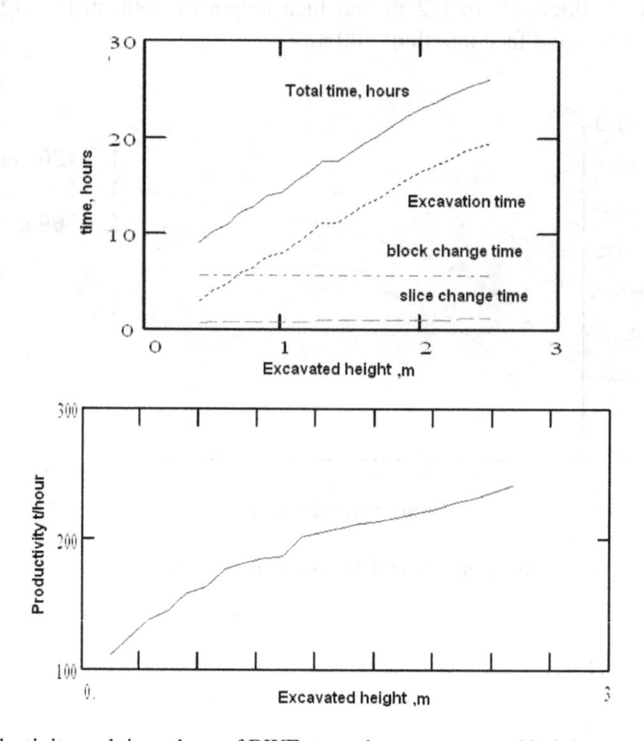

Fig. 6 Productivity and time-share of BWE operation on excavated height

From the figure, we can notice that technically, the BWE is able to excavate a face of minimal height of 0.4 m with a productivity significantly reduced compared with the nominal one, (112.2 m^3/h relative to 350 m^3/h which is guaranteed by the manufacturer). The productivity increases with the increase of the height of cut being close to the nominal value at a height of ca. 3.0 m. From the diagram of times, it results very clearly the dependence of total time with the times to change from one slice to another, and a constant value of change from one block to another.

Based on productivity diagrams shown below, we derived the total annual output as a function of face length and seam thickness, as shown in figure 7.

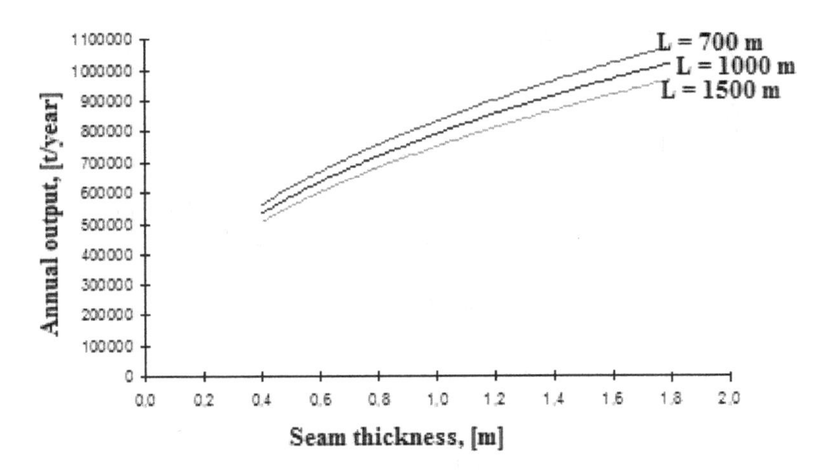

Fig. 7 Total annual output as a function of face length and seam thickness

We notice that the output slowly decreases with the increase of the face length, due to the contribution of downtimes imposed by the change of slice, respectively the block and the increase of time for translating the belt conveyor with the increase of the face length.

The regression equation of the dependence of the output with the face length and the seam thickness is given in (6):

$$Q_{an} = 2067235\, L^{-0.1383}\, m^{0.4159}, [t/an] \tag{6}$$

Taking into account an investment for the SRs 130 - 9/0,5 BWE at 2000000 \$, we derived the specific costs in the first year and the following years after recovery of investment, and a 12 % dilution, (1625 kcal/kg) for the incomes, which are presented in figures 8 and 9.

The regression equation for the specific costs is given in (7).

$$C_u = 0.4278\, L^{0.1535}\, m^{-0.4444}, [\$/t] \tag{7}$$

Analyzing the results above presented, we notice that the specific costs are in the range of 3.69 – 7.82 \$/t in the first year, when the investment is recovered, and in the future years the specific costs are between 1.19 – 2.53 \$/t.

If we add the transportation expenses, we obtain a range of 7.69 – 11.82 \$/t in the first year and a range of 5.19 – 6.53 \$/t in the following years.

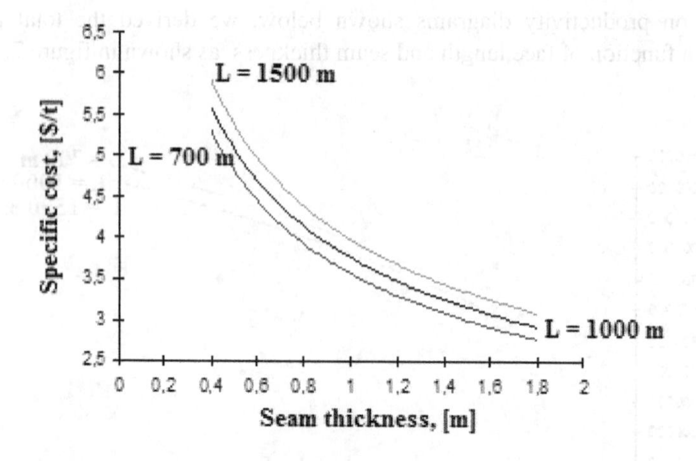

Fig. 8 Specific costs in the first year

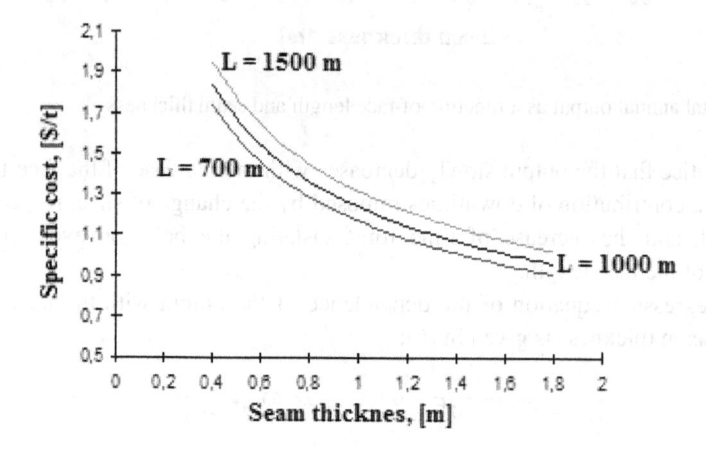

Fig. 9 Specific costs in the following years

3 Establishment of the Optimal Extraction Technology for Thin Seams

In order to obtain the optimal technology for a given range of face lengths and seam thicknesses, we used a multi-criteria – multi-attribute flexible optimization technique, based on weighted average maximization [7, 11].

This method implies the following steps:

- selecting a set of criteria and a set of variants
- normalization of data using:

$$a_{ij} = \begin{cases} \min\{a_{ij}\}/a_{ij} & \text{- for data to be minimized;} \\ a_{ij}/\max\{a_{ij}\} & \text{- for data to be maximized.} \end{cases} \quad (8)$$

given a weight p_j to each criterion (by heuristic reasoning, or consulting experts), the weight reflecting the importance of the criterion in the given performance indicator:

$$\sum_{j=1}^{m} p_j = 1 \quad (9)$$

deriving the optimum, for :

$$\max_i \sum_{j=1}^{m} p_j a_{ij} \quad (10)$$

In order to compare the three presented variants, we have chosen the selection criteria, which reflect technical, economical and qualitative attributes of the compared variants. The 4 chosen criteria are:

- annual output , t/year;
- the intensity of using available equipment time, %;
- the specific cost , $/t;
- the heating value, kcal/kg.

By using expert's opinion, we obtained the following hierarchy table of the four criteria using the opinion of 6 experts (table 3).

Table 3 Hierarchization of criteria (1-high importance; 4 low importance)

Criterion	Expert						Result	Weight, %
	E_1	E_2	E_3	E_4	E_5	E_6		
annual output , t/year	4	4	3	2	4	4	4	0,1
specific cost, $/t	1	1	1	1	1	1	1	0,4
intensity of using available equipment time, %	3	2	4	4	3	3	3	0,2
heating value, kcal/kg	2	3	2	3	2	2	2	0,3

Following the steps described above, we obtained the chart in figure 10 describing the optimal choice of solution, for the combinations of seam thickness and face length.

The results obtained following this sensitivity analysis on the described open pit model, lead to the conclusion that in the range of reduced seam thickness the most efficient technology is the use of Wirtgen SM. The area of application of this technology extends towards larger thicknesses, with the increase of the face length, as we can see in figure 10. The use of hydraulic shovel is efficient until the thickness value of 1.2 m, and the small BWE became efficient for thickness greater than 1.2 m, regardless the face length.

Seam thickness, [m]	Lenght of the exploitation front, [m]		
	700	1000	1500
0.2	V_1	V_1	V_1
0.4	V_1	V_1	V_1
0.6	V_2	V_1	V_1
0.8	V_2	V_2	V_1
1.0	V_2	V_2	V_2
1.2	V_2	V_2	V_2
1.4	V_3	V_3	V_3
1.6	V_3	V_3	V_3
1.8	V_3	V_3	V_3

V_1 – Variant of extraction using Wirtgen 2100 SM
V_2 – Variant of extraction using PROMEX – 3602 hydraulic shovel
V_3 – Variant of extraction using SRs – 130 BWE

Fig. 10 Chart for establishing the optimal extraction technology to be used

Based on the chart presented in figure 10, we have drawn up a nomograph, in which the optimality area can be determined for continuous ranges of thickness and face length, which is presented in figure 11. This nomograph uses the regression equations established before. The curves represent the border of separation between the areas of optimality of different technologies.

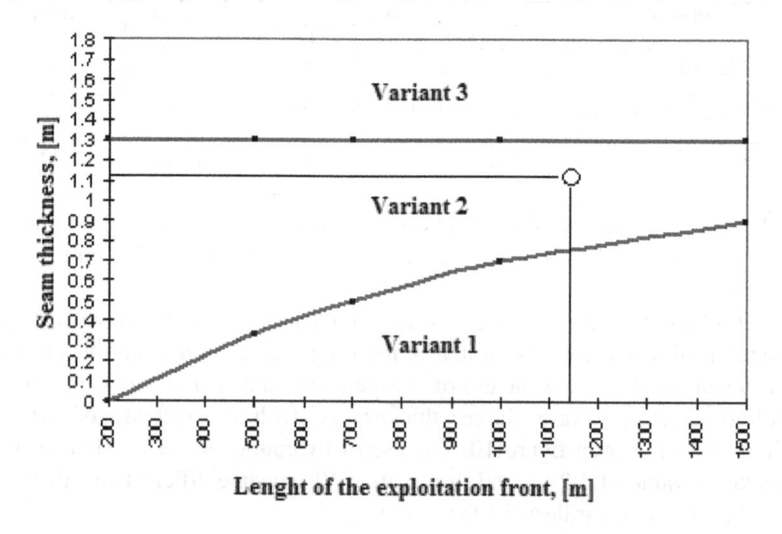

Fig. 11 Nomograph for establishing the optimal extraction technology to be used

The optimal technology for a given value of face length and thickness is given by the area in which the corresponding horizontal line of thickness crosses the vertical line of the face length. If the intersection point is situated on a borderline, then the both variants must be analyzed, using supplementary criteria. After selection of the optimal technology, the actual outputs and the costs can be derived using the corresponding regression equations.

4 Conclusions

In order to improve the quality of the throughput coal while reducing at the minimum possible the losses of coal, in the case of lignite deposits being mined out in Oltenia's open pits, two occurring situations, which require a high selective extraction must be considered.

The first case refers to the extraction of thin seams, which occurs relatively seldom in the overburden rocks, when the outcome of such an extraction is the avoidance of losses and reduction of the dilution. The second case is representative for the fascicular type seams, when many coal banks are separated by sterile intercalations.

Using the experience of a concrete situation, in which we encounter both these situations, three technological solutions are drawn-up, with different degrees of selectivity, analyzing the pros and cons of each. The three solutions envisaged are, apart from the actual , non selective method, using large BWE-s, with different selectivity degree, namely using Continuous Surface Miners, backhoe hydraulic shovels and small sized BWE.

In order to make the comparison possible, a simple surrogate model of open pit is presented, and the three technologies are analyzed, deriving the main technical, economical and qualitative results.

The areas of optimality of these technologies are derived using a multi-attribute-multi-criteria flexible optimization method, namely the weighted average maximization method.

In order to generalize the conducted study for various concrete exploitation conditions (front length and coal layer thickness), a nomograph has been drawn-out based on the extrapolation of the calculated values and on the regression equations of the curves representing separation limit between optimal zones.

References

1. Ciric, D., Niemann, C.: Abbauverfahren und Wirtschaftlichkeit von Continuous Surface Miner für zwei Braunkohlentagebaue in Jugoslavien am Beispiel des Krupp Surface Miners. Braunkohle (1991)
2. Fodor, D., Lazăr, M.: Determinarea productivității excavatoarelor cu acționare hidraulică în funcție de grosimea de excavare prin metode statistico-matematice. Revista Minelor, (3) (1998)

3. Fodor, D., Lazăr, M.: Posibilităţi de extragere a stratelor subţiri cu o tehnologie de tip Continuous Surface Miner. Revista Minelor, (4) (1998)
4. Fodor, D., Lazăr, M.: Posibilităţi de integrare în fluxul tehnologic al carierelor de lignit din Oltenia a utilajelor de tip CSM pentru extragerea stratelor subţiri. Revista Minelor, (5) (1998)
5. Golosinski, T.S.: Selective continuous surface miners and their possible applications in oil sands mining. CIM Bulletin (mai 1989)
6. Peretti, K.: Kombination von kontinuierlich und diskontinuierlich arbeitend Geräten und Anlagen in Tagebaubetrieben. Braunkohle (iulie1981)
7. Rădulescu, D., Gheorghiu, O.: Optimizarea flexibilă şi decizia asistată de calculator. Editura ştiinţifică, Bucureşti (1992)
8. Schimm, B.: Wirtschaftliche Gewinnung dünnmächtiger Flöze mit Surface Miner. Wirtgen GmbH (1995)
9. Vogt, W., Janecke, K.: Zuschnittüberlegungen für Kohletagebaue mit Schaufelradbagger und Continuous Surface Miner auf einer Sohle im Kombinationsbetrieb. Bergbau, (5) (1988)
10. Vogt, W., Strunk, S.: Hochselective Gewinnung geringmächtiger Schichten in Kohletagebauen - Begriff und Verfahren sowie Darstellung der Auswirkungen auf die Kohleverwendung. In: 4. Internationale Fachtagung für kontinuierliche Tagebautechnik, Aachen (1995)
11. Zorilescu, D.: Metode matematice de analiză şi decizie în geologie şi minerit. Editura Tehnică, Bucureşti (1972)
12. * * - Einsatzberichte. Wirtgen GmbH, Windhagen (1995)

Substantiation of Continuous Equipment Efficient Choice at the Selective Mining of Passing Minerals

Illia Gumenik and A. Lozhnikov

Department of open cast mine, National Mining University, Dnepropetrovsk, Ukraine

Abstract. Article deals with the actuality problem of continuous equipment efficient choice at the open cast mining when selective mining deposits. Analyses the advantage of bucket wheel excavators using at the selective mining of passing minerals. The parameters of face cut and bucket wheel that allow to increase the effectiveness of selective extraction technology by the bucket wheel excavator is calculated. The dependence of the rational use of bucket-wheel excavator types on it bucket width is determined. The results of the conducted researches show that the excavation of overburden layers with thicknesses that equal or more then selection degree is effective.

An integral feature of the mining process and reprocessing of minerals is waste and passing products, most of which is a raw material for various industries. The quality, technological and economic limits that separating raw materials from passing mining waste rock are conditional and have noticeable fluctuations in time. These changes depend on the needs of social production in each type of raw material, as well as technical and technological possibilities to satisfy it.

The industry transfer to intensive development level predetermines the need to implement resource management forms of multiply use of mineral resources and entering in the production passing minerals.

Effectiveness of selective extraction technique depends on the continuous mining conditions of attitude bad, physical and mechanical properties of mining rocks, the type of the equipment excavation, technological parameters of the mining and other factors.

The advantage of bucket wheel excavators in the selective extraction of passing minerals is that they provide a consistent removal of overburden layers from the crest to toe of bench.

Transportation of passing minerals are supplying to the console spreader, which allows one to selectively store the material in man-caused deposit to further use. For this process is necessary the consistency of movements between cutting and overburden equipments.

Effectiveness field of selective extraction using a bucket-wheel excavator are determined in the mining layer capacity (the height of the chip). If the thickness of the layer less than half of the bucket-wheel diameter it can lead to a decrease in cutting productivity.

C. Niemann-Delius. (ed.), *Proc. of the 12th Int. Symp Continuous Surface Mining - Aachen 2014*,
Lecture Notes in Production Engineering, DOI: 10.1007/978-3-319-12301-1_17

Selective extraction technology of passing minerals by the bucket-wheel excavator and its efficient is heavily dependent on the type of stope face, the form of chips, the parameters of bucket wheel. For selective mining more acceptable bucket wheel excavators low power with small diameter wheels.

To carry out the condition of full volume filling bucket E in the process of mining bench part (layer) of a given height H_C and loosening factor k_P chip thickness t_O at the wheel axis is

$$t_{\acute{\imath}} = \sqrt{\frac{E\xi_o}{k_P H_{\tilde{n}}}}, \qquad (1)$$

where ξ_O – 1,2 – 2,8 is attitude of layer thickness t_O toward its width b_O at the level of wheel ax.

Rational at the mining rock by bucket-wheel excavator there is a condition, when $t : b \approx 1$, where b is a bucket width. Development of passing minerals is made the vertical, horizontal chips and combined method depending on the fortress of the developed rocks. For example, the horizontal chips are develop dense rocks. Power intensity of selective mining, dynamics of loadings on a rotor and specific resistance digging at a mining vertical chips on 10 – 30% less than, than at a coulisse horizontal. During mining the vertical layers the productivity will be nominal, if power of one layer is equal (0,5 – 0,7)D [1]. Then $h_E \geq 0{,}5D$, where D is a diameter of bucket-wheel.

At development the horizontal chips variants are possible:

1 – when h_E accept the equal height of cutting edge of bucket $t = 0{,}08D$;
2 – when h_E accept equal to a few heights of cutting edge of bucket. Then

$$0{,}08D < h_E < 0{,}5D. \qquad (2)$$

On the pit when mining horizontal deposits of Nikopolskiy manganese ore, Dneprovskiy lignite, Pre-Carpathians sulfur pools for the development of overburden rocks the next bucket-wheel excavators are used: ERShR-1600-40/7, ERG-1600-40/10, SRS-2400-40/4, SRS-2400-35/9, ERG-1600-40/7, ShRS-1500-24/6, ER-1250, SRS-280-II/0,5 and RS-350.

For these machines, and the specific conditions of the development of pools below shows the procedure for determining the main parameters of selective mining.

It should be noted that the main indicator of the effectiveness of selective excavation is the lower limit of the effective recessed power (*the degree of selection*) – the minimum thickness of the layer of overburden, which is suitable for recess technical capabilities used excavators.

To establish it is necessary to calculate the change in productivity excavation equipment and slaughter coefficient depending on the height executed by layer for each type considered a bucket-wheel excavator.

Technical productivity of the bucket wheel excavator is

$$Q_T = Q_P \cdot k_R \cdot k_F, \text{ m}^3/\text{h}, \qquad (3)$$

where Q_P – theoretical productivity of bucket wheel excavator, m³/h; k_R – influence rock coefficient; k_F – coefficient of mining face.

To determine k_R the need to calculate a index of difficulties mining rock in face (P_{ER}), passport indicator difficulties mining (P_{EP}) and compare it with each other [2]. If $P_{ER} > P_{EP}$ then coefficient $k_R = 1$.

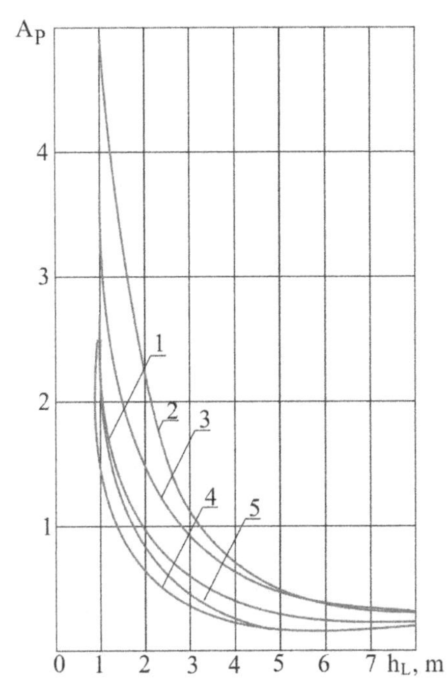

Fig. 1 Dependence of coefficient A_P on power of layer (h_L) for excavators with a nonprotractile boom: *1* – ERShR-1600; *2* – ShRC-1500; *3* – SRS-2400; *4* – ER-1250; 5 – SRS-280

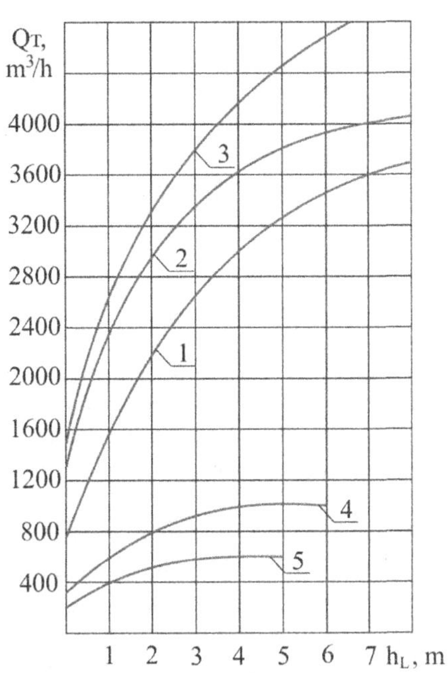

Fig. 2 Dependence of the technical productivity (Q_T) of excavators with a nonprotractile boom on thickness of mining layer (h_L): *1* – SRS-240; *2* – EPShP-1600; *3* – ShRS-1500; *4* – ER-1250; *5* – SRS-280

The determination of index k_R was made for all excavators were considered and found that for it $k_R = 1$, since the conditions of occurrence of approximately equal size, as well as other factors that characterize the rock developed.

Influence face coefficient k_F accounts the loss of time for assistive technology operations when developing this particular face; this quantity is determined design parameters for operating conditions and design parameters are applied excavators (A_P).

Estimation of A_P and k_F for excavators ERShR-1600-40/7, SRS-400-35/9, SRS-1500-24/6, ER-1250 and SRS-280-11/0.5 defined for specific conditions of pits in

Ukraine. When calculating the indexes thickness of layer change from 1 to 8 m. Input data and results of calculations A_P and k_F and technical productivity Q_T shown in Fig. 1 and 2.

For bucket wheel excavators with extendable boom type ERG-1600-40/10 , SRS - 2400 /4 and RS - 350 mining face coefficient is

$$k_F = \frac{t_E}{t_E + t_{AS}} \tag{4}$$

where t_{AS} – estimated time to perform additional operations when mining block; t_E – directly time of excavating at the mining block with next productivity

$$k_E = \frac{V_B \cdot k_P}{Q_T} \tag{5}$$

where V_B – block volume, m^3.

Schedule changes of technical productivity bucket wheel excavator with extendable boom is shown on Fig. 3.

From the graph on Fig 4.7 shows that the middle values of technical performance ΔQ_T at the increasing of height mining layers for all types of bucket wheel excavators are quite large and are, correspondingly, from from 14.2 to 18.6 %. For bucket wheel excavators with a bucket diameter more than 10 m the Q_T differ sharply on the areas elevation layers from 1 to 4 m and from 4 to 8 m. So, in the first case for excavators ERShR 1600, SRS-1500, SRS-2400 index Δk_F is 0.83 – 0.146, and at the increase thickness from 4 to 8 m Δk_F is 0,027 – 0,043. Similarly, changing the value of ΔQ_T: on the first section is 600 – 690 m^3/h, on the second – 253 – 308 m^3/h.

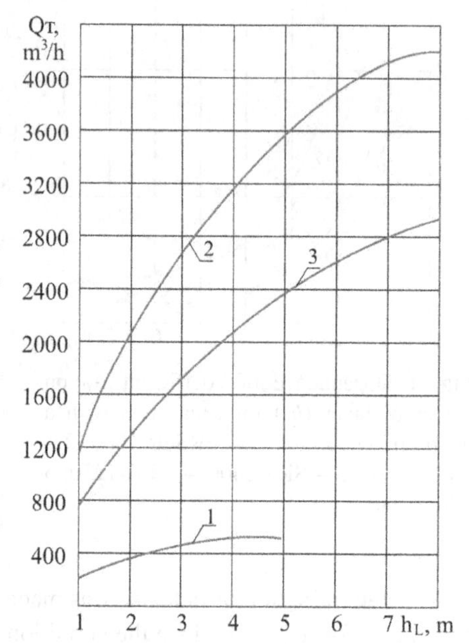

Fig. 3 Dependence of technical productivity (Q_T) excavators with extendable boom on thickness of mining layer (h_L): 1 – RS-350 2 – SRS-2400-40/4, 3 – ERShR-1600

Excavator with a small bucket (up to 6.5 m) values of Δk_F and ΔQ_T also significantly differ with increasing layer height. Thus, for the excavator ER-1250 the gap height from 1 to 2.5 m values of Δk_F and ΔQ_T constitute, correspondingly, 0,216 and 276 m^3/h, while increasing the height from 2.5 to 6 m – 0.085 and 66 m^3/h, correspondingly. Similarly set value Δk_F and ΔQ_T for excavator SRS-280 and RS-350, for which their sharp change in the (increase) was observed at a height of 2 m and 3 m, correspondingly.

Thus, based on the results of the analysis of changes in the values Δk_F and ΔQ_T executed by the height of the layer set the following next values of the degree of selection: for excavators ERShR-1600, SRS-1500, SRS-2400, ERG-1600 and CDS -2400 it is 4 m, for RS-350 – 3 m, ER-1250 – 2.5 m and SRS-280 – 2 m.

These values, as well as others previously set parameters of selective mining are summarized in Table. 1.

Table 1 Parameters of selective mining for bucket-wheel excavators applied in Ukraine

Type of power-shovel	Height of bench, *m*	Width of bench face, *m*	Mining method	Degree of selection, *m*	Block length, *m*	Volume of block, m^3	Coefficient of mining face, k_F	Technical productivity, Q_T, m^3/h	Time of working off a block, *h*
ERShR -1600 40/7	25	60	Multiserial vertical chips	4,0	35,0	63000	0,6885	3440	18,3
ShRS-1500 24/6	20	50		4,0	22,0	22000	0,5710	2860	7,7
SRS-2400 35/9	25	70		4,0	25,0	43750	0,6041	3980	11,0
ER-1250	15	25		2,5	6,7	2510	0,5842	730	3,4
SRS-280 11/0.5	10	16		2,0	8,0	1280	0,6101	420	3,0
ERG-1600 40/10	28	55		4,0	24,2	37270	0,4390	1975	18,9
SRS-2400 40/4	30	50		4,0	25,7	38550	0,4040	2909	13,2
RS-350	10	17		3,0	4,4	740	0,5500	412	1,8

Results in Table 1 show that the volume of overburden blocks, the mining of a one point distances, for powerful bucket-wheel excavators with the highest degree of selection are relatively large and the mining time change from 7.7 to 18.9 hours depending on the type of excavator. This indicates sufficiently the effectiveness of selective extraction of overburden by bucket-wheel excavator.

Conducted researches on determining the degree of selection confirmed the fact that the layers of overburden excavation capacity equal or biggest than degree of selection, is effective and does not involve decrease of productivity bucket-wheel complexes and give a real supposition for the subsequent dumping of separate passing minerals.

References

1. Baranov, E.G.: Experience the selective mining of complex fields deposits / E.G. Baranov, I.A. Tangail. - Frunze: Ilim, 346 p. (1869)
2. Rzhevskiy, V.V.: Processes of open cast mining / V.V. Rzhevskiy. 3rd ed., Rev. and add. - Moscow, Nedra, 541 (1978)

Mineable Lignite Reserves Estimation in Continuous Surface Mining

Christos P. Roumpos[1,2,*], Nikolaos I. Paraskevis[1], Michael J. Galetakis[2], and Theodore N. Michalakopoulos[3]

[1] Public Power Corporation SA, Mines Engineering and Development Department, Lenorman 195 and Amfiaraou, GR-10442 Athens, Greece
[2] Technical University of Crete, School of Mineral Resources Engineering, GR-73100 Chania, Greece
[3] National Technical University of Athens, Department of Mining Engineering, Laboratory of Excavation Engineering, Zographou Campus, GR-15780 Athens, Greece
c.roumpos@dei.com.gr

Abstract. The deposit modelling and evaluation and the mineable reserves estimation of multi-seam lignite deposits is an essential process for mine planning activities. The first approach of the original drill holes data evaluation of the exploration programs refers to the analysis of the in-situ lignite seams characteristics. The algorithm of compositing of the drill holes assay intervals should take into account specific criteria of evaluation based on the corresponding power plant specifications as well as on the applied mining equipment. Especially in the case of continuous surface mining, the accuracy of the mineable reserves and the run-of-mine lignite quality estimation is a very important subject of mine planning and design procedure, provided that the applied equipment is less flexible to operation changes regarding the short-term scheduling.

The objective of this paper is the investigation of the critical parameters which affect the lignite deposit evaluation procedure in relation to the algorithms of mineable reserves and run-of-mine lignite quality estimation in multi-seam continuous surface lignite mines. The analysis is based on a wide range of deposits, power plants and equipment parameters. Emphasis is placed on the spatial variability of the lignite deposit parameters and on the recommended design criteria as well as on mine development improvements.

Keywords: Mineable, Lignite, Reserves, Continuous surface mining, Algorithm.

1 Introduction

The electricity production system in Greece is mainly based, at a percentage of about 50%, on lignite exploitation in surface mines. The majority of the lignite deposits has a multiple-layered structure (Fig. 1) and is located in the Ptolemais - Amynteon basin, in the area of Western Macedonia, in northern Greece.

* Corresponding author.

C. Niemann-Delius. (ed.), *Proc. of the 12th Int. Symp Continuous Surface Mining - Aachen 2014*, Lecture Notes in Production Engineering, DOI: 10.1007/978-3-319-12301-1_18

Fig. 1 Typical multiple-layered structure of lignite deposits with data of a corresponding drill-hole (Barmpas, 2013).

The need for a high production rate and selective mining resulted in the main application of the continuous surface mining method, using large-scale equipment (high capacity bucket wheel excavators, conveyor belts and spreaders). In addition, non-continuous mining equipment (consisting of blast hole drills, dumpers, loaders, dozers and other auxiliary units) is utilized for mining the hard and semi-hard rock formations, which are encountered in the overburden strata, or for mining deposit formations that are not effectively mined by the continuous mining equipment. The non-continuous mining equipment is also utilized additionally in the cases that continuous mining equipment is not adequate for handling the required earthmoving works.

The lignite deposit modelling and evaluation and the mineable reserves estimation process is an essential and a very important subject for the mine planning and design phase of such surface mining projects. The deposit reserves have a particular effect on any surface mining project, affecting the life of the project, the optimal capacity of the corresponding power plant, the annual production rate of the mine, as well as the viability of the overall project (Galetakis et al, 2011, Roumpos and Papakosta, 2013). Quality parameters of run-off-mine lignite are also very important in strategic mine planning. They also affect the economic viability of the project, the efficiency of the power plant, the environmental performance of the project or the decision for the mining sequence

(Roumpos 2005, Roumpos et al., 2009). The accuracy of the mineable lignite reserves and the run-of-mine lignite quality estimation is, therefore, a very important subject of the deposit evaluation and mine planning and design procedure.

The scope of the present paper includes: (a) an investigation of the parameters that affect the lignite deposit modelling and the evaluation results, (b) a description of the suggested algorithm and procedure, (c) a presentation of a case study regarding the mid term development of a continuous surface lignite mine, and (d) conclusions.

2 Lignite Deposit Modelling and Evaluation

The modelling and evaluation procedure for the estimation of mineable reserves of multi layer lignite deposits could be divided into three main stages: (1) geologic modelling, (2) spatial block modelling, and (3) estimation of mineable lignite reserves. The flowchart of Fig. 2 shows the lignite deposit modeling and reserves estimation procedure, included the three stages that are discussed next in greater detail.

2.1 Geological Modelling

The frequent alteration of clastic sediments (normally consisting of soft sediments such as sands, clays and marls) and lignite beds is a characteristic of the largest lignite deposits of Greece (Papanikolaou et al, 2004). The majority of these deposits are characterized by high spatial quality variability. Modern commercial mining software packages, usually developed for covering all mine planning and design activities, are not ideally adapted to that type of multiple-layered stratified lignite deposits. They are based on a corresponding geology module which is designed to build a geological model from exploration drill-holes.

The first step in creating a geological model is the analysis and interpretation of the drill- holes data. This procedure includes the importing, editing, reporting and graphical representation of them. It goes through the various importing functions, checking and validating the data with queries and reports, and using various utilities to fix the problems encountered.

Commonly there are differences in the available exploratory data of drill-holes. These are attributed to the following reasons:

- The time period in which a drilling program is carried out is very long.
- The descriptions of the drilling data are carried out by many different geologists with different knowledge base.
- In the first years of drilling programs there was not a standardized logging procedure. Therefore, there is a lack of uniformity of the exploratory data.

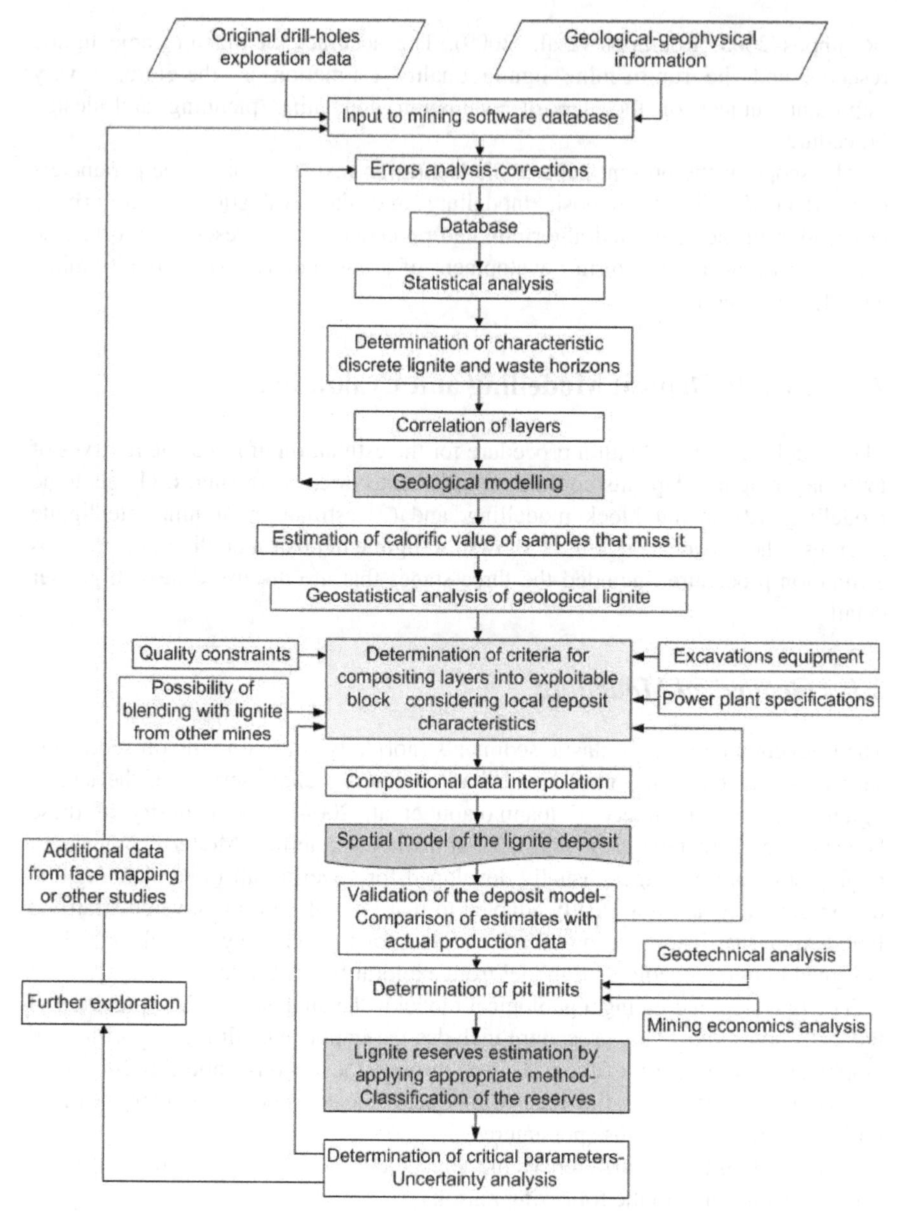

Fig. 2 Flowchart of lignite deposit modeling and reserves estimation procedure.

- The objective of some drill-holes is not the investigation of the whole deposit but of a certain problem (e.g. geotechnical, hydrogeological or other drill-holes).
- The descriptions of the drill-holes data may be done by geologists of incomplete experience. As a result, incorrect drill-hole locations, sampling and

analytical errors or errors in data input to databases may occur. Also, errors in the drilling procedure may result in core losses or in difficulties to discrete specific geological strata.
- Chemical analysis of drill-holes samples may not contain values of some important quality parameters.

Furthermore, large deviations from the actual data of face mapping, relating often to the lack of direct correlation of the lignite and waste strata, are often caused by the following conditions:

- The number of layers found in exploration drill-holes in many cases is extremely large.
- The quality characteristics of the lignite deposit present vertical and horizontal spatial variability, which is often reflected to the change in the number and thickness of lignite layers in the same seam.
- The deposits are characterized by many faults with different dip directions and ages. Besides, many of the faults were formed at the period of sedimentation, affecting the characteristics of each lignite seam.

In a first stage, the correlation procedure is mainly based on characteristic layers. In a second stage, the waste layers with specific characteristics or the lignite layers rich in carbon minerals could be separated. In this way, the interfaces between the characteristic lignite layers will be well defined in order to be used for the geological correlation and modelling Roumpos et al, 2011).

Furthermore, taking into account the data of the structural or other geological information, the locations of the characteristic layers can be determined. The effect of tectonics to quality variation of lignite in multilayered deposits is very clear (Galetakis et al, 2006). Additional data could also help to create sub-groups of layers between characteristic formations. Geological characteristics of the lignite deposit and interpretations of the formation of lignite seams can enhance the process of lignite deposit modelling (Leontidis et al, 2001).

2.2 *Spatial Block Modelling*

The evaluation of the mineable lignite reserves and quality is a process of compositing seams into blocks of exploitable lignite. The main criteria used for the formation of the exploitable compositing lignite blocks are (Kavouridis et al, 2000, Galetakis and Vasiliou, 2010)

- Minimum thickness of lignite blocks and of waste layers that can be excavated by selective mining. These values mainly depend on the technical characteristics of the excavator, the bench geometry and the type of cutting.

- Cut off quality of the run of mine lignite. This criterion is closely related to the specifications of the power station that will be fed by the excavated lignite. In the past, only the upper limit of ash water free content (included CO_2 in some cases) was used. However, for an accurate estimation of the exploitable block, the lower limit of calorific value should be used additionally.
- Dilution with waste and mining lignite loss at the top and bottom of each block. These parameters are used because the cutting geometry of the excavating machinery cannot be adjusted accurately to the geometry of the top and bottom surfaces of the lignite blocks.

The compositing of seams into blocks of exploitable lignite and waste material is a computer-aided iterative process (Karamalikis, 1992). The evaluation of the exploitable lignite is based on drill-holes data and the main steps of this process (described in Kavouridis et al, 2000) are:

- Initial raw seam coding. All seams of a hole are characterised as lignite or waste according to their quality characteristics (ash, water content and calorific value).
- Waste blocks formation. Successive non-lignite seams are grouped to from the initial waste beds.
- Initial determination of selectively excavated waste blocks. Waste blocks thicker than the defined minimum thickness for selective mining are classified as waste blocks and cannot contribute to the formation of lignite blocks.
- Evaluation of exploitable lignite blocks between waste blocks. First all lignite and waste (embedded dilution) layers between the above-mentioned selectively excavated waste blocks are examined. If the total thickness of these layers is greater than the minimum thickness for selective excavation and their compositing weighting quality values satisfy the cut off quality limit of the run of mine lignite, then all layers are grouped to form an exploitable lignite block. Otherwise, the algorithm examines the possibility of forming an exploitable block taking into account as many layers as possible. This evaluation process is repeated for all the remaining seams in the borehole and in this way the exploitable lignite blocks are determined. For the estimation of total thickness and average quality parameters the surface dilution and mining loss are also taken in account.

The compositing blocks of exploitable lignite may include thin waste layers while the blocks of waste material may include thin lignite layers (Fig. 3). The thickness h_e of an exploitable lignite block is estimated as the summation of all the individual layers, n layers of lignite, with thickness $h_{(l)i}$, *and* m layers of waste thickness $h_{(w)j}$ within the block (usually m<<n). In a similar way the value of a specific quality characteristic q_{bl} of the lignite block is determined as the weighted average of all q_i of the individual layers.

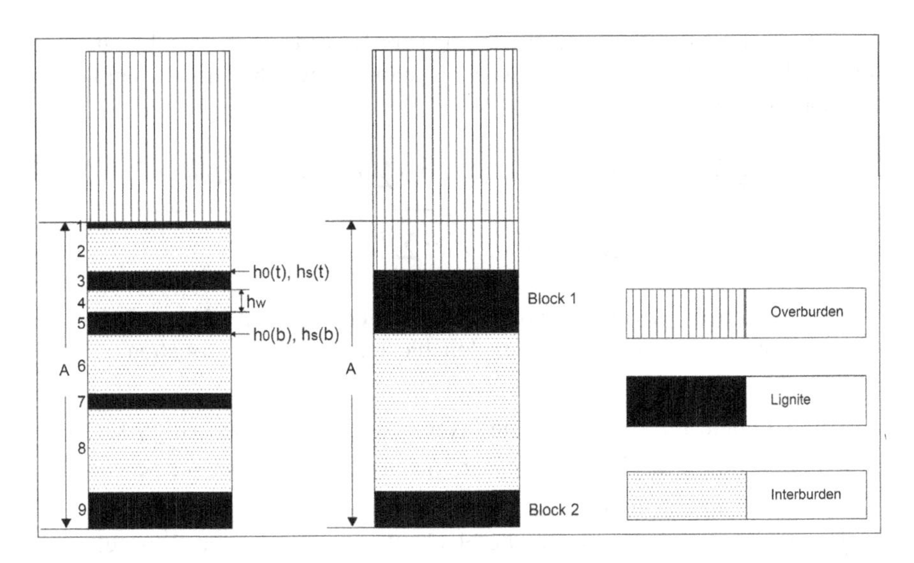

Fig. 3 Drill-hole data compositing. In the original drill-hole, dilutions and mining losses parameters are clarified.

The thickness and the quality of the exploitable block are corrected, as shown in equations 1 and 2, to take into consideration the surface dilution and mining losses at the top and bottom of the lignite blocks.

$$h_{bl} = \sum_{i=1}^{n} h_{(l)i} + \sum_{j=1}^{m} h_{(w)j} - (h_o(t) + h_o(b)) + (h_s(t) + h_s(b)) \tag{1}$$

$$q_{bl} = \frac{(h_{(l)1} - h_0(t))\, q_{(l)1}\rho_{(l)1} + \sum_{i=2}^{n-1} h_{(l)i}\, q_{(l)i}\rho_{(l)i} + \sum_{j=1}^{m} h_{(w)j}\, q_{(w)j}\rho_{(w)j} + (h_{(l)n} - h_o(b))\, q_{(l)n}\rho_{(l)n} + h_s(t) q_s(t)\rho_s(t) + h_s(b) q_s(b)\rho_s(b)}{\sum_{i=1}^{n} h_{(l)i}\rho_{(l)i} + \sum_{j=1}^{m} h_{(w)j}\,\rho_{(w)j} - (h_0(t)\rho_{(l)1} + h_0(b)\rho_{(l)n}) + (h_s(t)\rho_s(t) + h_s(b)\rho_s(b))} \tag{2}$$

Where:

h_0 (t) , h_0 (b): the equivalent thickness of a lignite layer corresponding to mining losses at the top and bottom of the lignite blocks

$h_s(t)$, $h_s(b)$ = the equivalent thickness of a waste layer corresponding to dilution at the top and bottom of the lignite blocks

$\rho_{(l)i}$ = the density of lignite layer i

$\rho_{(w)j}$ = the density of waste layer j

$\rho_s(t)$, $\rho_s(b)$: the density of waste layer at the top and bottom of the lignite blocks

For the estimation of the average ash water free content of the block, in eq. 2, dry densities should be used, while for the other quality parameters (calorific value, water content or other referring to the in situ material) density of wet samples should be used.

If, at a first approach, it is assumed that $\rho_{(l)i} = \rho_l$ for all lignite layers of the block (i=1,….,n) and $\rho_{(w)j} = \rho_w$ for all waste layers of the block (j=1,….,m), while

$\rho_w = \rho_s(t)=\rho_s(b)$, as well as that h_O (t) $=h_O$ (b)$=h_O$, and h_s (t) $=h_s$ (b)$=h_s$, then equation 2 is simplified to the following equation (3), which is usually applied in the compositing process:

$$q_{bl}=\frac{\rho_l\cdot(\sum_{i=1}^{n}h_{(l)i}\;q_{(l)i}-h_o\,(q_{(l)n}+q_{(l)1}))+\rho_w\cdot\sum_{j=1}^{m}h_{(w)j}\;q_{(w)j}+2\,h_s q_s\,\rho_w}{\rho_l\cdot(\sum_{i=1}^{n}h_{(l)i}-2h_0)+\rho_w\cdot(\sum_{j=1}^{m}h_{(w)j}+2h_s)} \tag{3}$$

From the above described compositing procedure, it can be deducted that the most important parameters of the exploitable lignite blocks formation in each drill-hole are:

(1) The minimum thickness of waste layers that can be excavated by selective mining ($h_{(w)\min}$). This parameter mainly depends on the excavating equipment and usually is higher for non-continuous mining equipment.
(2) The minimum thickness of lignite blocks that can be excavated by selective mining ($h_{bl\min}$).This parameter also mainly depends on the excavating equipment and usually is higher for non-continuous mining equipment.
(3) The quality characteristics of waste layers ($q_{(w)j}$) included in the compositing blocks. In most cases they are not known and must be determined.
(4) The quality characteristics of lignite layers ($q_{(l)i}$) included in the compositing blocks. The problem arises when some data are missing and should be estimated. This usually happens with the calorific value of the layers. The calorific value is an important quality characteristic of lignite since there is a strong relation between the calorific value and the mineable reserves. The calorific value of lignite layers often is not measured for all drill-hole samples taken and should thus be estimated from other measured parameters. In that case, the following geochemistry-based equation (4) can be applied:

$$LCV = Cwaf\cdot(1-w)\cdot(1-a)+Ca\cdot(1-w)\cdot a\cdot l+Cw\cdot w \tag{4}$$

Where:

LCV (kcal/kg) is the lower calorific value of the lignite layer, $Cwaf$ (kcal/kg) is the calorific value of the ash-free and water-free combustible matter of the lignite, w is the water content of the lignite, α is the ash content adjusted for the CO_2 content, Ca is the decomposition energy of the calcium carbonate ($CaCO_3$), l is the content of the ash in $CaCO_3$ and Cw is the energy of vaporization of the water contained in the lignite. The value of $Cwaf$ can be assumed as approximately constant for the whole mining area ($Cwaf\approx 5800$ kcal/kg for the lignite of West Macedonia Lignite Center), while $Ca=$ -437 kcal/kg and Cw =-583 kcal/kg.

Alternatively, a multiple linear regression method seems to produce more accurate estimations (eq. 5) (Pavlidis et al, 2013):

$$LCV = a_1 \cdot w + a_2 \cdot co_2 + a_3 \cdot ash + a_4 \tag{5}$$

Where:

LCV is the lower calorific value of the lignite layer, w is the water content of the lignite, ash is the ash water free content and CO_2 is the carbon dioxide content, while $\alpha_1,\dots, \alpha_4$ are the coefficients of the multiple linear regression model which must be determined.

(5) The thickness of dilution with waste (surface dilution) and the thickness of mining lignite loss at the top and bottom of each block. These parameters have a significant effect not only in the thickness of the lignite blocks (eq. 1) but also they affect the average quality characteristics of the block (eq. 2), in a way that they may not satisfy the corresponding cut off quality limits. A reduction of the surface dilutions will be a reason for a corresponding higher increment of the mining losses. The determination of these parameters (that in many cases should be different at the top and at the bottom of the block) in relation to cutting geometry of the excavators and to structural geometry of the deposit is a very important problem of lignite reserves estimation procedure.

(6) The densities of lignite layers ($\rho_{(l)i}$), of waste layers ($\rho_{(w)j}$) included in the blocks, and of waste layer at the top ($\rho_s(t)$), and bottom ($\rho_s(b)$) of the lignite blocks. These parameters affect the reserves estimation results in two ways:

- The calculated reserves (tonnage) of the lignite deposit from the corresponding volumes depend on the density of the lignite (tonnage factor).
- The average quality characteristics of the blocks (eq. 2) depend on these parameters (weighting factor). They may also affect the exploitability of the blocks in a way that they may not satisfy the corresponding cut off quality limits.

The above mentioned densities may vary within wide limits for lignite and waste according to their lithology, porosity, petrographic composition, quality or other mineralogy characteristics. Therefore, the spatial determination of the material densities is critical for the accuracy of lignite reserves estimation.

(7) The cut off limits of the quality of the run of mine lignite. The main quality parameters of the lignite are the ash content and the calorific value. The cut off limits of their parameters affect significantly the quantity and the quality of lignite reserves, since they are the main criteria besides block thickness for the exploitability of a block. The determination of that limits is a very important and critical subject of lignite reserves estimation procedure. In the process of the validation of the deposit model that is derived from the compositional data, the estimation data should be compared to available actual production data (Fig. 2). Furthermore, other quality parameters should also be taken into account, e.g. sulphur, nitrogen or other elements.

From the above mentioned parameters it is obvious that the determination of the criteria of the compositing layers into exploitable blocks is an iterative algorithmic process that should be revised very often according to local mining conditions.

After the determination of the exploitable lignite blocks the compositional data interpolation follows, by applying appropriate interpolation method and by integrating the geological structure information of the lignite deposit. Based on that interpolation process, the spatial model of the lignite deposit is developed. The modelling process should focus on drillhole database structuring, fault modelling as well as on stratigraphic correlation (Kapageridis and Kolovos, 2009). The validation of the model could be based on the available actual data of the deposit.

2.3 Estimation of Mineable Lignite Reserves

The estimation of the mineable lignite reserves should be based on the spatial model of the lignite deposit and also on the defined pit limits. For the determination of that limits, mining economics parameters should be considered and mainly the marginal mining ratio, as well as information of geotechnical analysis that will affect the pit slopes inclination.

The reserves estimation methods include conventional (mainly the inverse distance weighting) or geostatistical methods. A simulation-based approach may also be applied offering many advantages (Peattie and Dimitrakopoulos, 2013).

Uncertainty analysis and determination of critical parameters may lead to changes of the criteria of compositing of layers into exploitable blocks or to suggestions for further exploration. The new data should be integrated into the model and interpolations should be reviewed periodically, incorporating the new information.

From the above described procedure, it can be deducted that the results of the estimations of minable lignite reserves of multiple-layered deposits may vary within wide limits, depending mainly on the following factors:

- the algorithm for the determination of lignite and waste blocks in each drill-hole
- the parameters (criteria) for compositing layers into exploitable lignite blocks in relation to the type of excavators, to the specifications of power plant or to other constraints.
- the accuracy of the quality parameters and density of lignite and waste layers included in lignite blocks

Emphasis should be placed on the suitable integration of geological and local deposit characteristics information into the lignite deposit spatial model as well as on the lignite reserves estimation method. Benndorf (2013) investigates in situ variability and homogenization of key quality parameters in continuous mining operations and demonstrates the benefits of integrating conditionally simulated deposit models and mine planning and scheduling issues to optimize the mining operation, adjusting that to the production requirements.

3 Case Study

An application of the described process for the estimation of mineable lignite reserves, concerning the mid term development of Kardia continuous surface lignite mine is presented in this section.

3.1 Mine Location – Description of the Kardia Lignite Deposit

The Kardia mining field is located in the central part of the western boundary of Ptolemais mining area (Figure 4). The area of the mine covers approx. 20 km^2.

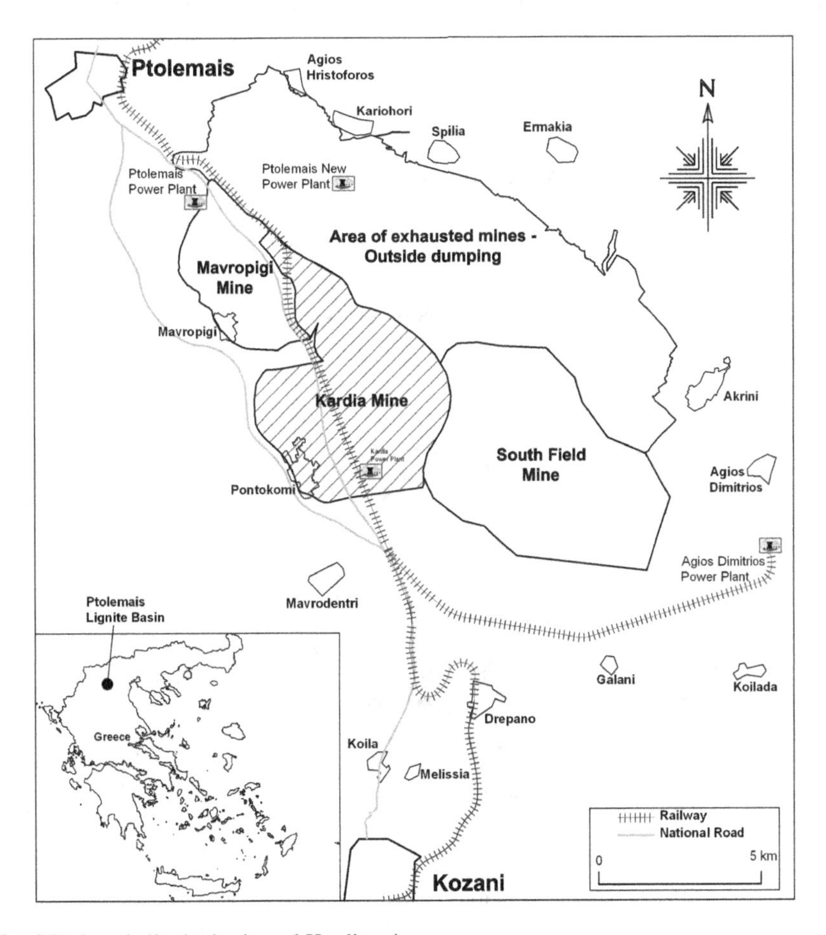

Fig. 4 Ptolemais lignite basin and Kardia mine

Faults striking from WNW-ESE to NW-SE dissect the mining field into several fault blocks. The western rim of the mining field is located near the mountain front. The thickness of the lignite-bearing layers, including intercalations, mainly ranges between 80 and 140 m. In a narrow part of the south-western boundary of the mining field, the lignite-bearing layers increase to a thickness of more than 210 m. In the central and north-western parts of the mining field the thickness of the individual split seams, which are partly replaced by intercalations, varies within wide limits. The overlying strata thicknesses vary between approx. 20 and 60 m. In the above mentioned south-western boundary zone, they amount to approx. 150 m. (Rheinbraun Engineering and Public Power Corporation of Greece, 1996).

The remaining exploitable lignite reserves amount to 300 Mt. The main mine equipment consists of 7 bucket wheel excavators and 4 spreaders, and the annual lignite production of the mine is approximately 10-12 Mt. Figure 5 shows the mining area, the final mine limits and the areas of continuous and non continuous operation of the excavators.

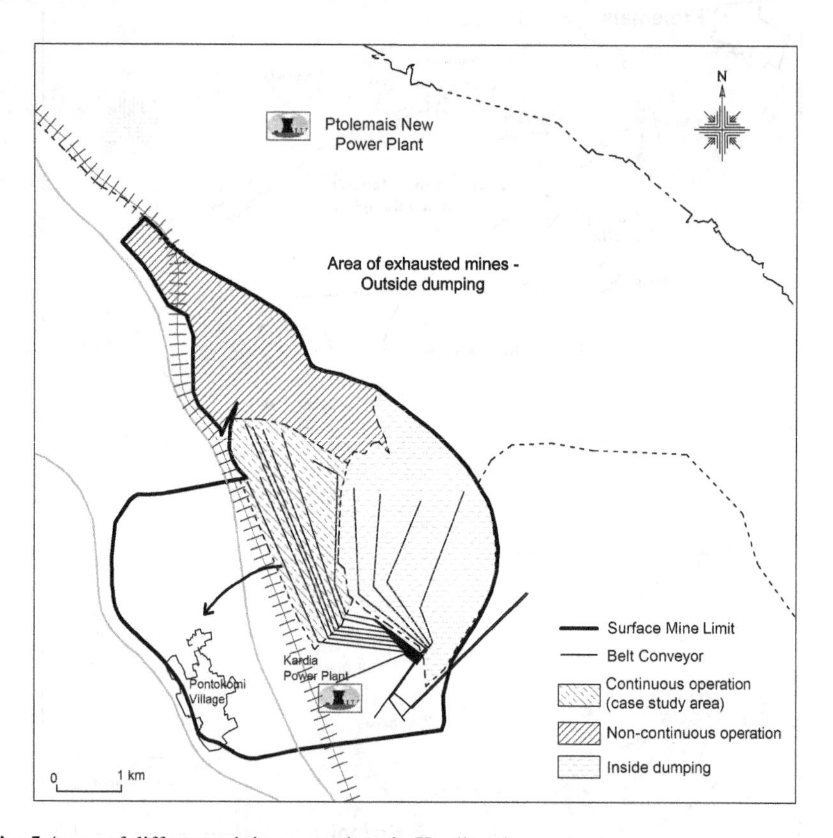

Fig. 5 Areas of different mining operations in Kardia mine

3.2 Lignite Deposit Modelling and Evaluation of a Mid-term Planning Sector

For the modelling and evaluation of the whole lignite deposit, the process described in section 2 was applied. The geological modelling was based on the drill-holes data and the available geological information. The case study is concentrated on a mid-term development of the mine, where for the lignite production only bucket wheel excavators were used, covering a period of two years. Figure 6 shows the area of case study, which in the first years of mine exploitation in the Ptolemais mines was used as an outside dumping area of waste material, the locations of drill-holes, as well as the main faults. In that area, major deviations between predicted an actual data were derived by applying the criteria of compositing layers into exploitable blocks that used for the previous mining sectors. A characteristic phase mapping view is shown in the picture of Fig. 1, which also includes the data of a drill-hole (201/205). The location of the drill-hole is also shown in Fig. 6.

For the geological correlation of the lignite seams, a face mapping program has been applied, in accordance with the available drill-holes data. In addition, new drill-holes in that area added new information for the continuity of mineralization and lignite spatial variability (Barmpas, 2013). The results of correlation process and the geological modelling are shown in the characteristic cross-section A-A' (Fig. 7). The location of this cross section is also shown in Fig. 6.

By applying the above described process for modelling of the lignite deposit, the effect of the critical parameters that were described in section 2.2 was investigated. From the analysis results it was concluded that the most critical parameter for the compositing layers into exploitable lignite blocks was the cut-off limit in relation to the missing data of calorific value from the most of lignite layers in that area. The main reason for this result is the high content of the calcium carbonate ($CaCO_3$) in the layers included in the compositing lignite blocks. This content affects significantly the calorific value of the layers. Figure 8 shows the comparative average lower calorific value- lignite reserves curve for the mined out area (case study area) as a function of ash water free cut off limit. The actual production data corresponds to the value of cut off limit (ash water free) 26%. The values of the calorific value were estimated by applying eq. 4.

Figure 9 shows the Deviation from actual data of main quantity and quality parameters for the case study area as a function of the ash water free cut off limit. Major deviations concern the tonnage of lignite reserves, which affect the mining ratio. From Fig. 8 and 9 it is concluded that the effect of the ash water free cut off limit is critical.

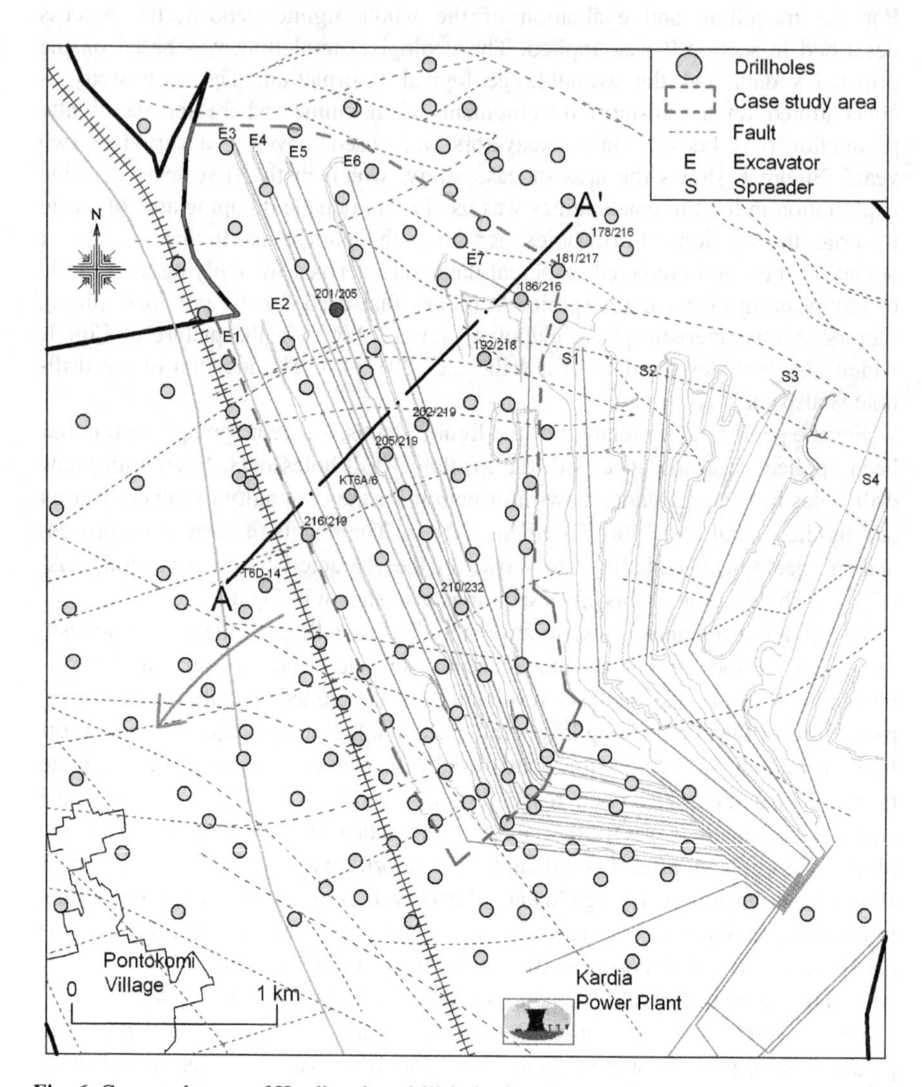

Fig. 6 Case study area of Kardia mine, drill-holes locations and main faults

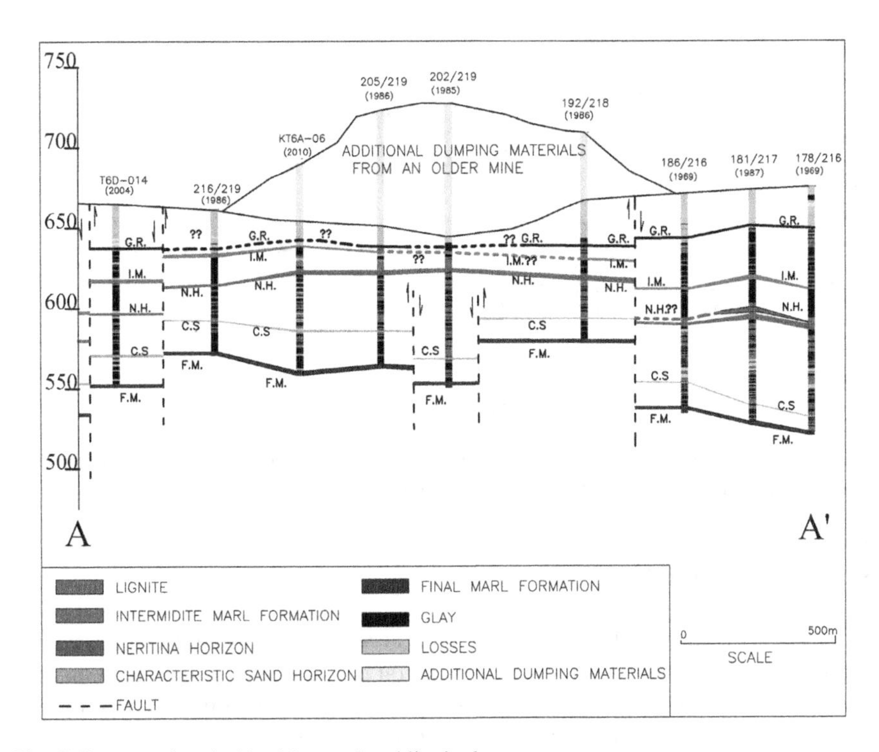

Fig. 7 Cross section A-A' with correlated lignite layers

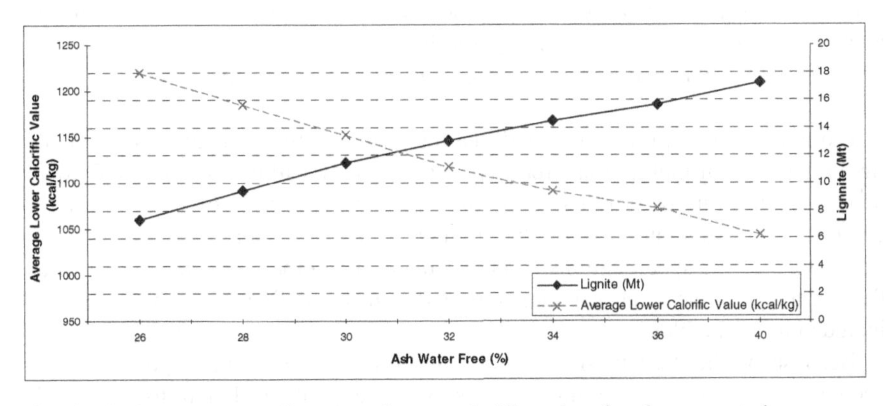

Fig. 8 Lignite reserves and average lower calorific value for the case study area as a function of ash water free cut off limit.

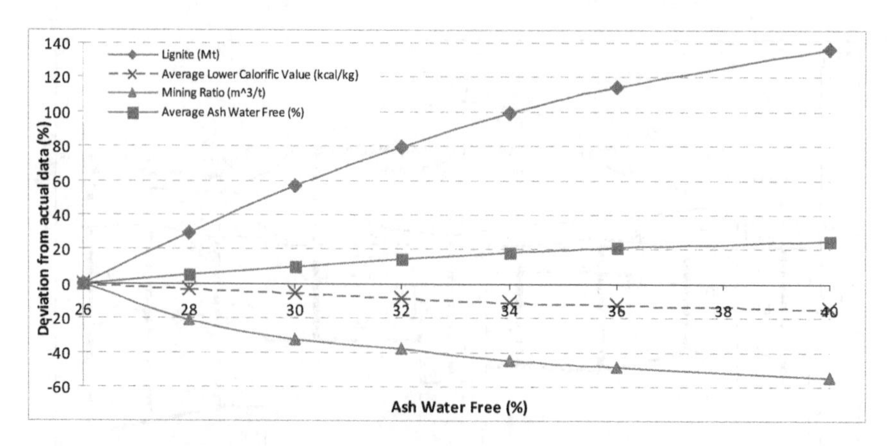

Fig. 9 Deviation from actual data of main quantity and quality parameters for the case study area as a function of the ash water free cut off limit.

4 Conclusions

The accurate spatial model of a lignite deposit as well as the mineable reserves estimation process have a significant impact on the successful mine planning and design phase of the corresponding mining project. In multiple-layered lignite deposits this process is more complicated because of the increased requirements for selective mining.

In this paper the iterative process of mineable lignite reserves estimation in continuous surface mining is defined. The discrete steps, from data input into the database to the application of the appropriate method of reserves estimation are described. The lignite modelling and evaluation stages are discussed in greater detail.

From the analysis, it is concluded that the results of the estimations of minable lignite reserves of multiple-layered deposits may vary within wide limits. Critical factors include the determination of the parameters (criteria of evaluation) for compositing layers into exploitable lignite blocks, in relation to the type of excavators, to the specifications of power plant or to other constraints as well as the accuracy of the quality parameters and density of lignite and waste layers included in lignite blocks.

Because of the high in situ spatial variability of the lignite deposits, for an accurate estimation of the minable reserves the geological and local deposit characteristics information should be integrated into the lignite deposit spatial model. In addition, emphasis should be placed on the appropriate lignite reserves estimation method.

The integration of the spatial model of the lignite deposit and mine planning and scheduling optimization techniques could lead to the adjustments in mine design and scheduling of key equipment to the production requirements.

References

1. Barmpas, T.: Face mapping in Kardia mine. Unpublished data (2013) (in Greek)
2. Benndorf, J.: Investigating in situ variability and homogenisation of key quality parameters in continuous mining operations. Mining Technology. Transactions of the Institutions of Mining and Metallurgy. Transactions: Section A 122(2), 78–85 (2013)
3. Galetakis, M., Vasiliou, A.: Selective mining of multiple-layer lignite deposits. A fuzzy approach. Expert Systems with Applications 37(6), 4266–4275 (2010)
4. Galetakis, M., Roumpos, C., Alevizos, G., Vamvuka, D.: Production scheduling of a lignite mine under quality and reserves uncertainty. Reliability Engineering and System Safety 96, 1611–1618 (2011)
5. Galetakis, M., Kavouridis, C., Kouvata, A., Roumpos, C., Pavloudakis, F., Kouridou, O., Stamoulis, K.: The effect of tectonics to quality variation of lignite mined from Southern field mine of Ptolemaes – Greece. In: Proc. 15th International Symposium on Mine Planning and Equipment Selection, Torino, Italy, September 20-22, pp. 1106–1111 (2006)
6. Kapageridis, I., Kolovos, C.: Modelling and Resource Estimation of a Thin-Layered Lignite Deposit. In: 34th International Symposium on the Application of Computers and Operations Research in the Minerals Industries (APCOM), Vancouver, pp. 95–103 (2009)
7. Karamalikis, N.: Computer software for the evaluation of lignite deposits. Mineral Wealth 76, 39–50 (1992) (in Greek)
8. Kavouridis, C., Leontidis, M., Roumpos, C., Liakoura, K.: Effect of dilution on lignite reserves estimation - Application in the Ptolemais multi-seam deposits. Braunkohle 52(1), 37–45 (2000)
9. Leontidis, M., Roumpos, C., Dadswell, J.: Planning problems posed by unusual geological sequences and explore a possible solution (A question of geology). World Coal 10(8), 49–50 (2001)
10. Papanikolaou, C., Galetakis, M., Foscolos, A.: Quality characteristics of greek brown coals and their relation to the applied exploitation and utilization methods. Energy and Fuels 19(1), 230–239 (2004)
11. Peattie, R., Dimitrakopoulos, R.: Forecasting recoverable ore reserves and their uncertainty at Morila Gold Deposit, Mali: An efficient simulation approach and future grade control drilling. Mathematical Geosciences 45, 1005–1020 (2013)
12. Pavlides, A., Hristopulos, D.T., Galetakis, M., Roumpos, C.: Geostatistical analysis of the calorific value and energy content of the Mavropigi multi-seam lignite deposit in Northern Greece. In: Proceedings of Fourth International Symposium on Mineral Resources and Mine Development, Aachen, Germany, May 22-23, pp. 199–209 (2013)
13. Rheinbraun Engineering, Public Power Corporation of Greece, Technical Mine Master Plan, unpublished report, 600 p. (1996)
14. Roumpos, C.: Optimisation of the combined project of a lignite mine exploitation and power plant operation. Mining-Metallurgical Annals 1 (1), 45–64 (2005)
15. Roumpos, C., Pavloudakis, F., Galetakis, M.: Optimal production rate model for a surface lignite mine. In: Proced. of the 3rd AMIREG International Conference: Assessing the Footprint of Resource Utilization and Hazardous Waste Management, Athens, Greece, pp. 360–365 (2009)

16. Roumpos, C., Liakoura, K., Barmpas, T.: Estimation of Exploitable Reserves in Multilayer Lignite Deposits by Applying Mining Software. In: The Significant Impact of Geological Strata Correlation. 17th Meeting of the Association of European Geological Societes (MAEGS), Belgrade, Serbia, September 14-18, pp. 247–250 (2011)
17. Roumpos, C., Papacosta, E.: Strategic mine planning of surface mining projects incorporating sustainability concepts. In: Proceedings, 6th International Conference on Sustainable Development in the Minerals Industry (SDIMI 2013), Milos Island, Greece, June 30 - July 3, pp. 645–651 (2013)

Application of Particle Swarm Optimization to the Open Pit Mine Scheduling Problem

Asif Khan[1] and Christian Niemann-Delius[2]

[1] Lehrstuhl und Institut für Rohstoffgewinnung über Tage und Bohrtechnik
[2] RWTH Aachen University Germany

Abstract. This paper proposes a procedure for applying Particle swam algorithm (PSO) to long term production scheduling problem of the open pit mines, which is a large combinatorial optimization problem involving large datasets and multiple hard and soft constraint. Constraint programming technique have been used to produce the initial population of solutions (particles) and then the continuous version of the PSO algorithm have been implied to search the feasible solution space for finding better solutions. The performance of different variants such as global, local and Multistart of the PSO algorithm have been studied and the results are presented.

1 Introduction

Mine design and production scheduling of open pit mines is a large scale complex optimization problem, which commonly aims to determine the extraction sequence of the mineralized material from the ground that produces maximum possible discounted profit from the operation while satisfying a set of physical and operational constraints such as precedence or slope constraints, processing capacity constraints, mining capacity constraints etc. Block model representation of the ore body is commonly used for the planning and production scheduling of the open pit mines. The block model discretize the ore body and the surrounded rock into a three dimensional array of regular size blocks. Each block has set of attributes attached to it such as: grade or metal content, tonnage, density etc. estimated using drill holes data. Using a fixed cutoff grade strategy the blocks are divided into two groups i.e. ore or waste blocks. If the prospective profit from a block exceeds its processing cost it is categorized as an ore block to be sent for processing once mined while the rest are the waste blocks. Depending on the location of the waste blocks they are either left in the ground or mined and sent to the waste dump to get access to the underlying high value ore blocks. An economic value is assigned to each block by taking into account its group i.e. its ore or waste block, the commodity price, mining, processing and marketing cost. The final contour of the pit is determined by solving the ultimate pit limit (UPL) problem. The UPL can be defined as the extent to which it is economically

C. Niemann-Delius. (ed.), *Proc. of the 12th Int. Symp Continuous Surface Mining - Aachen 2014*,
Lecture Notes in Production Engineering, DOI: 10.1007/978-3-319-12301-1_19

feasible to mine. Graph based Lerchs and Grossmann algorithm[1] or Max- flow algorithms [2] are commonly used for this purpose. Depending on the size of the individual blocks and of the ore deposit the UPL may contain thousands to millions of blocks that may have to be scheduled over a time horizon typically ranging from 5 to 30 years while considering different physical and operational constraints which makes it a large combinatorial optimization problem.

Several modelling and solution techniques have been proposed since 1960to solve this problem. Some of these techniques includes parametric or ultimate pit limit (UPL) based approaches [3-5], integer programming[6, 7], dynamic programming [8, 9], blocks aggregation / clustering[10]etc.. The main limitation of these techniques is their high computational cost when applied to real sized problem. In the recent years a new class of computationally less expensive algorithms i.e. metaheuristic techniques have attracted the attention of several researchers to solve the mine design and production scheduling problem such as Genetic Algorithms[11-13], Simulated Annealing [14, 15] , Ant Colony optimization [16, 17] etc.. Though metaheuristic techniques do not guarantee the optimality of the final solution that they produce but their low computational cost makes them attractive alternative choice over the computationally expensive exact optimization algorithms. In this paper a relatively new Metaheuristics technique known as Particle swarm optimization (PSO) has been applied to the open pit production scheduling problem. Ferland et al. [18] has proposed a procedure to apply particle swarm optimization to the capacitated open pit mining problem considering only one production capacity constraint during each period of the mine life. They used a genotype representation of the solutions based on the priority value encoding and proposed a decoding scheme to generate feasible phenotype solutions from the genotype solutions. A GRASP procedure was suggested to produce the initial population of genotype solutions that evolve through the solution space using global version of particle swarm algorithm. Different properties of the procedure were studied by its application to a group of two dimensional problems. This paper proposes a totally different procedure to apply particle swarm optimization to more general open pit production scheduling problem. It proposes to use constraint programming to generate the initial population of particles whose positions will be updated using the continuous version of particle swam algorithm.

The remainder of the paper is organized as follows. In section 2 the general integer programming formulation of the ultimate pit limit and production scheduling problem of the open pit mines are presented. In section 3 the standard PSO algorithm is discussed. In section 4 the proposed procedure for applying PSO algorithm to open pit production scheduling problem is presented in detail. Results are discussed in section 5 followed by conclusion in section 6.

2 Problem Formulation

The planning process of the open pit mine usually starts with definition of ultimate pit limit. The mathematical formulation of which can be defined as:

$$Maximize \sum_{i=0}^{N} V_i x_i \tag{2.1}$$

s.t.

$$x_i \leq x_j i = 1, 2, \ldots \ldots \ldots \ldots . N; \; j \in P_i \tag{2.2}$$

$$x_i \in [0,1]$$

Where V_i represents the economic value of block i, N represents the total number of blocks in the block model, x_i represents a binary variable corresponding to block i and P_i represents the predecessor group of block$i$. The solution to this problem just returns the collection of blocks that will result into maximum possible undiscounted profit, but it does not define the sequence of extraction of the blocks in each period while considering different processing, production capacity constraints and the discount rate, which is a more complex problem to solve. The general integer programming formulation of the open pit sequencing problem with the objective to maximize the net present value of the operation while satisfying a set of physical and operational constraints similar to the one defined by Gaupp in [19] can be described as follows:

Objective Function:

$$\textbf{Maximize Z} = \sum_{i=1}^{N} \sum_{t=1}^{T} v_{it} \, x_{it} \tag{2.3}$$

$$v_{it} = \frac{V_i}{(1 + d)^t}$$

Where v_{it} is the discounted economic value of block i if mined in period t and d represents the discount rate.

Subject to:

Reserve Constraint: A block cannot be mined more than once

$$\sum_{i=1}^{T} x_{it} \leq 1 \quad \forall \, \mathrm{i} = 1, 2 \ldots . \mathrm{N} \tag{2.4}$$

Slope constraints: Each block can only be mined if its predecessors are already mined in or before period ***t.***

$$x_{it} - \sum_{\tau=1}^{t} x_{j\tau} \leq 0 \tag{2.5}$$

$\forall$ i = 1,2 ….N ; t = 1,2 ….T; Where $j \in$ (set of predecessors blocks of block i)

Processing Capacity: The total ore processed during each period should be within the predefined upper and lower limits.

$$\sum_{i=1}^{T} o_i * w_i * x_{it} \geq \underline{O} \,\forall\, t = 1,2 \,....\, T \tag{2.6}$$

$$\sum_{i=1}^{T} o_i * w_i * x_{it} \leq \overline{O} \,\forall\, t = 1,2 \,....\, T \tag{2.7}$$

Mining Capacity: The total material mined during each period should be within the predefined upper and lower limits.

$$\sum_{i=1}^{T} w_i * x_{it} \geq \underline{M} \,\forall\, t = 1,2 \,....\, T \tag{2.8}$$

$$\sum_{i=1}^{T} w_i * x_{it} \leq \overline{M} \,\forall\, t = 1,2 \,....\, T \tag{2.9}$$

$$x_{it} \in (0,1) \forall\, i = 1,2 \,....\, N;\; t = 1,2 \,....\, T; \tag{2.10}$$

Where

N Number of blocks in the block model
T Number of periods
v_{it} The Net present value of the block$\boldsymbol{i}$ if mined in period $\boldsymbol{t}$

$$x_{it} = \begin{cases} 1 & \textit{if block } \mathbf{i} \textit{ is mined in period } \mathbf{t} \\ 0 & \textit{otherwise} \end{cases}$$

$\underline{O}$*and* $\overline{O}$are the upper and lower limits of the processing capacity, $\underline{M}$and $\overline{M}$are the upper and lower limits of the available mining capacity. As the UPL may contain thousands to millions of blocks that have to be scheduled over a time horizon typically ranging from 5 to 30 periods or may be more, the resulting integer programme may contain millions of integer variables and constraints, which can be extremely difficult and expensive to solve.

3 Particle Swam Optimization

Particle swarm optimization (PSO) is a stochastic population based optimization technique first presented by Kennedy and Eberhart in 1995[20, 21]. PSO is a nature inspired algorithm, based on the social interaction of individuals living together in groups e.g. bird flock, fish schools, animal herds etc. PSO algorithm performs the search process by using a population (Swarm) of individuals (Particles). Each individual (Particle) is a potential solution to the optimization problem. A random starting position and random velocity is assigned to each particle of the swarm. The velocity represents the speed of "*flying*" of the particle through the solution space. Each individual particle i possesses the following three pieces of information:

$\vec{x}_{i,t}$: Position of the particle i during iteration t,$\vec{x}_{i,t}$ can be considered as a set of coordinates specifying the position of a point in the solution space.

$\vec{v}_{i,t}$: Velocity of the particle i during iteration t, specifying the step size along every coordinate of the solution space.

$\vec{p}_{i,t}$: The personal best position experienced by particle i until iteration$t$. The corresponding personnel best value is stored in a variable called$Pbest_i$ (***Personal best***)which simplify the comparisons in the later iterations. This position is replaced if a better solution is found out by particle i in the later iterations.

The position of the best particle in the neighborhood of particle i until iteration t is stored in a variable called $\vec{g}_t$ *or* $\vec{l}_t$(depending on the topology being used) the corresponding value is stored in a variable called $gbest_t$(***global best***)*or* $lbest_t$(***Local best***) which simplify the comparisons in the later iterations. The position and the velocity of the particles is updated by using equation 3.1 and equation3.2 respectively.

$$\vec{v}_{i,t} = \vec{v}_{i,t-1} + c_1\vec{r}_1 \cdot (\vec{p}_{i,t-1} - \vec{x}_{i,t-1}) + c_2\vec{r}_2 \cdot (\vec{g}_{t-1} - \vec{x}_{i,t-1}) \quad 3.1$$

$$\vec{x}_{i,t} = \vec{x}_{i,t-1} + \vec{v}_{i,t} \quad 3.2$$

Where $\vec{v}_{i,t-1}$and $\vec{x}_{i,t-1}$are the velocity and position of particle i in the previous iteration, C_1 and C_1 are called the acceleration coefficients which are used to control the influence of cognitive and social term on the particle's velocity. $\vec{r}_1$ and $\overrightarrow{r_1}$are vectors of random real numbers in the range of (0, 1)[22].

The PSO algorithm can be described as:

i. Initialize a population (swarm) of particles with random initial positions $\vec{x}_{i,0}$ and random velocities $\vec{v}_{i,0}$ in the searchspace.

ii. Initialize each particle's personal best position$\vec{p}_{i,0}$ to its initial position $\vec{x}_{i,0}$

iii. Calculate the fitness value of each particle at its initial position $\vec{x}_{i,0}$and determine the initial global best position $\vec{g}_0$

iv. ***For all*** particles ***ido***

v. Update the particle's velocity and position using equation (3.1) & (3.2) respectively.

vi. Calculate the fitness value of each particle at its current position $\vec{x}_{i,t}$

vii. ***Iffitness***$(\vec{x}_{i,t})$ is better than the $fitness(\vec{p}_{i,t-1})$***then***

viii. $\vec{p}_{i,t} = \vec{x}_{i,t}$

ix. ***end if***

x. ***Iffitness***$(\vec{p}_{i,t})$ is better than the $fitness(\vec{g}_{t-1})$***then***

xi. $\vec{g}_t = \vec{p}_{i,t}$

xii. ***end if***

xiii. If the stopping criterion is met:***end loop***

xiv. Return the global best position: $\vec{g}_t$

Clerc and Kennedy proposed a constriction coefficients based PSO algorithm after analyzing the trajectory of a single particle[23]. These coefficients were introduced to prevent velocity explosion, ensure convergence and eliminate the need to set problem based maximum velocity parameter. The most simple and frequently used method to incorporate constriction coefficients into the velocity update equation also known as Constriction type1" can be defined as:

$$\vec{v}_{i,t} = \chi(\vec{v}_{i,t-1} + c_1\vec{r}_1 \cdot (\vec{p}_{i,t-1} - \vec{x}_{i,t-1}) + c_2\vec{r}_2 \cdot (\vec{g}_{t-1} - \vec{x}_{i,t-1})) \quad 3.3$$

Where $\varphi = c_1 + c_2 > 4$ *and*

$$\chi = \frac{2}{\varphi - 2 + \sqrt{\varphi^2 - 4\varphi}} \quad 3.4$$

$\varphi$is commonly fixed to 4.1,$c_1 = c_2 = 2.05$,which results into constant multiplier χ'svalue to be 0.7298 and each of the two terms i.e. $(\vec{p}_{i,t-1} - \vec{x}_{i,t-1})$and $(\vec{g}_{t-1} - \vec{x}_{i,t-1})$multiplied by 1.49445 times random number between 0 and 1. These parameters will be used throughout this study.

4 Applying Particle Swarm Algorithm to Open Pit Mine Production Scheduling Problem

In this study the continuous version of particle swarm algorithm as described in section3 (with a few modifications) will be applied to open pit mine production scheduling problem. The reason for using the continuous version of particle swarm algorithm is twofold: First the block model may contain thousands to millions of regular sized blocks; making the scheduling decision on the blocks level may be computationally expensive. Secondly the special structure of the block model, precedence relationship between the blocks in the same column and the one dimensional nature of the PSO algorithm [24] dictates that the scheduling problem can be turned into depth determination problem for different periods, which makes the PSO calculations faster and the implementation simple. In the following sections a few requirements for applying the PSO algorithm to open pit production scheduling problem will be discussed in more details.

4.1.1 Initial Solutions

To generate an initial population of random feasible solutions, the open pit production scheduling problem was turned into a constraint satisfaction problem which was then solved using the CP optimizer of the IBM ILOG CPLEX Optimization Studio V12.5 to get multiple feasible solutions. A Constraint satisfaction problem needs the following three pieces of information to define a problem:

- A set of decision variables
- A finite set of possible values for each variable
- Constraint defining the inter variable relationship

The solution to this problem is to determine the value of every variable from its domain such that all the constraints are satisfied. This is achieved by using different search algorithms such as arc and path consistency, simple backtracking and forward checking and different heuristics to guide the search [25].

4.1.2 Solution Encoding

In order to apply the continuous version of particle swarm algorithm to open pit production scheduling problem an encoding scheme is being used. The proposed encoding scheme determines the deepest block along a certain column to be mined in a certain period. This value (depth) is stored into a variable corresponding to that column and period. These depths are them updated using standard PSO algorithm in each iteration

4.1.3 Back Transform

After each iteration a back transform scheme is used to determine the period in which a block is scheduled. The procedure takes the depth variable for a certain column and period and determines all the blocks that are lying above that depth and below the depth defined by the PSO algorithm for the same column for the prior period. There is no upper limit for the first period. This process progresses from the first to the last period. During the optimization process it is insured that depth for the successive periods along a column does not cross each other.

4.1.4 Constraint Handling

Particle swarm algorithm like other evolutionary algorithms does not have an explicit mechanism for constraint handling. Considering the special structure of the constraints the following two techniques have been used for constraint handling in this study:

- Due to the one dimensional nature of the PSO algorithm, the mining depths along each column for each period are determined independently from each other which may result in infeasible solution in terms of the required slope angles. A normalization technique (a constraint repair technique)is being applied to each back transformed infeasible solution after each iteration to turn it into a feasible solution in terms of the required slope angles as shown in Figure 4.1 and Figure 4.2.(SGeMS software have been used for visualization[26]).

- A constant penalty method[27] is used to deal with the Mining and Processing capacity constraints, where a constant penalty is added to the objective function for per ton violation of the capacity constraints, to decrease the quality of the infeasible solutions .The value of the penalty is problem dependent.

Fig. 1 Infeasible back transformed solution while scheduling for three periods

Fig. 2 Feasible solution in terms of the required slope angles (45^0in this case) after Normalization

4.1.5 The PSO Algorithm for Open Pit Production Scheduling Problem

The proposed procedure to apply PSO algorithm to open pit production scheduling problem can be summarized as (The proposed modification have been highlighted by using bold text):

a. Initialize algorithm's parameters and a population (swarm) of particles with random initial positions $\vec{x}_{i,0}$ and random velocities $\vec{v}_{i,0}$ in the searchspace.

b. Initialize each particle's personal best position$\vec{p}_{i,0}$ to its initial position $\vec{x}_{i,0}$

c. Calculate the fitness value of each particle at its initial position $\vec{x}_{i,0}$and determine the initial global best position $\vec{g}_0$

d. ***For all*** particles ***ido***

e. **Encode each particle's current position (x_i), its personnel best position ($Pbest_i$), and the populations' global / local best position (Gbest or Lbest)**

f. Update the particle's velocity and position using equation (3.1 or 3.3) & (3.2) respectively.

g. **Back transform the solutions**

h. **Normalize the current solutions**

i. Calculate the fitness value of each particle at its current position $\vec{x}_{i,t}$

j. $\boldsymbol{If} fitness(\vec{x}_{i,t})$ is better than the $fitness(\vec{p}_{i,t-1})\boldsymbol{then}$

k. $\vec{p}_{i,t} = \vec{x}_{i,t}$

l. ***end if***

m. $\boldsymbol{If} fitness(\vec{p}_{i,t})$ is better than the $fitness(\vec{g}_{t-1})\boldsymbol{then}$

n. $\vec{g}_t = \vec{p}_{i,t}$

o. ***end if***

p. If the stopping criterion is met:***end loop***Return the global best position: $\vec{g}_t$

4.1.6 Population Topology

The learning process of the individual particles of the Particle swarm algorithm is based on the social interaction among the particles within the same neighborhood. Considering the neighborhood structure / Population Topology the following two different versions of the PSO have been proposed in the literature [20, 28]:

- **gbest:** In the gbest neighborhood every particle is influenced by the best solution found by any member of the entire population. In this topological structure every particle compares its performance with every other member of the entire population and follows the best one.
- **lbest:**In the lbest topology a particle is influenced by the best solution found by its k immediate neighbors. The index based neighbourhood will be used in this study.

Particle swarm algorithm using the gbest neighborhood converges much faster than the one using the lbest neighborhood, with much higher likelihood to get trapped into a local optimum position[22]. A detailed description of different Population Topologies and their influence on PSO algorithm's performance can be found in [29].

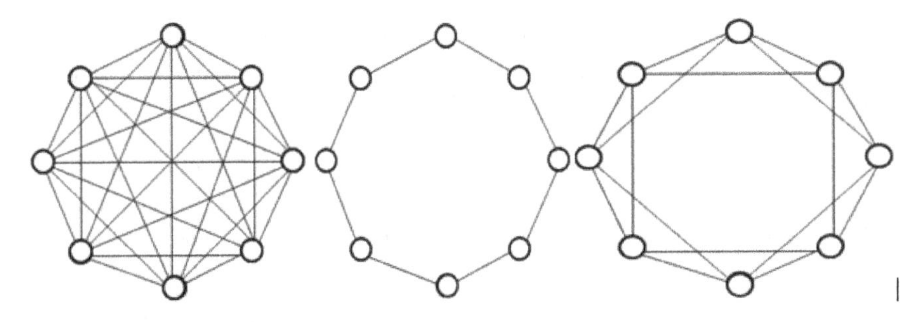

Fig. 3 Topologies gbest (left), lbest with k = 2 (middle) and k=4 (right)

5 Numerical Results

To check the capabilities of the proposed procedure for applying PSO algorithm to long term production scheduling problem of the open pit mines, a computer programme has been developed in Microsoft Visual studio 2010 (C++) programming environment. A hypothetical 3D block model has been used for performance analysis of the proposed algorithm. The ultimate pit limit was determined by solving the ultimate pit limit problem as defined in section 2.The required slope angles were supposed to be 45^0 in all directions. The general information about the blocks within the predefined UPL and optimization parameters are given in Table 1.

Table 1 General information about the blocks within UPL and optimization parameters

DESCRIPTION	VALUE
Total No of blocks	7836
Ore blocks	4707
Waste blocks	3129
Mine Life (Years)	3
Discount rate (%)	10

The length of each scheduling period was supposed to be 1 year. The upper and lower limits for Processing and Mining Capacity were set to be within ± 10% and ± 15% of the average available quantity of ore and rock within the UPL for each period respectively. All the numerical experiments have been completed on AMD Phenom IIX 4 945 (3 GHz) and 4 GB Ram running under windows 7. An exact solution of the problem was determined using CPLEX, which will be used as a benchmark to assess the performance of the PSO algorithm in terms of computational time and solution quality. The details about of the solution found by CPLEX are given in Table 2.

A series of experiments were carried out to find out the optimal values for certain parameters such as Population size and maximum number of iteration for the algorithm to converge for the problem under consideration. These values will remain constant in the subsequent experimentation.

- **Population size:** In comparison to other evolutionary algorithms e.g. Genetic Algorithms, PSO needs smaller population [30]. After conducting a series of experiments a population size containing 50 particles was found out to be performing well for the problem under consideration. A population of 50 feasible solutions (particles) was generated using constraint programming technique; these particles will be used throughout this study. The quality of the initial population was found out to have significant effects on the performance of PSO algorithm. The objective function values (NPV) of the generated particles are given in Figure 5.1
- **Maximum Iterations:** The global variant of the PSO algorithm generally needs less time / iterations to converge than the local variant of the algorithm. To get an upper bound on the number of iteration required by both the variants before convergence, a series of preliminary tests were carried out, where it was found out that the maximum numbers of iterations required by the global and local variants of the PSO algorithm to converge are approximately 2200 and 5000 iterations respectively for the problem under consideration. A general idea about the convergence behavior of local variant with different neighbourhood sizes vs. the global variant can be observed in the Figure 5.2, it shows that the local variant of the PSO algorithm shows a relatively slower rate of convergence and needs relatively more time to converge than the global variant. The convergence behaviour of the global

variant of PSO algorithm can be seen in the Figure 5.3 where the iteration best continuously shows an improvement in terms of objective function quality as particles keep on moving towards better solution areas as the process progresses.

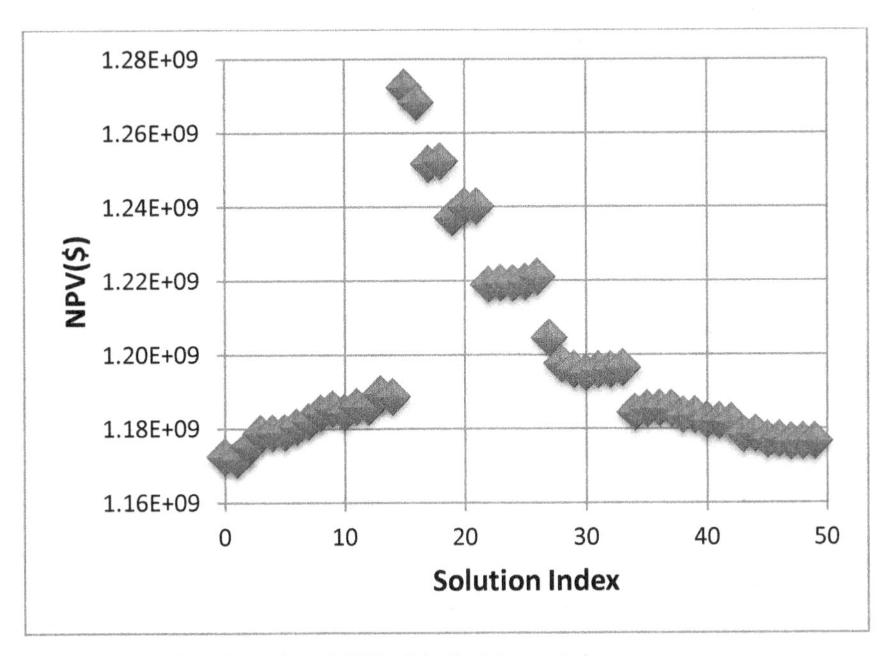

Fig. 4 Objective function values (NPV) of the initial population

To further analyze the performance of the proposed procedure a series of experiments were carried out using the following algorithmic settings.

- Global variant of the PSO algorithm where each particle is influenced by the best particle of the entire population.
- Multistart Global variant of the PSO algorithm, where while using the global variant once the algorithm is converged, as a diversification strategy the particles' position and velocities are reinitialized to their initial values while keeping their current personal best positions.
- Local variant of PSO algorithm with different neighbourhood (K) sizes, where K represents the total number of indexed based neighbours of each particle of the population.
- Due to the stochastic nature of the PSO algorithm the problem was solved 10 times using each of the above mentioned settings. In Table 3 the average computational times and average relative %Gap between best solutions generated by the PSO algorithm i.e.$Z_{PSO}$and the optimal solution found by CPLEX i.e. Z_{CPLEX} and their standard deviations are given.

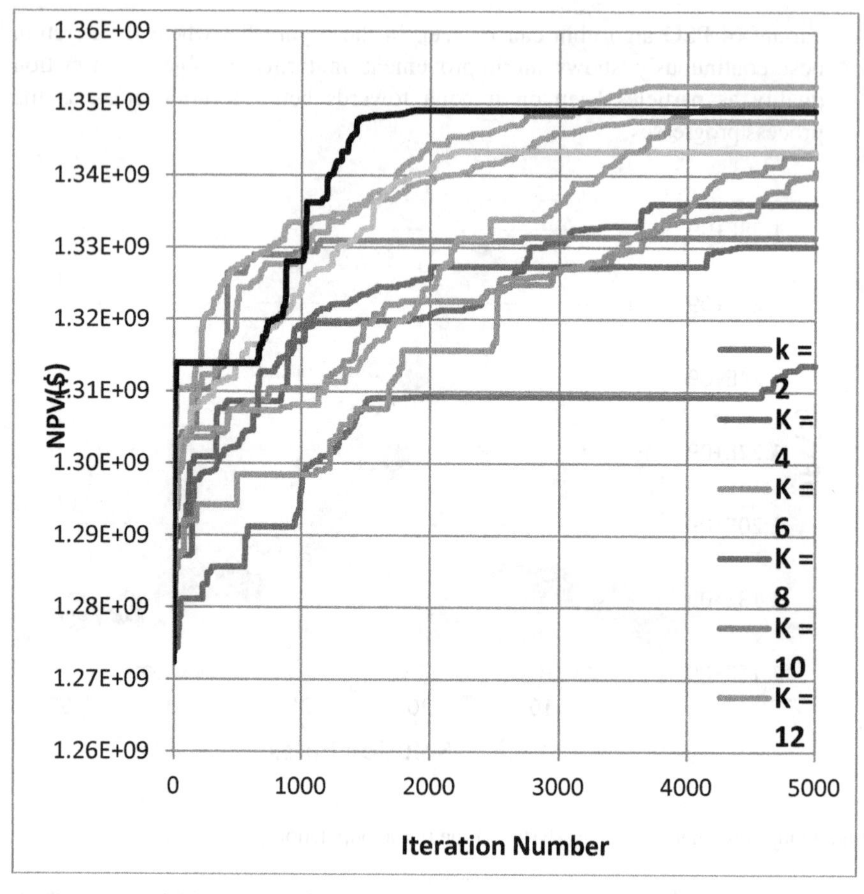

Fig. 5 Convergence behaviour of local variant with different neighbourhood sizes vs. global version of PSO algorithm for a single run.

$$\% \, Gap = \frac{Z_{CPLEX} - Z_{PSO}}{Z_{CPLEX}} * 100$$

As described before local version of PSO algorithm with different neighbourhood sizes were allowed to run for a fixed number of iteration i.e. 5000 in the current case. The CPU time column shows that the average computation time generally increase with increasing neighbourhood size except in the case when K = 2, whose performance both in the terms of solution quality and computation time is not satisfactory with relatively higher % Gap and computational time. By considering the quick convergence behaviour of the global version of the PSO algorithm, it was allowed to run for 2200 iteration. The global versions always found better quality solutions in comparison to the local version of the PSO algorithm in relatively less time. Multistart strategy almost always performed better than the local and global variants showing its ability to escape the local optima. The standard deviation of % Gap in almost all the cases has very small values showing the robustness of the procedure. The proposed procedure in all the cases can produce better quality results in relatively shorter period of time in comparison to the exact optimization algorithms implied by CPLEX.

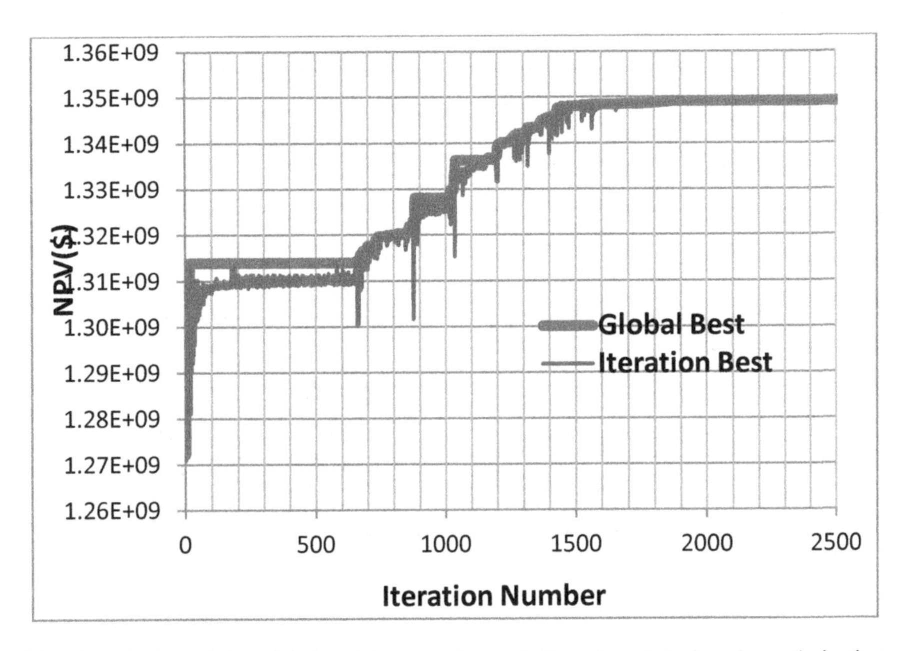

Fig. 6 Evolution of the global and iteration best solutions found during the optimization process

Table 2 General information about the solution found by CPLEX

DESCRIPTION	VALUE
Objective Function value ($)	1.37854e+009
CPU time (Seconds)	79740
Optimality Gap (%)	4.5 %

Table 3 Numerical results of the PSO algorithm with different population settings

LOCAL	% GAP		CPU TIME (SECONDS)	
	Mean	Standard Deviation	Mean	Standard Deviation
K = 2	4.91	0.6	621.5	18.17
K = 4	3.61	1.1	609.4	4.02
K = 6	3.27	0.7	601.4	4.65
K = 8	2.89	0.7	600.2	2.70
K = 10	2.41	0.5	604.4	4.29
K = 12	2.51	0.1	608.6	6.21
K = 14	2.46	0.3	615.4	7.41
K = 16	2.11	0.2	616.9	4.35
K = 18	2.18	0.2	626.2	5.23
K = 20	2.20	0.4	630.3	5.49
K = 22	2.21	0.3	635.3	4.56
K = 24	2.21	0.1	636.1	2.57
K = 26	2.34	0.4	642.0	2.98
K = 28	2.39	0.4	655.1	18.84
K = 30	2.43	0.1	656.9	18.54
Global	3.10	0.8	271.6	16.69
Multistart Global	2.00	0.43	595.245	16.4209

6 Conclusion

In this paper, a procedure has been proposed to apply a metaheuristic technique based on Particle Swam Algorithm to long term open pit scheduling problem. Constraint programming technique has been used to produce the initial population of feasible solution (particles). This population was kept constant to study the effects of different algorithmic settings on the performance of the procedure i.e. global, local and Multistart variants of the PSO algorithm. By making comparison with the results obtained using CPLEX in terms of computational time and solution quality it was learned that the proposed procedure can produce better quality solutions in relatively shorter period of time with smaller % Gap and standard deviation showing the robustness of the procedure. The proposed procedure is quit flexible and can easily accommodate additional constraints, variable slopes angles and grade and market uncertainties.

References

1. Lerchs, H., Grossmann, I.F.: Optimum design of open pit mines. Canadian Institute of Mining Trans. 68, 17–24 (1965)
2. Johnson, T.B.: Optimum open pit mine production scheduling, in University of California, Berkeley, CA (1968)
3. Dagdelen, K., Johnson, T.B.: Optimum open pit mine production scheduling by Lagrangian parameterization. In: Proceedings of 19th APCOM Symp., Littleton, Colorado, pp. 127–142 (1986)
4. Caccetta, L., Kelsey, P., Giannini, L.: Open pit mine production scheduling. In: Proceedings of Third Regional APCOM Symp., vol. (5/98). The Australasian Institute of Mining and Metallurgy Publication Series, Kalgoorlie (1998)
5. Whittle, J.: Four-X user manual. Whittle Programming Pty. Ltd., Melbourne (1998)
6. Gershon, M.E.: Mine scheduling optimization with mixed integer programming. Mining Engineering 35(4), 351–354 (1983)
7. Caccetta, L., Hill, S.: An Application of Branch and Cut to Open Pit Mine Scheduling. Journal of Global Optimization 27(2-3), 349–365 (2003)
8. Onur, A.H., Dowd, P.A.: Open-pit optimization — part 2:production scheduling and inclusion of roadways. Transactions of the Institute of Mining and Metallurgy Section A, p. 102, A105 – A113 (1993)
9. Wang, Q.: Long - term open-pit production scheduling through dynamic phase- bench sequencing. Transactions of the Institute of Mining and Metallurgy Section A 105, A.99–A.104 (1996)
10. Ramazan, S.: The new Fundamental Tree Algorithm for production scheduling of open pit mines. European Journal of Operational Research 177(2), 1153–1166 (2007)
11. Denby, B., Schofield, D.: Open pit design and scheduling by use of genetic algorithms. Trans. Inst. Min. Metall (Sec. A: Min. Industry) 103, A21–A26 (1994)
12. Denby, B., Schofield, D.: Genetic algorithms for open pit scheduling-extension into 3-dimentions. In: Proceedings of MPES Conference, Sao Paulo, Brazil, pp. 177–185 (1996)
13. Denby, B., Schofield, D., Surme, T.: Genetic algorithms for flexible scheduling of open pit operations. In: Proceedings of 27th APCOM Symp., London, pp. 473–483 (1998)
14. Kumral, M., Dowd, P.A.: Short-Term Mine Production Scheduling for Industrial Minerals using Multi-Objective Simulated Annealing. In: Proceedings of 30th APCOM Symp., Fairbanks, Alaska, pp. 731–742 (2002)
15. Kumral, M., Dowd, P.A.: A simulated annealing approach to mine production scheduling. Journal of the Operational Research Society 56, 922–930 (2005)
16. Sattarvand, J., Niemann-Delius, C.: Long Term Open Pit Planning by Ant Colony Optimization, Institut für Bergbaukunde III, RWTH Aachen University (2009)
17. Sattarvand, J., Niemann-Delius, C.: A New Metaheuristic Algorithm for Long-Term Open-Pit Production Planning/Nowy meta-heurystyczny algorytm wspomagający długoterminowe planowanie produkcji w kopalni odkrywkowej. Archives of Mining Sciences 58(1), 107–118 (2013)
18. Ferland, J.A., Amaya, J., Djuimo, M.S.: Application of a particle swarm algorithm to the capacitated open pit mining problem. In: Mukhopadhyay, S., Sen Gupta, G. (eds.) Autonomous Robots and Agents, pp. 127–133. Springer, Heidelberg (2007)

19. Gaupp, M.P.: Methods for Improving the Tractability of the Block Sequencing Problem for Open Pit Mining. In: Economics Business Div., Colorado School of Mines (2008)
20. Eberhart, R., Kennedy, J.: A new optimizer using particle swarm theory. In: Proceedings of the Sixth International Symposium on Micro Machine and Human Science, MHS 1995 (1995)
21. Kennedy, J., Eberhart, R.: Particle swarm optimization. In: Proceedings of the IEEE International Conference on Neural Networks (1995)
22. Engelbrecht, A.P.: Fundamentals of Computational Swarm Intelligence. John Wiley & Sons (2006)
23. Clerc, M., Kennedy, J.: The particle swarm - explosion, stability, and convergence in a multidimensional complex space. IEEE Transactions on Evolutionary Computation 6(1), 58–73 (2002)
24. Angeline, P.: Evolutionary optimization versus particle swarm optimization: Philosophy and performance differences. In: Porto, V.W., Waagen, D. (eds.) EP 1998. LNCS, vol. 1447, pp. 601–610. Springer, Heidelberg (1998)
25. Barbara, M.S.: A Tutorial on Constraint Programming. Division of Articial Intelligence University of Leeds School OF Computer Studies, Research Report Series Report 95.14 (1995)
26. Remy, N., Boucher, A., Wu, J.: Applied Geostatistics with SGeMS: A User's Guide. Cambridge University Press (2009)
27. Michalewicz, Z., Schoenauer, M.: Evolutionary algorithms for constrained parameter optimization problems. Evol. Comput. 4(1), 1–32 (1996)
28. Kennedy, J.: Small worlds and mega-minds: effects of neighborhood topology on particle swarm performance. In: Proceedings of the 1999 Congress on Evolutionary Computation, CEC 1999 (1999)
29. Mendes, R.: Population Topologies and Their Influence in Particle Swarm Performance, Universidade do Minho, Portugal (2004)
30. Kennedy, J.F., et al.: Swarm Intelligence. Morgan Kaufmann Publishers (2001)

Application of ARENA Simulation Software for Evaluation of Open Pit Mining Transportation Systems – A Case Study

Ali Saadatmand Hashemi[1] and Javad Sattarvand[2]

[1] Queen's University
a.saadatmand.hasehemi@queensu.ca
[2] Mining Engineering Faculty,
Sahand University of Technology
sattarvand@sut.ac.ir

Abstract. ARENA simulation software as one of the most comprehensive tools of discrete event simulation modeling has been used in this research for evaluation of the transportation system of the Sungun copper mine. Its user friendly environment makes it easy for the users to construct a representative model of complex systems and freely make diversity of modifications to monitor the upcoming results. Using simulation modeling, different management systems of the open pit mining equipment including fixed and flexible assignments of trucks for loaders have been studied. Developed model in ARENA has the capability of considering detailed features of both loading and hauling equipment. Productivity assessment scenarios have been established on the constructed model for the current fixed assignment managing procedure. Furthermore, a dispatching simulation model with the objective function of minimizing truck waiting times have been developed and a 7.8% improvement obtained by applying a flexible assignment of the trucks for the loaders compared to the fixed assignment system.

Keywords: ARENA, Simulation modeling, truck assignment, dispatching, open pit mining.

1 Introduction

Truck and loader systems are widely used in open pit mines due to their flexibility and productivity features in comparison with the other haulage systems. Although these systems require the least infrastructure to be settled in mining operations [1], but still haulage costs in earthmoving projects constitutes the main portion of total expenditures [2]. Lizotte et al. believe that material handling represents more than 50% of whole operating costs and some others estimate this amount to reach 60%

C. Niemann-Delius. (ed.), *Proc. of the 12th Int. Symp Continuous Surface Mining - Aachen 2014*,
Lecture Notes in Production Engineering, DOI: 10.1007/978-3-319-12301-1_20

in some cases [3]. Hence the mining operation is a highly expensive industry; even a small reduction in these costs would be noteworthy for both managers and mine planners. In other words, reducing haulage costs just for a few percent would cause remarkable savings in both operating and capital costs.

As the need for raw material by the industry increases, open pit mines go deeper and low grade deposits look economic for exploitation [4]. Though trying to design bigger equipment was considered as the first step to respond to this query, that's why today we encounter some gigantic trucks with about 350 tons capacity and some big shovels to serve these dump trucks. But it seems that the trend for capacity increase cannot keep up with the high demand for productivity, so the next step in coping with this highly increasing demand is to try to find the best solution for the management of the operating fleet in the mine.

Transition of fleet management paradigm from fixed to flexible assignment of trucks for loaders has been discussed for more than 35 years. Application of first dispatching system in a limestone quarry mine in Germany showed a great improvement in equipment efficiency [5] and since then different dispatching packages with diversity of optimization criteria have been introduced to mining industry.

In 1988 Eduardo Bonates and Yves Lizotte worked on evaluation of applying a dispatching system by use of a computer simulation model. In this model, they categorized fleet management systems in 3 main groups including: Manual Dispatching, Semi-Automated Dispatching and Automated Dispatching [5]. The resulted model which was written in FORTRAN, attempts to take into account the real features of the mine and optimize the utilization of the trucks and shovels as an LP objective function [5]. Z. Li introduced a methodology for optimum control of shovel and truck operation in 1989 [6]. The mentioned methodology was developed by applying 3 indicating questions and trying to find the best answer to each:

- How much load in each route should be handled inside the pit and which route network is the optimum?
- How are the trucks assigned to the shovels?
- What is the best number of trucks to achieve the excavation target?

Later F. Soumis et al. developed a procedure for solving the dispatching problem [7]. Their research was based on a nonlinear objective function with 3 parameters including: deviation of shovel's operational excavation from its objective excavation, deviation of trucks' real working hours from their scheduled operating hours, the third factor consists of some penalties assigned for deviation from desired ore blending [7]. In 1993 B. Forsman et al. designed a computer simulation model for Aitik open pit mine [8]. Using a discrete event modeling microcomputer called METAFORA, they could build a graphical model of the mine. Simulation results dictated the optimum haulage network and number of trucks in order to reach desired excavation target [8]. D. Gove and W. Morgan worked on truck-loader matching and the influential parameters using CAT's fleet production and cost (FPC) software [9]. B. Kolonja and J.M. Mutmansky developed a simulation model for analyzing dispatching strategies using SIMAN [10]. A dispatching strategy is the way a dispatching system encounters with optimization problem [10]. Heuristic strategies try to find local optimums while in combined strategies an LP model takes the production objectives into account and then a heuristic procedure assigns the truck to the shovels [10]. M. Ataeepour and E. Y. Baafi worked on a simulation model for truck-shovel operation using Arena [11]. The first step is to monitor the effect of number of trucks in system utilization and then applying a dispatching rule with objective function of minimizing trucks waiting times. In other attempt in 1999 A. J. Basu proposed a simulation model for Kalgoorlie Consolidated Gold Mines (KCGM) using GPSS/H. Applied strategy in this model is to assign a truck to a shovel with least queue length [12].

S. Alarie and M. Gamache conducted a research on solution strategies used in truck dispatching systems for open pit mines [3]. Based on their research, dispatching problem can be solved using two major approaches: a single stage approach tries to dispatch the trucks without applying any excavation constraint or blending while in a multistage approach an excavation target using LP or heuristic methods is set to the operation as the upper stage. In the lower stages the assignments are handled in a way that the deviation of operational excavation from the targets that are suggested by upper stage to be minimum. Qiang Wang et al. studied truck real-time dispatching from a macroscopic perspective [13]. By evaluating the flow rates of the trucks and introducing some base nodes, the proposed model showed better results in comparison with the conventional dynamic programming methods.

In 2007 C. N. Burt and L. Caccetta proposed some approaches for the calculation of a match factor for heterogeneous truck and loader fleets [14, 15]. Based on their methods, a match factor of a fleet consisting different types of trucks and shovels is attainable considering their cycle times. The proposed method is also capable of considering different haul road features.

A. Jaoua et al in 2009 developed a framework for realistic microscopic modeling of surface mining transportation systems [16]. They believed the conventional macroscopic models lacked considering microscopic features of the system under study. Therefore a simulator called SuMiTSim is developed by A. Jaoua et al using SIMAN. An important indicator in this simulator is the occupancy level of the road segments [16]. Real time study of the mine traffic using the strategy of optimizing the occupancy level of road segments makes it easy to deal with unwanted traffic congestions and strengthens the flexibility of the system. Comparisons of SuMiTSim with other macroscopic simulators concluded that the microscopic approach has the ability of considering detailed features of mining operation like interaction between equipment and operators. Thus, the dispatching model based on microscopic approach has higher reliability level in contrast with other macroscopic models [16]. E. K. Chanda and S. Gardiner compared different methods for prediction of truck cycle times [17]. The results proved that by using multiple linear regression method and artificial neural networks (NN), one can estimate the truck cycle time with the least deviation from reality while using simulation software like TALPAC leads to more deviated outputs.

2 Sungun Mine Features and Model Construction

Sungun copper deposit is the second largest copper mine in Iran. It is located in a mountainous area between Sungun and Pakhir rivers in Eastern Azerbaijan province of Iran, by the border of Azerbaijan and Armenia and 130 km from Tabriz, the capital of the province. The geological reserve of the deposit is estimated at 828 million tons with an average copper grade of 0.62 percent. Mining operations are handled using rigid frame rear dump trucks and hydraulic shovels, backhoes or front end loaders. Annual excavation target for the mine is 30M tons of ore and waste. The fleet is managed in a non-dispatching mode. Operating fleet consists of Komatsu HD 785 and HD 325 dump trucks, Liebherr R9350 Hydraulic shovel, Komatsu PC 800 hydraulic backhoes, Komatsu WA 600 and CAT 988 front end loaders. There are 3 main dumping destinations for trucks: crusher, low grade stockpile and waste dump. Each working day consists of 3 shift rosters. The mine operates 24 hours a day and 363 days per year.

The required data for analyzing the haulage system utilization is the record time periods for trucks' and loaders' cycle, queue time at loading point and loaders' idle time. These time sets are then used as input data for the simulation software. For recording cycle times, a recording procedure was designed for each of the operating faces. A chronograph instrument at each face recorded the time data. For recording of the each truck's cycle time, a reference time point was assigned for it which was the moment that the loader pours its first bucket load into the truck tray. This moment was recorded in the report paper in front of the truck's ID which was unique for each truck. When after a period of time the same truck appears in loading point, again the mentioned reference point was recorded to the truck, therefore, each truck cycle time would be attainable via deducting two following reference points. Similarly another reference point was assigned for recording of the loading times which was the moment that the empty truck at the loading point started to maneuver for positioning till it starts leaving the loading station. In this case the latter would be considered as loading with exchange time instead of just loading time. Truck waiting time was considered the time period from getting a truck to the loading point till starting to maneuver for loading. Gathered raw data were then inserted into a spreadsheet to conduct some statistical analysis. At first point some data were considered as outlier and omitted from following calculations [8]; the logic applied was considering a significance level of α=0.05 so that accepted data are as followed:

$$\bar{X} - 1.96S < x < \bar{X} + 1.96S$$

Where S is the standard deviation and $\bar{X}$ is the average of the sample data; then x would be accepted region for data. Once all data sets were approved through this procedure, each set was statistically analyzed to fit a distribution function. Using Kolmogorov-Smirnov and Chi Square tests, the best function with the least square error from the empirical data was selected.

Table 1 Specifications of the loading stations

No.	Loading Station	Dumping Station	Distance (m)	Loader	Hauler	No of Hauler
1	2037.5 S	Crusher	1000	CAT 988	Komatsu HD 325	3
2	2100	Crusher	2700	Komatsu WA 600	Komatsu HD 785	2
3	2087.5	Dump	1200	Hyundai 500	Komatsu HD 325	3
4	2025	Dump	1975	Komatsu PC 800	Komatsu HD 785	3
5	2087.5	Stockpile	2200	Hyundai 500	Komatsu HD 325	3

Table 2 Distribution Functions for Collected Data

Working Bench	Data Set	Best Fitted Model
1	Truck interarrivals	0.5+10*BETA(0.84,1.12)
	Waiting in queue	-0.5+7*BETA(1.08,1.43)
	Loading times	NORM(3.22,0.443)
	Hauling times to destination	2.19+3.29*BETA(1.32,1.82)
	Dumping times	0.4+ERLA(0.0579,4)
2	Truck interarrivals	7.5+ERLA(0.728,3)
	Waiting in queue	0
	Loading times	TRIA(4.39,4.68,6.26)
	Hauling times to destination	6+1.76*BETA(1.18,1.04)
	Dumping times	NORM(0.647,0.0973)
3	Truck interarrivals	2.5+GAMM(0.369,6.38)
	Waiting in queue	-0.5+ERLA(0.525,5)
	Loading times	TRIA(2.82,4.02,4.53)
	Hauling times to destination	3+1.66*BETA(2.48,2.5)
	Dumping times	0.4+WEIB(0.263,2.36)
4	Truck interarrivals	NORM(6.28,2.15)
	Waiting in queue	NORM(4.97,1.92)
	Loading times	NORM(5.29,0.67)
	Hauling times to destination	NORM(4.32,0.784)
	Dumping times	0.4+GAMM(0.0527,4.52)
5	Truck interarrivals	0.5+WEIB(6.4,2.01)
	Waiting in queue	-0.5+LOGN(0.912,0.733)
	Loading times	2.48+ERLA(0.233,5)
	Hauling times to destination	6.3+GAMM(0.246,4.56)
	Dumping times	0.4+GAMM(0.0527,4.52)

As it was screened inside the mine, there were 5 major loading faces, which were considered as 5 loading stations. Table 1 shows destinations for each station, distances from loading to dumping points, and specifications of fleet in each subsystem. (Contents in loading station column refer to elevation of the corresponding operating bench inside the pit). Table 2 shows the best fitted models for each data set. All times are in minutes in this table.

Once the required input data were prepared for all operating faces, a simulation model should be constructed to represent real characteristics of loading and hauling operations. To this end, *Arena* simulation software which is one of the most powerful discrete event simulation tools was applied for building a model of haulage system.

Since two of the five operating subsystem's destination is crusher, a distinct model should be considered for representation of these two subsystems in order to realistically take the effect of crushing process into account. Figure 1 shows the layout of the constructed model.

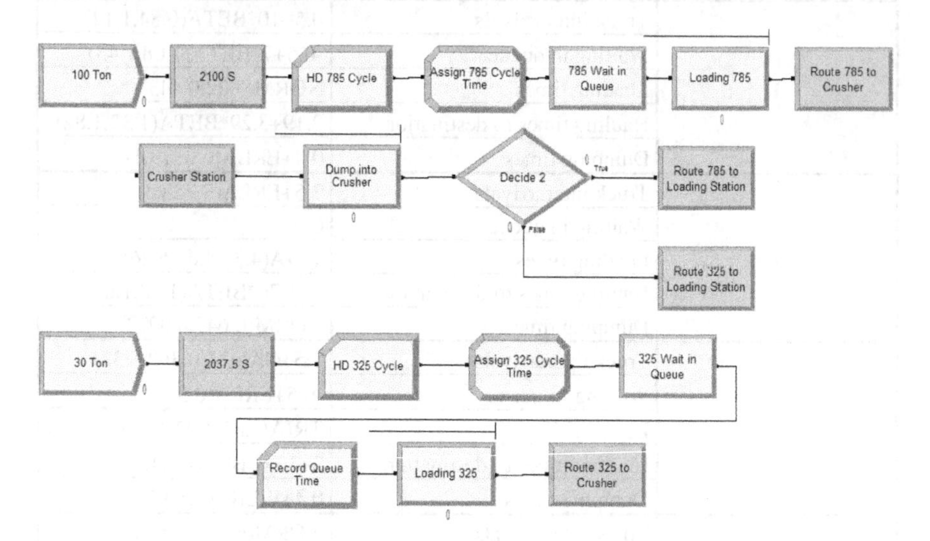

Fig. 1 Layout of Non-Dispatching Model in Arena

To have an acceptable reliability of the model and to ensure that the model truly represents the real system, there should be an indicating factor for comparing model outputs with the real system [9]. The most important output of the model is trucks cycle time, because it involves all time variables in it; therefore it was applied as an indicating factor for validation of the model. As it can be seen from Figure 2, constructed model represents the characteristics of the system under study with a high level of confidence.

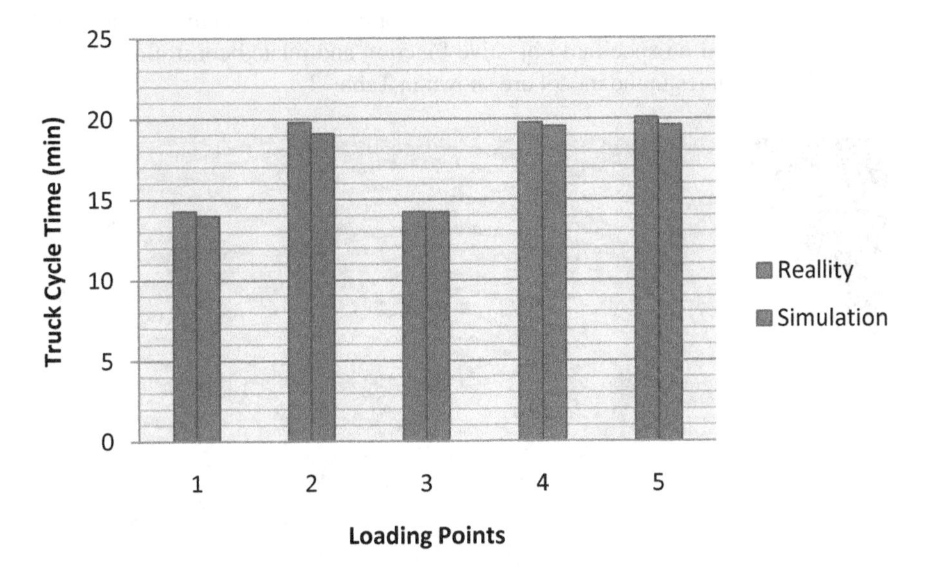

Fig. 2 Comparison of real data with simulation results

3 Modelling of Dispatching System

Once the system evaluated by match factor index, it seems all independent sub fleets should operate in a way that there should not be any waiting time for loaders and no queue of trucks at the loading faces, but due to the inherent variability in mining operations, in some cases queue lines of trucks or idle loaders have been observed. This variability which can occur because of unwanted traffic congestion or a special loading condition, leads to disarrangement of the fleet. Therefore, applying a flexible managing system for trucks which monitors the dynamic condition of in pit operation seems inevitable. Such a system dispatches every truck to a loader in a way that an optimization criteria be satisfied and leads to a condition that waiting times are set to minimum.

Unique geologic condition of Sungun copper mine forces the mine managers to apply front end loaders rather than shovels due to their high operational flexibility. Considering this condition and also by comparing operational costs of loaders with trucks, it is decided to set the dispatching objective function on minimizing truck wait times and applying of the strategy of dispatching one truck to *n* shovels.

Dispatching model was applied for the 5 working benches of the study, though for loading and hauling times, the data used from fixed assignment model and applied to have proper situation for future comparison of these two managing systems. It was approved that using dispatching systems in a condition that the system is over-trucked or under-trucked makes no efficient improvement in system utilization, though the important factor while setting the input data is the number of trucks. Therefore it is crucial to have the correct number of trucks while using dispatching system. Consequently the number of trucks in dispatching model was set to those applied in the modified fixed assignment model.

Dispatching decisions are made in two dumping sites, including crusher and waste dump based on specified objective function and related constraints (Figure 3). Applied variables in the model are shown in Table 3.

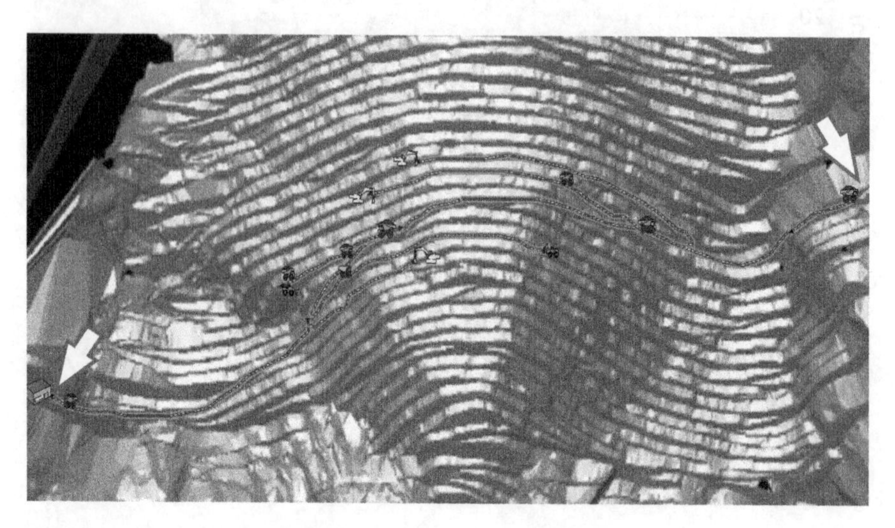

Fig. 3 Layout of dispatching points

Table 3 Applied Variables in Arena for dispatching model

Arena Variable	Definition
ECT(i,j)	Expected capture time of (i^{th}) loader by (j^{th}) truck
ACT(i,j)	Assigned capture time of (i^{th}) loader by (j^{th}) truck
ERT(i,j)	Expected release time of (i^{th}) loader by (j^{th}) truck
LOADING T(i,j)	Loading time of (j^{th}) truck by (i^{th}) loader
DEL(i,j)	Waiting time of (j^{th}) truck in (i^{th}) loader's queue
TRAVEL T(i,j)	Travel time of (j^{th}) truck from dump point to (i^{th}) loader
TARGET	Specified loader that satisfies dispatching criteria
TNOW	Current time

In this table index *i* refers to the number of loaders while *j* refers to the type of the truck, in other words there can be desired types of trucks based on their capacity and mechanical features.

Following calculations are carried out every time a truck dumps its load and asks for new assignment:

ECT(i,j)=ACT(i,j)
ERT(i,j)=ECT(i,j)+LOADING T(i,j)
DEL(i,j)=ERT(i,j)-[TNOW+TRAEL T(i,j)]
TARGET=Min{DEL(1), DEL(2),…DEL(5)}

Once the truck assigned for a loader, the target loader's variables should be updated for the next calculations. Figure 4 shows the layout of the designed model for dispatching system.

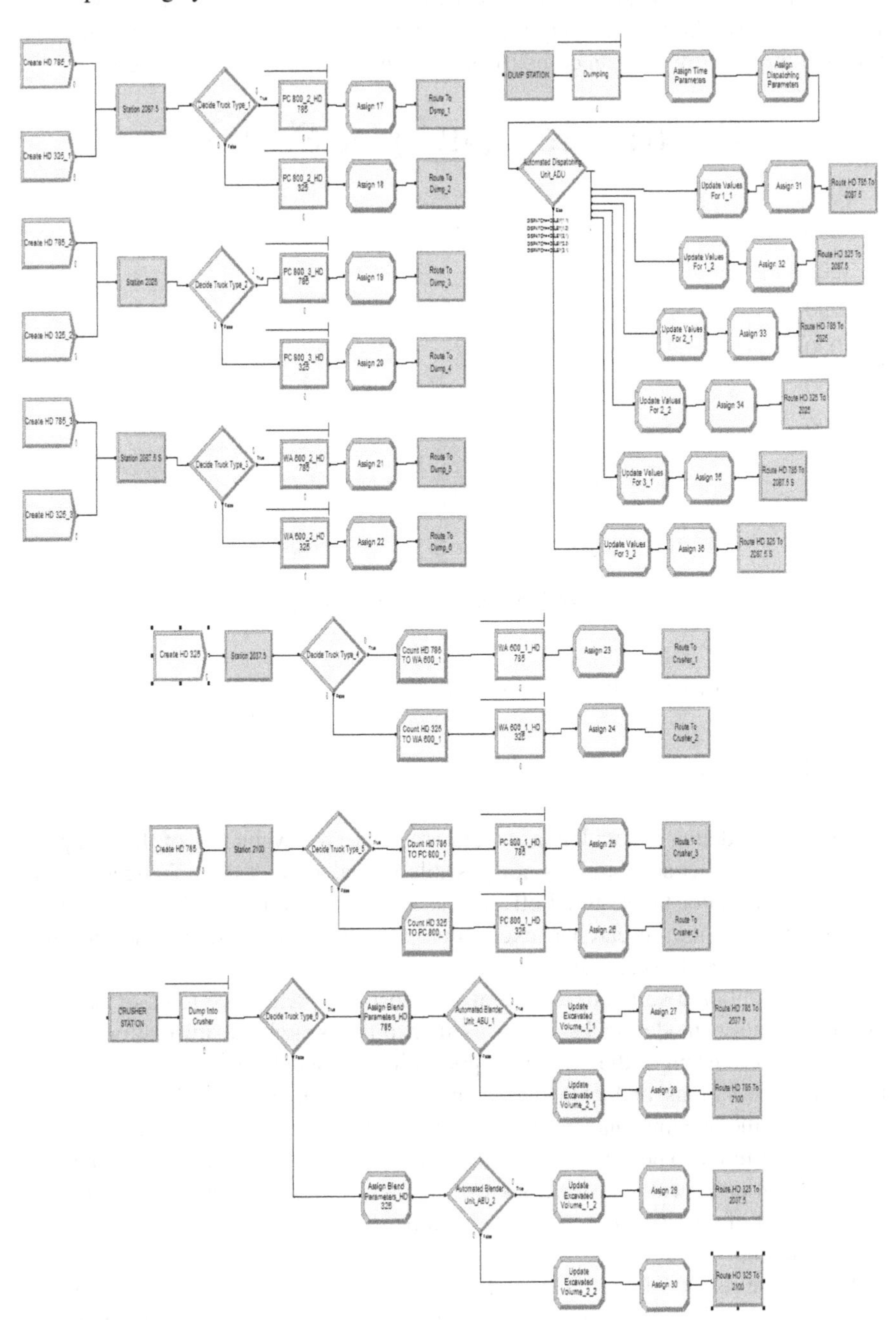

Fig. 4 Layout of Dispatching Model in Arena

The dispatching model after validation and assurance of the correctness of its framework is run for a period of one shift in order to be able to compare the outputs with fixed assignment model. Results of the simulations are shown in Figure 5. It reveals that by applying dispatching system for trucks the production rate will improve by 7.8% in comparison with the fixed assignment system.

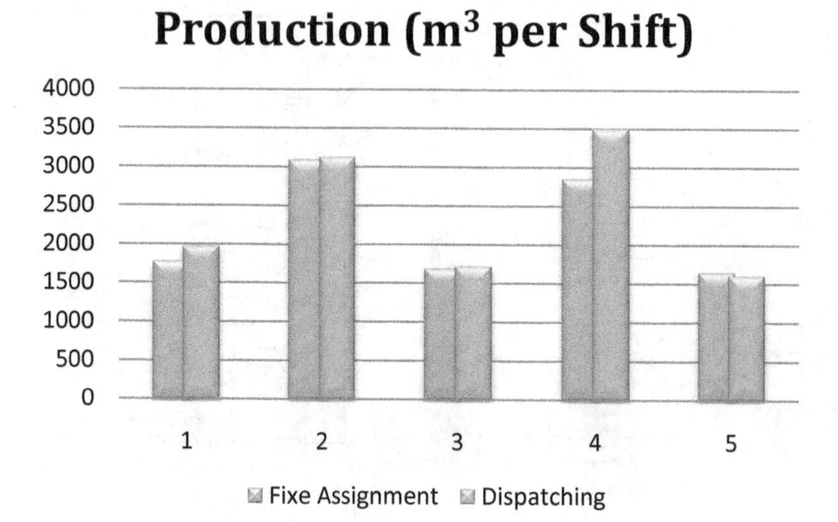

Fig. 5 Effect of Dispatching in Production Rate

4 Conclusion

Simulation modeling is a widely accepted procedure for evaluation of the complex systems. In some conditions it is not possible to make the desired changes in real system and monitor the resulted effects; it's where simulation modeling comes in handy and lets the researcher exert any possible modifications and get the feedbacks with the lowest costs.

Arena simulation software as one of the most comprehensive tools of discrete event simulation modeling has been used in this research for evaluation of the transportation system of the Sungun copper mine. A model was designed based on the site observations and collected empirical data representing the operational features of the loading and hauling units which currently function on a fixed assignments mode. Then the model is validated by comparing the model outputs with the real empirical data for truck cycle times, it was approved that the model in 95% level of confidence is valid and can be relied on for future assessments of the system under study.

Simulation results of operating haulage and match factor analysis approved that the current system is severely under-trucked. Therefore, optimization scenarios based on modification of the number of the trucks simulated on the model. Resulted outputs from these scenarios approved that by modifying the match factor of the haulage system, the production rate in the working benches of the research will increase by 40%.

Searching for productivity improvement, a dispatching model based on the strategy of dispatching one truck to several loaders and objective function of minimizing truck waiting times have been developed for the transportation equipment of Sungun copper mine. Simulation results showed a 7.8% improvement in the total production rate in comparison with the previous fixed assignment system.

References

1. Rodrigo, M., et al.: Availability-based simulation and optimization modeling framework for open-pit mine truck allocation under dynamic constraints. International Journal of Mining Science and Technology 23, 113–119 (2013)
2. Oraee, K., Asi, B.: Fuzzy model for truck allocation in surface mines, pp. 585–591. Taylor & Francis Group (2004)
3. Alarie, S., Gamache, M.: Overview of Solution Strategies Used in Truck Dispatching Systems for Open Pit Mines. International Journal of Surface Mining, Reclamation and Environment 16, 59–76 (2002)
4. Ahangaran, D.K., et al.: Real –Time Dispatching Modelling For Trucks With Different Capacities In Open Pit Mines. Arch. Min. Sci. 57(1), 14 (2012)
5. Eduardo Bonates, Y.L.: A computer simulation model to evaluate the effect of dispatching. International Journal of Surface Mining, Reclamation and Environment 2, 99–104 (1988)
6. Li, Z.: A methodology for the optimum control of shovel and truck operations in open pit mining. Mining Science and Technology 10, 337–340 (1989)
7. Soumis, F., Ethier, J., Elbrond, J.: Truck dispatching in an open pit mine. International Journal of Mining, Reclamation and Environment 3, 115–119 (1989)
8. Forsman, B., Ronnkvist, E., Vagenas, N.: Truck dispatching computer simulation in Aitik open pit mine. International Journal of Mining, Reclamation and Environment 7, 117–120 (1993)
9. Gove, D., Morgan, W.: Optimizing Truck-Loader Matching. In: SME Annual Meeting (1994)
10. Kolonja, B., Mutmansky, J.M.: Analysis of truck dispatching strategies for surface mining operations using SIMAN. In: SME Annual Meeting (1995)
11. Ataeepour, M., Baafi, E.Y.: ARENA simulation model for truck-shovel operation in dispatching and non-dispatching modes. International Journal of Mining, Reclamation and Environment 13, 125–129 (1999)
12. Basu, A.J.: Simulation of a large open pit mine operation in Australia. Mineral Resources Engineering 8, 157–164 (1999)
13. Wang, Q., et al.: Open pit mine truck real-time dispatching principle under macroscopic control. In: First International Confernce on Innovative Computing, Information and Control (2006)

14. Burt, C.N., Caccetta, L.: Match Factor for Heterogeous Truck and Loader Fleets. International Journal of Surface Mining, Reclamation and Environment 21, 262–270 (2007)
15. Burt, C.N., Caccetta, L.: Match Factor for Heterogeous Truck and Loader Fleets. International Journal of Surface Mining, Reclamation and Environment 22, 84–85 (2008)
16. Jaoua, A., Riopel, D., Gamache, M.: A framework for realistic microscopic modelling of surface mining transportation systems. International Journal of Mining, Reclamation and Environment 23, 51–75 (2009)
17. Chanda, E.K., Gardiner, S.: A comparative study of truck cycle time prediction methods in open-pit mining. Engineering, Construction and Architectural Management 17, 446–460 (2010)

Discrete-Event Simulation of Continuous Mining Systems in Multi-layer Lignite Deposits

Theodore N. Michalakopoulos[1], Christos P. Roumpos[2], Michael J. Galetakis[3], and George N. Panagiotou[1]

[1] National Technical University of Athens, Department of Mining Engineering, Laboratory of Excavation Engineering, Zographou Campus, GR-15780 Athens, Greece
[2] Public Power Corporation SA, Mines Engineering and Development Department, Lenorman 195 and Amfiaraou, GR-10442 Athens, Greece
[3] Technical University of Crete, School of Mineral Resources Engineering, GR-73100 Chania, Greece

Abstract. Continuous surface mining systems, employing bucket wheel excavators, belt conveyors, and stackers, are used in the exploitation of most of the large lignite mines in northern Greece. One particular characteristic of these mines is that the deposits consist of a series of lignite layers of thickness varying from just a few centimeters up to several meters, with interbedded layers of sandy and clayey waste material. This multi-layer geology dictates frequent changes of the material excavated on each bench, which adds to the complexity of the inherently stochastic mining system and makes material flow a critical performance parameter.

In this paper an animated discrete-event simulation of such an operating continuous mining system is presented. An extensive statistical analysis of recorded operational data for a period spanning a full calendar year, as well as of the deposit's spatial variability, provides the empirical distributions used to model input variables. The distribution of material flow at the belt conveyor hub, production estimates, equipment availability and utilization are model outputs. The simulation model is evaluated in terms of its suitability for decision making under risk during mine planning and design.

Keywords: Continuous surface mining systems, Discrete-event simulation, Multi-layer, Lignite, Arena.

1 Introduction

The continuous excavation system and the strip mining method are widespread in Europe for mining lignite deposits. Such systems consist of bucket wheel excava-

C. Niemann-Delius. (ed.), *Proc. of the 12th Int. Symp Continuous Surface Mining - Aachen 2014*,
Lecture Notes in Production Engineering, DOI: 10.1007/978-3-319-12301-1_21

tors (BWE), belt conveyors (BC), and stackers operating in series and forming a network of continuous excavated material flow.

The serial operation makes the production levels that can be achieved very sensitive to both reliability and proper matching of the various equipment units. Limited production capacity or frequent overloading and spillage, resulting in additional costs for secondary materials handling, characterize inadequately designed systems. Respectively, high owning and operating costs in relation to actual production levels characterize over-designed systems [1].

Material flow rate distribution is a critical performance parameter in designing and operating such continuous excavation systems. Given the stochastic nature of unit breakdowns, loaded material on the intersecting belts and material arrival at the transfer points, of particular interest is the amount of spilled material at transfer points and hub bottlenecks.

Because of this inherent stochastic nature, simulation is well suited for designing such systems. Walkley and Hutson [2], Watford and Greene [3], Yingling and Kumar [4], Agioutantis and Stratakis [5], Lebedev and Staples [6], and McNearny and Nie [1] have published simulation methods and applications for designing continuous excavation systems. Yingling reviews such methods in reference [7].

Sturgul et al. present an interesting aspect regarding the simulation of continuous material flow systems in reference [8]. Modeling *continuous* material flow with *discrete-event* system simulation is one of the difficult problems in mining systems simulation. They suggest a method where the continuous material flow is divided into *finite excavated material units* placed onto the conveyors.

Discrete-event system simulation is one in which the state of the model, and consequently of the system being simulated, changes only at a discrete, but possibly random, set of time points, known as event times [9].

In this paper, the exploitation of a multi-layer (zebra-like) lignite deposit with a continuous surface excavation system consisting of six BWEs of three different types, three stackers, and approximately 40 km of conveyors is simulated. Multi-layer lignite deposits are very common in Greece and lignite is still the main resource for power generation providing 48% of the electric power. These deposits consist of a series of lignite layers of thickness varying from just a few centimeters up to several meters, separated by interbedded layers of sandy and clayey waste materials. Figure 1 depicts a typical multi-layer stratigraphy of the Greek lignite deposits. Geology and the fact that lignite layers with ash content more than 30-40% are considered waste for quality and environmental reasons, make the application of selective mining with BWEs the principal method for the exploitation of Greek lignite deposits.

ARENA™, a product of Rockwell Automation, Inc. [11], has been used as the simulation software in this paper. Recorded historical data regarding the geology of the lignite deposit, the configuration of the mine, the operational state of equipment and production figures formed the basis for setting up and validating a discrete-event system simulation model. The model output is the distribution of the material flow rate at the transfer points, production estimates, equipment utilization etc.

In order to facilitate easy communication of the simulation model and the results of this investigation, animation of the continuous excavation system operation is provided as well.

Fig. 1 Typical multi-layer stratigraphy of Greek lignite deposits [10]

2 Modeling a Continuous Excavation System

2.1 System Configuration

The simulated mining system is based on the Kardia Field mine as it was during 2006. This mine is owned and operated by the Public Power Corporation of Greece SA (PPC). It is part of the Western Macedonia Lignite Centre, which is a mine complex located in northwest Greece. Lignite excavated from the mine feeds the 1250 MW Kardia thermal power station. The configuration of the simulated mine is shown in figure 2.

The simulated excavation systems are organized in six branches so that each mining bench has its own BWE, face conveyor and main conveyor. Each BWE selectively excavates either lignite or waste in terrace cuts and transfers the material to the face conveyor, which conveys it along the bench to the main conveyor. All material excavated at the six benches finally arrives at one single conveyor hub, where waste is distributed to one of the three stackers for dumping, and lignite is forwarded either to the power station or to an intermediate lignite bunker. Table 1 shows the technical specifications of the BWEs on the simulated branches.

The height of each terrace cut is governed by the thickness of the interbedded lignite and waste layers and the need for selective mining. Due to the multi-layer nature of the deposit, changes from lignite to waste and vice versa are very frequent, and dilution of lignite with waste or losses of lignite in waste occur. Each time a material change takes place, the BWE stops excavating while the face conveyor continues conveying its load, so that a void is created on the belt. This void facilitates the adjustments at the BC hub, which are required for the proper distribution of the excavated material.

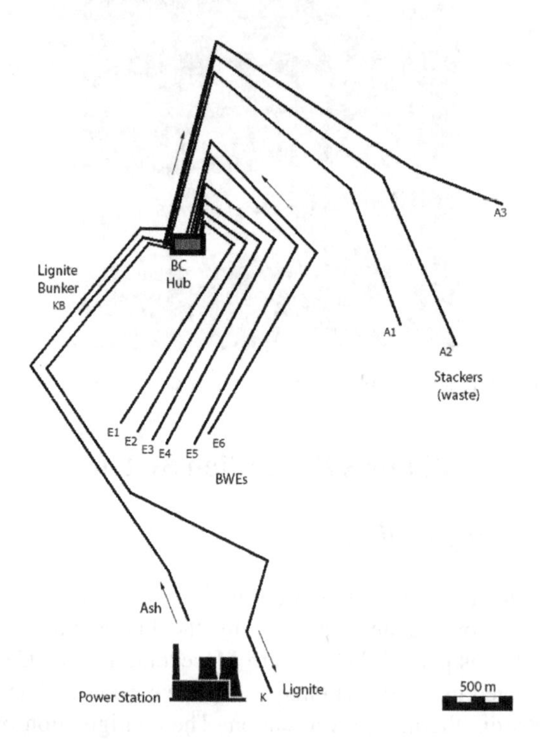

Fig. 2 Simulated mine configuration

Table 1 Technical specifications of the BWEs

Bench	Branch	BWE Model	Discharges/min	Bucket volume (m^3)
OBC	E1	Krupp 1337 SchRs 2300	44	2.3
1st LC	E2	Krupp 1264 SchRs 630	54	0.6
2nd LC	E3	Krupp 1264 SchRs 630	54	0.6
3rd LC	E4	Krupp 1264 SchRs 630	54	0.6
4th LC	E5	Krupp 1339 SchRs 2300	44	2.3
5th LC	E6	Krupp 1264 SchRs 630	54	0.6

Note: OBC - overburden cut. LC - Lignite cut (multi-layer).

The mine operates 24/7. Regular maintenance is carried out on a weekly, monthly, and annual schedule. The annual maintenance has a duration of one month and is scheduled in off-peak power demand periods. During annual maintenance the production on the given mining bench ceases. Breakdowns of the equipment units are random and independent. Due to the “in series” system configuration, equipment units feeding a failing unit are blocked and set out of operation while the failure is being repaired.

The system is of a characteristic strong stochastic nature that is based on the randomness of significant system parts and operations. The combined effect of random equipment breakdowns, frequent material type changes, and cuts of varying heights, make the prediction of the exact material flow rate at any given future time point a difficult task of great uncertainty.

The mine is closely monitored and all events are recorded in a digital log. Each entry holds information for the affected branch, the event type and description, its start and duration. Data from the events logbook have been used for input modeling. Production output has been used for model validation and verification. The data used cover one calendar year (2006).

2.2 Model Assumptions

The model represents logically and numerically the flow of excavated material from the point of excavation by a BWE to its final destination. The smallest quantity of excavated material taken into consideration is the *finite bucket load*. Each excavated bucket load is represented in the model as an individual *entity*, which after creation is routed through the conveyor network on the basis of whether it represents lignite or waste.

At simulation initialization a single production control entity is created for each BWE and forwarded to an infinite loop. There it cycles producing bucket loads by seizing and releasing a *resource* representing the bucket wheel. The process is shown graphically in figure 3. The excavation rate is controlled by the discharges per minute, the bucket volume in m^3, a dimensionless bucket fill factor, and the state of the BWE. Material type (i.e. lignite or waste), volume in m^3 for both waste and lignite, and weight in t for lignite is saved in *attributes* of each bucket load entity and used for calculating statistics regarding production output and material flow rate.

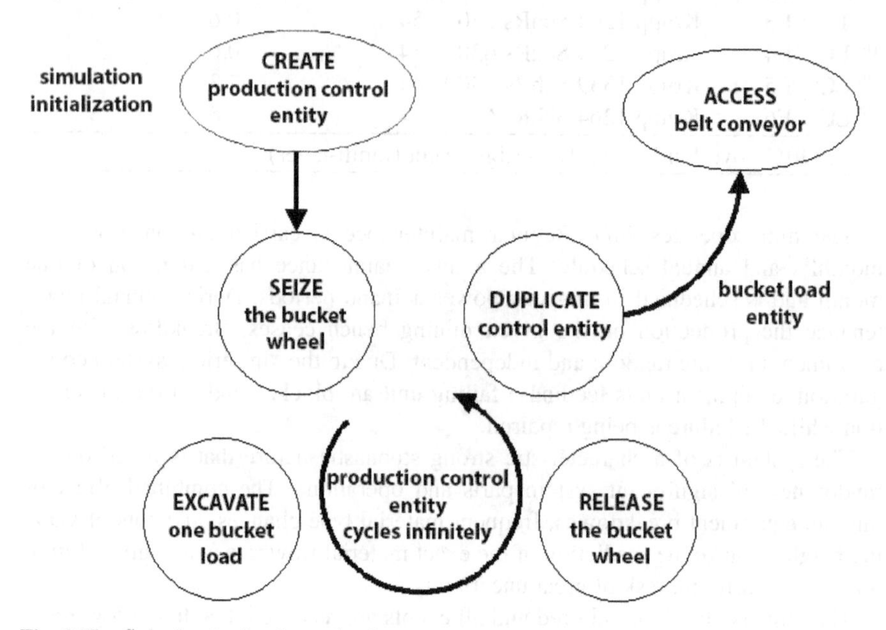

Fig. 3 The finite bucket loads production process.

In the excavating state the BWEs advance along their bench with a velocity of 0.7 m/h, thus changing location. Each bench is divided in 40 m long blocks of known waste to lignite ratio. These blocks correspond to the geological model of the mine. The location of the BWE combined with the geological model defines the excavated material's type and its final destination. The material type is sampled in random time points. If a material change is experienced, the corresponding destination of the excavated bucket loads is changed as well, according to the distribution of the final destinations for the respective branch.

On a material change the BWE stops to create a void on the belt of the conveyor, as mentioned above. Other reasons which stop a BWE are its annual maintenance, scheduled maintenance, and random unscheduled failures. Except for annual maintenance that takes place according to a calendar schedule, all other events follow their own interarrival time and duration distributions.

In general, equipment stops are random and independent in regard to their interarrival times. However, there are situations where overlapping reasons may exist for a unit to be halted. For example, if during a BWE failure its downstream conveyor fails as well, and the BWE repair finishes before the repair of the conveyor, the BWE becomes available only after the repair of the conveyor has been completed. In other words, until the conveyor is available again the BWE is *blocked.* Such situations are an inherent characteristic of the "in series" equipment operation.

In order to manage blocking and keep track of the number of causes that set a certain equipment unit unavailable, a control *storage* is defined and used for each BWE and conveyor. Each time a unit changes its state, the state of all upstream units is updated as well, by using appropriate HOLD modules either waiting for a signal or scanning for a specific condition to fulfill, which usually is a certain state of the storage.

The simulation model has been organized using submodels, in a way that provides a template for its easy adjustment to other continuous mining system configurations. A view of this submodels structure is shown in figure 4. Additionally, a different random number stream is used for each branch, so that in simulation runs with alternative configurations no interaction exists between the experimental conditions of the branches.

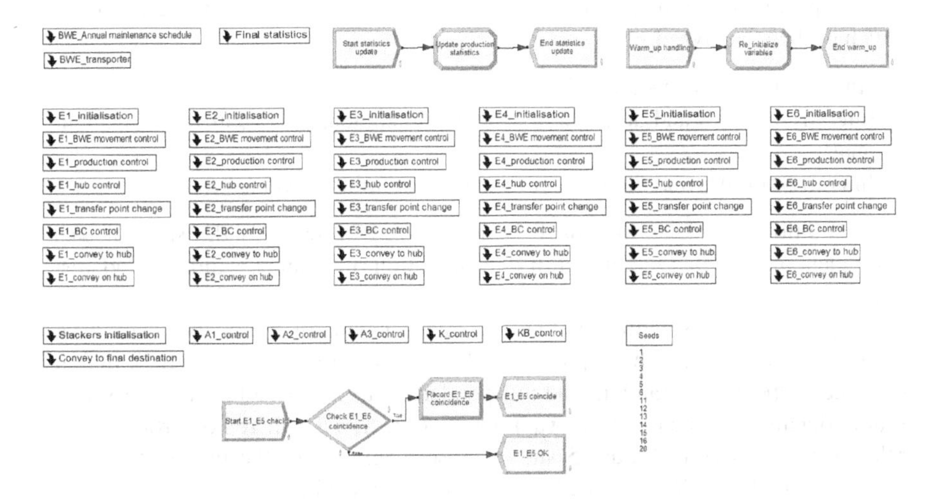

Fig. 4 The submodels structure

2.3 *Input Modeling*

The basis of input modeling has been the digital events logbook of the Kardia Field mine for the entire calendar year 2006. A total of 189364 entries have been grouped according to the affected branch, equipment unit and event type.

For the branches ingoing to the BC hub, i.e. the BWE branches E1 through E6, the following data have been analyzed:

- Interarrival times and duration for belt conveyor events (up/down).
- BWE annual maintenance start and end.

- Interarrival times and duration of BWE scheduled maintenance events (other than annual).
- Interarrival times and duration of BWE unscheduled failures events.
- Interarrival times and duration of vacations, strikes, and other special events. For all branches these events were found to be negligible and were not taken into further consideration.
- Interarrival times and duration of destination changes, i.e. the event of the BWE stopping in order to create a void on the belt conveyor that will facilitate the change of destination at the BC hub. Material changes are always associated with such events, though it is not necessary to experience a material change on every occasion. In the model these events trigger the sampling of material type. Their duration has been found to be consistently equal to four minutes.
- The distribution of destinations for waste and lignite.
- The distribution of excavated material.

For the branches outgoing from the BC hub, i.e. the stacker branches A1, A2, A3, the power station branch K, and the lignite bunker branch KB, the following data have been analyzed:

- Annual maintenance start and end.
- Interarrival times and duration of scheduled maintenance events (other than annual).
- Interarrival times and duration of unscheduled failures events.

In addition, all log entries have been analyzed in terms of the state of each equipment unit, so that the output of the simulation runs can be validated and verified.

To gain an understanding of the dynamic behavior of the analyzed time variables their values have been plotted in the sequence they were recorded. In general, they appear to be random and independent. This has been further assessed by scatter diagrams of the pairs (X_i, X_{i+1}) for $i = 1, 2, \ldots, n-1$, where no particular correlation was observed. Figure 5 shows those diagrams for the interarrival times and duration of the BWE unscheduled repairs of branch E3.

Summary statistics have been estimated as well, with emphasis on specific percentiles. In Table 2 the summary statistics for the interarrival times and duration of the BWE unscheduled repairs of branch E3 are given.

The estimated percentiles formed the basis for modeling input to the model. They were used to define *piecewise-linear empirical distributions* for all time variables. In order to determine how representative the fitted distributions are, they were plotted on the same graph with the observed data themselves. In general, a very good fit was observed, which can be attributed to the relatively large number of percentiles used to define the model. A typical example can be seen in figure 6.

In total, 62 empirical distributions have been defined for interarrival times and durations.

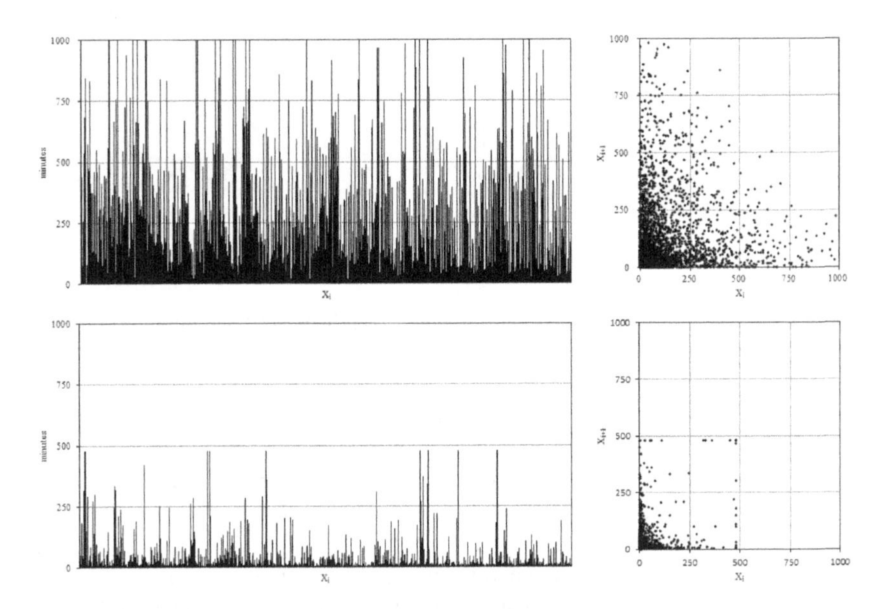

Fig. 5 Sequence plots (left) and scatter diagrams (right) for the interarrival times (top) and duration (bottom) of the BWE unscheduled repairs of branch E3. Time is measured in minutes.

Table 2. Summary statistics for the interarrival times and duration of the BWE unscheduled repairs of branch E3. Time is measured in minutes.

Statistic	Interarrival	Duration
data points	3450	3451
sum	454512	69862
average	132	20
standard deviation	294	48
coefficient of variation	2.23	2.38
min	0	2
10th percentile	6	4
25th percentile	24	6
median	58	6
75th percentile	152	14
90th percentile	332	40
95th percentile	482	70
99th percentile	860	279
max	13110	480

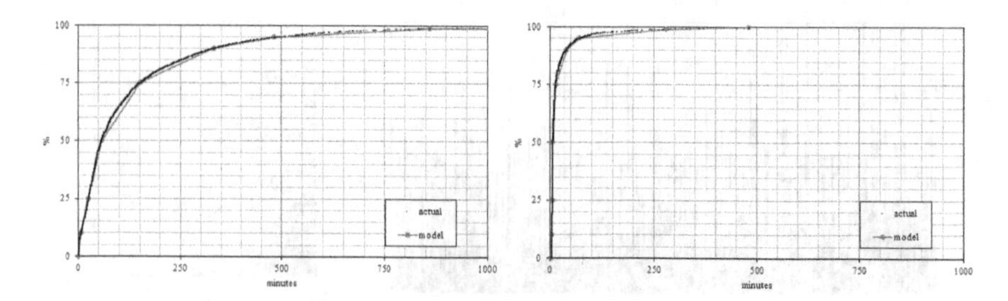

Fig. 6 Distributions of the actual and modeled interarrival times (left) and duration (right) of the BWE unscheduled repairs of branch E3. Time is measured in minutes.

As mentioned above, each mining bench has been divided in 40 m long blocks, which correspond to the blocks of the mine's geological model. For each such block the waste to lignite ratio has been used to define a discrete Bernoulli distribution that describes the probability of excavating either waste or lignite. During a simulation run the location of the BWEs is monitored and each time a destination change event takes place the associated Bernoulli distribution is sampled.

Regarding the distribution of destinations, for all ingoing branches a discrete distribution has been defined for each of the two cases of material excavated (lignite or waste). When a destination change event takes place the appropriate distribution is sampled, and the hub settings are changed accordingly.

Furthermore, the technical specifications of the BWEs and the BCs are used as model input. The most significant BWE characteristics are the bucket capacity in m^3, and the discharges per minute. The latter is the reciprocal of the duration of each bucket discharge and therefore it is used in the model to control the production rate of bucket loads. For the BCs, the length in meters and the velocity in m/s are used to setup the conveyors network model.

3 Simulation Runs and Output

Ten replications of the simulation model have been executed. Each replication started with a system warm-up period of one day and simulated a full year of operation.

The replications were run in five antithetic pairs. *Antithetic variates* are a variance reduction technique that is applicable in simulations of single systems [9]. The paired replications were partially synchronized through the dedication of specific random number streams to each branch.

During the simulation runs, data regarding the excavated material and system state were recorded. Analysis of these data provided the required production indices.

The most significant output variables of interest are material flow rate, and lignite and waste volumes excavated and conveyed on each branch. For the purpose of model validation and verification, emphasis was given on the state of the system as well.

In figure 7 the distribution of simulation time in regard to the state of the E3 BWE in each replication is shown. The average of all replications for BWE E3 is shown in figure 8. The advance of the BWEs on the benches can be seen in figure 9. The production yielded is given in Table 3, where the simulated average is compared to the actual one

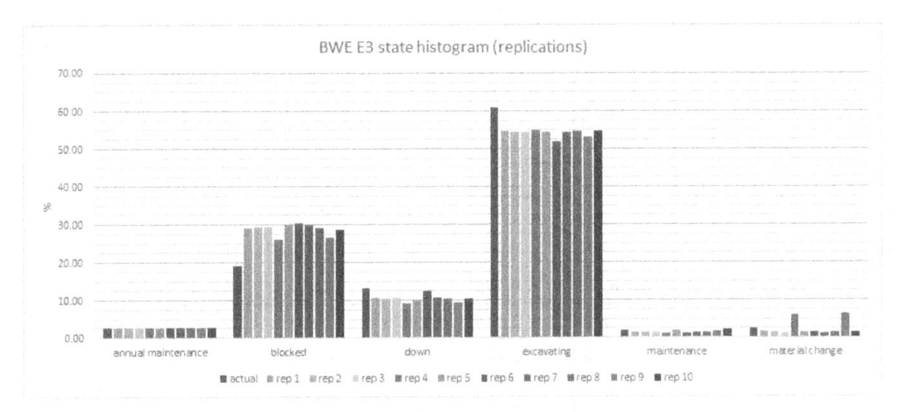

Fig. 7 BWE E3 state histogram for all replications

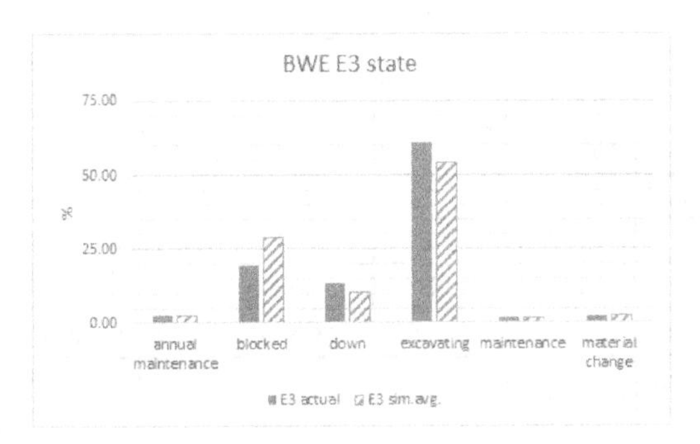

Fig. 8 BWE E3 state histogram: actual vs. simulated average

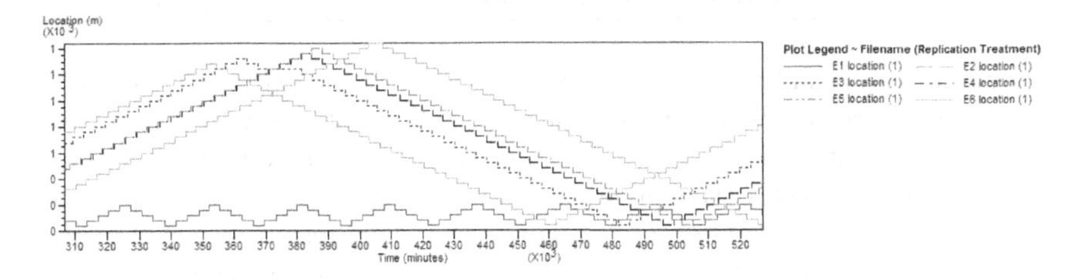

Fig. 9 BWE advance in replication #1 for the period August – December

Table 2 Ingoing material (all replications).

branch	material		actual	simulated/ actual ratio	simulated				
					min	avg	max	stdev	c.v.
E1	Lignite	m^3	69750	1.424	12852	99300	243489	91334	0.92
	Lignite	t	83700	1.424	15423	119161	292186	109601	0.92
	Waste	m^3	4869400	0.965	4379550	4698310	5047788	265528	0.06
	Sum E1	**m^3**	**4939150**	**0.975**	**4394973**	**4817470**	**5339974**	**243642**	**0.05**
E2	Lignite	m^3	1059667	0.715	697769	757166	786697	35596	0.05
	Lignite	t	1271600	0.715	837327	908601	944039	42714	0.05
	Waste	m^3	4584100	1.064	4652413	4876494	4948231	126258	0.03
	Sum E2	**m^3**	**5643767**	**1.025**	**5489740**	**5785095**	**5892270**	**135766**	**0.02**
E3	Lignite	m^3	712000	1.372	926232	976946	1082969	63120	0.06
	Lignite	t	854400	1.372	1111474	1172335	1299563	75744	0.06
	Waste	m^3	5554900	0.908	4914150	5042267	5141685	100073	0.02
	Sum E3	**m^3**	**6266900**	**0.992**	**6025624**	**6214602**	**6441249**	**74703**	**0.01**
E4	Lignite	m^3	804833	2.114	1469261	1701032	1922630	167030	0.10
	Lignite	t	965800	2.114	1763116	2041240	2307155	200435	0.10
	Waste	m^3	5536300	0.773	4082544	4279184	4424454	124392	0.03
	Sum E4	**m^3**	**6341133**	**0.997**	**5845660**	**6320424**	**6731609**	**72603**	**0.01**
E5	Lignite	m^3	1568833	2.021	2122902	3171047	3968567	668741	0.21
	Lignite	t	1882600	2.021	2547495	3805255	4762265	802478	0.21
	Waste	m^3	9227300	0.808	6913054	7451528	7948229	371462	0.05
	Sum E5	**m^3**	**10796133**	**1.043**	**9460549**	**11256782**	**12710494**	**340127**	**0.03**
E6	Lignite	m^3	2361917	0.857	1925950	2023694	2232056	124993	0.06
	Lignite	t	2834300	0.857	2311139	2428428	2678462	149992	0.06
	Waste	m^3	2620200	1.039	2517803	2721876	2997331	178153	0.07
	Sum E6	**m^3**	**4982117**	**1.034**	**4828942**	**5150304**	**5675793**	**106223**	**0.02**
TOTAL	Lignite	m^3	6577000	1.327	7531785	8729185	9802579	839160	0.10
	Lignite	t	7892400	1.327	9038159	10475019	11763082	1006977	0.10
	Waste	m^3	32392200	0.897	28062313	29069659	30311628	824453	0.03
	Total	**m^3**	**40284600**	**0.982**	**37100472**	**39544678**	**42074710**	**219828**	**0.01**

In general, the simulated average of the *combined lignite and waste volume* is in very good agreement with the actual one. The differences range from -4.3 % to 2.5 %. The same applies for the coefficient of variation, with very low values for the combined lignite and waste volume. The agreement is not good for the particular material types excavated on each bench, which calls for a refinement of the model in regard to the interaction among the processes for the selection of material type, the BWE advance, and the selection and implementation of the excavated material conveying sequences.

Figure 10 is a time plot for the transfer rate demand, i.e. the rate at which lignite and waste arrive from the excavating branches at the BC hub. The plot highlights the dynamic, stochastic characteristics of this demand, which results from the combination of the random and independent performance of the individual branches.

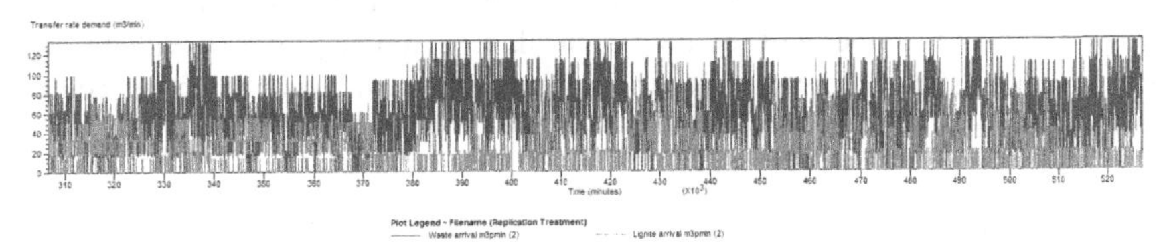

Fig. 10 Time plot of the transfer rate demand of lignite and waste in replication #2 for the period August – December

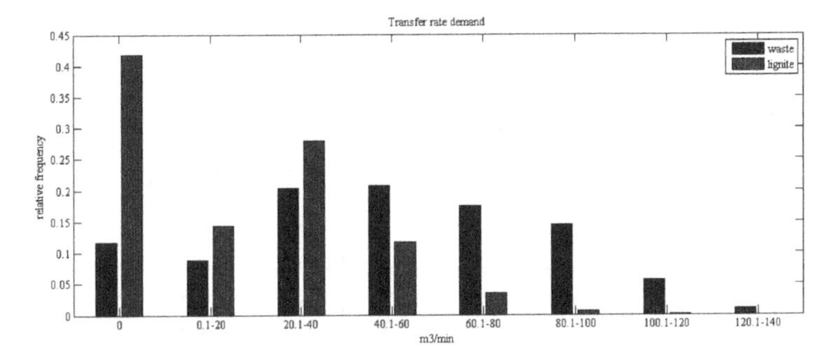

Fig. 11 Transfer rate demand histogram for replication #1. Relative frequency is percentage of time

The continuous recording of the transfer rate demand provides the data for the estimation of its distribution over time. In figure 11 a typical histogram of the demand for waste and lignite transfer rate can be seen. Such distributions, in combination with the respective distributions of the capacity of the downstream lignite and waste handling subsystems, can be used to estimate the probability of experiencing bottlenecks at the BC hub or material spillage. This information forms the basis to move from *decision-making under uncertainty* to *decision-making under risk* in the design and dimensioning of continuous mining systems, where a rational balance between owning and operating cost on the one hand, and system capacity on the other, has to be reached.

4 System Animation

Animation was used to verify the simulation model and to facilitate easy communication of the underlying assumptions and the results.

In figure 12 a typical screenshot of the system animation is shown. On the left hand side the flow of the excavated material through the conveyor network is visualized. The main information is dynamically displayed in the state and production matrix, where data can be found regarding to the current state of all units composing the system, the destination setting on the BC hub, the production achieved to-date and its distribution to the various destinations, as well as the location, direction and velocity of the BWEs. In addition, a time plot is provided to visualize the transfer rate demand for waste and lignite for the last month, as well as histograms of the to-date transfer rate of all outgoing branches.

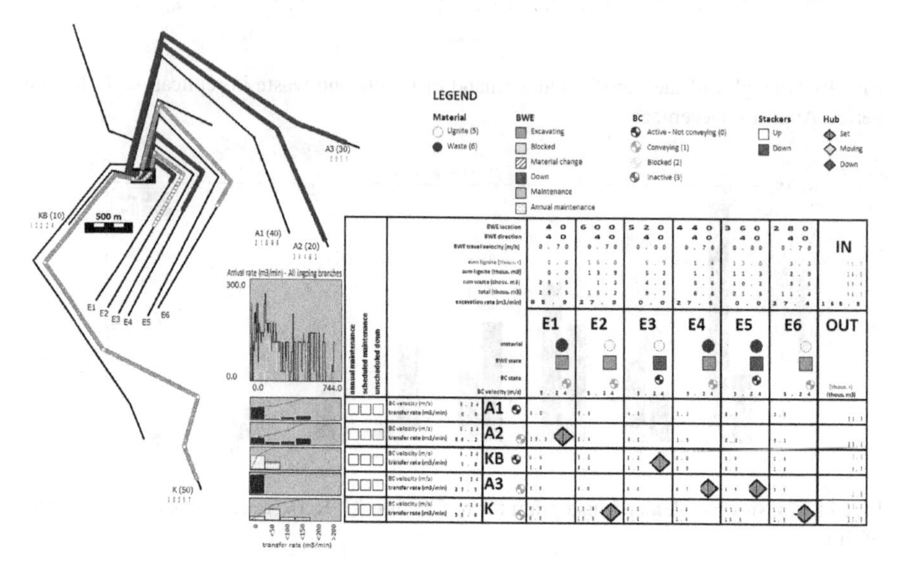

Fig. 12 Typical system animation screenshot

5 Conclusions

A discrete-event simulation model of a continuous mining system used to mine a multi-layer lignite deposit has been presented in this paper. The model simulated the configuration of PPC's Kardia Field mine. Extensive statistical analysis of historical data was the basis for input modeling and verification of the simulated output. Detailed piecewise-linear empirical distributions were used for all random time variables modeling the interarrival time and duration of the events. The geological model was used to define the excavated waste to lignite ratio at each location of the mining benches. Continuous material flow has been modeled by using a finite bucket load approach.

Five antithetic pairs of simulation replications have been executed. Preliminary output shows that the simulated average of the *combined lignite and waste volume* is in very good agreement with the actual one. The differences range from -4.3 %

to 2.5 %. Still, a refinement of the model in regard to the interaction among the processes for the selection of material type, the BWE advance, and the selection and implementation of the excavated material conveying sequences, is needed.

Animation was used to verify the simulation model and to facilitate easy communication of the underlying assumptions and the results.

The continuous recording of the transfer rate demand provides the data for the estimation of its distribution over time. This information forms the basis to move from *decision-making under uncertainty* to *decision-making under risk* in the design and dimensioning of continuous mining systems, where a rational balance between owning and operating cost on the one hand, and system capacity on the other, has to be reached.

References

1. McNearny, R.L., Nie, Z.: Simulation of a conveyor belt network at an underground coal mine. Mineral Resources Engineering 9, 343–355 (2000)
2. Walkley, R.D., Hutson, N.D.: Simulation of the operation of the coal supply system at a 2000 MW generating station. In: Third Conference on Applications of Simulation, Winter Simulation Conference, Los Angeles, pp. 213–225 (1969)
3. Watford, B.A., Greene, T.J.: Simulation software for bulk material transportation systems: a case study. In: 19th Annual Symposium on Simulation, Tampa, FL, pp. 85–103 (1986)
4. Yingling, J.C., Kumar, A.: A new framework for simulation of material flows on mine conveyor networks. Mineral Resources Engineering 4, 265–295 (1995)
5. Agioutantis, Z.G., Stratakis, A.: Simulation of a continuous surface mining system using the Micro Saint visual simulation package. In: Panagiotou, Michalakopoulos (eds.) Information Technologies in the Minerals Industry, Balkema, Rotterdam, p. 85 (1998)
6. Lebedev, A., Staples, P.: Simulation of coal mine and supply chain to a power plant. Mineral Resources Engineering 7, 189–202 (1998)
7. Yingling, Y.C.: Cycles and systems. In: Hartman, H.L. (ed.) SME Mining Engineering Handbook, 2nd edn., vol. 1, pp. 783–805. SME, Littleton (1992)
8. Sturgul, J.R., Walter, C., Masik, S., Bayrhammer, E.: The design of complex conveyor belt systems for large surface mines. In: Singhal, Singh (eds.) Mine Planning and Equipment Selection, pp. 1005–1009. Oxford & IBH, New Delhi (2001)
9. Law, A.M., Kelton, W.D.: Simulation modeling and analysis, 2nd edn. McGraw-Hill, New York (1991)
10. Public Power Corporation, Power for Greece. Public Power Corporation, Athens (2010) (in Greek)
11. Kelton, W.D., Sadowski, R.P., Sturrock, D.T.: Simulation with Arena, 4th edn. McGraw-Hill, New York (2007)

Optimizing of Sampling of Lignite Deposit Using Geostatistical Methods – A Case Study

Katarzyna Pactwa

Institute of Mining Engineering,
Wroclaw University of Technology, Poland

1 Introduction

Sampling on each stage (D, C2, C1, B) aims at documenting quality parameters of the fossil in a given location (obtaining discrete data) so that it is possible to obtain information with increased reliability on the deposit and the manner of arrangement of collected samples may cause significant consequences in the aspect of economics (Carrasco at al., 2004). Wrong net of holes is included as a geological risk (one of many existing in the chain of values' creation (Jurdziak, Wiktorowicz, 2008a). Therefore, a key issue is optimization of recognition and sampling of deposits and, as a consequence, estimating resources of deposits intended for exploitation so that the estimated values are encumbered with the smallest error possible. Simultaneously, it should be remembered that the costs resulting from sampling (recognition) are adequate to the benefits involving a possibility to steer the quality of the winning directed to the power station (Naworyta, 2008). Geostatic methods are used to solve the abovementioned problems (Clark 1987, Clark, Harper, 2001; Journel, Huijbregts, 2003; Mucha, 1994; Namysłowska-Wilczyńska, 2006). The reasonability for using them in order to reduce costs of geological costs is confirmed by Mucha and Dolik (1999) to describe, inter alia.:

- optimization of placement of samples in mine excavations – calculations allowed to establish an optimum sampling interval which turns more beneficial due to the reduction in costs connected with sampling and chemical analyses.
- selection of proper size of samples – on the basis of applied calculations with the use of geostatic methods it is stated that the influence of the size of a sample on the exactness of estimation is visible only in the case when a random component significantly overweighs the random one
- selection of optimum shape of sampling net – it was proved that the value of error in contents assessment of one of the elements depends on the shape of the net of sampling

Large meaning of the arrangement of sampling places towards the deposit batch assessed is indicated by Mucha (2001). He also pays attention to the fact that better effects in the assessment of parameters may be assured by proper calculation procedures (probability kriging, „soft" kriging) to recognize simultaneously that further extension of mathematic procedures loses its meaning

C. Niemann-Delius. (ed.), *Proc. of the 12th Int. Symp Continuous Surface Mining - Aachen 2014*,
Lecture Notes in Production Engineering, DOI: 10.1007/978-3-319-12301-1_22

in connection with little improvement in estimation exactness which can be offered (similar opinion of the topic, which is followed by approximate conclusions presented by Kokesz (2006a) and Mucha and Stala-Szlugaj, (2002)). The issue of influence of initial information on effective usage of geostatistic procedures in estimating resources is described also by Kokesz (2006b) to emphasize the necessity to recognize the variability structure of the deposit in an exact manner, to pay attention to anisotropy and heterogeneity in estimating the deposits' resources.

For the last couple of decades, the optimisation of the deposit sampling has been a problem that interested geostatisticians and mining engineers alike. Walton and Kaufman (1982)[1], describing the way of choosing a localisation for additional drilling holes, state that the variance of estimation can be limited the most by drillings in the areas of the highest uncertainty. The analysed region was divided into blocks, calculating the variance of estimation for each of these. Next, the blocks with the highest variance were chosen in order to localise additional holes within those chosen blocks and then evaluate the influence of the drilled holes on the global variance of kriging. This process was repeated until the acceptable global value was found for the variance of estimation. Gershon (1987) also dealt with the problem of localisation of drilling holes and compared various approaches to this problem with his own, based on theoretical geostatistics. He proposed an algorithm allowing the choice of such localisations for the drilling, that minimise the variance of estimation and at the same time limit the number of holes and reduce the cost of drilling. Szidarovszky (1983) presents two models allowing the choice of the optimal localisation for additional exploratory drillings. The first model minimises the variance of estimation and limits the number of additional measuring points or lowers the additional measurement costs. The second model minimises the number of additional points or lowers the costs and preserves the upper bounds accepted for the variation of estimation. These models are based on kriging theory and the solutions are based on the appropriate algorithm. The solution proposed by Szidarovszky does not take into account the three-dimensional geometry of the deposit and the drilling holes. This aspect was taken into account by Soltani and Hezarkhani (2013) when dealing with the problem of optimisation of localisation of additional exploratory drilling holes.

In Poland, a lot of attention has been paid to the problem of determining the optimal lignite sampling network by analysing the location of the existing holes. The researchers looked into the variability structure of qualitative parameters of the deposit and the question of estimation of their values using block kriging and varying the size of the blocks (Naworyta, Mazurek, 2007). The analysis was made taking into account the two following parameters: sand and sulphur content in the lignite. The conclusion is that it is sufficient to focus on the analysis of the sulphur content in lignite. In the case of sand content, there also is no justification found

[1] Walton D, Kauffman PW (1982) Some practical considerations in applying geostatistics to coal reserve estimation. SME-AIME,Dallas Information for (Soltani and Hezarkhani 2013)

for increasing the network density. This is influenced by a greater variability of the parameter, the extreme values and low autocorrelation. The level of financial expenses, a consequence of drilling additional holes, is shown to be disproportionate to the benefits of the improved estimation of the parameter. In his subsequent work Naworyta (2007) tries to answer the following question: is it possible to choose the sampling network in such a way, so that it is possible to estimate the average values of all the interesting parameters of lignite deposits with the same level of precision. Comparing the results of estimation after the two-fold and three-fold reduction of the drilling holes network, the changes in the values of the block kriging standard error were observed. It turs out that the network with the lowest number of holes is sufficient to sample the sulphur variability structure. Whereas, the estimation of sand content – even in the case of the optimal network – was always characterised by a large error.

This article compares the results reported in literature related to the problem of choosing the sampling network for Polish lignite deposits with the results of statistical and geostatistical analysis of qualitative data describing one of the parameters (the calorific value (Q_i^r)). The solutions presented in the abovementioned works serve here as guidelines. The main aims were to minimise the error by modifying the estimation procedure (by changing the block sizes in block kriging and the range of sample searching in the three-dimensional space) and to evaluate the estimation effects after the change in the density of drill holes has been made in the analysed deposit area.

2 Description of the Experimental Material

The lignite deposit near Bełchatów is located in the Łódź Upland, the Kleszczowa graben. The graben is oriented latitudinally. The deposit basis consists of Zechstein, Triassic, Jurassic and Upper Cretaceous sediments. The Zechstein deposits are made of salt domes, anhydrites and gypsum. On top of them one finds the Tertiary compositions with the following complexes: sub-coal, coal and super-coal. The layout and deformations of the Tertiary sediments, including the lignite deposit are influenced by the tectonic deformations present in the Mesozoic basis. An entire area of the deposit is covered by the Quaternary sediments. The deposit is a foredeep deposit. Its average volume is more than 50m. It is divided into three fields: Szczerców, Bełchatów and Kamieńsk (Konstantynowicz, 1994; Ciuk, 1970).

We analysed the data made available by the "Bełchatów" Lignite Mine coming from the chosen region of the Bełchatów deposit (Szczerców Field). The data consist qualitative information from 722 holes, with the total number of samples being 15406. Drilling holes orifices located in the area of $12.5 \cdot 10^6$ m^2, are on average 132m apart. The drilling holes network is not uniform; the west part contains more sample holes (average distance between orifices equals ca. 97m), whereas the middle and east part contain noticeably less sample holes (average distance between orifices equals ca. 160m), moreover several separated points are located on the boundaries. Fig. 1 presents of drillholes.

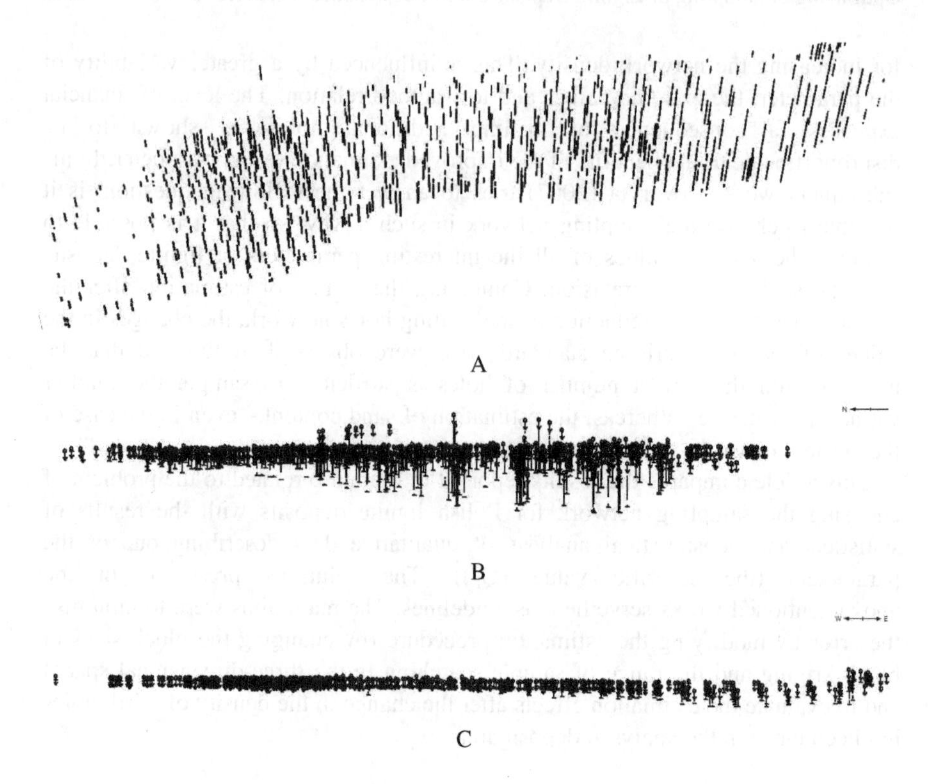

Fig. 1 Localization of drillholes A) 3D B) North-South C) East-West view

3 Research Methodology and Results

The geostatistical methods used in cases of regionalised variables (spatial, localised) that is those characterised by randomness and structure (non-random) were also utilised in our research. Semivariogram is the main function used to describe the relationship between geostatistical deposit parameters. Its classic form is presented below (Davies, 2002; Mucha, 1994; Clark 1982; Clark, Harper 2001; Webster, Oliver 2001):

$$\gamma(h) = \frac{1}{2n_h} \sum_{i=1}^{n_h} (z_{\overrightarrow{h+1}} - z_i)^2$$

where:

$\gamma(h)$ –semivariogram value

$z_i, z_{\overrightarrow{h+1}}$ –are measurements separated by a distance h

n_h –is the number of data pairs separated by a distance h

Semivariogram is normally defined in reference to geological data for 2D and 3D spaces. The calculation of semivariograms is done until such moment, when the distance h between the points is equal to half of the maximum distance between observations. The limit may also be represented by the number of data pairs which ought not to be smaller than 30-50. Empirical semivariogram is presented in the form of a scatter diagram which, in order to be useful in solving geological exercises, requires applying one of the analytic functions (Mucha 1994).

The mathematical model of this function is the basis of kriging and is used to decide the feasibility of this approach in a given case.

Kriging is the most popular geostatistical tool applied for estimating average values of parameters of resource deposit (Mucha Dolik, 1999; Nieć, Mucha 2007).

It is defined as the best linear unbiased systematicness estimator (ang. *BLUE – Best Linear Unbiased Estimator*). Within kriging an unknown value of the parametr in the deposit is settled as the weighted average from the pattern (Davies 2002; Namysłowska-Wilczyńska 2006):

$$z_k^* = \sum_{i=1}^{N} w_i \cdot z_i$$

where:

z_k^* – value of parameter estimated kriging method

z_i – value of parameter in i-th point of sampling

w_i – weighting factor of kriging

N – number of points of sampling used within interpolation

Weighting factors are selected in such method so as to ensure unbiased perception and maximum efficiency of the estimator. Fulfilling the initial postulate requires that:

$$\sum_{i=1}^{n} w_{ik} = 1$$

Whilst, second postulate signifies that the variance of difference between the real and the average value of the m parameter and its assessment z_k^* ought to be smaller. Weighting factors are assigned from the equation system of kriging which may be written in simpler terms as follows (Davis 1977):

$$\left|\frac{\bar{\gamma}(S_i, S_j)}{1}, \frac{1}{0}\right| \cdot \left|\frac{w_i}{\lambda}\right| = \left|\frac{\bar{\gamma}(S_i, A)}{1}\right|$$

where:

$\bar{\gamma}(S_i, S_j)$ –value of semivariogram (taken from the model) for the distance between sampling points S_i i S_j

$\bar{\gamma}(S_i, A)$ – value of semivariogram between sample S_i and field (point) A, for which the average value of parameter is estimated

λ – Lagrange multiplier

Semivariograms and estimation of the deposit parameter had been preceded by statistical data analysis.

3.1 Statistical Analysis

The basic statistics are presented in Table 1; these are the values of arithmetic mean, variance, median and the minimal and maximal values of the analysed parameter. The calculations were made on the basis of the sets of samples after standardisation (regularisation), that is after normalising for length. The length of the standardised samples is chosen on the basis of the mode. In the remainder of the article, the standardised samples are referred to as composites.

Table 1 Basic statistics calculated for the calorific value

Calorific value Q^{i}_{r}, kcal/kg	
Mean	1983
Variance	135993
Median	1950
Maximum	4111
Minimum	763

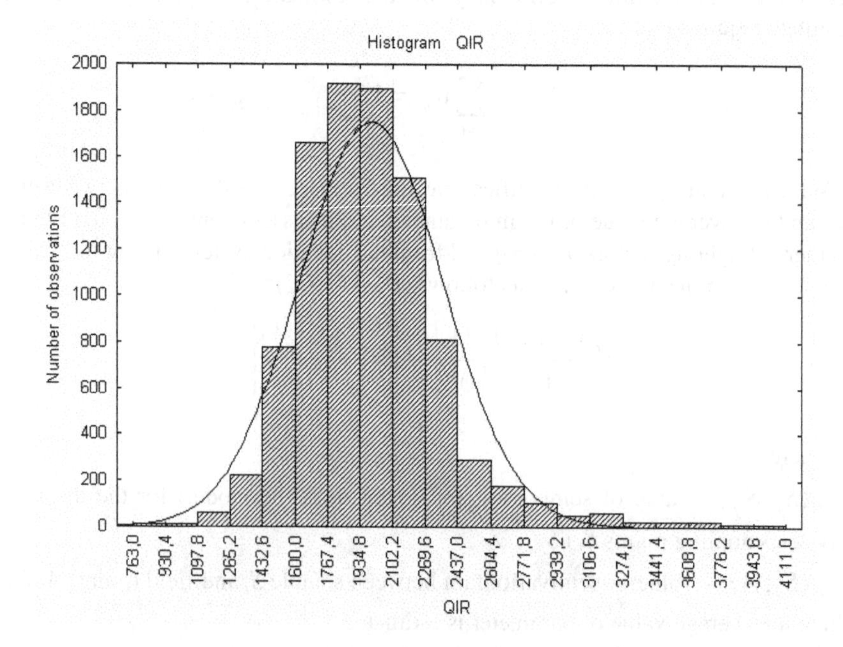

Fig. 2 Histogram for the calorific value of the set of composites

The value of the variability coefficient of the analysed feature equals 18.6%, which, according to the five-tier classification of deposits by Baryszew (Mucha 1994), classifies the analysed deposit region as belonging to Group I, the small variability deposits. This is also confirmed by the Figure 2. The distribution of the calorific value shows semblance to the normal distribution. On the basis of the characteristics of probability density distribution a decision was made that there was no necessity to transform the variable.

3.2 Geostatistical Analysis

We begin the analysis of the spatial variability of the parameter by creating a semivariogram. A theoretical model is first fitted to the empirical function calculated on the basis of all available data from the set of composites (an averaged and isotropic variogram), and then the correctness of the model is verified using the cross-validation method. Figure 3 presents the semivariogram graph and the parameters describing the function.

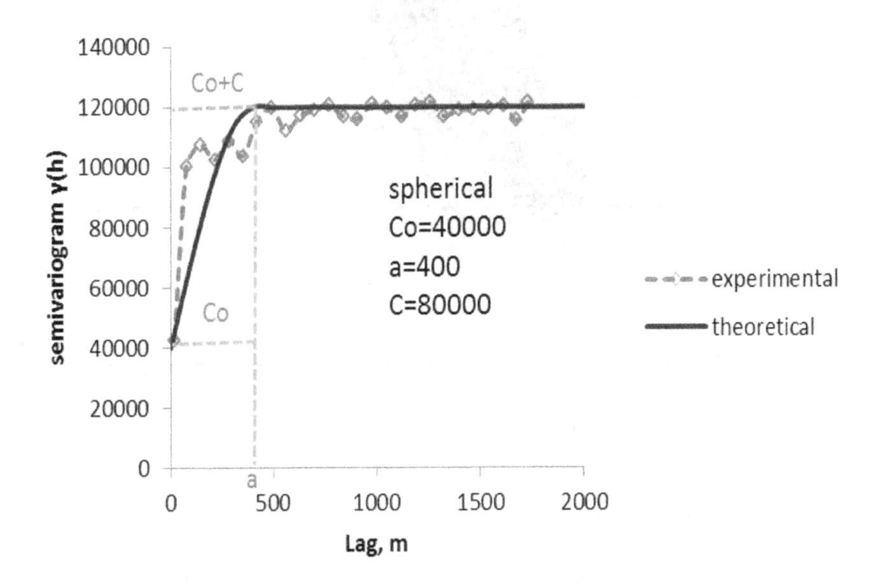

Fig. 3 Experimental semivariogram and the fitted theoretical model

The results of cross-validation in the form of diagram of dependence of real values of the parameter and the estimated values are presented on fig. 4. The linear correlation factor between the presented values equates to 0,789 and was classified as high.

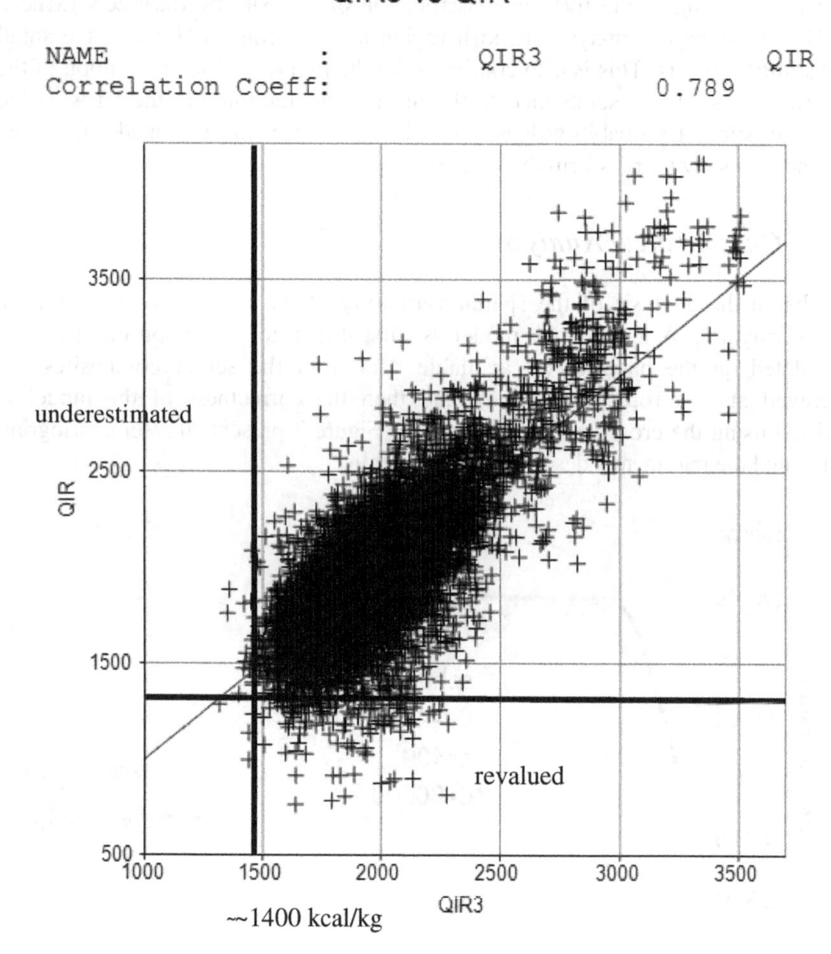

Fig. 4 The results of cross-validation

Then, the distribution models of the average estimates of calorific value. To that end, the ordinary kriging was used. The values of the parameter are estimated in blocks. Since the size of blocks influences the estimation errors (Naworyta, Mazurek 2007, Jurek et al., 2013), the variability of variance values with the increase of block size is evaluated. The analysis is conducted for the following 4 sizes of blocks (in the XY plane): 50x50, 100x100, 150x150 and 300x300 metres. Estimation is conducted for two values of minimal number of samples required in estimation: 20 and 30 and evaluating the influence of these values on the estimation precision. The results are depicted in the graphs below (Fig. 5) and Table 2.

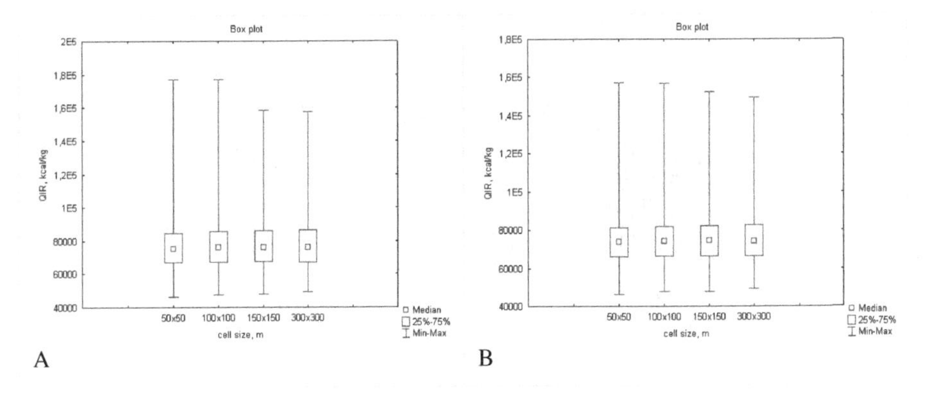

Fig. 5 Box-plot depicting the variability of estimation error statistics for 4 block sizes, with minimal number of samples required in estimation equals A) 20 B) 30

Table 2 The estimation variance for the parameter Q^i_r

Min number of samples	Block size	Mean	Median	Min	Max
20	50x50	80084,52	76165,05	46400,80	176822,5
	100x100	80191,57	76213,38	47584,08	177116,5
	150x150	80154,66	76308,01	47783,46	158378,0
	300x300	79932,22	75976,19	49317,48	157783,8
30	50x50	76334,95	73721,16	46400,80	157046,4
	100x100	76214,24	74334,13	47584,08	156892,8
	150x150	76520,34	74428,51	47783,46	152627,0
	300x300	76292,62	74252,46	49317,48	149652,8

The presented results show that the greater the block size, the smaller the range of variance values. More precise parameter estimates can be obtained by increasing the required number of samples/attempts. The best results (the lowest values of mean, median and range) were obtained for the 300x300 m blocks.

The next step is to evaluate the impact of sampling density in the area on the precision of estimates. In order to allow such an evaluation, nearly 50% of the holes are first removed (339 out of 722). The drilling hole network before and after the selection is presented on Fig. 6.

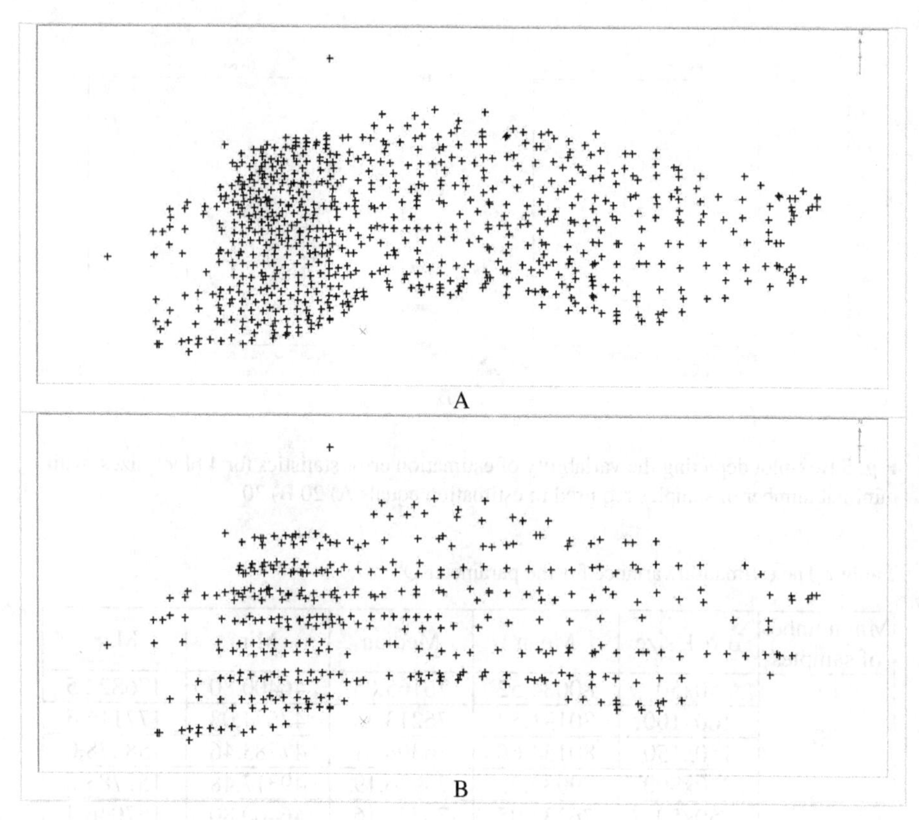

Fig. 6 Drilling hole network presented in the XY plane A) original B) after the selection

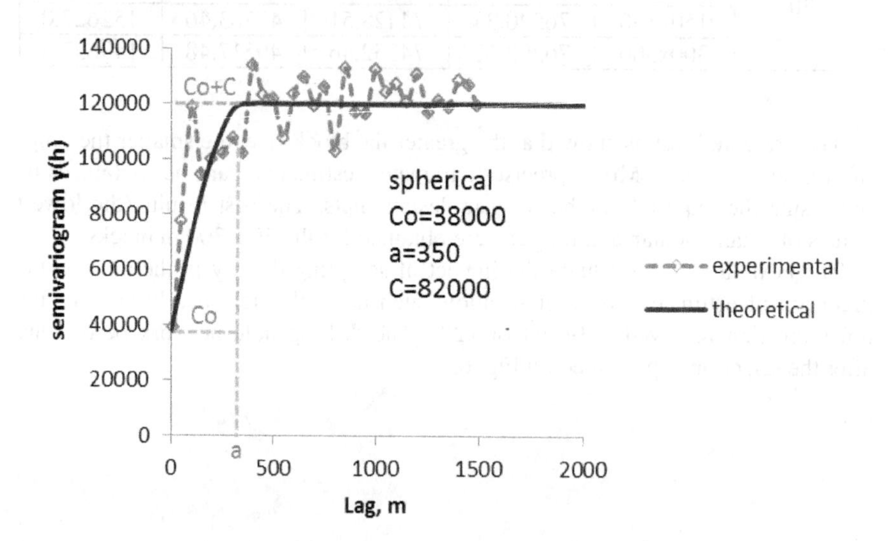

Fig. 7 Experimental semivariogram and the fitted theoretical model (after selection)

Prior to parameter value estimation in blocks, a semivariogram model is constructed, taking into account the network changes. Again, it is the Matheron spherical model that turned out to be the model describing the variability of the parameter in the analysed region. Figure 7 and 8 presents the semivariogram graph and results of cross-validation.

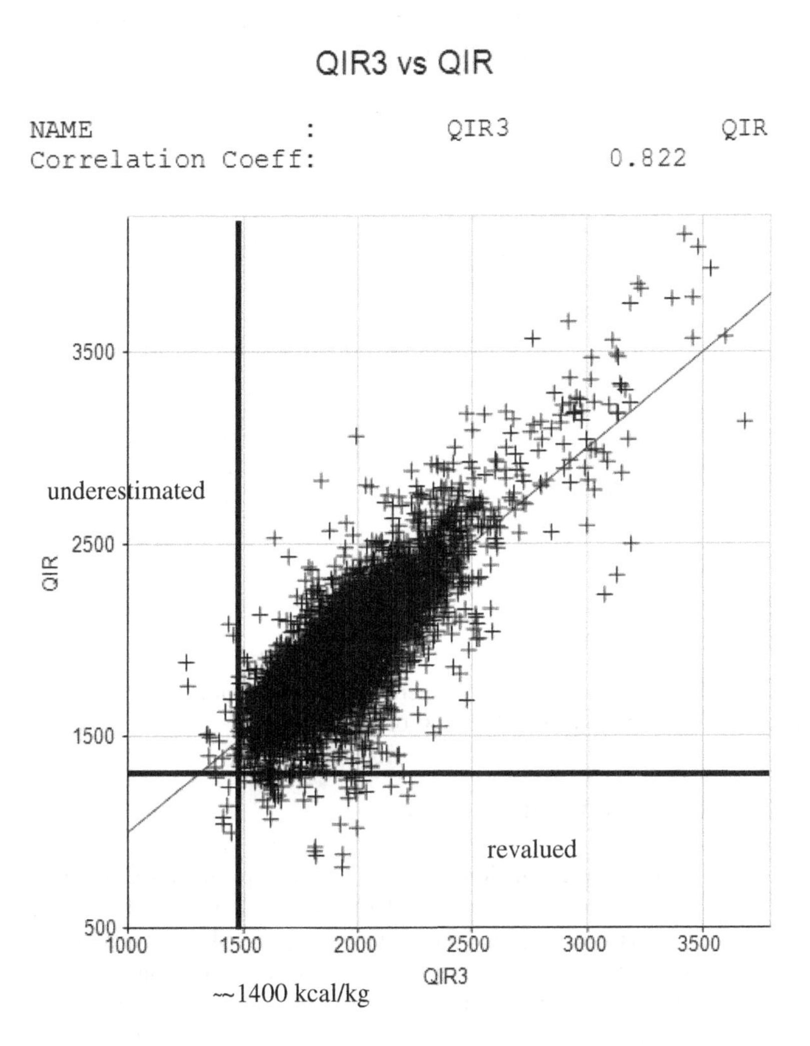

Figure 8 The results of cross-validation

The values of the parameter are estimated and the mean value of variance is then calculated, yielding: 85088 $kcal^2/kg^2$ with at least 20 samples required for the process and 78009 $kcal^2/kg^2$ with 30 samples, respectively. Comparing these results with those presented in Table 2, the differences are clearly visible. Additionally, apart from the higher variance values, it is also observed that the estimation does not take place in the entire sampled region. Figure 8 presents two models: the first one – for the chosen points – the model with the highest value of average variance, the second one – for the original drilling network – with the lowest value of average variance. The smallest differences between the estimated values are visible in the western part of the area, which is more densely sampled. Nevertheless, the second model is clearly better in terms of estimation precision.

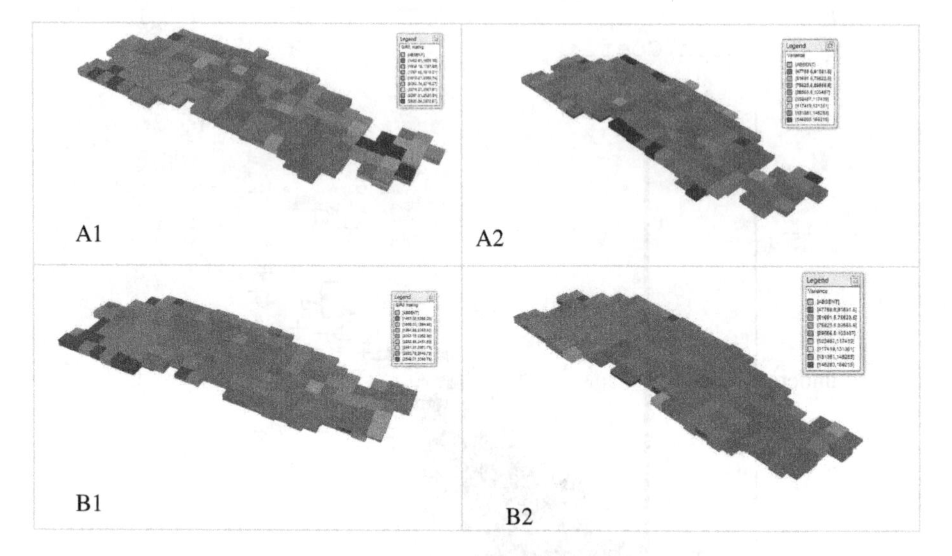

Figure 9. Block model describing average estimated parameter values (1) and variance (2) A) selected drilling holes B) original network

4 Summary and Conclusions

The problem of resource estimation of Polish lignite deposits has been present in the literature for no more than two decades. It seems however, that in the last couple of years it has considerably gained currency. Most likely, it is related to the universal access to specialised software, allowing one to analyse large sets of data. The simple analysis presented in this article can be a source of information on the level of resources, but also on the quality of the obtained estimation values. These can also serve as an indication on how to obtain information allowing one to make the decision regarding the location and density of the drilling hole network. Given the high cost of drilling, it seems worthwhile, prior to making the decision whether to perform the drilling, to first update the qualitative model to see if the decision is justified.

The conditional simulation can be used in resource estimation. It is a method recommended in risk analysis related to the problem of exceeding critical deposit values and evaluation of the forecasted prediction error (Naworyta, Benndorf, 2012) and one that uses geostatistical parameters to construct a number of equally probable qualitative models of the deposit (Jurdziak, Wiktorowicz, 2008b, Jurdziak, Kawalec 2011).

In the case of mineral resource estimation, there are hardly any universally applicable solutions but it seems justified to analyse these, which have the potential of being implemented and can potentially improve the profitability of mining.

References

1. Davies, J.C.: Statistics and data analysis in geology, p. 638. John Wiley & Sons (2002)
2. Carrasco, P., Carrasco, P., Jara, E.: The economic impact of correct sampling and analysis practices in the copper mining industry. Chemometrics and Intelligent Laboratory Systems 74, 209–213 (2004)
3. Clark, I.: Practical geostatistics. Elsevier Applied Science, p.129 (1987)
4. Clark, I., Harper, W.V.: Practical geostatistics 2000, p. 342. Ecosse North America Llc, Columbus (2001)
5. Ciuk, E.: Lignite deposits in: Geology and Polish mineral resources. Instytut Geologiczny, Biuletyn 251, 661–688 (1970)
6. Gershon, M.: Comparisons of geostatistical approaches for drillhole site selection. In: APCOM 1987. Proceedings of the Twentieth International Symposium on the Application of Computers and Mathematics in the Mineral Industries. Geostatistics, vol. 3, pp. 93–100. SAIMM, Johannesburg (1987)
7. Journel, A.G., Huijbregts, C.J.: Mining geostatistics, p. 600. Blackburn Press (2003)
8. Jurdziak, L., Wiktorowicz, J.: Identification of risk factors in a bilateral monopoly of a mine and a power plant,Prace Naukowe Instytutu Górnictwa P. Wr. Nr 123, Studia i Materiały (34), 97–111, (2008a) (in Polish)
9. Jurdziak, L., Wiktorowicz, J.: Conditional and Monte Carlo simulation - the tools for risk identification in mining projects. Economic evaluation and risk analysis of mineral projects, pp. 61–72. Taylor and Francis (2008b)
10. Jurdzik, L., Kawalec, W.: Geological risk assessment in the lignite surface mining industry with the use of the conditional simulation method. Przegląd Górniczy (12), 72–82 (2011) (in Polish)
11. Jurek, J., Mucha, J., Wasilewska-Błaszczyk, M.: Overviev of geostatistics applications for estimation of parameters of Polish lignite deposits. Zeszyty Naukowe, Instytut Gospodarki Surowcami Mineralnymi i Energią PAN (85), 143–153 (2013) (in Polish)
12. Kokesz, Z.: Dificulties and limitations in geostatistical modelling of mineral deposit variabilities and resources/reserves estimaton by kriging. Gospodarka Surowcami Mineralnymi-Mineral Resources Management 22(3), 5–20 (2006a) (in Polish)
13. Kokesz, Z.: Application of linear geostatistics to evaluation of Polish mineral deposit. Gospodarka Surowcami Mineralnymi- Mineral Resources Management 22(2) (2006b)
14. Konstatntynowicz, E.: Geology of mineral deposits. Energy minerals, Wydawnictwo Uniwersytetu Śląskiego, Katowice, p.495 (1994) (in Polish)

15. Mucha, J.: Metody geostatystyczne w dokumentowaniu złóż Geostatistical method, Kraków, pp. 1–155 (1994) (in Polish)
16. Mucha, J., Dolik, M.: Geostatystyka jako narzędzie obniżenia kosztów działalności geologicznej w kopalni. In: Materiały Szkoły Eksploatacji Podziemnej 1999, Szczyrk 22-26 lutego (1999) (in Polish)
17. Mucha, J.: Bariery i ograniczenia geostatystycznej oceny parametrów złożowych. Geologia 27(2-4), 642–658 (2001) (in Polish)
18. Mucha, J., Stala-Szlugaj, K.: Struktura zróżnicowania i dokładność szacowania zawartości Pb i Zn w złozu Cu-Ag Lubin. Gospodarka Surowcami Mineralnymi-Mineral Resources Management 18(3), 29–51 (2002) (in Polish)
19. Namysłowska-Wilczyńska, B.: Geostatistics. Theory and applications, Wydawnicza Politechniki Wrocławskiej, Wrocław, p. 356 (2006) (in Polish)
20. Naworyta, W., Mazurek, S.: Verification of correctness of borehole grid density selection for defined degree of depodit parameter recognition using geostatistical methods. Górnictwo Odkrywkowe 49(5-6), 139–145 (2007) (in Polish)
21. Naworyta, W.: Impact of borehole net density on the accuracy of deposit parameters recognition considering its variability. Górnictwo Odkrywkowe 49(7), s.46–s.51 (2007) (in Polish)
22. Naworyta, W.: Variability analysis of lignite deposit parameters for output quality control. Gospodarka Surowcami Mineralnymi- Mineral Resources Management 24(2/4), 97–110 (2008) (in Polish)
23. Naworyta, W., Benndorf, J.: Accuracy assessment of geostatistical modelling methods of mineral deposit for the purpose of their future exploitation – based on one lignite deposit. Gospodarka Surowcami Mineralnymi- Mineral Resources Management 28(1) (2012) (in Polish)
24. Soltani, S., Hezarkhani, A.: Proposed algorithm for optimization of directional additional exploratory drill holes and computer coding. Arab. J. Geosci. 6, 455–462 (2013)
25. Szidarovszky, F.: Multiobjective observation regionalized variables. International Journal of Mining Engineering 1, 331–342 (1983)
26. Webster, R., Oliver, M.: Geostatistics for environmental scientists, p. 271. John Willey & Sons (2001)

A Real-Time Regulation Model in Multi-agent Decision Support System for Open Pit Mining

Duc-Khoat Nguyen[1,*] and Xuan-Nam Bui[2]

[1] Department of Automation for Mining and Petroleum,
Hanoi University of Mining and Geology, Vietnam
nguyenduckhoat@humg.edu.vn

[2] Surface Mining Department, Hanoi University of Mining and Geology, Vietnam
buixuannam@humg.edu.vn

Abstract. This paper deals with the real-time regulation of traffic within a disrupted transportation system of a large open pit mine. We outline the necessity of a decision support system that detects, analyzes, and resolves the unpredicted disturbances. Due to the distributed aspects of transportation systems, we present a multi-agent approach for the regulation process. Moreover, this approach also includes a genetic algorithm for a programming model of the real time rescheduling problems. The genetic algorithm then treats the rescheduling problem as an optimization and provides the regulator with relevant decisions that can result in a partial reconfiguration of the transportation system.

Keywords: Real time rescheduling, Dispatch system, Genetic Algorithm.

1 Introduction

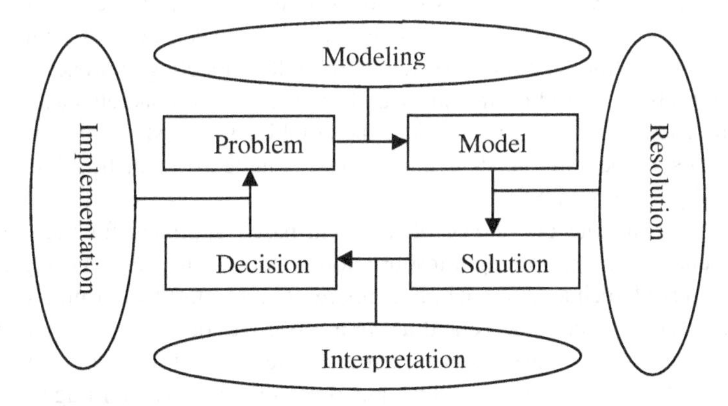

Fig. 1 DSS decision-making process

Since the early 1970s, decision support systems (DSS) technology and applications have evolved significantly. Many technological and organizational

* Corresponding author.

C. Niemann-Delius. (ed.), *Proc. of the 12th Int. Symp Continuous Surface Mining - Aachen 2014*,
Lecture Notes in Production Engineering, DOI: 10.1007/978-3-319-12301-1_23

developments have made an impact on this evolution. Initially, DSSs possessed limited database, modeling and user interface functionality, but technological innovations enabled the development of more powerful DSS functionality [1]. DSSs are, in fact, computer technology solutions that can be used to support complex decision making and problem solving. Decision making is the study of how decisions are actually taken, and how they can be better, or more successfully taken [2].

In a DSS decision-making process (Fig. 1), once the problem is recognized, it is defined in terms that facilitate the creation actors and of the concerned entities, the definition of the decision horizon, of the parameters and the constraints, and also the criteria formalization. The resolution stage imposes a choice of an exact or a heuristic algorithmic approach. A set of decision proposals is then established through the interpretation stage and presented to the concerned actors. The final implementation stage consists in applying the operational decisions, supervising their impacts, taking corrective actions, and validating the decisions. Carlsson and Turban in [3] state that modern support systems research is focused on the theory and application of intelligent systems, and soft computing in management. This includes processes of problem solving, planning, and decision making. The context for this research ranges from strategic management, business process re-engineering, effective collaboration, improved user-computer interfaces, and mobile and electronic commerce to production, marketing, and financial management. The methodologies that are used may be analysis or system-oriented, action research or case-based, or they may be experimentally or empirically focused. An emerging common denominator for many field studies, favored in DSS, is the design and use of intelligent (expert systems, multi-agent systems, etc.) and/or soft computing systems (evolutionary algorithms, fuzzy logic, etc.). Moreover, in the architectures of DSSs, the complexity reduction tools should not curb the combinatorial capabilities of the system [4]. For instance, when dealing with a DSS, such as on production scheduling systems (PSS), the modeling approaches and the resolution tools are based on the study and the analysis of concrete cases coming from real problems. Hence, we consider the combined task which includes "satisfaction needs cooperation needs computational complexity reduction," as the major capability of such a DSS.

In order to validate the choice of an agent-based approach for the real time management of open pit mining transportation systems, it is necessary to grasp the characteristics of such an approach. For this reason, we start by defining agents as conceptual entities that perceive and act in a proactive, or reactive manner within an environment where other agents exist and interact with each other based on shared knowledge of communication and representation. A multi-agent system (MAS) can then be defined as a loosely coupled network of problem solvers interacting to solve problems that are beyond their individual capabilities or knowledge. MASs constitute a powerful tool for handling open, complex, and distributed systems since they offer modularity and abstraction. They are in fact being used in an increasingly wide variety of industrial, commercial, medical, and entertainment applications, such as air traffic control. Accordingly, an agent-based approach seems the most appropriate for studying the real-time control of fleet control and dispatch within mining transport networks. In fact, transportation in open pit mining systems are complex and therefore, require a set of interacting distributed entities. In addition, the global behavior of these systems emerges from the individual behavior of the entities and also from their interactions. Hence, traffic disturbance detection, analysis, and resolution are held by a set of cooperating agents within a MADSS that assists the regulator in order to provide a better LOS for customers by reducing their waiting and route time, and especially by ensuring better conditions for connections between the network lines. That is, the MADSS has to optimize the different regulation criteria since it can have a more global view on the network than the regulator. The present MADSS for open-pit mining transport network consists of the following two modules:

The supervision module, responsible for the supervision of the network traffic and the disturbance detection. It is composed of the agents VEHICLE and STOP;

The regulation module, responsible for the disturbance analysis and the generation of the appropriate rescheduling measures. It is composed of the agents INCIDENT, ZONEPERT, and ZONEREG. The agents of the two modules communicate with each other in order to cooperate in the real-time treatment of the different incidents (Fig. 2). The regulation module has a hierarchical organization with horizontal and vertical communication.

The roles of the agents will be explained in Sections 2, then some conclusions are finally shown in Section 3.

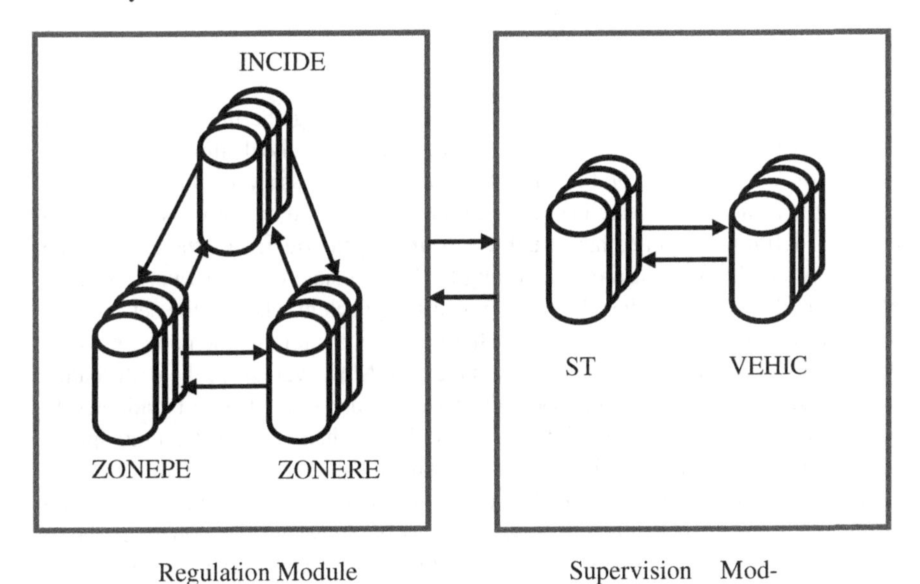

Fig. 2 MADSS modules and agents

2 Regulation Module

This module contains the agents INCIDENT, ZONEPERT, and ZONEREG. It operates in disturbed conditions. It is responsible for the identification, analysis, and resolution of the incidents. This rescheduling process needs several simulations in order to forecast the impact of the incidents and the regulation decisions on the network.

2.1 Agent Incident

An agent STOP, associated to, creates, at, an agent INCIDENT when a disturbance caused by appears. Being responsible for the considered disturbance, this agent first identifies its characteristics (disturbed haul truck, stop, delay, cause, etc.). Then, it creates an agent ZONEPERT for the analysis and the first-level regulation of the incident. According to the importance of the disturbance, INCIDENT can decide to create an agent ZONEREG that will generate several possible rescheduling solutions through a genetic approach. This agent has then a coordination role in the rescheduling process. Moreover, it must supervise the arrival of the others to avoid regulation conflicts when simultaneous incidents are related either in time (haul truck) or in space (stops). These agents propose the relevant final rescheduling measures to the regulator.

2.2 Agent Zonepert

This agent has a diagnosis role. It is responsible for the gathering and analysis of the information related to the space-time zone affected by the disturbance. In order to control the evolution of the disturbances, it is necessary to define first the space-time limits of the search space. That is, a space-time horizon has to be identified by defining the set of haul truck fleets and stops affected by the disturbance and the rescheduling measures, according to the real state of the network. The set of haul truck fleets describes the time axis of the problem whereas the set of stops describes the spatial one. If we assume that is the set of stops of the network and is that of all the haul truck fleets, then we have and. Moreover, since the disturbance evolves according to time and space, the considered horizon has to be adapted to the real changing conditions of the network. It has then to be a dynamic space-time horizon or window. The schedules that are situated beyond this horizon should be equal to the theoretical ones. Consequently, the starting and ending points of the haul truck routes in mine have to be respected. It cooperates therefore, with a society of agents BUS and STOP, called ZonePert representing the horizon. If a connection node belongs to the stops managed by ZONEPERT, then ZonePert should include agents representing the other line(s). So, in order to calculate the disturbed schedules, a simulation of the future state of the network is effected by calculating the estimated passage times of the haul truck fleets at the stops of ZonePert. The impact of the incident can be evaluated through a comparison between the theoretical and the disturbed schedules, via the regularity, punctuality, and transfer criteria. This impact is then estimated through the computation of, and. Moreover, ZONEPERT generates, at a first strategic level, some regulation decisions through a rule-based approach that describes the nature of the rescheduling measures adapted to the type of the incident.

2.3 Agent ZONEREG

This agent is created by INCIDENT. It operates by an anytime evolutionary regulation approach that takes into account the several rescheduling criteria and the solutions proposed by ZONEPERT. Through a comparison between the situations before and after regulation, ZONEREG considers the regularity, transit, and punctuality criteria that have been previously stated.

The MADSS regulation module has in fact, a hierarchical organization that can be considered as an expert community where each agent is specialized for performing a particular task and the solutions are constructed through a mutual adjustment (Fig.3).

2.4 Fleet Control and Dispatch

Over the last 10 years computerized dispatch of haul truck fleets has developed into highly sophisticated an efficient tool for control and optimization of mine performance. In its control function the existing system provide accurate, real time information on equipment performance, while a variety of customizable algorithms help with optimization of mine performance. The most popular of the dispatch systems, DISPATH [5], is used in over 100 mines worldwide. In its most recent form it monitors location of mining equipment using a combination of GPS and proximity detectors, and collects information on equipment status using a variety of electromechanical sensors as well as operator inputs. Both are then processed by a computer to develop haul truck assignment that optimize the objectives of operation as set by the mine operator. The system is also used to generate a variety of site specific reports that accurately reflect mine performance, facilitate performance analysis and simulation of operations under changed operating scenarios, and the like.

While the advanced dispatch system are able to define optimum equipment assignments that optimize a pre-defined set of production objectives, the last need to be defined by a mine planner or operator. Manual customization of dispatch application is required if the objectives change, done through customization or adjustment of dispatch proprietary algorithms, some of which are heuristic while others use sophisticated operations research methods.

2.5 Equipment Condition and Performance Monitoring

The current vision in the present work offer monitoring and reporting of three sets of parameters: those related to equipment condition, those related to its performance (load carried, speed…) and those derived by digital manipulation of the first two. The first are of help with failure prevention and diagnostics, the second allow accurate production and performance reporting, while the last are normally designed to facilitate performance analysis.

2.6 Information Module

This module contains a set of programs for the creation of databases on the pit and also for continuous updating of the geological, survey, and technological data.

The geological section of the database is intended for the storage of primary geological information and integration of the data. The primary geological information includes core-sample data on bed intersections and intersections of rock lithotypes, obtained from geological rifts and trenches. The survey section of the database includes information from performance monitoring module. The technological section of the databases is intended for the storage of information on the technical potential of the pit, the parameters of all the technological systems used in the pit or considered as options at the design stage.

2.7 Design Module

This module contains the basic programs required for determining the order and boundaries of the pits, the output level, and scheduling. The reserves and rate of removal are calculated, with differentiation over the levels, terraces, technologically uniform parcels, and beds and appropriate parameters of the production system are determined.

2.8 Planning Module

This module addresses the following geological, survey, and technological problems: taking account of the movement of resources, preparing accounting documents, analyzing tachometric survey recordings, calculating production volumes, automatic recoding of the current position of the workings on the survey documentation, calculation of the coal reserves in storage, planning the output volumes and the contours of the mine working, and so on.

2.9 *Analyze Problem Module*

This module basically consists of programs for graphical work: making charts and plans of the workings. Graphical operations may be performed both in a graph plotter and using a printer.

The six components of the hierarchical organization described above lays a firm foundation of the regulation module for open pit mining, but to make it a reality several technologies need further development. Between those are:

- The need for all the monitoring, fleet control, dispatch systems to be integrated in a one coherent whole;

- Eliminate or reduce the need for operator interference into overall system operation;

- Effective and efficient processing of voluminous data.

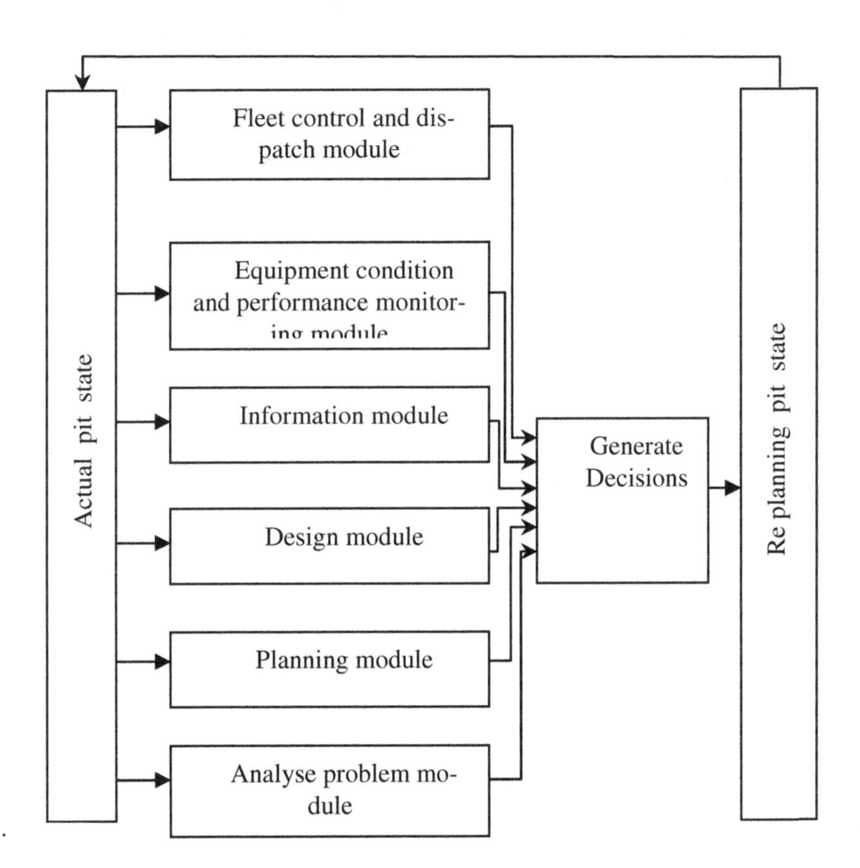

Fig. 3 Hierarchical organization of Regulation Module

3 Conclusions

In our paper, we present a regulation module in multi-agent decision support system for open pit mining that provide mine equipment operator with information required to optimize mine performance in terms of mine efficiency and effectiveness. Shows for the development of the regulation module for open pit mining, some work are needed for integrated all components in a one coherent one.

References

1. Shim, J.P., Warkentin, M., Courtney, J.F., Power, D.J., Sharda, R., Carlsson, C.: Past, present, and future of decision support technology. J. Dec. Support Syst. 33(2), 111–126 (2002)
2. Roy, B., Bouyssou, D.: Aide Multicritère à la Décision: Méthodes et Cas: ECONOMICA (1993)
3. Carlsson, C., Turban, E.: DSS: directions for the next decade. J. Dec. Support Syst. 33(2), 105–110 (2002)
4. Meystel, I.A.: The tools of intelligence: Are we smart enough to handle them? In: Proc. European Workshop on Intelligent Forecasting, Diagnosis Control, Santorini, Greece, June 24-28, pp. 2–4 (2001)
5. Modular Mining System (2000), Company information at http://www.mmsi.com

Operating a Large-Scale Opencast Mine in the Rhenish Lignite-Mining Area – Tasks and Challenges in Operating the Hambach Mine

Hans-Joachim Bertrams

RWE Power, Niederzier, Germany

1 Introduction

The Hambach opencast mine is located in the Rhenish lignite-mining area and, together with the Garzweiler and Inden opencast mines, ensures the supply of lignite for the power plants and the refining facilities operated by RWE Power AG. In 2013, electricity from Rhenish lignite accounted for some 13% of Germany's total power consumption, equivalent to an absolute electricity volume of 82TWh, for which about 86 million tonnes of coal were used. Including the deliveries to the refining facilities, total output was 98 million tonnes.

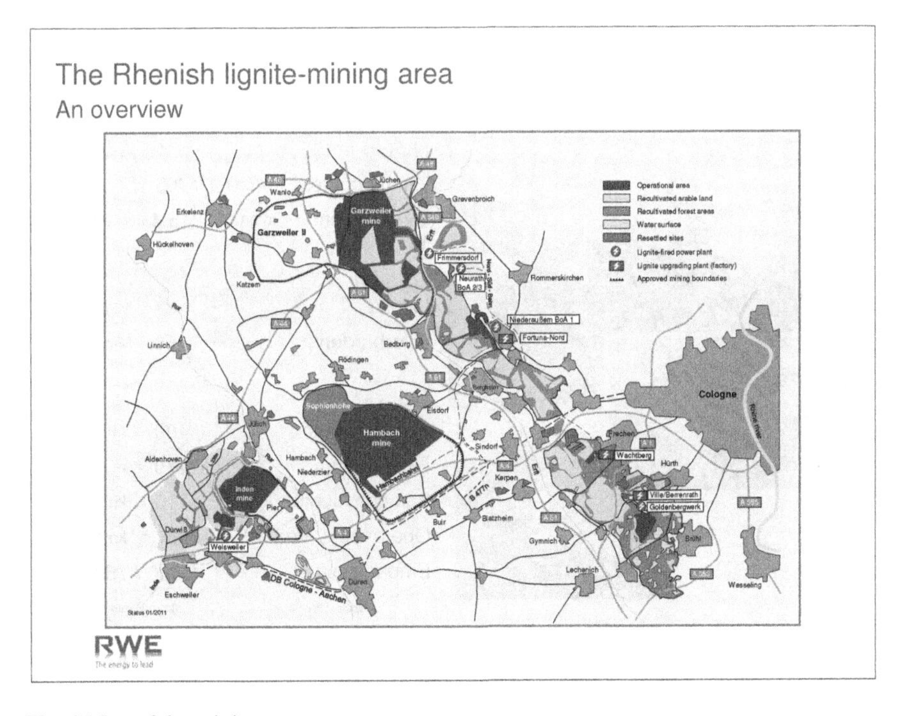

Fig. 1 Map of the mining area

C. Niemann-Delius. (ed.), *Proc. of the 12th Int. Symp Continuous Surface Mining - Aachen 2014*, Lecture Notes in Production Engineering, DOI: 10.1007/978-3-319-12301-1_24

In this period, the Hambach mine produced roughly 31 million tonnes of coal for power generation. Another 12 million tonnes were used in the refining facilities to make products like briquettes, pulverised lignite and coke. To expose the coal, it was necessary to remove 236 million m³ of overburden and to raise more than 280 million m³ of water.

Besides the economic and entrepreneurial aspects involved, the operation and steering of a large-scale opencast mine involves numerous other activities. Thus, regular day-to-day busi-ness has to take account of environmental and public acceptance issues in addition to the actual mining work, along with approval-law and, increasingly, political questions as well.

2 Geology and Mining Field

The mining field of the Hambach mine has a surface of approx. 8,500 ha and stretches between the parishes of Niederzier and Elsdorf, being bounded in the south by the Aachen-Cologne railway line. The total quantity of mineable coal located in the planning area amounted originally to some 2.5 billion tonnes, of which about 1 billion tonnes have already been mined since the start of extraction. The total overburden to be moved amounted to 15.4 billion m³, which translates into an overburden-to-coal ratio (O:C) of 6.2 : 1 (m³:t).

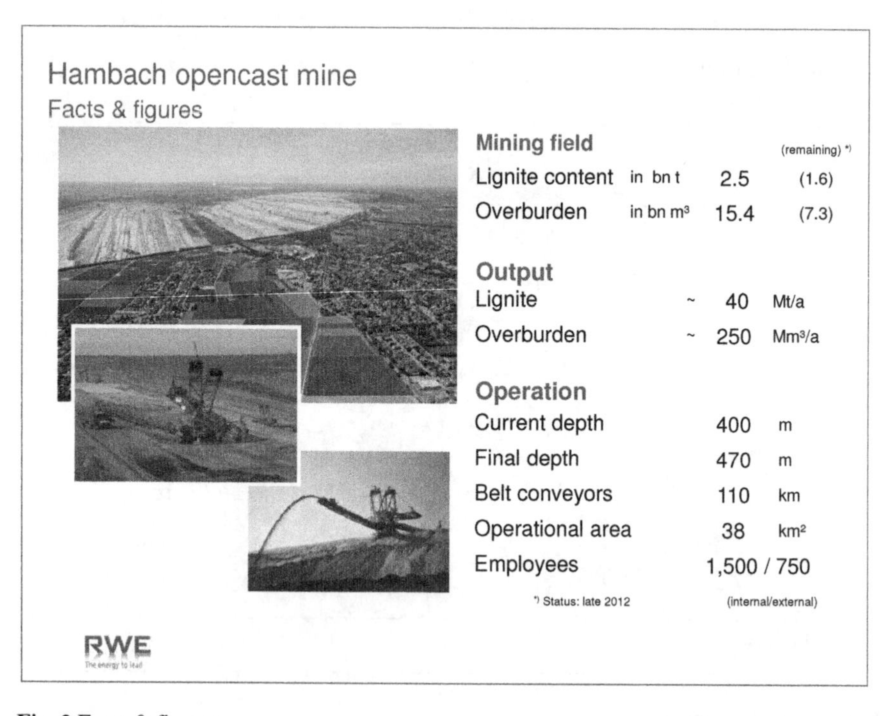

Fig. 2 Facts & figures

3 Underlying Energy-Policy and Energy-Management Conditions

3.1 Lignite and the Energy Transition

The energy sector in Germany is currently caught up in a phase of fundamental change. The growing feed-in of renewable energy is leading to long-term price erosion on the power ex-changes. The reason is the daily course of the feed-in, with a peak in the consumption-heavy midday hours, above all from photovoltaics. This ousts peak-load power plants with their priceforming effect. New hard-coal power stations or lignite-based power plants come to form the price. This effect has now made inroads into the prices of forward electricity trading as well.

On the other hand, at times of low feed-in of renewable electricity, virtually the entire peak consumption still has to be supplied by power stations firing conventional energy sources. This still means constant high requirements to be met by the availability of lignite power plants – coupled with high flexibility – in order to offset the volatility of the renewable feed-in. In this way, lignite power stations are becoming important, dependable and flexible partners for renewables in the energy transition.

3.2 Regional Economic Importance

Lignite is not only an important guarantor of secure and low-cost power generation in Germany: with its value-add, it is also a major economic factor of both regional and nationwide importance.

In 2012, RWE Power had a direct workforce of some 13,000 in lignite mining, power generation and refining. A total of around 41,000 employees depend directly or indirectly on Rhenish lignite.

In addition, the Rhenish lignite-mining area places orders, mainly for maintenance material and replacement investment, amounting to approx. €1 billion and – year in, year out – trains about 750 young people in some 20 different vocations.

4 The Opencast Mine and Its Environs

4.1 Large-Scale Mining in a Conurbation

The Hambach opencast mine is located in one of Germany's most densely populated regions, with over 50,000 people living in the mine's immediate environs, some of them in townships directly adjacent to the mining field.

People in the neighbourhood of the mine perceive its operations in the most various of ways. One of the main challenges facing management – besides the organisational and technical measures for emission control and noise abatement –

is to stay in touch with its neighbours, to have an open ear for people's concerns, and to nurture partnership-based relations.

4.2 *Resettlement and Relocation Facilities*

Lignite mining in opencast operations is necessarily associated with complete use of the land affected for extraction purposes. In addition to nature and infrastructure, there are townships, too, to be relocated. For local residents, this involves extraordinary burdens, so that the main focus of attention must lie on appropriate, socially compatible conditions.

2012 saw the start of the programme to resettle the Kerpen-Manheim parish, and resettlement of the Morschenich district belonging to the Merzenich municipality commences in 2014.

In the course of the further development of the mine in an easterly direction and to the south, there are additional infrastructural facilities to be relocated. Most of the planning and approval procedures this requires began already in the early 1990s.

One substantial measure is the partial relocation of the Company's own coal railway, the so-called Hambachbahn. This two-track, electrified railway line handles the transport of lignite from the Hambach mine to the power plants and refining facilities, and crosses approx. 9km of the pre-mining area. The new route runs east around the mining field in parallel with the German Railways Cologne-Aachen line and with the new route of the A4 federal autobahn.

The existing A4 runs for some 9km through the mining field of the Hambach opencast mine be-tween Düren and Kerpen. Around 2017, the opencast mine will reach the existing autobahn route. The switchover of the A4 to the new route will take place as early as 2014.

4.3 *Approval-Law Stipulations*

The Hambach opencast mine is operated under the supervision of the Arnsberg regional government on the basis of approved operating plans (*Betriebspläne*). The master operating plans (*Rahmenbetriebspläne*) are produced and applied for under the responsibility of a central unit for all three opencast mines in the mining area. Applications for main and special operating plans, by contrast, are usually worked out by the opencast mine concerned and/or by the technical department in charge. After approval, adherence to any ancillary provisions it contains is the responsibility of management.

In the course of the mining operating-plan procedure (*bergrechtliches Betriebsplanverfahren*), numerous subjects are dealt with – specifically those of relevance for the environment. In particular, in the wake of the amendment of Germany's Federal Nature Conservation Act in 2007, there is now an even stronger focus on species protection.

To create new habitats for endangered animal species, on the one hand, and to enable the animals to migrate to further adequate habitats – here, specifically, old forest stocks in the areas of Kerpen and Nörvenich – on the other, so-called "guide structures" have been created since 2011. For this purpose, surfaces previously used for farming are usually bought by RWE Power and ecologically upgraded or partially afforested, or old forest stocks ecologically upgraded.

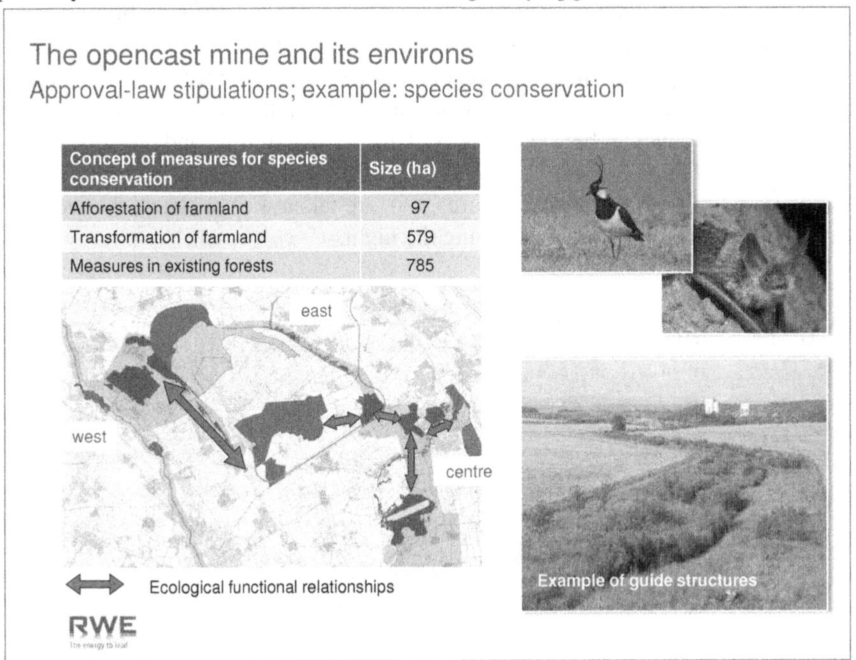

Fig. 3 Guide structures

A further approval-law stipulation for operating an opencast mine is the investigation and analysis of archaeological finds in the mining area. Here, work is in close collaboration with the Rhenish Office for the Preservation of Historical Monuments (*Rheinisches Amt für Denkmalpflege*). The archaeological work is financed by a foundation jointly set up by RWE Power and the State government. Also, the opencast mine provides hands-on support during excavations by providing staff and equipment. In this way, the conflict between land use for mining and the claims of the preservation of historical monuments can be minimised.

4.4 Neighbourhood Protection, Emission Control and Noise Abatement

A mining project in a conurbation necessarily has points of contact with the neighbourhood. The extraction and dumping sides of the Hambach mine, for instance, brush the outskirts of the town of Elsdorf some 300m away.

Operation of an opencast mine is inevitably associated with dust and noise emissions. To minimise nuisance levels in the adjacent townships, the Hambach opencast mine takes numerous planning, technical and organisational mitigation measures. The basic aim here is to operate the mine in such a way that despite the encumbrances due to the mine operation the highest possible public acceptance is achieved and the trust of the people in the region is safeguarded.

Emission control and noise abatement are thus accorded great significance.

To protect direct residents, embankments and/or walls are erected in the areas of parishes. The opencast-mine equipment and facilities are operated according to the state of the art in noise abatement. In addition, the drives of excavators, spreaders and conveyor belts are encased to contain noise, and noise-lowered conveyor-belt idlers are used. Within the scope of what is operationally possible, organisational noise-abatement measures, too, are taken. For instance, the work of auxiliary equipment is minimised during the night.

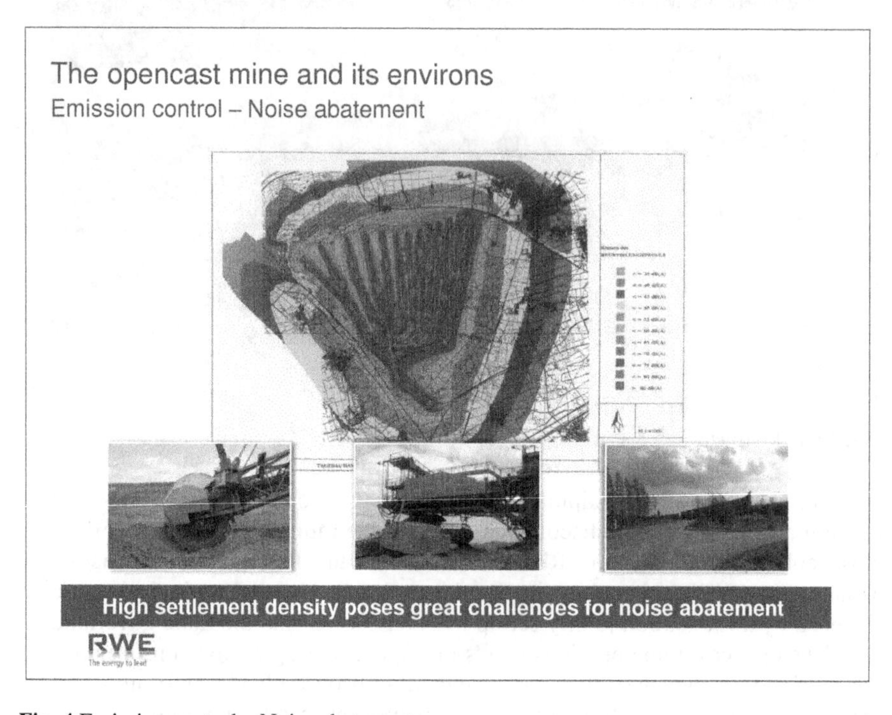

Fig. 4 Emission control – Noise abatement

The exposed opencast-mine surface at Hambach measures approx. 38km². This circumstance, if weather conditions are conducive, can give rise to coarse-dust emissions in the environs. To minimise the dust load emitted from the opencast mine, a multitude of dust-combating measures have been and are being developed, operated and refined in the course of the mine's development.

As a planning measure, the embankments created in locations close to townships are also making a contribution here. In wind-critical areas, they are equipped with sprinklers and/or sprayer masts throughout. These measures have proved their worth, especially in mitigating dust emissions from the opencast mine. The primary aim, however, is to avoid the emergence and whirlup of dust in the first place.

To this end, bark mulch is spread on longer lying surfaces and slopes in the opencast mine, or they are greened by sowing. The surfaces first covered and stabilised by plant growth in this way have lower coarse-dust emissions and are also less prone to erosion.

Fig. 5 Emission control – Dust prevention

Technical emission control measures on large-scale equipment units and opencast-mine facilities, like belt sprinklers or spraying systems at the transfer substations of the belt routes, are now standard at Hambach. Further water-related pollution-control facilities are also used, like stationary or mobile rotary sprinklers on conveyor belts and/or in open-air spaces. Roads, as well as surfaces subject to the rough travel action of auxiliary equipment, are continuously sprayed with water, especially if the weather is critical, in order to minimise dust emissions from moving vehicles. In addition, each staff member can make a contribution to successful airpollution control. The Hambach workforce are regularly trained and instructed in behaviour with emission control in mind. Moreover, personnel are regularly informed about emission control issues by suitable messages on information displays in the washhouse area.

The coarse-dust burden in the environs of the Hambach mine is continuously measured by an independent institute at a total of 29 measuring points. To assess the dust precipitation, the Technical Instructions on Air Quality (*TA Luft*) set an annual mean value. The dust-precipitation values read at the mine rim have always been well below this value. Although severe storms may in principle cause dust whirl-ups of exposed surfaces despite the measures taken, such events are extremely rare and usually do not last long.

Besides the visible coarse dust, it is the fine dust which is not visible to the human eye, which has increasingly been in the spotlight in recent years. Fine dust has numerous sources, for many of which human activity is answerable. Fine dust emerges both in combustion processes in industry, households and traffic, and, in opencast-mine operations, also in grain crushing, eg during materials handling. According to the World Health Organisation (WHO), fine dust from terrestrial sources is less toxic than fine dust from combustion or metallurgical processes.

Since the beginning of fine-dust measurements in Niederzier in 2004, the annual mean value of the fine-dust burden (40μg/m³) set by the EU has been reliably met. However, the permissible daily mean value for PM10 (50μg/m³ on more than 35 days) has been exceeded in 3 years since 2005. Hence, in a first step, an action plan for clean air was drawn up and, in 2012, a so called "clean-air plan" produced for the environs of the Hambach mine.

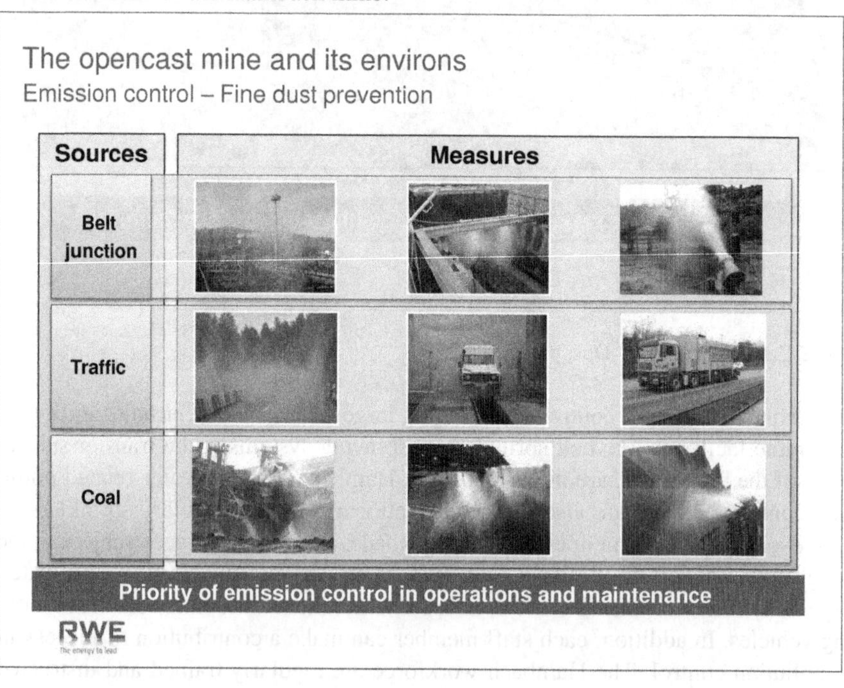

Fig. 6 Emission control – Fine dust prevention

In the course of the procedure, the share of the Hambach mine in the overall concentration was first calculated at 25%. The rest stemmed from diffuse sources and could not be assigned to specific emitters. In cooperation with independent institutes, numerous measures were worked out and taken by the Hambach mine to lower the fine-dust concentration. These included, eg, the installation of high-pressure spraying systems on the bucket wheels of the coal excavators, the operation of mobile dust-consolidation machines, the building of a tyre-washing facility as well as many more technical and organisational measures.

Since 2011, all threshold values have been met again at the Niederzier measuring point. In addition, the findings of the State environment office have confirmed the efficacy of the measures taken. For instance, the share of the Hambach mine in the region's fine-dust volume has fallen to a mere 11%. This development is also due not least to the fact that existing emission control measures have been steadily improved, with additional measures being developed that go beyond the original action plan, while all staff members have been called upon to display environmentally compatible behaviour.

5 Opencast-Mine Operation

The Hambach mine has a workforce of approx. 1,500, along with some 750 employees (full-time equivalents) of partner companies. In terms of function, Hambach is organised in a production, a mechanical-engineering and an electrical-engineering department. For administrative tasks, the so called "Bergtechnik" department reports directly to management.

5.1 Health & Safety

The safety and security of the personnel employed has been writ large in the operations of Rhenish lignite mining for decades now. In the last 15 years this point – true to the motto "No work is so important and urgent that we may jeopardise the health of our employees" – has been prioritised by management. Executives unreservedly subscribe to the principle: "We do not want accidents!" and are also prepared to act accordingly. The success of the efforts made on behalf of industrial safety depends strongly on the personal commitment of the executive staff, a presence on the spot and on the systematic action of supervisors. Industrial safety tolerates no compromises. The success of the action taken is reflected in Hambach's accident figures.

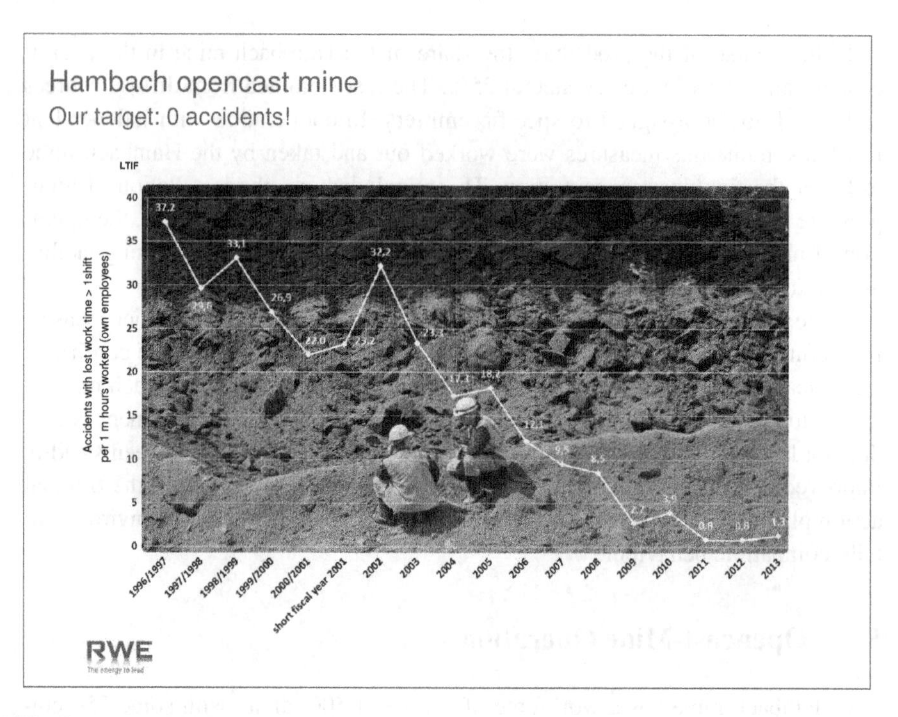

Fig. 7 Accident figures

In the nature of things, improvements become harder and harder to obtain, the closer you get to the zero line. A crucial breakthrough was achieved here when the efforts on behalf of industrial safety focused on staff behaviour, and employees became involved in solving safety problems in a spirit of eye-level partnership. This approach involves "behaviour-oriented visits" by senior executives and "brief safety talks" with supervisors on the spot.

5.2 *Continuous Improvement*

The feed-in of apparently low-cost electricity from wind and solar is leading to a tougher com-petitive situation on the energy market. The energy source 'lignite' has to face more and more competition. Here – for market-related reasons – mining only has levers on the cost side if it wants to improve its result. The improvement processes required to maintain competitiveness are compelling a constant analysis of the processes and the organisation of the Company's fields of action. For this purpose, the Hambach mine has established a continuous improvement process (CIP) going by the name "*immer:besser*" ("getting better and better") that is firmly integrated into its structures. Principals, as "users of ideas", are continually on the lookout for improvement potentials. In their efforts, they can harness the experience and knowledge of their staff. Administrative support comes from so called "idea coaches". At the start of 2013, RWE Power also launched a programme for the sustainable underpinning of competitiveness under the heading "NEO".

5.3 Energy and Environmental Management

Prudent and effective use of the available resources is a fundamental precondition for economic action and for achieving a long-term secure position in a competitive situation. The most recent developments on Germany's energy market, too, are reason enough for thematising the effective use of resources in lignite mining and in electricity generation from lignite.

Extracting lignite in opencast operations requires a not inconsiderable energy input, chiefly electric energy. In 2013, the energy input in the three opencast pits in the Rhenish mining area amounted to just under 3.1TWh (incl the raising of water throughout the mining area), equivalent to approx. 4% of the electricity produced there. Effective monitoring and pinpointed controlling of the specific energy consumed by the mine, supplemented to include a focus on energy efficiency in procurement and operating processes, are the aims of energy management in lignite mining. As a consequence of the - at present - unstable electricity market prices maintenance activities are rescheduled to periods with high prices. Operational costs can be reduced by running mining processes at full capacity during periods characterised by low electricity prices.

In 2013, Hambach was the first location in the mining area to introduce an energy-management system certified by an independent body. One module of energy management in operations is a pinpointed focus on energetic subjects in the CIP. Employees and executives are informed and sensitised on energy-efficiency topics both in regular talks and via the location's internal communications.

In view of the substantive links between energy and the environment, the Hambach mine's environmental management was certified at the same time. This mainly involved the relevant processes resulting from approval procedures, official stipulations and action in the public relations area; these have, in principle, been practised in mining for decades without leading to complaints and have been documented and, in this way, made transparent. Certification of the environmental management has confirmed the standard of work and activity at Hambach.

5.4 Partner-Company Management

Operations at Hambach are supported by numerous partner companies. If the goals of industrial safety and health are to be achieved, a number of measures have to be taken together with the partner companies. To this end, Hambach engages in partner-company management.

This starts with the selection of firms and the award of contracts. The rule is that the opencast mines now only employ partner companies that are certified in industrial safety. Before beginning any work at Hambach, all partner-firm employees receive instructions on the location's special features and on fundamental safety issues. Only after a successful check of knowledge levels and upon submission of a safety pass may partner firms' personnel commence work in the opencast mine. Moreover, risk assessments must be drawn up and reviewed with the client. In the execution of their work, the partner firms are monitored by supervisors and by the location's Industrial Safety unit. This is done by random sampling in workplace inspections, by regular talks with management and, in the wake of unsafe actions or of accidents, by appropriate consequences in the form of an escalation programme.

The overall picture of a partner firm in industrial-safety terms is documented within the scope of a supplier and/or industrial-safety assessment.

5.5 Operations Management Systems

One key precondition for an economic design of processes is the systematic improvement of the efficiency of plants and personnel using modern operations control instruments. RWE Power has been engaged in their development and rollout ever since the end of the 1990s. Besides opencast-mine and extraction planning, the chief focus here is on the main processes, like excavation, transport and dumping and/or bunker management and coal logistics.

Starting out from the long- or medium-term opencast-mine planning based on a digital deposit model, complex applications for pinpointed short-term and equipment planning have been created in recent years. The underlying aim has been to optimise the mine's material scheduling, which is subject to numerous dependencies and restrictions. The core task here was the combination and continuous calibration of the deposit model using data from the actual situation at the working face as well as an orderly dump structure based on dumping schemes. Building up on this, it should be possible to simulate potential scenarios and to outline optimisation processes.

The basis of such a system is real-time data from the mining equipment units on their position and movements. The development of the satellite-assisted excavator and spreader operation control (SABAS/SATAS) in 2002 created the basis for a downstream production- and material-management system. To this very day, the applications – some of which were developed and successfully deployed by the Hambach mine – have been and are still being continually optimised. Besides these advances, systems for the control of auxiliary processes in the opencast mine were also developed and optimised. These mainly include applications for scheduling auxiliary equipment or the monitoring of water-management systems.

5.6 Scheduling and Mass Distribution

The planning of mining operations in an opencast mine covers various periods of time, with a basic distinction being made between long-, medium- and short-term planning horizons. Strategic long-term planning affects an opencast mine across its entire life cycle, all the way to the end of mining supervision after the land surface has been reinstated. Medium-term planning covers each next five-year period.

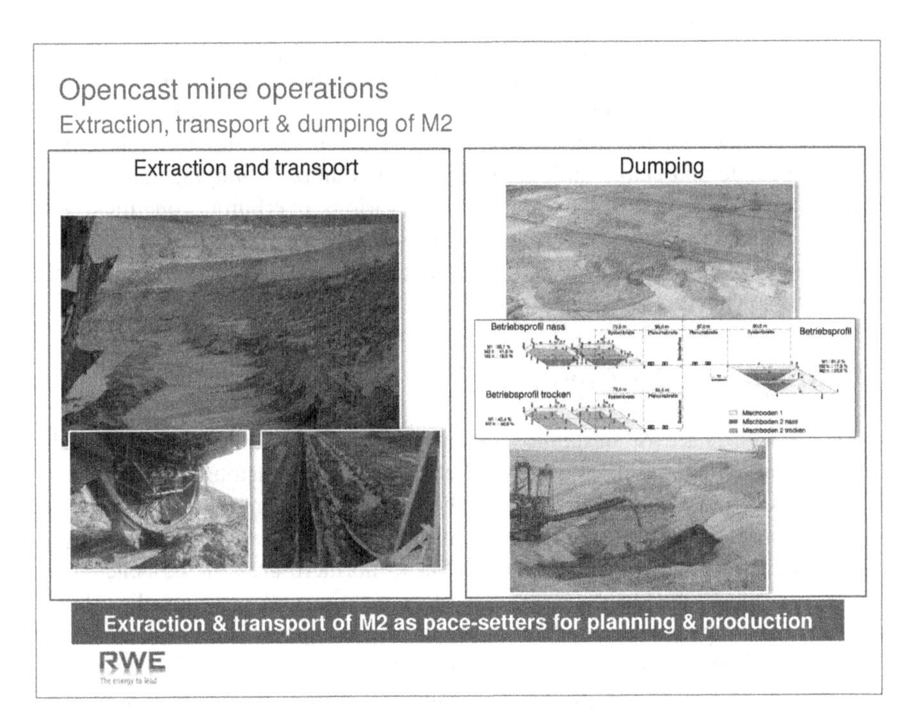

Fig. 8 M2 extraction, transport and dumping

An opencast mine's planning department schedules one-year periods all the way to seam-focused equipment use and standstill planning. The various planning horizons are interlinked in the form of control loops. Planning the coming shifts, days and weeks is a daily iterative task at an opencast mine. Besides mine operational planning and maintenance aspects, the factors impacting a mine's operation and, hence, the variables in such rolling mining equipment planning, specifically include further mining-related subjects like geology and hydrology, geomechanics and equipment technology.

For periods of up to one year, a computer-aided extraction plan is drawn up to take account of all important factors that affect an opencast mine's operations. This tool can map, block by block, the operation of the mining equipment units with position-dependent output data, maintenance and preparatory-work dates. In this way, it is possible to steer the excavator's digging operations, especially as regards the scheduling of difficult-to-dump overburden masses, and to precisely forecast a mine's output in the period under review.

The operation of the mining equipment and the daily steering of operations at the Hambach mine are marked in particular by its geology and hydrological conditions. Besides sands and gravels, the overburden which sits atop the coal mainly consists of a not inconsiderable share of clays and silts. The safe discharge of just this overburden from the extraction side on an inside dump with 7 spreader benches and a height of just under 600m are the constant features of the deployment of the mining equipment. The daily scheduling of the overburden removal is mainly determined by the limited availability of dump space for unstable mixed soil. For the safe and stable build-up of the inside dump, a dumping scheme was developed for the controlled dumping of the various overburden qualities. The dumping principle is based on a system of pinpointedly created dams of stable material (so called "M1") that can be used to form polders in which unstable mixed soil (so called "M2") can be discharged.

5.7 *Mining of Clay Ironstone*

Besides a high share of unstable interstratifications containing clay-silty sediments, it is above all highly consolidated clays and clay ironstone deposits on the 5th and 6th bench that continue to impede the whole extraction process (incl. transport and dumping). Most of the irregular, in places also thickly bedded, clay ironstone deposits are encountered in the central to northern area of the bench concerned. The clay ironstones mainly consist of siderite with varying clay shares and, together with the surrounding clays, form a compact, very hard structure. Precise spatial exploration of the clay ironstone deposits is possible only with qualifications owing to their irregular distribution. It must be assumed, however, that – in the medium to long term – these compacted layers must still be expected on the 5th and 6th benches.

The compacted clay ironstone strata cause serious problems during excavation. For one thing, only low extraction volumes can be achieved where the bucket-wheel technology is used. For another, the material leads to higher wear and tear, specifically in the case of the bucket wheels and conveyor belts and of the transfer substations.

Ever since clay ironstones were encountered for the first time, numerous processes and measures have been trialled at Hambach to ensure output-optimised and wear-minimising extraction. These included blasting trials or advance loosening of material by auxiliary equipment. These approaches were not pursued any further for technical and cost reasons.

Besides the structural reinforcement of transfer substations and the steel construction of the conveyor belts and the large-scale units, it was the development of buckets specially optimised for clay ironstone that brought a sustainable improvement. Ever since their first deployment in 2009, these specialised buckets have been continuously further developed. In fact, deployment of bucket-wheel excavators in clay ironstone is receiving in-depth engineering support beyond the usual measure.

Opencast mine operations
Bucket-wheel technology in clay ironstone
▪ Performance restrictions due to difficult-to-mine material
▪ Wear-prone clay-ironstone extraction requires in-depth maintenance
▪ Steady optimization of the equipment deployed (clay-ironstone buckets, etc)

Fig. 9 Bucket-wheel technology in clay ironstone

5.8 Coal Qualities in the Hambach Opencast Mine

Another pace-setter in daily operations – specifically on the lower extraction benches – is the range of coal varieties occurring there or about to be accessed by the excavator, in conjunction with the quantities and qualities demanded by the coal logistics.

The coal in the Hambach opencast mine can mostly be extracted in a thickness of some 70m and a bench length of about 3.5km. The quality of the coal is marked by high calorific values, in part by a high sodium and potassium content and fluctuating iron values. A high calorific value and the chemical composition of Hambach's coal mean that, in the older boilers of lignite-fired power plants, there is a risk of ash caking on the boiler tubes. For this reason, a system for feeding the power stations with coal tailored to the boiler was developed, which means that, in daily scheduling, a distinction must be made between 7 different boiler coal varieties. For some years now, the Hambach mine has been the sole supplier of the refining facilities, and has made available 2 further coal varieties in quantity and quality for the dry-lignite market.

Opencast mine operations

Coal qualities

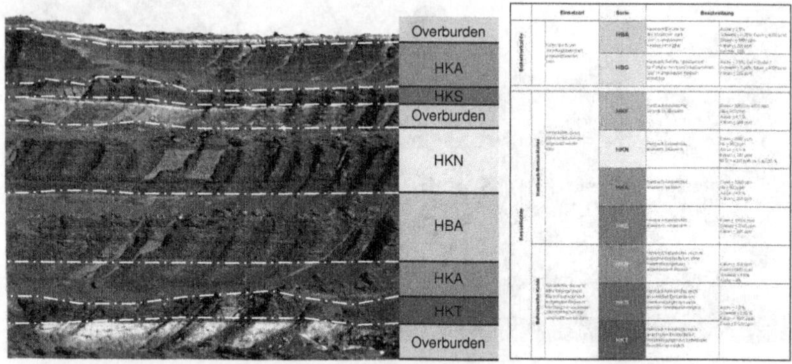

- Timely provision of qualities and quantities demanded
- Selective extraction in large-scale equipment operations
- Extraction of coal for upgrading solely from Hambach opencast mine

RWE

The energy to lead

Fig. 10 Coal qualities at Hambach mine

5.9 Opencast-Mine Drainage

Digging deep-lying seams in an opencast mine is not possible without comprehensive drainage measures. The deep wells required for this are operated on the extraction benches, in the premining areas and on the opencast mine's rim areas.

The aim of effective dewatering operations is the maximum utilisation of the well infrastructure. This aim conflicts with the obstacle-free operation of the mining equipment. So, what matters is having in-pit wells operated until just before the arrival of mining operations and then, as soon as possible, putting the ploughed-under wells back into operation. Timely decommissioning and recommissioning requires close daily review and coordination with the Production department and the specialist units involved.

For the dewatering measures, Hambach currently uses some 200 in-pit wells. Roughly the same number is operated in the pre-mining area. This raises about 280 million m³ of water annually. Besides the raising and discharging of the well water, all benches maintain an extensive infrastructure for surface dewatering, consisting of trench systems for rainwater and residual rock waters, temporary dewatering sumps and discharge systems. In the stationary area are water-retention basins and sedimentation tanks as collection systems from which the free waters are pumped via mine-water purification plants to the higher-level piping system. Both the well network and its discharge pipes and the surface-dewatering system are each monitored and controlled by computer.

5.10 Morschenich Colliery

With the aim of achieving a high degree of self-sufficiency in raw materials for Germany in the 1940s, various projects were launched at the time to trial underground lignite mining. One of these projects was the Morschenich colliery (underground lignite mine "Union 103") in the municipality of the same name. As the opencast mine progresses, the underground structure has to make way for mining operations.

The mining activities at the Morschenich colliery were confined to sinking two shafts to a depth of some 330m and a roadway system in the lignite seam along a total distance of some 12km. Owing to recurrent water penetration, the underground activities were discontinued as early as 1950 without any commercial extraction having taken place. In 1960, the underground structure was flooded after protective measures had been taken to enable later re-commissioning.

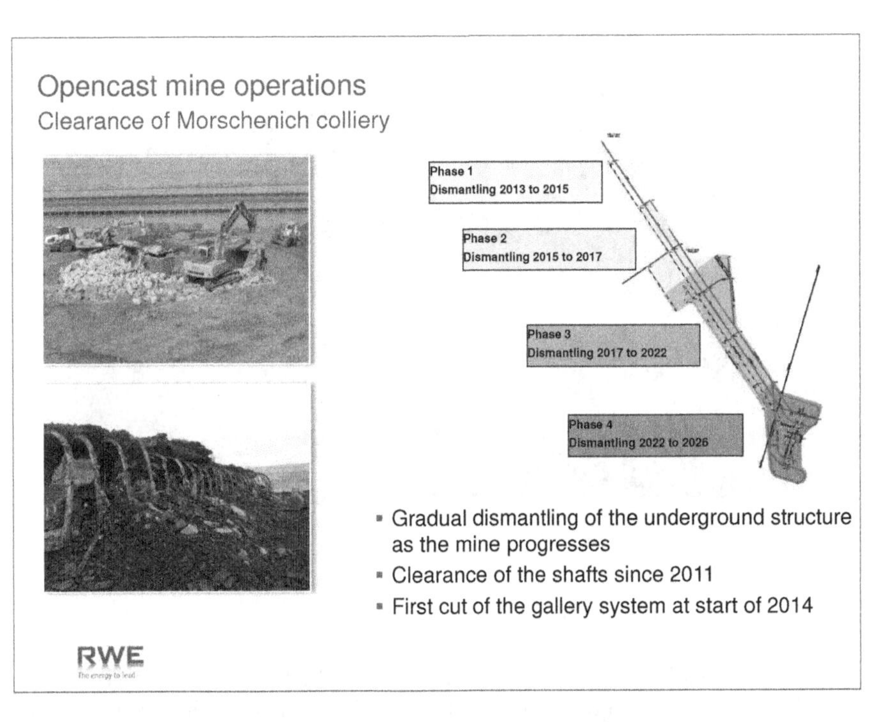

Fig. 11 Clearance of Morschenich colliery

The opencast mine's first bench reached the first section of the underground structure in 2011 with shaft 2. In a costly procedure, the first 40m of the shaft column were dismantled in special operations down to the excavator's operating level. At the start of 2012, in a second dismantling step, shaft 1 was used up by the first bench and dismantled. The opencast mine first cut through the roadway system on the 6th bench in early 2014.

5.11 Recultivation

High-quality recultivation of depleted areas is writ large at the Hambach opencast mine. Here, the mine is not only building on its own experience of several decades of successful recultivation, but – for the sake of continuous improvements in this field – is also collaborating with research institutes, universities and independent experts from environmental protection, forestry and agriculture.

In recultivation, what counts basically is not merely restoring the landscape used by an open-cast mine the way it was prior to mining. Rather, the aim is to offer Nature a basis for a land-scape that is at least of the same high quality, so that the recultivated areas integrate quickly into the existing cultivated landscape.

Fig. 12 Recultivation at Hambach mine

The "Sophienhöhe" heights, the outside dump of the Hambach mine, rises some 300m MSL and is a major landscape feature. There and on the adjacent inner dump of the mine, ten million trees have been planted as saplings since the mine was first developed. The oldest populations are now aged over 30. The variety of the trees, shrubs and wild flora, either especially planted or else the result of voluntary settlement, is in line with the typical vegetation of the Rhineland. In addition, the recultivated Hambach opencast mine now also offers a home to numerous protected species of domestic flora and fauna.

On the various terraces of "Sophienhöhe", numerous water surfaces and biotopes have been created, but some surfaces have been deliberately not afforested, being left to free succession.

"Sophienhöhe" is now a popular outing destination for quiet recreation. A hiking trail network of some 100km has been created and is open for public use.

The recultivation of the Hambach opencast mine is making an important contribution toward retaining and increasing the acceptance of lignite mining in the region and of work in the mine.

6 Upshot

Operating a large-scale opencast mine involves a whole host of tasks and challenges. In addition to the commercial aspects that must be considered, it is above all technical mining issues that determine the day-to-day business of mine management. Still, account also has to be taken of the underlying conditions set by geology and the deposit as well as environmental-policy and approval-law issues.

What is more, decisions arising on a daily basis increasingly include the side effects of Ger-many's energy transition. The fiercer competitive situation on Europe's energy market has an effect on Hambach, all the way to excavator and spreader.

Besides economic factors, all the processes in an opencast mine, starting with dewatering and clearing the pre-mining area, via overburden removal and dumping, coal extraction and recultivation, as well as all concomitant processes, always have to be in harmony with higher-ranking interests and stipulations, like neighbourhood protection, air-pollution control and noise abatement.

Besides these aspects, which are concerned more with planning and operation, there are leadership and management tasks, which are further important components in the day-to-day business of operating an opencast mine. These include, in particular, the promotion and active implementation of industrial safety. This is true both of the company's own workforce and of all partners. Our fundamental goal, by having a corresponding operational organisation as well as unequivocal delegation of tasks and responsibilities, is to ensure safe, economic and proper mine operations.

The creation of transparency in opencast mining operations by stepping up public-relations work in the neighbourhood and by an in-depth dialogue with local politicians and institutions will go on being of undiminished importance. Not least also owing to the energy transition and developments on the energy market, extending and stabilising the public acceptance of the work in an opencast mine and of the 'lignite system' in the direct environs are tasks of growing significance.

References

[1] Kulik, L., Hempel, R.: Planung und Steuerung des Tagebaus Hambach. Festschrift 25 Jahre Tagebau Hambach (2003)
[2] Gärtner, D., Hempel, R.: Überwachung und Steuerung der Prozesse in den Braunkohletagebauen im Rheinland. In: Der Braunkohlentagebau, pp. 391-408. Springer (2007)
[3] Strunk, S., Houben, B., Schollmeyer, P.: Gewinnung von verfestigten Toneisensteinschichten mittels kontinuierlicher Fördertechnik - Betriebliche Erfahrungen und technische Lösungen im Braunkohletagebau Hambach. World of Mining 65(4), 237–248 (2013)

"Best Practice" Concepts for a Fast-Track Lignite Mine Opening

Christos J. Kolovos

Mines Strategic Development Unit,
Mines Division, Public Power Corporation of Greece SA, Greece

Abstract. The paper presents the unusually rapid opening of the Mavropigi lignite mine in the Western Macedonia area, Northern Greece. Three mines, employing totally 9-10 small to medium sized Bucket Wheel Excavators (BWEs) and producing annually 7-8 million tons of lignite, were scheduled to close between 2006 and 2010 and be replaced by a new bigger mine. The transition from the three closing mines to the new one had to be planned carefully, in order to maintain the total annual lignite production level. Proper short-term mine planning and design of both the initial box-cut and of the first benches, permitted the deployment of four BWEs within a 13-month period, an achievement acknowledged as "best practice".

1 Introduction

Public Power Corp. of Greece SA (PPC), the biggest utility company in Greece, is a major lignite surface mine operator. The total PPC lignite production for 2012 reached 61.74 million tons (Mt) in the 2 major lignite-mining complexes in Western Macedonia (Northern Greece, 52.14Mt) and Peloponnese (Southern Greece, 9.6Mt). Lignite is used for power generation, contributing to almost 50% of the national demand. Three small local minor producers in the Western Macedonia area also sell mostly to PPC Power Plants.

The size of the deposit in the Western Macedonia area, its energy importance and the oil crises in the 1970's led to further development of mining operations. In 2002, a peak production of 55.8Mt was achieved by PPC's Western Macedonia Lignite Center (WMLC). The main activity takes place in the basin between the towns of Kozani and Ptolemais, (Fig. 1), where three main and some minor mines are currently developed, while another mine is developed further north in the Amynteon basin, between the towns of Ptolemais and Amynteon.

The multilayer type deposit necessitates selective excavation. Mining in PPC mines employs bucket wheel excavators (BWEs) for both the soft overburden and the multi-layer lignite deposit. Harder overburden formations are handled with blasting and shovel + heavy truck systems or contractor operations. Contractor backhoes have also excavated an increasing proportion of the lignite deposit during recent years.

C. Niemann-Delius. (ed.), *Proc. of the 12th Int. Symp Continuous Surface Mining - Aachen 2014*,
Lecture Notes in Production Engineering, DOI: 10.1007/978-3-319-12301-1_25

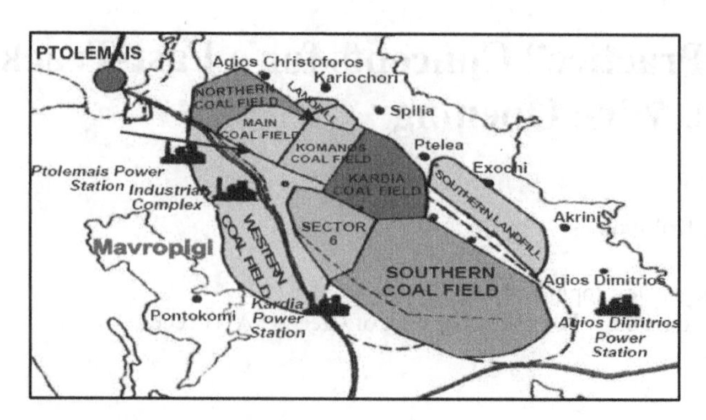

Fig. 1 The Kozani-Ptolemais lignite basin.

2 The Mavropigi Lignite Mine Case

2.1 The Mine Concept

The Main Field Mine was the first mine opened in the area in the late 50's and its reserves were mined out by the mid 80's. The main equipment (BWEs-conveyor belts-stackers) was utilized to open new operations at the adjacent Northern Field and the Komanos Field Mines (Fig.1). By the mid 90's, the East Komanos operation had been developed separately. These three mines, employing totally 9-10 small to medium sized BWEs of different types and producing annually 7-8 million tons of lignite, were scheduled to close between 2006 and 2010.

PPC had been planning the Komnina Mine at the Amynteon area, north of the Ptolemais town, for several years in the late 80's, and both a detailed mine study and an environmental study had already been prepared by 1991. However, the decision was never implemented due to the very promising borehole results at the Western Field of the Kozani-Ptolemais basin; the decision to develop the Mavropigi Mine was made in 1995.

Rheinbraun Engineering (RE), -a subsidiary of the German energy company RWE-, undertook the initial design and planning of the new mine as part of the Western Macedonia Technical Mine Master Plan Project (PPC SA-RE, 1996). The Western Field was divided by RE into two parts, the Northwestern and the Southwestern Field. The Northwestern Field was preferred over the Komnina Field as a new mine, due to the favorable deposit conditions, whereas the Southwestern Field is planned to be exploited in the forthcoming years as an extension to the Kardia/Sector 6 mine. The original mine name "Northwestern Field" was changed in June 1996 to honor the nearby Mavropigi village.

According to the RE planning (1995-96, Fig. 2), the Mavropigi Mine was initially scheduled to open from the SE side and advance in a clockwise slewing operation to the north, producing up to 6Mt per annum. The main advantage of this solution was the capability for an early development of the internal dump; low production and the early disruption of the Mavropigi village connecting road were the main disadvantages.

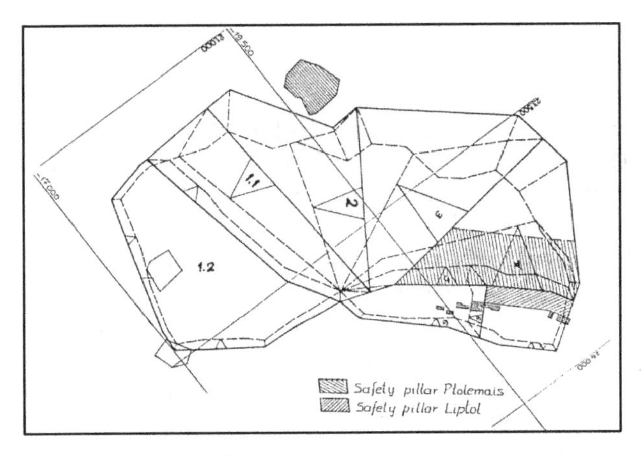

Fig. 2 Mavropigi Mine: RE initial design

After an extensive discussion with PPC, an alternative solution was adopted in 1996 (Fig. 3); to open the mine from the northern side and advance counter-clockwise to the south, in order to obtain a more stable production rate for a longer period. The excavation of the Mavropigi village connecting road area was also postponed for some years. The key point in adopting the second solution was the result of the geotechnical stability studies, which indicated that a much smaller than the originally anticipated pillar could be abandoned towards the nearby Ptolemais Power Plant (Kolovos C., 2013).

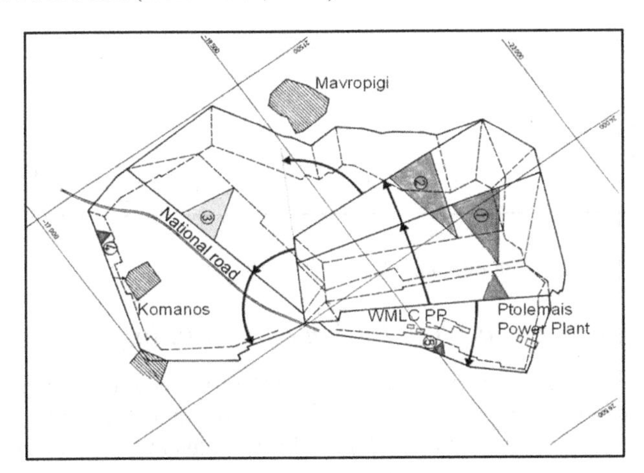

Fig. 3 Mavropigi Mine: RE-PPC solution 2

The deepest part of the mine is in sectors 1 and 2, with a maximum depth of about 180m and the mine design includes 8 benches/BWEs. Due to the fact that the mine would be opened with already existing equipment, it was possible to pre-assign every available BWE, according to their excavating capacity and their suitability for overburden or lignite deposit excavation, to a specific bench of the new mine.

The expropriation of the land for the first 5-year mine life was completed at the end of 1999. However, the residents of the nearby Mavropigi village insisted that the village had to be relocated and refused to allow entry of PPC equipment into the expropriated land; a court decision to expel the villagers was obtained at the end of 2000. The environmental permit was issued in September 2001 and the excavations for the initial box-cut commenced in October 2001 (contractor backhoes + on-highway trucks).

The final pit limit, which was originally drawn by RE at a distance of 38m from the village cemetery, had to be redrawn in 1997 and moved further away from the village, at the distance permitted by the Greek Mining Regulation, (minimum 250m from the village and cemetery perimeter). In January 2002, in an effort to assure the safety of the village, PPC moved the final pit limit at a distance of 500m away from the village perimeter, double than the minimum distance permitted by the Greek Mining Regulation (Fig. 4). The Komanos village, shown inside the Mavropigi pit limit, had been relocated years ago.

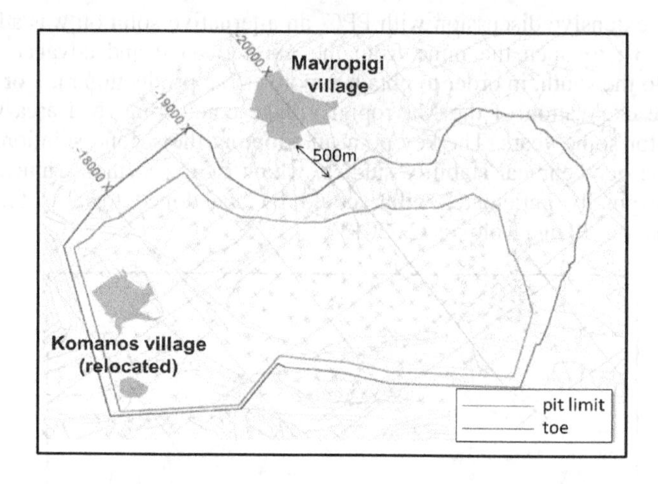

Fig. 4 Mavropigi Mine: final pit limit - 2001

2.2 *Modifications in Mine Design and Planning*

There was an almost 3-year delay in the opening of the new mine, which resulted in production issues for the entire WMLC. The operations at the three closing mines (Northern Field, Komanos Field, East Komanos, all under the local administration of the Main Field Mine) had to be extensively rescheduled and redesigned, so that no BWE would be without assignment and stand still for months, waiting to be moved to the new mine. The transition from the three closing mines to the new one had to be planned carefully, in order to maintain the total annual lignite production level.

Adding contractor equipment to accelerate box-cut excavations was not feasible at the time: in 2001-2002, almost every piece of contractor equipment in Greece was employed in Athens, preparing the 2004 Olympic Games facilities and it proved difficult to find available contractors to accelerate the opening of the Mavropigi Mine. Low supply and high demand for contractor equipment had also significantly increased the price per cubic meter excavated. As an aftermath, "unorthodox" ways to accelerate operations had to be examined, which resulted in an extended redesign of the initial phase of the mine (Kolovos, 2004a). It is also mentioned that since the BWEs are rather old (initial deployment year between 1958 and 1970), an extended maintenance program was planned to be applied to the BWEs, whenever a BWE would be moved to the new mine. This program was planned to last a minimum of three months and be applied just before it would start operations to the new mine. The maintenance program could not run during the winter months, due to heavy winter in the area, therefore it had to be scheduled to run within the remaining months.

All the necessary revisions and amendments to the original mine planning affected the short and mid-term production planning, and, as a result, they were assigned to the Main Field internal mine planning unit.

2.2.1 The Rolling 5-Year Monthly Timetable Tool

A rolling 5-year monthly timetable tool (Fig. 5), in the form of a Gantt chart, had already been developed by November 2001, from the very beginning of the new mine, in order to monitor the progress of each bench of all four mines, the assignment of the BWEs to exploitations and benches, the rapid opening and development of the new Mavropigi Mine, the possibility for increased lignite production with fewer excavations from any exploitation, the potentiality to reduce expenses for contractor operations, the ensuring of the necessary period for the BWE extended maintenance program, etc. (Kolovos, 2004b).

PUBLIC POWER CORPORATION OF GREECE SA
WEST MACEDONIA LIGNITE CENTER
MAIN FIELD MINE - MINE PLANNING DEPT.

BUCKET WHEEL EXCAVATORS OPERATING PER BENCH
SCENARIO 1

NORTHERN FIELD MINE (NF)

KOMANOS FIELD MINE (KO)

EAST KOMANOS EXPLOITATION (E.KO.)

MAVROPIGI MINE (MA)

KARDIA MINE(KA)

BWE PER EXPLOITATION	YEAR x												YEAR x+1											
NF	3	3	3	3	3	3	3	3	3	3/2	2	2	2	2	2	2	2	2	2	2	2	2	2	2
KO	5	5	5	5	5	5	5	5	5	5/5	5	5	5	5	5	4	4	4	4	4	4	4	4	4
E.KO.	1	1	1	1	1	1	1	1	1	1	1	1	1	1	1	1	1	1	1	1	1	1	1	1
MA	0	0	0	0	0	0	0	0	0	0	0	0	0	0	0	1	1	1	1	1	1	1	1	1
KA	0	0	0	0	0	0	0	0	0	0/1	1	1	1	1	1	1	1	1	1	1	1	1	1	1
TOTAL	9	9	9	9	9	9	9	9	9	9	9	9	9	9	9	9	9	9	9	9	9	9	9	9

Fig. 5 Example of the rolling timetable tool

The remaining volumes of overburden and lignite for every sector and every bench till the end of operation of the three exploitations under exhaustion had to be most accurately calculated. Based on the assignment of the BWEs to the different exploitations (or even in some cases, to other nearby mines), the mining capacity of the different types of BWEs, allowed a preliminary hypothesis on the time the BWE would have completed its remaining volumes and would be either available to be moved to the new mine or assigned to another operation.

The BWEs were preferably assigned to the lignite deposit benches instead of the overburden ones, in order to keep production quantity and quality under strict control. In case a bench lacked a BWE, the respective volumes would be moved by contractor diesel equipment, and a provision to the mine budget and the relevant procedures could be made in time.

The BWE serial number was preferred over the BWE type to appear on the timetable, so that it could assist the programming of the maintenance departments. Different colors for every BWE allowed an easy visual follow-up of each BWE's movement around the different exploitations within the rolling 5-year period covered by the timetable. The bottom part of the timetable included a verification of the total number of BWEs dedicated every month to every mine, to ascertain that a BWE was neither omitted nor used twice.

Priority was given to the opening of the new mine over operations in the existing ones, so, as soon as a bench floor was ready at the new mine, the BWE would be moved there, even if there was still an unexploited area at the closing down mine. Every time a new bench floor was prepared at the Mavropigi Mine, the appropriate BWE type had to crawl some kilometers to the new mine, passing over a small stream and both the Kozani-Ptolemais national road and railway. Two high voltage and one medium voltage electricity lines had to be also taken down every time a BWE was moved to the new mine (Fig. 6).

Fig. 6 Two BWEs crossing the Kozani-Ptolemais national road

This procedure called for a detailed programming of the BWEs exact travelling dates, since many activities and departments had to coordinate. This timetable proved to be a most valuable tool for the mid-term production planning, as well as the communication with the various departments involved in the new mine project and the coordination of operations in all four mines/exploitations. Five different scenarios for the assignment of BWEs to operations and benches had already been produced by November 2001, allowing the respective lignite production scenarios. By November 2005, 6 BWEs had been deployed to the new mine and 27 different 5-year scenarios had been produced, covering the development of the mine to the end of 2010.

2.2.2 Redesigning the Box-Cut

The box-cut was redesigned in order to accelerate both the rapid mine opening and the early lignite production. The initial design aimed at preparing the ground surface for the control tower of the mine, the floor for the belt distribution point and the floor of the 1st bench (Fig. 7). The new design, (Fig. 8), altered the shape of the box-cut, decreasing it to the west and extending it to the north, so that the first bench could be assembled on the original ground topography and not within the box-cut, decreasing the volume of the necessary contractor works.

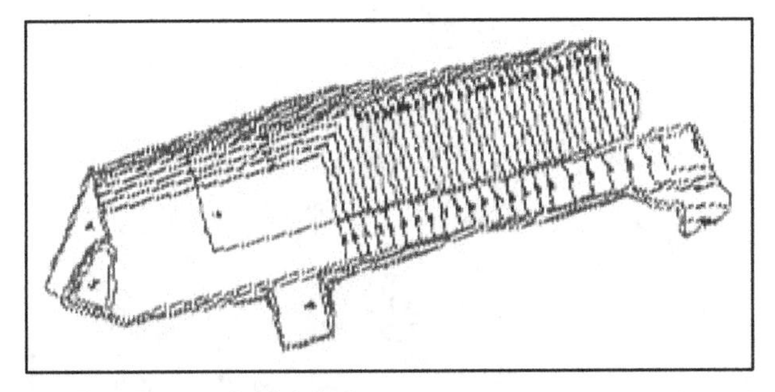

Fig. 7 Mavropigi Mine: Box-cut initial design

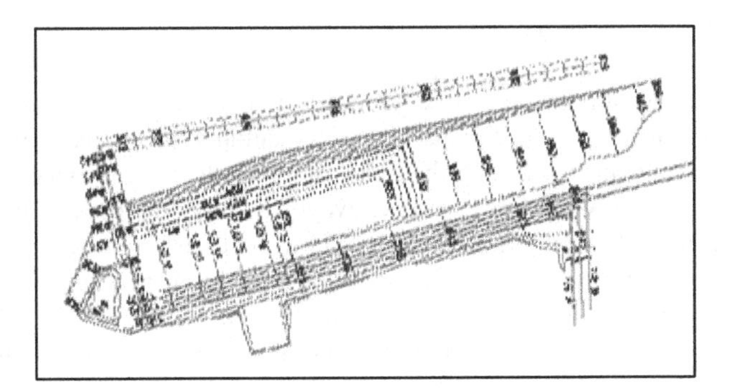

Fig. 8 Modified box-cut design

The narrow excavation, originally dimensioned just to host the mine distribution point area, was elongated to the north, allowing deeper box-cut levels down to the lignite deposit. The alteration to the shape and levels of the box-cut resulted in a lignite production of 1Mt in the very first year (2002) of the mine life (box-cut excavations 4.6 Mm^3). The box-cut was excavated via diesel equipment (backhoes + loaders) and the lignite was transported by on-highway trucks to the nearby Ptolemais Power Plant, since neither a run-of-mine (ROM) lignite stockyard nor lignite conveyor belts had been prepared yet.

The possibility for an in-pit dump does not exist in a new mine, only an ex-pit dump is possible during the opening phase. The ex-pit dump for the Mavropigi Mine was supposed to be the mining voids at the already existing exploitations (Northern & Komanos Fields). However, at the time the box-cut started, there was no available infrastructure that would permit the crossing of the Kozani-Ptolemais national road and railway. Mining trucks certainly cannot cross a national road and a new, wide enough, bridge for a road and the three stacking side conveyor lines was ready only two years later. As a consequence, all the soft overburden excavated from the box-cut was temporarily dumped nearby, in order to speed up the opening phase (Fig. 9).

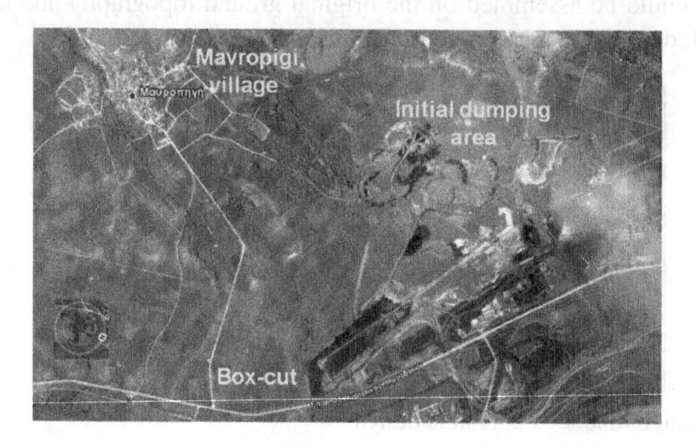

Fig. 9 The Mavropigi Mine box-cut in 2003

The BWEs would then be able to remove this loose overburden in small quantities and at a later stage. This temporary dump was also used for covering some streams and either levelling the topography for the bench floors or preparing the floors to assembly the bench side conveyor belts. Some hard formations that were found in the box-cut, along a fault, had to be transferred further away, at a small external dump outside the projected mine pit limit, near a PPC settlement. The BWEs were relieved from excavating hard material and this dump also worked as a small noise barrier, in order to protect the settlement from mine noise and dust.

2.2.3 Rapid Deployment of Four BWEs

At first, the initial box-cut was redesigned in order to accelerate the deployment of the first BWE. The redesigning of the initial box-cut (Fig. 8) altered the shape of the excavation quite enough to allow a different approach to the deployment of the BWEs.

An investigation of the possibility for an early deployment of BWEs proved that it was possible to deploy the second BWE right after the first one, thus allowing significant savings in contractor works and the relevant expenditure (Kolovos, 2004a).

In June 2003, the elongated and deepened box-cut, combined with the favorable landscape morphology, permitted the deployment of the first BWE, directly inside the box-cut and at the second bench, connected to conveyor belts 2-12 and one stacking conveyor belts line (Fig. 10).

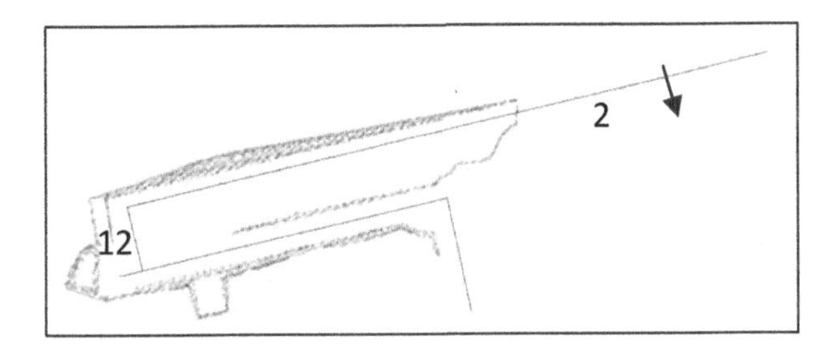

Fig. 10 Belts 2-12 & 1st stacking line

A versatile Krupp C700 was selected to be the first BWE deployed, for its very convenient capability of working easily at either a high or a low step.

The key element in accelerating the opening of the mine was the ability to assemble the connecting conveyor bench 11, of the first bench, at the southwest part of the box cut, on a ramp that, due to favorable landscape morphology, could be easily extended outside the box-cut. The first bench, formed at the next stage, was divided into two segments, and a BWE was deployed to each segment. The two segments were developed at a later stage in the first two benches of the mine.

This action permitted the acceleration of the overburden excavation with the deployment of two BWEs (Krupp SchRs 600x21/3.3) in November and December 2003. These BWEs had separate bench side conveyor belts (named 1 & 1a - belt width 1200mm), but they were both connected to only one connecting conveyor belt (named 11 - belt width 1600mm, Fig. 11 & 12).

The upper BWE (upper segment of the first bench) was then able to advance rapidly and create the necessary space for the slewing point of the mine. While the minimum initial assembly length of the connecting conveyor belt 12 was 178m (which included the 5-position shunting head, the belt tail and just two fitting frames), the adjacent connecting belt 11 was designed with an initial assembly length of 350m and could be easily lengthened by another 60m, if a piece of land could be expropriated: fortunately it took place with only minor delays.

The space of 172m, between the initial lengths of connecting belts 11 and 12, provided just enough room to deploy and operate three BWEs; their close proximity called for the utmost attention.

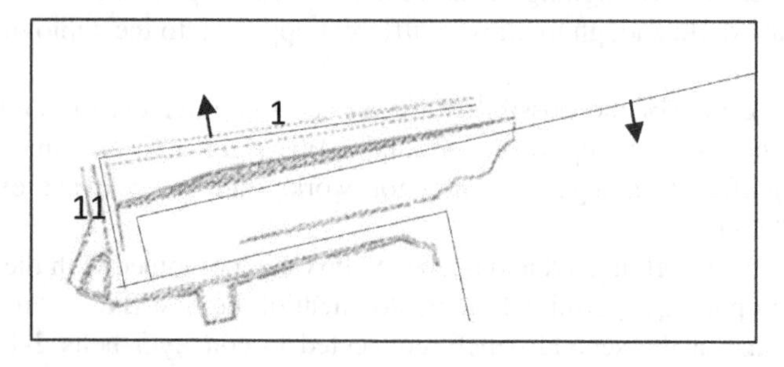

Fig. 11 Assembly of Belts 1-11

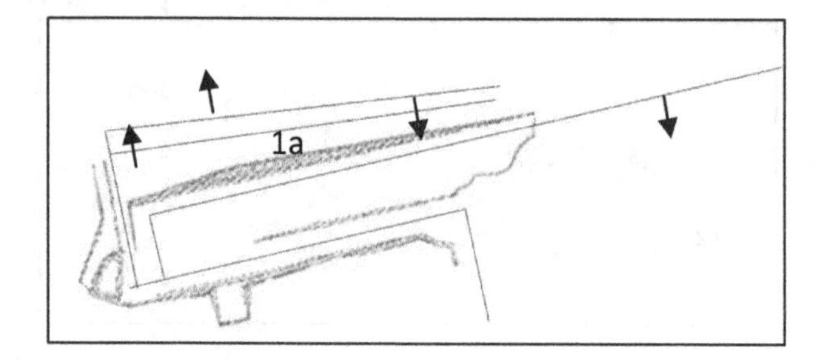

Fig. 12 Assembly of Belt 1a behind Belt 1

The upper BWE worked at first in parallel mode, with a shortened excavation side belt length (belt 1-1.0km), then in left-hand slewing mode, so that it would create the necessary space for the lower benches. After 4 successive elongations it reached a length of 2.1km. Then the slewing operation of the bench belt (belt 1) stopped, and it was transformed into a connecting belt, whereas a new bench belt (belt 1b-0.9km) was assembled, to deal with a hilly terrain. Belt 1b continued a counter-clockwise slewing operation until it fell into line with belt 1 and the two belts were joined into one, with a total length of 2.7km (Fig. 13). The upper bench conveyor belt gained that way its full mine length and continued the left-hand slewing operation.

While the upper BWE worked with a shortened bench belt in high step, the second BWE, (lower segment of the first bench), had to work in low step. The bench belt (belt 1a) was assembled with an initial length of 0.9km, but it was almost immediately, in January 2004, lengthened to 1.7km by being joined with part of the conveyor belt of the second bench (belt 2), which was the first to be

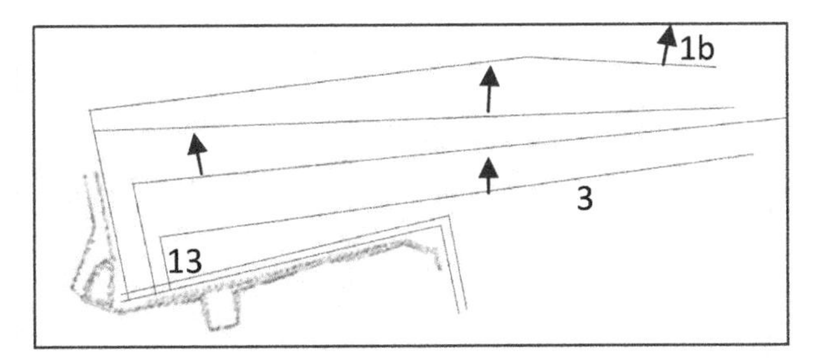

Fig. 13 Assembly of Belt 1b, Belts 3-13 and 2nd stacking line

assembled in June 2003. In the beginning, belt 1a followed a clockwise slewing operation, by transferring the belt head close to belt 1, in order to create space at the slewing point of the mine; after that it was soon elongated to 2.3km. With this operation, it soon gained the full bench length and worked mostly at the periphery of the mine, so that the advance between the benches would be monitored and kept as close as possible. The water from the small lake shown in Fig. 9 was pumped and the void was covered by a mix of overburden, moved by the contractor working at the box-cut, plus fly ash, acting as drying agent, so that the BWE could pass over the lake area and continue its operation beyond the lake to the final pit limit.

From the beginning one stacker conveyor belt line was constructed for conveying the overburden, whereas the lignite conveying lines were intentionally delayed; priority was given to moving the overburden. This was taken into account when designing the floors of the upper benches, so that the BWEs met only minor quantities of lignite near their slewing point and it was very easy to dump it on the bench floor. As soon as a lignite belt line was constructed, the initially deployed BWE was able to deepen the mine and prepare the floor of the fourth bench.

Meanwhile, the first deployed BWE at the second bench was moving overburden in a trench shaped excavation, to prepare the rest of the bench floor. Contractor works prepared the remaining part, in trench-like shaped excavations, so that the full bench length could be attained as soon as possible.

The fourth bench started operations in July 2004 (Fig 13). The advance of the upper benches and the widening of the box-cut permitted the construction of just the minimum necessary length of the connecting conveyor belt 13 (178m as in belt 12, including the 5-position shunting head, the belt tail and two fitting frames).

The bench floor was prepared by the C700 excavator, working in low step. A Buchau Wölf SchRs 660x21/3 was deployed at the 3rd bench (belts 2-12) and the C700 was immediately moved to the fourth bench, to continue working in low step and deepen the mine. The Mavropigi Mine continued its development and is today one of the largest lignite mines in Greece (Fig. 14).

Fig. 14 The Mavropigi Mine in 2009

The usual case in similar worldwide BWE lignite operations is to deploy one BWE per year, because every BWE has to work in low step back and forth, to prepare the floor for the next bench. However, in the case of the Mavropigi mine, four BWEs were deployed in 2003-2004 within a 13-month period. This accomplishment was acknowledged in PPC as an international "best practice" and was proudly mentioned in the 2004 Annual Report (PPC SA, 2004).

3 Conclusions

- The opening of a new mine is never a simple project: it has to be planned carefully and in great detail. In the case of opening a new mine with already existing equipment, which is not idle and ready to be used, but already working at other mines, the transition phase can be extremely challenging.
- The role of the internal mine planning unit is indispensable in revisions and amendments to the long-term studies, as well as in the coordination of the departments responsible for the construction of the mine infrastructure.
- The 5-year rolling timetable, developed for the Mavropigi case, proved to be a valuable tool; not only in short and mid-term production planning, but also in setting and rescheduling targets, in easing the communication and establishing a basic understanding between the mine departments and all external supporting units, in setting the mine budget, in programming on time procurements and the necessary procedures for contractor equipment available on time. It was a valuable mine management tool, investigating the best possible arrangement of the BWEs and critical in clarifying what needs to be done, when it needs to be done, why it needs to be done and by whom needs to be done.
- A close co-operation of BWEs with diesel equipment is very significant for a fast-track preparation of bench floors.

- The long term mine study is indispensable in general guidance, however, there is always enough room for modifications, amendments and innovations in the short and mid-term mine planning.

References

1. Kolovos, C.: Mavropigi mine: Deployment of four BWEs by May 2004 and operation in the 2004-2006 periods. PPC, Ptolemais (2004a) (in Greek)
2. Kolovos, C.: Coal Mining Technology, 349 p. ION Publishing Group, Athens (2004b) (in Greek)
3. Kolovos, C.: Corporate social responsibility and the future of mining in Greece. In: Proceedings of the 6th International Conference on Sustainable Development in the Minerals Industry, Milos island, Greece, June 30-July 3, pp. 160–166 (2013)
4. PPC SA-RE, Technical Mine Master Plan for PPCs LCPA and LCM, Final Draft Report. PPC, Athens (1996)
5. PPC SA, Annual Report 2004, Athens: PPC (2004) (in Greek)

Optimisation of Ugljevik Basin Open Pit Mines with Regards to Long-Term Coal Supply of the Thermal Power Plants

Cvjetko Stojanović[1] and Bojo Vuković[2]

[1] ZP RiTE Ugljevik, Republika Srpska, BiH
[2] ZP RiTE Gacko a.d. Gacko, Republika Srpska, BiH

Abstract. Capital mining projects are very complex due to the influence of various internal and external techno-economic and natural factors and constraints arising from social and economic environment. By investing in the development of surface mining the company „Rudnik i Termoelektrana Ugljevik" („ Mine and Thermal Power Plant Ugljevik") in the Republic of Srpska (Bosnia and Herzegovina), implements previously defined goals of growth, development policy and strategy of the company, as well as the national energy strategy. This investment project represents a capital project for the company, since it requires extensive financial assets, significant amount of other resources as well as time. The realization of the project, with regards to deposit conditions, is a very complex process consisting of multi-dimensional examination activities of all relevant determinants of future conditions and changes that the project entails. In addition to these internal resources, the complexity of the project realisation is increased by external influences of the State as the Company owner. The consequences of external influences to the investment project are strategic changes to the project scope, concept, work schedule and budget. To address the long-term coal supply Thermal Power Plants Ugljevik was optimised open pits Ugljevik basin, using modern software tools.

This paper gives a brief overview of the methodology and results of the optimisation.

Keywords: optimisation, open pit mine, efficiency and effectiveness.

1 Introduction

Ugljevik coal basin is located in the north-eastern part of Republic of Srpska, and Bosnia and Herzegovina. According to the long-term development program of the coal basin, a planned successive exploitation of certain deposits has been foreseen for the supply of four power generating units, each providing 300 MW of power.
The first unit was designed to be provided with coal from Bogutovo Selo coal deposit. The deposit contained about 50 million tons of coal reserves. The coal for the second unit would have been provided from the Ugljevik East (Ugljevik Istok)

C. Niemann-Delius. (ed.), *Proc. of the 12th Int. Symp Continuous Surface Mining - Aachen 2014*,
Lecture Notes in Production Engineering, DOI: 10.1007/978-3-319-12301-1_26

deposit, containing about 55 million tons of coal reserves. The coal for the third and the fourth unit would have been provided from other coal deposits, whose level of research was significantly lower in the time the long-term program was made, i.e. in the mid-seventies.

2 History of Past Activities in the Development of Thermal Power Complex Ugljevik

In late seventies began the construction of the first unit of Ugljevik Thermal Power Plant. The unit was completed and put into operation in 1985. The construction of the unit 1 of the Thermal Power Plant was followed in parallel with the activities on preparing the open pit and of Bogutovo Selo coal deposit. After completion of this phase, the construction of the second unit of Thermal Power Plant Ugljevik began alongside with the activities on the opening and preparing the Ugljevik East coal deposit for exploitation.

Complete project documentation was made for this project, defining the technology, external landfill sites with the required volume of storage space for overburden, industrial site facilities, etc. The above activities were performed continuously until the beginning of year 1992, when, due to well-known events, they were completely stopped.

The following period was marked by efforts to resume the previously initiated construction. However, mostly due to lack of funds, the project was not implemented. Meanwhile, a reconstruction and modernization of the first unit was performed, thereby extending its lifetime. Additionally, a long-term loan for the desulphurisation project has been provided, with the last instalment of the said loan expiring in 2039.

Based on studies and planning documents, while taking into consideration the remaining service life of Thermal Power Plant Ugljevik I, it was concluded that it is necessary to provide a total of about 45 million tons of coal, i.e. 1.7 million tons per year. Accordingly, the need arose for additional quantities of coal for the operation of this unit, as the available coal reserves in the coal deposit Bogutovo Selo, according to the remaining reserves and mining projects, enabled plant operation until the year 2021.

The following figure contains a final contour of the open pit mine, Bogutovo Selo, 2011 mining project.

Meanwhile, a new investor appeared with the intention to build a new unit of 300 MW. Among other things, it was required to define a coal deposit which would provide sufficient quantities of coal for the new unit as well.

In the search for optimal solutions that meet both requirements, the project team joined the detailed analysis of both the total mineral resources and the assessment of opportunities and capabilities for their exploitation.

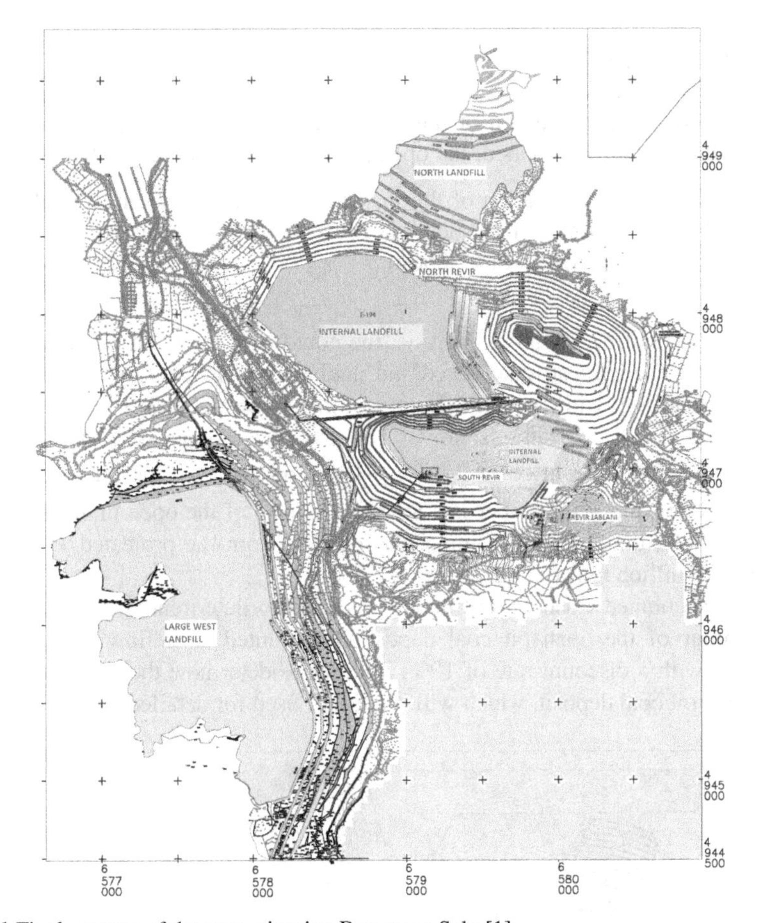

Fig. 1 Final contour of the open pit mine Bogutovo Selo [1]

3 A Brief Description of the Methodology and Results of the Optimisation

The first step was the high-quality estimate of coal deposits. The estimate was made by applying the standards and by using a modern 3D software for this purpose, as well as the software for optimisation and long term mine planning and design. Such software allows users easier, faster and more effective approach to the design and production planning in comparison with traditional calculations that are based on numerous assumptions.

Therefore, the first task set before the project team was to determine the final contours of the existing open pit coal deposit Bogutovo Selo, which will provide the necessary amount of coal for the uninterrupted operation of Thermal Power Plant until the rest of its operational lifetime. Consequently, the process of determining the final contours of the open pit coal deposit Bogutovo Selo was made in two steps, as follows:

Option 1 - Exploitation of the maximum possible amount of coal within the existing contours of the open pit coal deposit Bogutovo Selo

Option 2 - Exploitation of the maximum possible amount of coal within the expansion of the open pit coal deposit Bogutovo Selo onto the section of Ugljevik East deposit.

Based on the multiple-iteration analysis, the contour of the open-pit coal deposit that generates the highest (discounted and undiscounted) cash flows was adopted as the optimal one, with predefined input optimisation parameters. Since one of the key parameters for optimisation is the coal price, expressed via its lower heating value, the base price of 2.2 € / GJ was adopted for the purpose of this analysis. The result of this paper is a new Contouring of the open pit coal deposit Bogutovo Selo, thus increasing mineable reserves from the projected 19 million tons to 26.5 million tons of coal.

DCF (Discounted Cash Flow) model was used to determine and optimise the final contour of the open pit coal deposit. Discounted cash flow analysis was performed with a discount rate of 10%, in order to determine the optimum limits of the open pit coal deposit, which will be further used for detailed design.

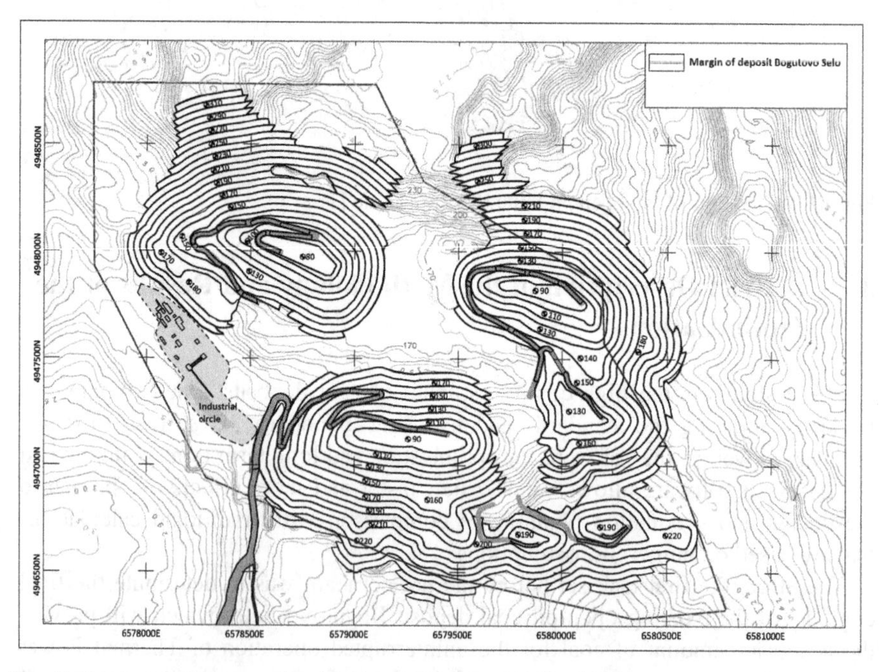

Fig. 2 Contour of the open pit mine Bogutovo Selo, Option 1 [2]

Table No.1 shows a portion of the cash flows (discounted and undiscounted) for individual contours of the mine with the corresponding quantities of coal and overburden and indicated an optimum open pit mine limit and optimisation technical parameters of the open pit mine for Option 1 is shown in Table 2.

An optimal pit mine is the one whose contour generates the highest cash flow (discounted or undiscounted) with predefined input optimisation parameters. In this case, the approved final contour of the mine corresponds to the mine contour no. 17.

The following figure shows a final contour of the open pit mine, Bogutovo Selo, Option 1.

Table 1 Economic parametres of optimisation [2]

Final pit	Open pit cashflow best € disc	Open pit cashflow worst € disc	Open pit cashflow best € no disc	Coal tonne	Waste tonne	Mine life years
1	195.438.654	195.438.654	327.439.375	16.373.237	52.293.873	9,36
2	210.965.049	196.942.343	327.439.375	20.287.122	82.166.791	11,87
3	213.237.277	194.816.810	327.439.375	20.924.045	87.811.219	12,23
.						
.						
16	223.311.077	161.070.085	327.439.375	26.220.524	149.323.904	15,39
17	**223.340.703**	**158.113.385**	**327.439.375**	**26.505.339**	**153.986.487**	**15,55**
18	223.302.418	151.866.603	327.439.375	26.872.574	159.979.252	15,80
.						
.						
50	215.529.702	62.204.711	327.439.375	31.015.997	259.293.879	20,35
51	215.517.455	62.073.305	327.439.375	31.019.648	259.408.148	17,1

In order to provide the required quantity of coal for Thermal Power Plant Ugljevik to the end of its lifetime, year 2039., optimisation of the open pit coal deposit Bogutovo Selo was performed in the next step, thus expanding onto the open pit coal deposit Ugljevik East, with the aim of providing the lacking approximately cca. 20 million tons. Optimisation of Bogutovo Selo open pit coal deposit onto a section of the Ugljevik East deposit - Option 2 (Figure 3) was performed using the same methodology as in the previous version. The results of economic optimisation have been provided in Table 3, and the results of technical optimisation provided in Table 4.

Table 2 Technical parametres of optimisation[2]

Pit	Rev Ftr	Waste Tonnes	Coal Tonnes	DTE, MJ/t	Strip Ratio, t/t	Strip Ratio, m^3/t
1	0,5	52.293.873	16.373.243	11.950	3,2	1,6
2	0,53	82.166.791	20.287.128	11.896	4,1	2,0
3	0,56	87.811.219	20.924.050	11.908	4,2	2,1
.						
.						
16	0,95	149.323.904	26.220.529	11.928	5,7	2,8
17	**0,98**	**153.986.487**	**26.505.344**	**11.932**	**5,8**	**2,9**
18	1,01	159.979.252	26.872.579	11.930	6,0	3,0
.						
.						
50	1,97	259,293,879	31,016,001	11,958	8,4	4,2
51	2	259,408,148	31,019,652	11,958	8,4	4,2

Table 3 Economic parametres of optimisation [2]

Final pit	Open pit cashflow best € disc	Open pit cashflow worst € disc	Open pit cashflow best € no disc	Coal tonne	Waste tonne	Mine life years
1	211,643,105	211,643,105	432,848,714	22,636,818	89,764,839	13.00
2	217,222,374	209,836,432	453,920,115	24,269,869	103,472,165	13.94
3	220,050,468	208,884,509	465,601,043	25,249,322	111,824,105	14.50
.						
.						
17	232,116,904	96,126,892	541,115,603	37,821,819	243,321,797	22.37
18	**232,118,110**	**91,316,240**	**541,154,828**	**38,135,344**	**247,394,545**	**22.56**
19	232,079,878	85,913,114	540,869,496	38,586,623	254,885,090	22.91
.						
.						
44	226,508,269	-64,536,923	462,287,418	48,464,728	433,617,504	31.26
45	**226,401,609**	**-66,777,775**	**460,188,066**	**48,603,484**	**436,612,444**	**31.34**
46	222,870,214	-131,868,935	385,728,488	53,186,613	541,463,157	37.34
.						
.						
50	222,452,769	141,400,458	370,781,228	53,942,469	561,928,095	37.99
51	222,404,730	-142,570,002	368,966,255	54,033,234	564,287,300	38.10

From the given analysis, it can be seen that, in this option, maximum profit is achieved with the exploitation of the mine deposit no.18. The results of the economic optimisation show that it would be economically justified to excavate about 48.6 million tons of coal (mine deposit no. 45) with an average stripping ratio of 4.5 m^3/t, meaning that such contour of the mine should provide the necessary amount of coal of about 45 million tonnes for a smooth and safe operation of the power plant by the end of its operational lifetime.

Table 4 Technical parametres of optimisation [2]

Pit	Rev Ftr	Waste Tonnes	Coal Tonnes	DTE, MJ/t	Strip Ratio, t/t	Strip Ratio, m^3/t
1	0,5	89,764,839	22,636,807	12,077	4.0	2,0
2	0,53	103,472,165	24,269,858	12,090	4.3	2,1
3	0,56	111,824,105	25,249,310	12,083	4.4	2,2
.						
.						
17	0,98	243,321,797	37,821,807	11,571	6,4	3,2
18	**1.01**	247,394,545	38,135,330	11,555	**6,5**	**3,2**
19	1,04	254,885,090	38,586,610	11,557	6,6	3,3
.						
45	1.82	436,612,444	48,603,470	11,110	9.0	4.5
50	1,97	561,928,095	53,942,455	10,909	10,4	5,2
51	2	564,287,300	54,033,220	10,906	10,4	5,2

4 Design of Final Open-Pit Coal Deposit Contour and Exploitation Reserves

The final contour of open-pit coal mine was designed based on the optimal pit contour (shell no. 45, Table 4) by predefining all necessary geometric elements of the open pit coal mine: final berm width, angles of inclination of working floors and final pit slope, width and longitudinal incline of roads, radius of the road curvature, etc. Likewise, final verification of the stability of final open-pit coal mine slopes was performed at this design stage, in order to provide a safe and secure execution of mining operations from the current state until the final open-pit coal mine outline. After the verification of the slope stability in all directions of the mine, the final contour of open-pit coal mine was made, with the expansion onto a section of Ugljevik East open-pit coal mine, shown in Figures 3.

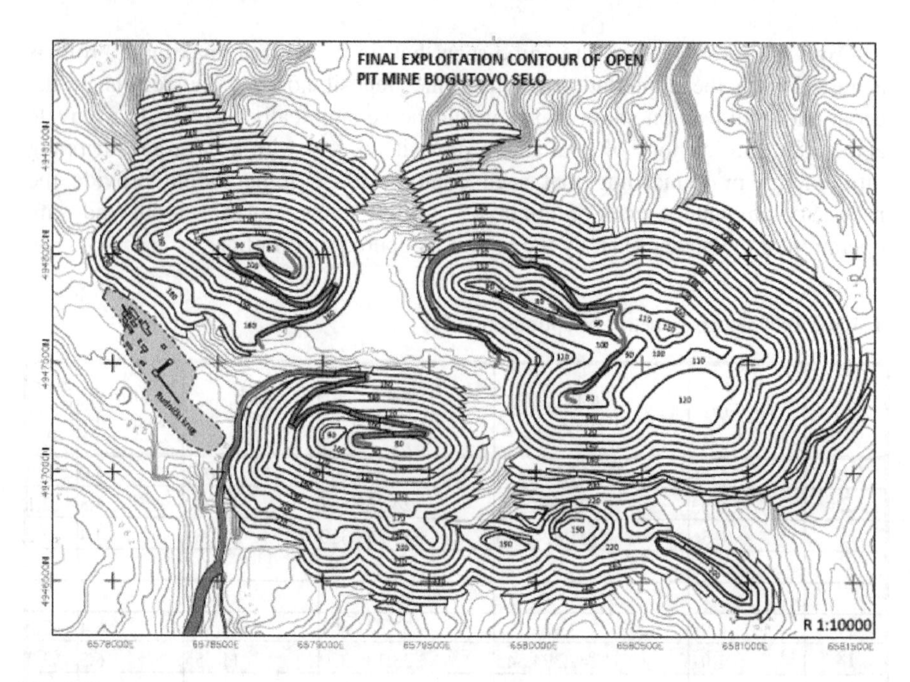

Fig. 3 Final contour of the open pit mine *Bogutovo Selo* with the expansion onto a section deposit of Ugljevik East [2]

Mineable quantities of coal that are affected by the final contour or the open pit mine Bogutovo Selo, with the expansion onto a section deposit of Ugljevik East with the average exploitation ratio amount to:

$$K_0 = \frac{199.118.000}{46.795.000} = 4{,}26 \text{ m}^3/\text{t}$$

By comparing the coefficients of overburden from the previous project and the results of the optimisation described in this paper, it can be concluded that the average ratio of overburden until the end of the exploitation was about 4.5 m^3/t for the open-pit coal deposit Bogutovo Selo, according to the previous project designs, while the same ratio for the Ugljevik East was 6.5 m^3/t.

From the results of technical optimisation, it is clear that a significantly lower ratio of overburden was obtained for the selected final contour of the open-pit coal deposit Bogutovo Selo, with the expansion onto the section of open-pit coal deposit Ugljevik East.

The result of the analysis described is an increase of mineable reserves in the open pit coal deposit Bogutovo Selo to about 7 million tons as a result of including those reserved not being included in the earlier design proposals. The outstanding quantities of coal required for the operation of the first unit would have to be provided from the open pit coal deposit Ugljevik East. Such development would actually represent a continuation of the exploitation from Bogutovo Selo deposit, thus meeting the coal needs of the first unit.

The next step was to perform the optimisation of the remaining section of the open-pit coal deposit Ugljevik East in order to determine the remaining mineable reserves for the second unit of Thermal Power Plant Ugljevik. The optimisation was performed according to the same criteria as in the previous cases, with the analysis result amounting to approximately 40 million tons of mineable reserves. That way, most of the coal demand required for the operation of the new unit was met, provided that the missing amount of coal is provided from other deposits whose mining-geological conditions are somewhat less favorable for their exploitation. General final economic parametres of the optimisation have been shown in Table [5].

Table 5 General final economic parametres of the optimisation [3]

Parametre	Description	Value in the project	Desired value	Accepted
Income	Average anual income according to the Balance statement	77.297.947		YES
Net profit	Average anual net profit according to the Balance statement	21.779.264		YES
Net profit ratio*	Net profit / Total investment	12,44%	10%	
Cost efficiency*	Net profit / Total income	28.18%	15%	
Reproducibility*	(Net profit+wear & tear cost)/Total investment	15,72%	5%	YES
Liquidity	Based on financial flow, the project is liquid		liquid	YES
NPV	Sum of discounted net income of economic flow	160.396.022	>0	YES
IRR	Discount rate which brings the current value of the investment project to 0	34,63%	Higher than discounted (8%)	YES
Return period	Time period (number of years) required for the investment to be repaid to the Investor.	10 years	Shorter than project lifetime	YES

According to newly designed structure of the final contour of the open-pit coal deposit Bogutovo Selo, a remapping of concession boundaries in the coal basin was performed. The coal exploitation concession boundary for the open-pit coal deposit Ugljevik Istok 1, foreseen for the use with Unit 1 of the Thermal Power Plant, is shown in Figure 3; As for the Figure 4, it shows the boundaries of the open-pit coal deposit concession for Ugljevik East 2, whose reserves have been provided for the requirements of the second unit of Thermal Power Plant Ugljevik.

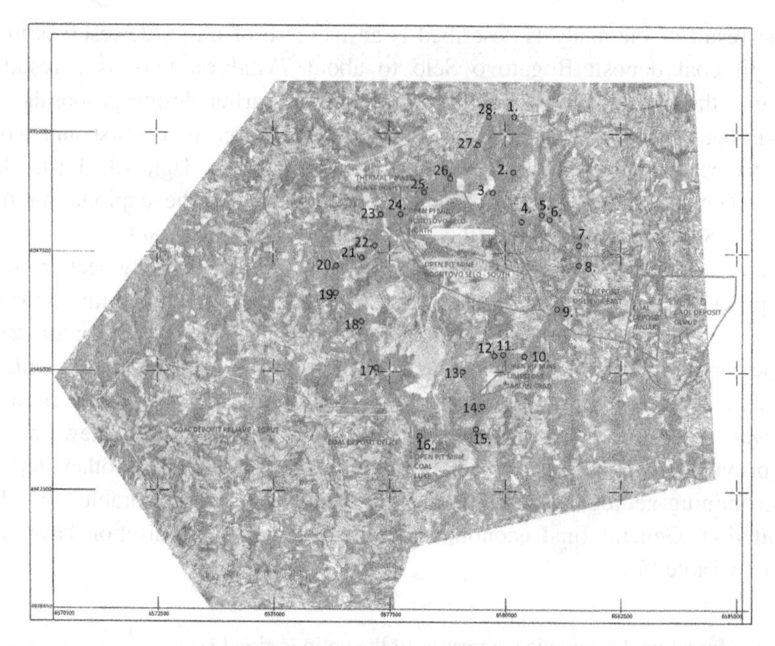

Fig. 4 Concession limit of coal excavation within Bogutovo Selo coal deposit and within a secion of Ugljevik Istok 1 coal deposit [2]

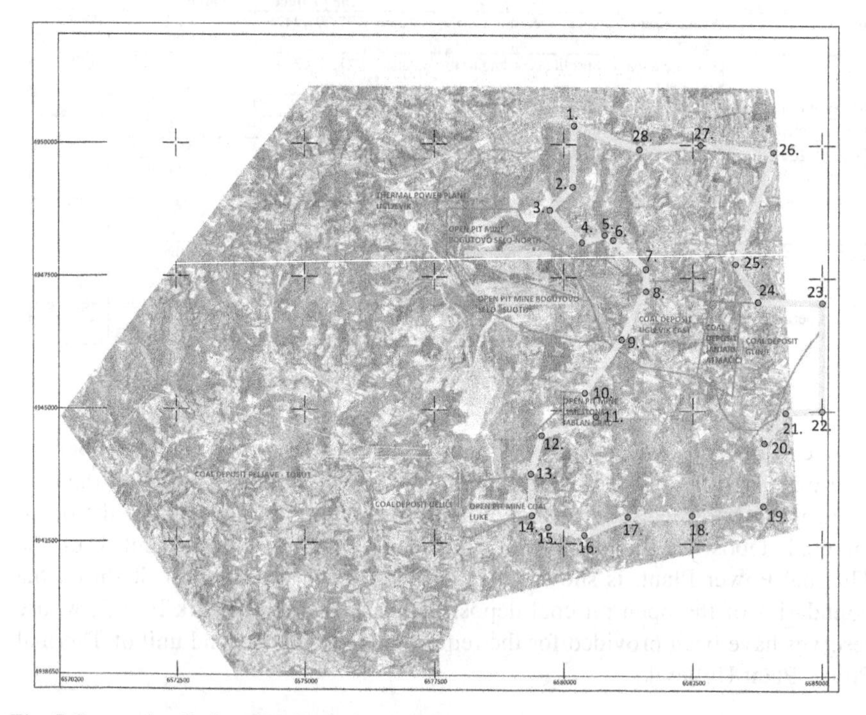

Fig. 5 Concession limits of Ugljevik Istok 2 deposit [2]

Conclusion

Advantages of the optimisation in comparison to the earlier designs:

1. Better utilisation of deposits
2. Balancing the overburden ratio, thus balancing operating costs
3. Significantly reduced the amount of investment overburden
4. No need for external landfills since almost all overburden from the future open-pit coal deposit Ugljevik Istok is to be deposited in the excavated area of the open-pit coal deposit Bogutovo Selo
5. Significantly less investment in infrastructure facilities, because of using the existing facilities
6. The ability to use existing equipment without additional investment of purchasing new equipment in the first years of exploitation, etc.

References

[1] Mining project of the open pit mine Bogutovo Selo, Mining Institute in Tuzla (2011)
[2] Expert analysis on the status of coal reserves within Bogutovo Selo and Ugljevik Istok deposits, Faculty of Mining and Geology - University in Belgrade (2013)
[3] Feasibility Study for the concession award for the exploitation of open-pit coal deposits Bogutovo Selo and Ugljevik East 1, Centre for surface mining in Belgrade (2013)

Results of In-lake Liming with a Underwater Nozzle Pipeline (UNP)

Michael Strzodka[1] and Volker Preuß[2]

[1] GMB GmbH, Knappenstraße 1, 01968 Senftenberg, Germany
michael.strzodka@gmbgmbh.de
[2] Faculty of Environmental Sciences and Process Engineering,
Brandenburg University of Technology, Siemens-Halske-Ring 8, 03046
Cottbus, Germany
preuss@tu-cottbus.de

Abstract. In fall 2011, one of the largest pit lakes of Eastern Germany was limed with a novel on-site process. Within 16 weeks of operation, its 110 Mm^3 of water was shifted from the iron buffer to circumneutral pH-values. Due to a thorough consideration of the chemical and hydrodynamic parameters the method obtained 80 % process efficiency. In this paper we will present the details of this novel UNP-process.

1 Introduction

Usually, after open pit mining ceases, the residual hole is filled and a pit lake is created. In the Eastern German Lusatian lignite mining area, this will result in Europe's largest artificial lake district. Many of these lakes comprise large water volumes and surface areas and are therefore amongst the largest lakes in Germany (Nixdorf et al. 2001). The inflow of potentially acid groundwater from the adjacent overburden dumps results in sulphate dominated acidic conditions. Many of the newly developed pit lakes have a pH of around 3.0, in the range of the iron buffer (Geller et al. 1998).

Back in the 1970s and 1980s methods were developed to treat lakes acidified due to acid atmospheric depositions (Nyberg 1988, Sverdrup 1985). However, the acidity of the acid sulphate pit lakes exceeds those of the Scandinavian softwater lakes by 2…3 orders of magnitude (Geller 2009). As a result a substantially larger amount of neutralizing agents is needed to treat the pit lakes. To minimize the costs for creating and keeping pH-neutral conditions it is essential to apply the neutralising agents as efficiently as possible. Current procedures, such as sprinklers (Benthaus and Weber 2012) or ships (Pust et al. 2010) that spread the suspension over the water surface have weaknesses that are inherent in the procedures and are mainly a result of the hydraulic and logistical circumstances.

C. Niemann-Delius. (ed.), *Proc. of the 12th Int. Symp Continuous Surface Mining - Aachen 2014*,
Lecture Notes in Production Engineering, DOI: 10.1007/978-3-319-12301-1_27

This paper presents our work that aimed at developing and testing a highly efficient procedure to lime the pit lakes.

2 Methods

Usually, lime products are used to neutralize acidified lakes. On-site they are mixed with water to produce a suspension which is then injected into the water body. To successfully inject the neutralizing agent as efficiently as possible, it is necessary to disperse the suspension evenly in the lake volume using a minimum amount of energy.

This requirement is best met by applying the free jet principle (fig. 1). Velocity differences between the free jet and the ambient fluid generate exchange processes at the jet boundary (Schlichting and Gersten1997). Fluid particles of the ambient fluid near the jet boundary are incorporated into the eddies and accelerated. The fluid particles are decelerated within the jet as a result of turbulent conditions and eddies in the jet direction. Due to the incorporation of ambient fluid into the jet, the jet volume increases with length and the jet velocity decreases while its momentum stays constant. Based on the investigations of Kraatz (Bollrich et al.1989) it is possible to describe the jet's velocity distribution and special development as a function of its initial velocity and length.

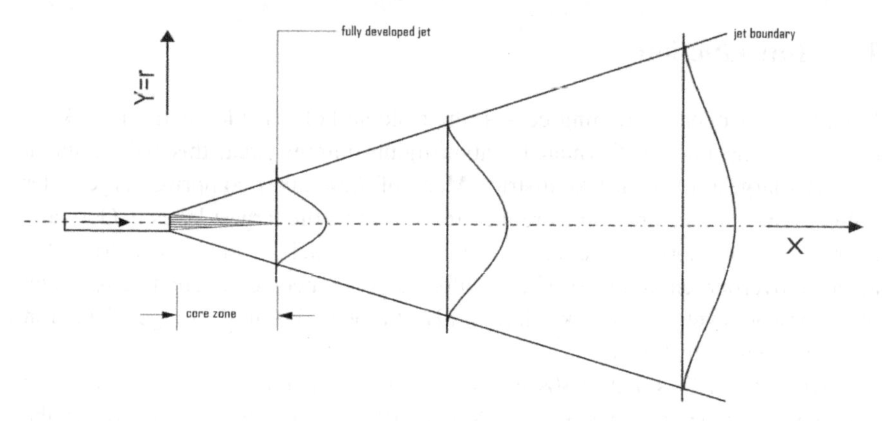

Fig. 1 Velocity distribution inside a free jet flowing into a stagnant and homogenous surrounding fluid

By using the free jet for injecting and mixing the lime suspension, density differences between the particle loaded jet beam and the ambient fluid are induced. Permanent mixing of the ambient fluid over the length of the jet causes a continuous dilution and consequently a reduction in the suspension's density. Those density effects superimpose the spreading of the free jet and determine the beam's trajectory (fig. 2).

The free jet's spread is either limited by the jet reaching the lake's floor or a layer of water with a higher density. Those layers might be a result of thermal stratification during the summer stagnation (thermocline) and the suspension will then spread horizontally along this boundary layer.

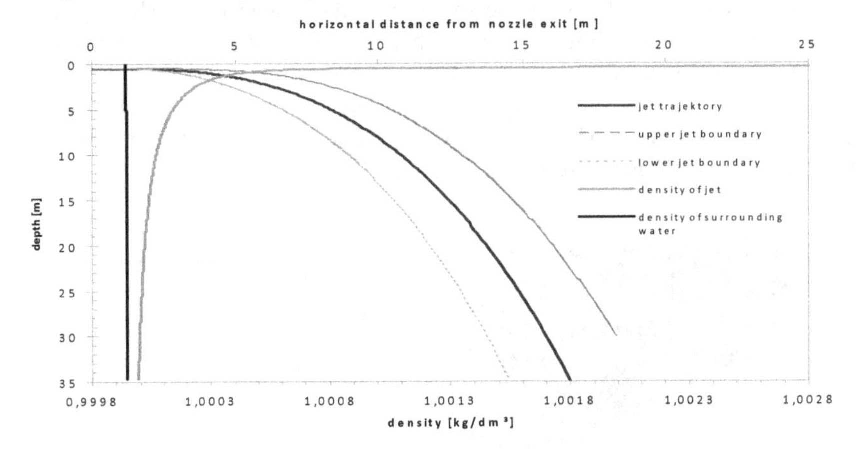

Fig. 2 Jet trajectory and change in density over the whole jet run length

In order to neutralize the acid pit lake, the lime suspension's dilution at the end of the jet beam's length must possess the chemically necessary application rate. The best neutralisation results will be obtained if the liming is conducted during the lakes full vertical circulation periods as the complete length of the jet beam can be used for the mixing of the lime suspension (fig. 2). An adequate number of nozzles ensure that the ambient lake water mixed into the free jet spreads throughout the entire volume of the lake during the application period at least once. The method works on a 24-7 basis and consequently even in the case of large water bodies a relatively short period of time is necessary for the liming.

Each pit has a characteristic chemical composition and morphology. Consequently, the UNP-process requires a configuration specifically designed for each water body. The parameters to be considered include the maximum concentration of the lime suspension, the best treatment period and the location of the nozzles to produce the free jet.

3 Area of Investigation

A first pilot test of the UNP liming process was conducted in the waters of the pit lake *Scheibe*, which has a volume of 110 Mm³ and a water surface of 6.8 Mm^2. With its length of 5.2 km and a maximum width of 1.7 km it is one of the largest pit lakes in the Lusatian lignite mining area. As a result of the lignite mining technology used, the lake's morphology is characterised by two distinct features: the eastern part of the lake consists of the former pit's inner dump with a shallow water area of 2…6 m depth and the western part with a water depth of 35 m (fig. 3). Lake *Scheibe* is characterised by dimictic conditions with full circulation phases in spring and fall.

A determining aspect for the water composition of lake *Scheibe* are the ground water inflows into the lake: from the south, from the mother rock a slightly acidic ground water with an acidity of approximately 1.0 mmol/L and from the inner dump water with an acidity of 9.0 mmol/L.

The initial state determined of the lake *Scheibe's* water to be treated was an acidity of 3.4 mmol/L, a pH of 2,9, as well as calcium and sulphate concentrations of 150 mg/L and 550 mg/L, respectively.

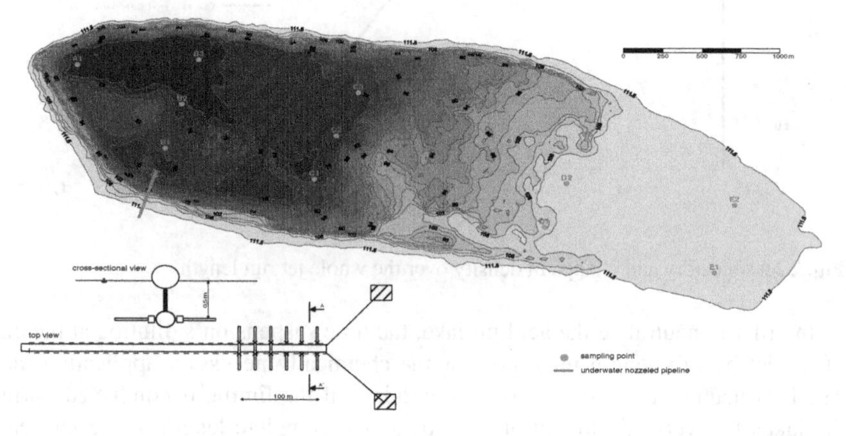

Fig. 3 Morphometrics of lake Scheibe, sampling points, and a detailed view of the UNP-process

4 Pilot Project Implementation

Prior to the pilot test, various potential lime products were investigated in the laboratory to determine if they can be used to neutralize Lake *Scheibe* with the UNP-process. The quicklime *(CaO)* provided by Fels-Werke GmbH proved to possess the best application properties with a application rate necessary determined of 150 g/m³ at an efficiency of 70 %. Consequently the necessary amount of lime to be added was 16.5 kt.

Technologically, the UNP-process has been kept simple: A submersible pump draws water from the lake and supplies the mixing station with this water by ways of a pipe. Two lime silos dose the neutralization agent into a mixing tank. From there, the lime suspension is pumped into a maturation tank and finally into a submerged floating pipe transporting the suspension into the lake. At the end of the pipe the lime suspension mixing nozzles are installed in pairs. Lake *Scheibe* had a nozzle configuration with 6 pairs at a distance of 20 m, which can be considered to be a punctiform injection in relation to the lake's size (fig. 3). On

October 4, 2011 the treatment of Lake *Scheibe* started and was successfully completed on January 25, 2012, after just 16 weeks of operation and two short operational interruptions of the liming installation (fig. 4). For monitoring the liming, 33 water samples were taken on a weekly basis at 12 sampling locations (fig. 3).

Based on the changes of the water quality, the specifications for the further operation of the neutralisation plant were determined. An additional monitoring of the hydraulic conditions of the lake provided the basis for validating a 3D lake model (MOHID-Water Modelling System). Both data were used to verify the design calculation algorithms previously used. In addition, the aim of the 3D modelling was to identify the fraction of the momentum input, density driven flow, and wind induced flow responsible for the overall water treatment.

5 Results

At the beginning of the treatment, Lake *Scheibe* was characterised by stratification with the thermocline being located at a depth of 12 m. Liming started initially with a 10.4 t/h mass flow, equivalent to 250 t/d. Such an application rate makes high demands on the logistics of the lime supply as up to 10 silo trucks were needed daily. As expected, the application of the suspension was restricted to the epilimnion, but wind induced currents during this phase of the injection assisted in the uniform distribution of the concentrations throughout the whole epilimnion. Consequently, a treatment effect could already be observed at the two farthest measurement points E1 and E2 in the first week of operation (fig. 3).

A certain proportion of the neutralising agent is stored as a result of its horizontal spreading along the thermocline. Since the samples were always taken from the same depths, part of the injected neutralisation agent is therefore not detected and thus, the average effect in the entire lake is temporarily underestimated. In view of a full lake circulation and the subsequent homogenisation of the lake's conditions the treatment effect is eventually correctly represented.

During the continuation of the water treatment, the thermocline gradually disappeared and the effect of the treatment could fully develop over the entire water depth. This phase of the treatment is purely controlled by momentum input and the density driven flow in the lake.

Figure 4 shows the temporal development of the average lake's conditions. As a measure of the acidity and alkalinity the modified neutralization potential NP of Schöpke (2008) is used.

As planned, at the end of the water treatment, Lake *Scheibe* exhibited pH-neutral conditions with a 0.16 mmol/L buffering capacity. The amount of lime used was 15.2 kt, which is less than calculated and the chemical efficiency of 80 % was above the pre-determined value. All project objectives agreed to with the client were met and the financial framework was not exhausted. With treatment costs of less than 0.01 €/mol the UNP-process is well below other lake treatment costs with lime.

Various boundary conditions for the lake treatment could be identified by the 3D modelling. Wind induced currents are supportive only within the epilimnion. Yet, the main treatment effect is controlled by the momentum input and the density driven flow (fig. 5). Moreover, the pre-determined parameters for predicting the process could be proved to be sufficiently accurate.

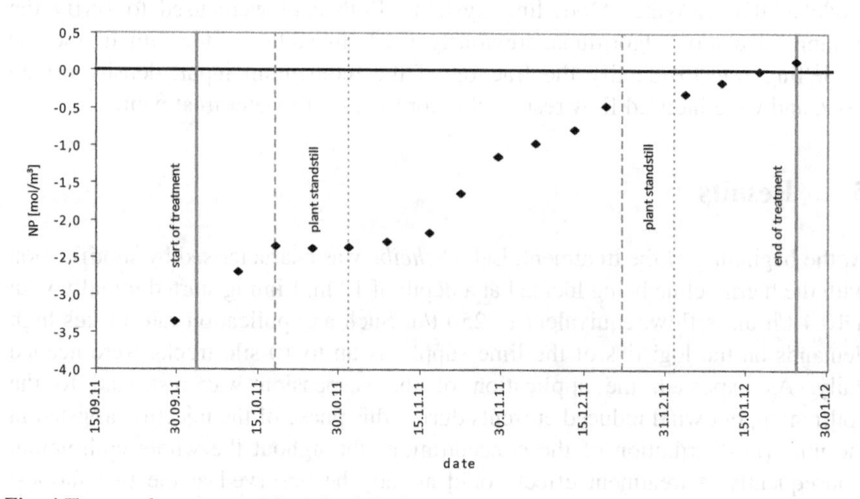

Fig. 4 Temporal progress of the lake *Scheibe* treatment

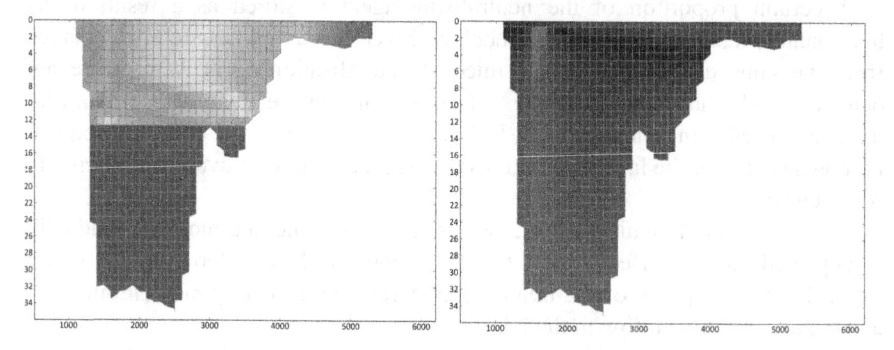

Fig. 5 Longitudinal section of Lake Scheibe with discharge of lime suspension under stratified conditions (left) and while full circulation is completed (right) [dimensions in m]

6 Extrapolation

The continuous inflow of acidity via the ground water leads to the lake becoming acidic once again, thus creating the necessity for further technical measures to maintain the neutral conditions. The calculation was made with the premise of an acid inflow of 27.1 Mio. mol/a with a follow up neutralization after one year with an introduction of approx. 1,000 t quicklime. Adding CO_2 leads to a hydrogen carbonization creating a buffer and significantly lengthening the time required before aftercare treatment is necessary.

Calculations were made based on the UNP pilot test of Lake *Scheibe* and discussed with the client. As a result aftercare follow up treatment is planned for the spring circulation phase of the lake waters March/April 2014. Thereafter hydro carbonization will be carried out of the lake with the addition of pulverized limestone and CO_2 simultaneously. The UNP process is technically suitable for this and as in the primary and first aftercare neutralization its reliability and economical benefits have been demonstrated.

The installation will be assembled as follows: For liming the UNP equipment remains unchanged as described. The gaseous CO_2 will be dissolved in a reactor on land. The dissolved CO_2 will be also introduced using the equivalent technical installation. The underwater nozzle pipelines with its nozzles will be placed approx 1 m above the bottom of the lake at its deepest point, >30m deep. In contrast to the primary neutralization, the CO_2 and lime will be introduced in the summer stagnation period. The underlying idea of the procedure is to use the entire body of water for the reaction. During the introduction phase the lime suspension is self-distributing in epilimnion, the dissolved CO_2 in hypolimnion. While the lime particles gradually sink they dissolve and have a hydrogen carbonate buffering effect first in hypolimnion. In autumn with the dissolution of thermal stratification both water bodies mix and lead to a buffering of the entire lake. 9,600 t pulverized limestone and 7.500 t CO_2 are intended for alkalinization. There will be a total application period of 16 weeks with a 6 week starting preparation time introducing CO_2 alone and then the remaining CO_2 and pulverized limestone are added simultaneously. According to the calculations the intervals between necessary follow up aftercare will increase to > 6 years at costs calculated as significantly under other procedures.

7 The Effect of Liming on the Downstream Ground Water Quality

The lake waters predominantly infiltrate the grounds on the northern side. The aquifer investigated is of sands and the permeability is $6.0*10^{-4}$ m/s. The 155 m distance from the shores of the Lake *Scheibe* to the ground water measuring point 6048 has a calculated flow time of 0.5 years. The local geological conditions were described in detail by Preuß and Koch in 2013.

Although a very rapid neutralization and the complete deferrization of Lake *Scheibe* took place –there was a substantial delay before the change in the water quality was detected (fig. 6). A clear decrease in acidity was first observed 1.5 years after the liming of the lake (three times the theoretical flow time). With the transformation of oxic lake water to anoxic ground water significant amounts of ferric ions are mobilized. Prior to liming the lake the groundwater had an iron content of approx. 70mg/l, more than double than the lake water approx. 28mg/l. Although at the ground water measuring point there was a significant decrease in iron content, the concentrations of 35mg/l were higher than any of those measured in the lake water prior to neutralization. (fig. 6 right)

Despite a slight re-acidification of the lake waters the dissolved iron content remains under 1 mg/l two years after liming was carried out.

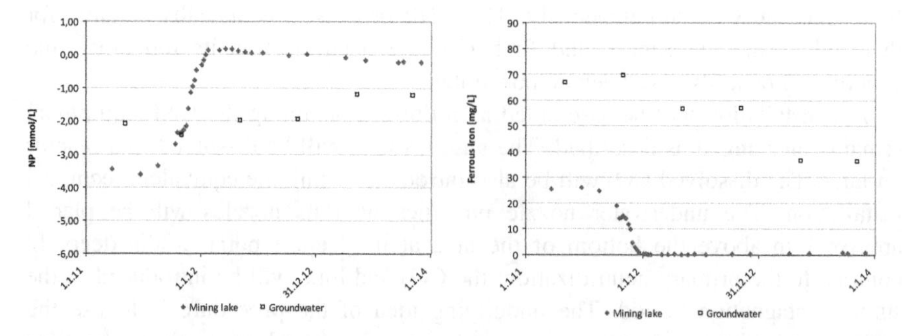

Fig. 6 Changes in the acidity (left) and lake dissolved iron concentrations (right) respectively as a result of liming the lake and the ensuing development over the following two years

8 Conclusions

As the LMBV pilot test for neutralising Lake *Scheibe* showed, a stationary, continuously working liming installation can treat large pit lakes within a relatively short period of time. The UNP-process described in this paper combines chemical and hydrodynamic conditions within its design calculation algorithm. In order to achieve an optimum treatment, the process is fitted into the natural circulation period of the water body. To our knowledge this was the first time that 15.2 kt of limestone were applied within a 16 weeks operation period. The chemical efficiency of 80 % exceeds the expected efficiency obtained during preliminary tests and the treatment costs of less than 0.01 €/mol extraordinarily prove that the UNP-process is a highly efficient treatment option.

Acknowledgements. This pilot project was conducted on behalf of the LMBV (Lusatian and Central German Mining Administration Company). The authors appreciate their financial contribution.

References

Benthaus, F.C., Weber, L.: Innovative Verfahren zur Verbesserung der Gewässerbeschaffenheit. In: LMBV-Flutungskonferenz 2012 am 08.03.2012 in Leipzig (2012)

Bollrich, G., Autorenkollektiv: Technische Hydromechanik Band 2, Spezielle Probleme. VEB Verlag für Bauwesen, Berlin (1989)

Geller, W.: Folgeseen des Braunkohletagebaus und deren Sanierung. WRRL-Seminar 30 am 19.01.2009 in Radolfzell (2009)

Geller, W., Klapper, H., Salomons, W.: Acidic mining lakes: acid mine drainage, limnology and reclamation. Springer, Heidelberg (1998)

Nixdorf, B., Hemm, M., Schlundt, A., Kapfer, M., Krumbeck, H.: Braunkohlentagebauseen in Deutschland - Gegenwärtiger Kenntnisstand über wasserwirtschaftliche Belange von Braunkohlentagebau¬restlöchern. In: Umweltbundesamt, UBA Texte 35/01 (2001)

Nyberg, P., Thørneløf, E.: Operational liming of surface waters in Sweden. Water, Air and Soil Pollution 41, 3–16 (1988)

Preuß, V., Koch, C.: Wechselwirkungen der Wasserbeschaffenheit zwischen Tagebaufolgeseen und dem Grundwasser am Beispiel des Scheibe-Sees. In: Proceedings des DGFZ e.V., Heft 49. Eigenverlag (2013)

Pust, C., Schüppel, B., Merkel, B., Schipek, M., Lilja, G., Rabe, W., Scholz, G.: Advanced Mobile Inlake Technology (AMIT) – An efficient Process for Neutralisation of Acid Open Pit Lakes. In: Wolkersdorfer, C., Freund, A. (eds.) Mine Water & Innovative Thinking, pp. 175–178. CBU Press, Sydney (2010)

Schlichting, H., Gersten, K.: Genzschicht-Theorie. Springer, Heidelberg (1997)

Schöpke, R.: Experimental Development and Testing of an in Situ Technology to Reduce the Acidity of AMD-laden Groundwater in the Aquifer. In: Rapantova, N., Hrkal, Z. (eds.) Mine Water and the Environment, pp. 337–340. VSB – Technical University of Ostrava, Ostrava (2008)

Sverdrup, H.U.: Calcite Dissolution Kinetics and Lake Neutralization. Lund Institute of Technology, Dissertation, Lund (1985)

Analysis of Sound Emissions in the Pit and Quarry Industry

Alexander Hennig and Christian Biermann

Department of Mining-Surface Mining and Drilling,
RWTH Aachen University, Germany

1 Introduction

The awareness of the population of harmful influences in their environment is increasing and the operations of the pit and quarry industry are coming to the fore of public interest. They are often discussed with regard to sound emissions, which arise from the mining, transport, processing and loading of the materials to excavate.

The goal of this analysis was to find out by way of two pit and quarry operations what the emissions are due to the operation and how they are to be evaluated on the basis of legal guide values and psychological threshold values.

The measurements were made during operational hours as well as non-operational hours so as to ascertain to what extent the operation is responsible for sound emissions.

In order to obtain an overview of the propagation of the emissions punctually measured results have been interpolated to the whole plant premises and adjacent areas. From these measurements sound maps have been generated making the special distribution of the emission values visible and forming a basis for the evaluation of the emission behavior of the plant.

The measurements cover a total period of about 150 hours. By saving the measurement values by the second, 540.000 measuring values were available for the propagation calculation.

2 What Is Sound

The human ear perceives sound at any time; unlike the eye it does not have the ability to close itself in order to so prevent the intake of sounds.

The ear takes in the sound which is processed in the brain as a sensatory stimulus. Only then, any person decides whether it is perceived as pleasant or unpleasant.

If the sound perceived and processed is considered unpleasant it is generally called noise. According to this definition sound is a subjective matter.

C. Niemann-Delius. (ed.), *Proc. of the 12th Int. Symp Continuous Surface Mining - Aachen 2014*,
Lecture Notes in Production Engineering, DOI: 10.1007/978-3-319-12301-1_28

Sound that is perceived as pleasant can also be noise in the sense of legal regulations and provisions if it exceeds levels that are hazardous to health. The table below provides these levels.

Table 1 Threshold values for health risks [1]

.	dB (A)
Psychological reactions, e.g.high blood pressure, increased heart frequency	>60
Inhibited behavior, symptom of first damage caused by noise	>65
Danger of permanent hearing impairment if perceived for a longer period.	>85

Sound is measured in Decibel (dB).

"Decibel denotes the measured sound pressure i.e. the sound output per area. To express the measured values in simple figures the decibel scale is logarithmic so that a difference of a few decibel can account for an over-proportional difference in sound intensity." [1]

In view of its effect on humans sound is usually measured in dB(A). This means that the characteristics of the human ear considered in sound measuring. As humans are significantly less sensitive for deep frequencies filters are used for the dB(A) measurement that attenuate these frequencies.

However, the difficulty remains that the dB values are abstract values that are difficult to interpret without reference values (see table 2).

Table 2 Orienting sound levels of various sound sources [2]

Sound level (dB)	Sound source	Sound sensitivity
20 to 30	ticking of clock, whispering of leaves	Sound hardly audible
40 to 50	Conversation, quiet residential street	Weak sound
60 to 70	Loud conversation, office noise, car from 10m away	Moderate sound
80 to 90	Traffic noise,noisy factory room	Strong sound
100 to110	Car horn 7m away, boiler shop	Very strong sound
120 bis130	Air hammer 1m away, jet engine	Deafening sound
140 bis150	Close-up range to explosion or jet engi-ne	Pain

3 Regulatory Legal Provisions

The determination of admissible emissions that may emanate from industrial operations or other technical installations and also the immissions to be measured at an immission point is regulated in Germany by the "Law on the protection from hazardous environmental effects by air pollution, noises, vibrations or similar activities (Federal Immission Protection Law)

This law prescribes that installations that due to their characteristics or operation are especially suitable to cause hazardous environmental effects require a special immission law approval, next to construction-law approval [2]

Even if an installation does not require approval it must nonetheless be operated in the way that all hazardous effects on the environment are avoided according to the state of the art.

The measurements and the evaluation of the sound emissions of industrial installations are effected according to the German guidelines of „Technical instructions for noise protection" (TI-Noise). The TI-Noise is the "Sixth General Administrative Rule of the Federal Immission Law (Technical instructions for noise protection).

Its purpose is to protect the general public and the neighborhood from hazardous environmental effects by noise as well as the prevention of hazardous environmental effects by noise. It applies for installations that require approval and are subject to the second part of the Federal Immission Protection Law..." [3]

The TI-Noise stipulates the immission target values. They indicate what noise immissions are allowed in different areas. These immission target values have been used in the sound maps.

The adherence to these values needs to be checked by the responsible state environmental agencies unless the installations are subject to mining law.

4 Measurements in the Gravel Pit

The examined pit produces 750.000 tons of gravel and sand per year using wheel loaders and conveyor belts.

The measuring points are located within the boundaries at significant points of emission generation and are further distributed around the gravel quarry. Within the quarry the measurements were carried out at nine points and outside the plant at six points. The location of the measurement points can be seen in figure 1.

At all measuring points non-operational and operational measurements were carried out. That was not possible at all points due to operations. In the case of the non-operational measurements outside the plant it has to be pointed out that the traffic was much lower during non-operational hours than at operational hours. Therefore, it can be assumed that the measurements do not give a true picture of the non-operational volume towards the bottom end.

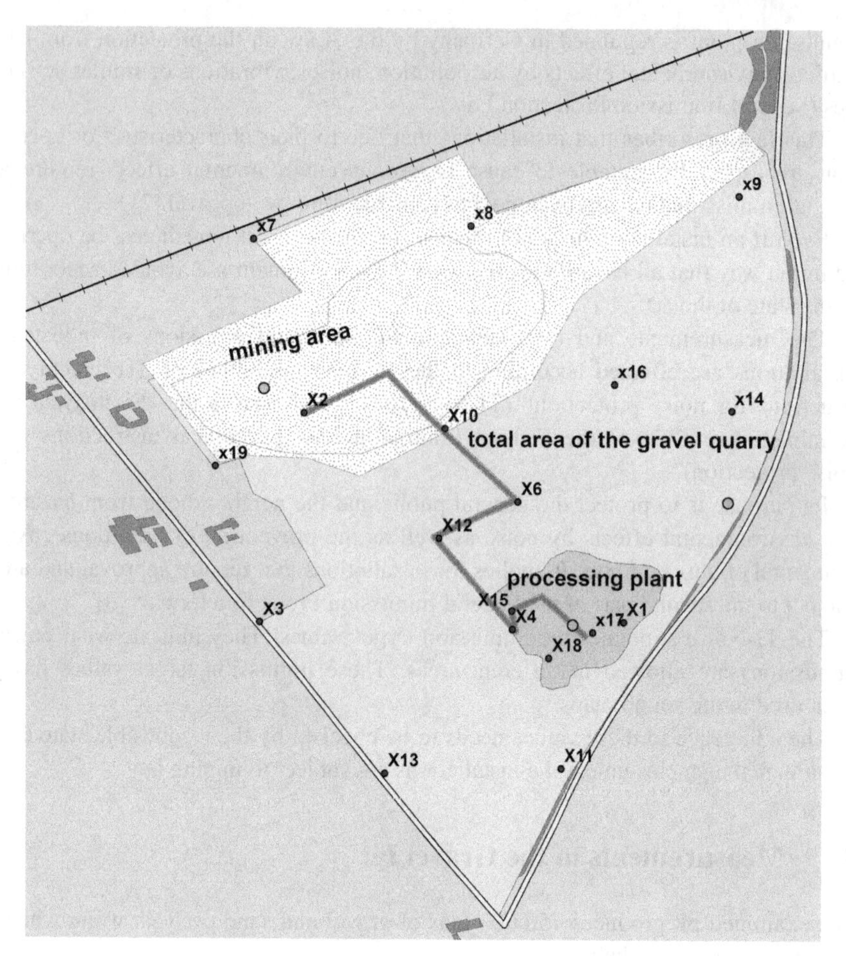

Fig. 1 Overview of the measuring area

The map displayed in figure 2 shows the distribution of the expected exposure approximately calculated with ArcGIS (Spline with Barriers) during the operational hours of the gravel plant. The boundaries of the individual zones were drawn according to the Immission guide values indicated in the TI-Noise as listed in the legend to figure 2. As there are emissions inside the plant too, which are above the values allowed for industrial zones and which are not defined as a class in its own right two more areas were added to this classification, one range between 70 dB(A) und 75 dB(A) and an another between 75 dB(A) and 80 dB(A). These two additional zones of 5 dB(A) steadily prolong the scale upward.

The map suggests that outside the boundaries of the plant no high immission values occur caused by its operation. The maximum values to be expected based on the calculations occur in two short boundary areas in the north- west and west of the premises, which are admissible for mixed zones. In the larger part of the direct surroundings of the plant the values are admissible for general, in most cases even all-residential areas. Further outside, the effect of the noise level generated by the gravel plant naturally wanes. The calculations here only resulted in values that are admissible for spa areas.

Here, it becomes very evident that the immissions to be expected outside the plant and caused by it are so low that a lasting impairment of the quality of living conditions can basically be excluded.

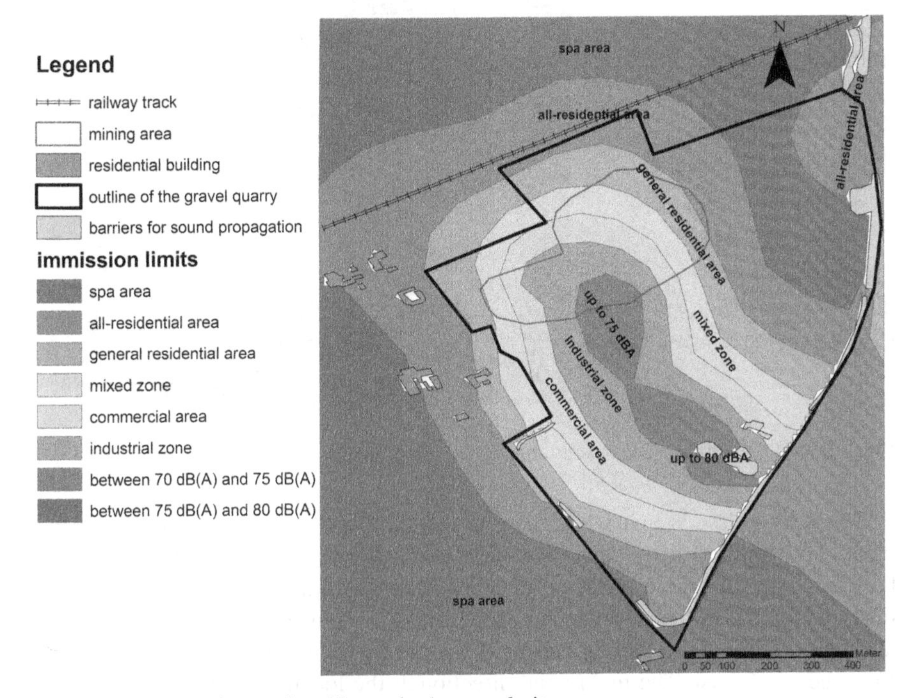

Fig. 2 Sound map of operational hours in the gravel pit

Inside the plant the highest strains to be expected are above the maximum values of 65 dB(A) are admissible for the industry These values may arise near the processing plant as well as at the installations near the conveyor belt. The values calculated in this area are in almost all cases still below 75 dB(A), only directly at the processing plant values of up to 80 dB(A) were measured. So they are still below the threshold of 85 dB(A) that is hazardous to human hearing. The load on the staff can be considered as low as the personnel do not stay in the open (production area) during normal operations. In these areas vehicles are used for work where the drivers stay in closed cabins. It can be assumed that the cabins lower the sound from outside to non-critical levels.

What is more, there are large noise exposures in the area near the scales. Since the staff here are in a closed building health risks can be widely excluded.

Figure 3 displays the calculated sound propagation during non-operational hours for comparison.

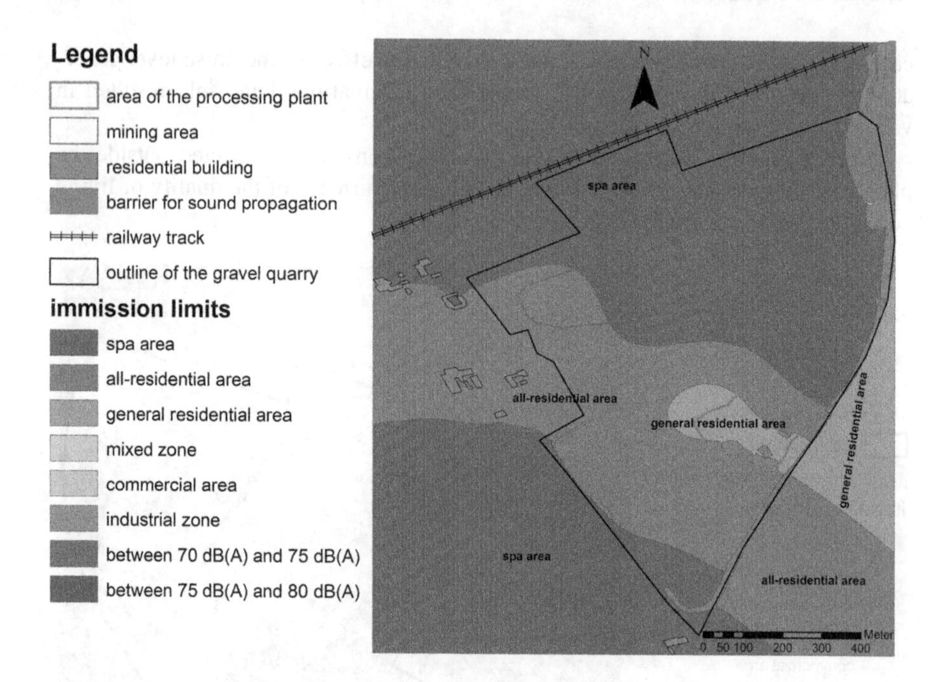

Fig. 3 Sound map during non-operational hours, gravel plant

The bottom line is that while there is nuisance caused by the operation of the plant in the areas directly adjacent to the plant it is on a low level. These findings, however, can only be transferred to other plants within limits because there are many factors of influence in sound propagation. So, nuisance on residents would also be possible in the case of this plant caused by an unfavorable wind direction from the south-east. The main wind direction at the location of the gravel quarry, however, is south-west and so a permanent change of the immission situation by wind can be ruled out.

5 Measurements Inside the Quarry

The plant produces basalt on a total of five levels using drillings and blasting, where only one is in operation at a single time. The loading is effected by a back-hoe onto a heavy-duty truck taking over the transport. The plant has a yearly output of 200.000t of material.

Figure 4 shows the area of the quarry with different production areas and the positioning of the measuring points. The plant is located directly next to a federal highway.

In this plant too measurements were carried out inside the plant and at its boundaries.

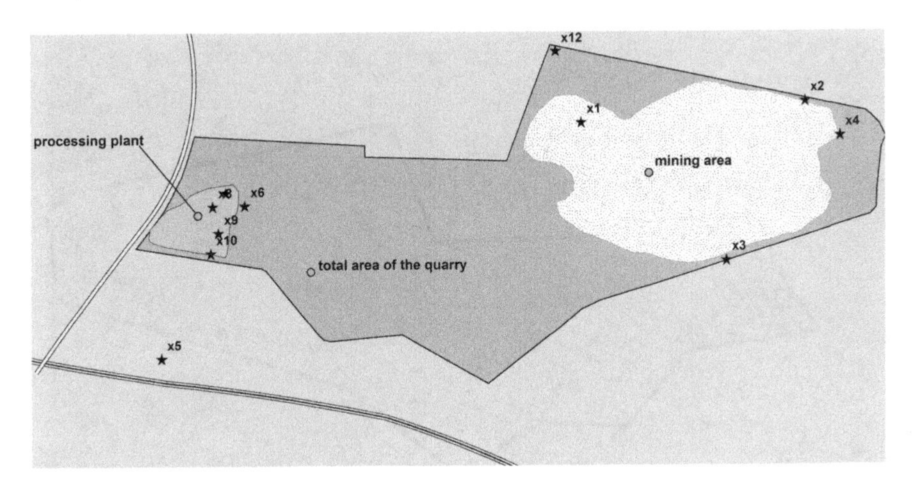

Fig. 4 Overview, quarry

In this plant, too, measurements were carried out during operational and non-operational hours.

The map displayed in figure 5 shows the distribution of the nuisance to be expected approximately calculated with ArcGIS during the operational hours of the quarry. The separation of the individual areas was done according to the immission values of the TI-Noise. Unlike the gravel plant, the presentation of the sound emissions in this map the scale had to be completed towards the top by four more ranges. In addition to the sound map regarding the gravel plant the categories between 80 dB(A) and 85 dB(A) as well as over 85 dB(A) were added. Concerning the category above 85 dB(A) it has to be underlined that this value may lead to non-reversible hearing impairments when humans are permanently exposed to it.[see table 1]. The numbers displayed in the marked-up areas are to be regarded as dB(A) maximum values.

In analysing the map it becomes evident that the highest emission values are measured in the zone around the processing unit. The values are partly above the range of 85 dB(A), which is hazardous in the case of permanent exposure. As can be seen from the map the value rapidly decrease as the distance from the plant decreases, yet this does not mean, however, that this is also the case in reality.

Here, a difficulty in drawing up the maps comes into play. The interpolation was calculated and regarded in 2-D, any barrier of sound propagation thus seems to be the same. In the case of a 3-D modelling that would take into account the actual height of the barriers themselves and their spatial position to the emission source, we could assume that the high values in the northern direction would not decrease so rapidly. The digital terrain model of the plant premises necessary to this end was not available.

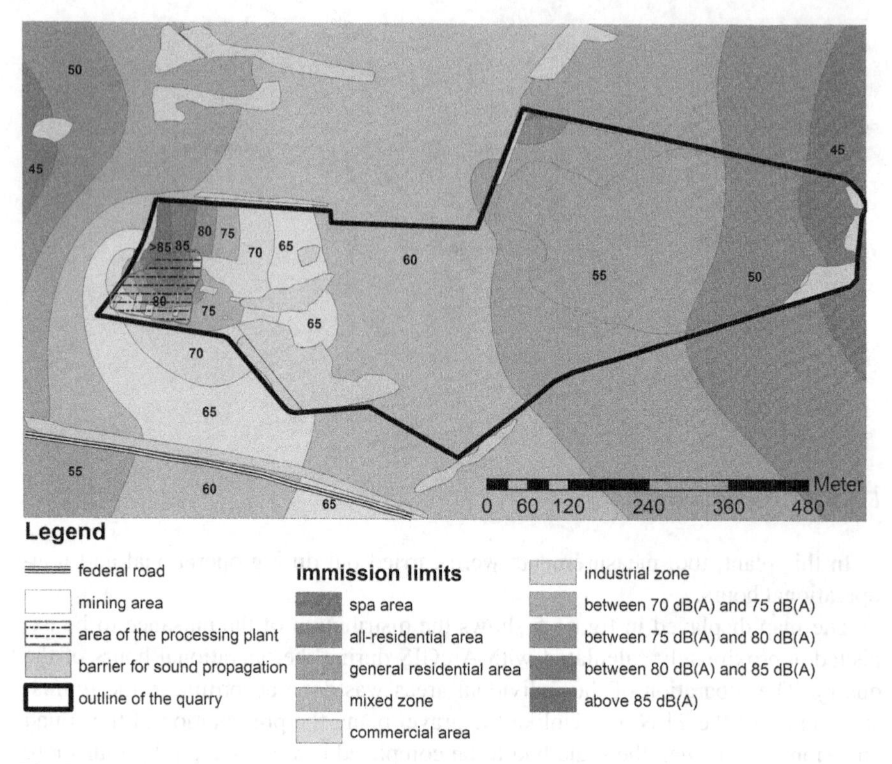

Fig. 5 Sound map during operation hours, quarry

Since there are no residents in the near vicinity of the plant any conflict over the emission peaks can be largely excluded. Nonetheless, this aspect should be taken into account in view of a transfer of the measuring results to other plants or the planning of other measuring campaigns.

In the mining area located in the eastern part of the premises the low calculated average values make for the fact that it hardly plays a role in the total average exposure. So, the mining area is always within one range admissible for residential areas. Towards the east it even fades into a range admissible for residential areas, further east even spa areas would be allowed. On the whole, the highest values that go beyond the boundaries of the plant are in a range admissible for the industry and rapidly decrease to a level admissible for mixed-use zones. If there are any problem zones regarding emissions leaking outside then they are towards the direction of the highway in the south of the premises. In this direction the calculated emission values are relatively high; they are in the value range of mixed-use zones.

To evaluate the exposure of staff inside the plant two areas mining and processing have to differentiated. In the mining zone their exposure is relatively low on average with values below 60 dB(A). They are thus in a range that can be called hazard-free according to table 1. There is only a problem regarding the maximum levels. Since the workers are either in the excavator or truck cabins during mining dangerous exposure can be widely excluded.

In the area near the processing plant significantly higher values are obtained on average, which may lead to hearing impairments during extended exposure. As no staff member stays in the open in this area for a longer time hazardous exposure cannot be fully ruled out but it is relatively unlikely. However, it is preferable that these levels are lowered in the area.

Figure 6 in turn shows a sound map during non-operational hours for comparison.

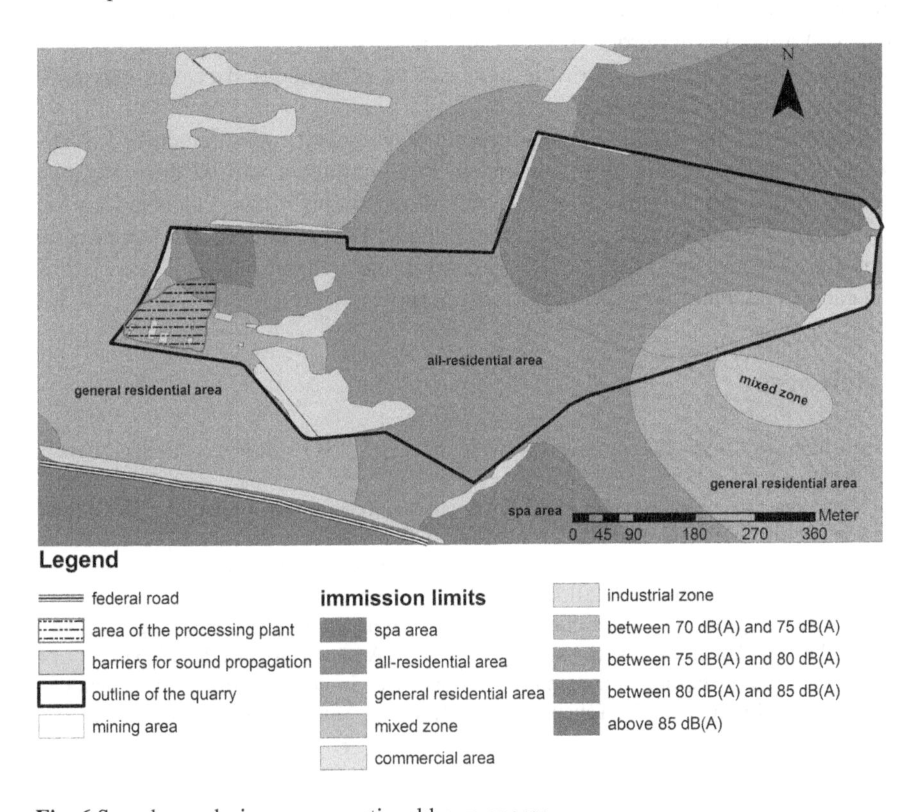

Fig. 6 Sound map during non-operational hours quarry

6 Summary

The studies conducted in the two operations proved considerable sound immissions outside the premises, especially true for the quarry. But already in a distance of 200 m the immissions decrease to a level which is acceptable for mixed-use zones. Main sound emitters in open pit operations are processing plants and loading terminals. Sound emissions of more than 85 dB(A) have been measured close to these facilities inside the quarry. Extended exposure to this sound level can lead to non-reversible hearing impairments. Hence hearing protection is required in these areas.

Relatively low sound levels are emitted at the loading point and by transport operation. Exceptions are blasting operations. Sound levels of more than 115 dB(A) occur. A specific measurement of sound emissions caused by blasting operations was not part of this project.

On one hand the results found cannot be directly transferred to other projects or operations as the sound transmission is directly linked to the specific site. On the other hand the effort for these measurements is rather small, as the measurements only have to be done during operational hours. At critical locations (close to settlements) a monitoring program is recommended. With the results generated by a computer based calculation critical areas can be identified and certain measures for noise reduction can be derived.

In addition to the data already gathered further acoustic measurements of single processing units, loading terminals and mobile equipment will be done at other quarries in the future. This set of data will allow computer based calculations for any terrain, presuming that a digital terrain model is available. Hence forecasts of acoustic emissions for green-field projects or project expansions will be possible and can be taken into account for planning purposes.

References

1. Aecherli, W.: Umweltbelastung Lärm, 1st edn., pp. 18–19. Rüegger Verlag, Zürich (2004)
2. Landesamt für Natur, Umwelt und Verbraucherschutz Nordrhein-Westfalen, Geräusche von gewerblichen und industriellen Anlagen, `http://www.lanuv.nrw.de/geraesche/gewerbe9.htm` (letzter zugriff December 3, 2012)
3. Sechste Allgemeine Verwaltungsvorschrift zum Bundes-Immissionsschutzgesetz (Technische Anleitung zum Schutz gegen Lärm – TA Lärm) (August 26, 1998)
4. Fuchs, H.: Reduzierung von Sprengschallemissionen in den Festgesteinstagebauen der Stein- und Erdenindustrie, Tabelle 4-4, p. 51. Verlag Mainz, Wissenschaftsverlag, Aachen (2003)

Short Method for Detection of Acidfying and Buffering Sediments in Lignite Mining by Portable XRF-Analysis

Andre Simon[1], M. Ussath[2], Nils Hoth[2], Carsten Drebenstedt[2], and J. Rascher[3]

[1] Freiberg Mining Academy (TUBAF),
Institute of Mining and Special Civil Engineering,
Gustav Zeuner Str. 1a, 09596 Freiberg/Germany
andre.simon@tu-freiberg.de
[2] TUBAF, Germany
{maria ussath,drebenst}@mabb.tu-freiberg.de
nils.hoth@tu-freiberg.de
[3] GEOmontan GmbH Freiberg, Am St. Niclas Schacht 13,
09599 Freiberg, Germany
j.rascher@geomontan.de

Abstract. The densely populated Central German region in the south of Berlin is rich in lignite. In addition to an energetic use of the lignite, the material use of lignite will become more important in the future. Due to the annual excavation of 1 billion m³ of sediments, valuable land will be lost. Therefore, it is important to use edited dumps for the generation of reclamation areas. These dumps should have good geotechnical as well as geochemical conditions, considering the surrounding waters and to protect the receiving waters. Unfortunately, the overburden above the coal includes pyrite, which causes problems such as groundwater acidification and the discharge of sulphate, iron and trace metals. As the groundwater level in the future tipping of the open pits will rerise, it is important to control the influence of weathering of pyrite depositions (acid mine drainage), buffering by glacial till and the resulting solutes on the surrounding water body. By superposing geochemistry and geology it is possible to mark main problem areas, improve buffer potentials by changing the technology and implement suitable technological countermeasures. Therefore a short method by using a portable XRF-analysis will be helpful to determine buffering and acidifying potentials. In the future, this will significantly reduce the impact to the surrounding waters by material loads from the dump bodies. The optimal use of the existing buffering potentials can save a large amount of money, that otherwise would be necessary for the subsequent rehabilitation.

Keywords: acid mine drainage, buffering, pyrite weathering, glacial till, geochemistry, dump structure, advanced mining technology, XRF-analysis, sulphate, iron, trace metals.

C. Niemann-Delius. (ed.), *Proc. of the 12th Int. Symp Continuous Surface Mining - Aachen 2014*,
Lecture Notes in Production Engineering, DOI: 10.1007/978-3-319-12301-1_29

1 Introduction and Objectives

In the area of lignite mining there are geological units which contain pyrite (FeS_2). So open-cast lignite mining induces sulphide weathering connected with Acid-Mine-Drainage-phenomena and the mobilisation of acidity, sulphate and cat ion metals (Wisotzky 1994). This partial weathering of sulphides is embedded in hydro-geochemical buffering reactions. Essential buffers are carbonates, aluminium or iron hydroxides and aluminium silicates. Especially the carbonate buffering is important (Hoth 2004). For sustainable strategic activities to reduce the acidification of groundwater around the lignite dumpsites of Peres, Schleenhain und Profen (Germany, South of Leipzig/Saxony), it is necessary to evaluate the acidification and buffer potentials of the overburden units. These investigations need to consider the applied mining technology and mass management (Rascher 2010).

For a sustainable development of strategic measures to avoid or reduce the acidification of groundwater in the surroundings of lignite mining, the evaluation of the natural geogenic potential in relation to the proposed mining technology is an essential issue.

A complete avoidance of the acidity generated through mining operations is impossible up to now. It is rather a question of minimising the negative effects by applying appropriate technology from the initial mine development to the final reclamation. Regarding this, primary, secondary and tertiary technical measures can be distinguished (Drebenstedt, Struzina 2008).

The first intention of the research was, to identify technological reduction measures to minimise groundwater acidification based on the hydrochemical situation within the mining field and by taking the mining technology into account. For this a complex system of research methods has been developed and will described below.

Because of the complex and long-term processing of the samples in the laboratory, it is beneficial to develop a short method to characterize the future overburden cheaper and quicker. So in operation, it will be possible to tilt the upcoming overburden in a structured and documented system.

In the context of the preparatory work on the project "ibi" (inovative lignite integration) at the Technical University Bergakademie Freiberg (Pfütze, Drebenstedt 2012) and the purchase of a handheld XRF-sensor in the ibi-project, the idea was born to develop a short method for the detection of acidifying and buffering sediments in mid German lignite mining. Furthermore, it is required to distinguish the overburden and valuable materials.

2 Solution of the Tasks

To solve the acidification problem a working scheme was created (Simon 2013). This includes three sections the geological model, the geochemical model and the technological production model. Each of these sections has been investigated with different processing methods. These methods will be explained in more detailed in the following. Figure 1 shows the basic concept and processing scheme of the investigations. The laboratory and field work is of particular importance for the processing (Simon 2012). Especially on-site eluates, weathering tests and buffering tests are useful and will described in these items.

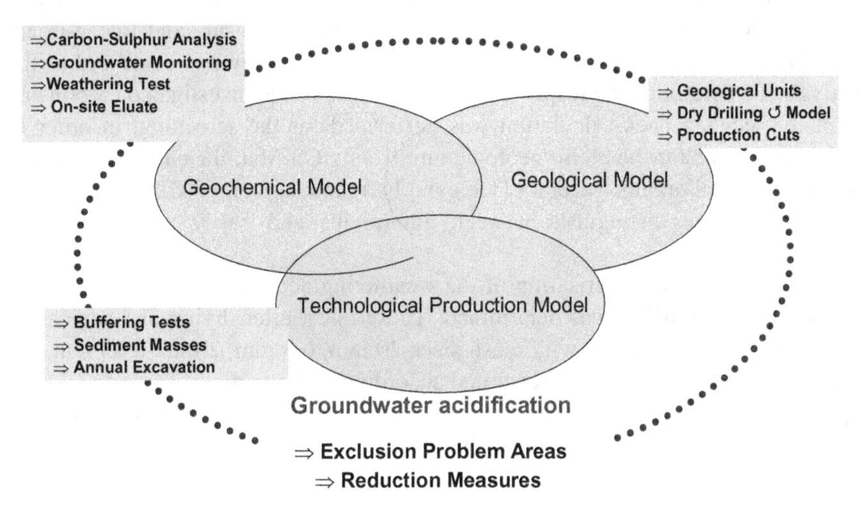

Fig. 1 Basic concept and processing scheme of investigation

2.1 On-site Eluates

The experiments for the determination of geogenic acidification and buffering potentials were carried out with the aim of identifying weathering-relevant units and relevant carbonate buffer potentials for the overburden profile. On-site eluates are a quick method to detect problematic material and buffer material. The sediments of the core boxes from the dry drilling were eluted and pH-value and electrical conductivity were measured. For elution, 10 mL of sediment and 25 mL of deionised water were mixed. pH-value and conductivity were measured by portable multi-parameter instruments (for example WTW 3430).

After that, the values could plotted depth-oriented. Test shows clearly low pH-values and higher conductivities in aquifer 2 and 3 sediments. In contrast, glacial till shows high pH-values and low electrical conductivities.

2.2 Solid Analysis - Hydrolytic Acidity, Carbon-Sulphur-Analysis, Grain Size Analysis and Water Content

To determine the **hydrolytic acidity**, as a measure for the tendency of acidification, 40 g field-moist sediment was filled in a NALGENE® wide neck bottle. It was loaded with 100 mL of a 0.1 molar calcium acetate solution and shaken for 1 hour in an overhead shaker. A centrifuge separation of the solids followed. If necessary, an additional filtration step was carried out on 0.45µm. After this, the solution was titrated with 0.1 M NaOH solution to a pH of 8.2 the end point of phenolphthalein. For a correct calculation the results were normalised by a blank value and water content to mmol/100g dry matter.

CS-Mat-analysis was used to determine the solids content of carbon and sulphur. The samples were sieved about a size cut out of 2 mm and freeze-dried. The fine material was ground by a ball mill and then finely powdered fed to the analysis. To determine the respective mass fractions of the investigated carbon and sulphur species a back-calculation was performed for the screening, in order to make statements for the entire geologic unit. Using CS-Mat, the carbon content in the form of total organic carbon (TOC) and inorganic carbon (TIC) was achieved. It is possible to distinguish between amorphous and easily available sulfur fractions as well.

As part of the characterisation of the weathering accessibility of the liners the **grain size distribution** was determined. These units, after drying and sonication in the sieve stack of sieves with mesh sizes 20 mm, 6.3 mm, 2 mm, 0.63 mm, 0.2 mm, 0.063 and 0.02 mm were separated according to the particle size.

Another important method is the determination of the **water content**. By weighing the mass of test sediment before and after drying at 105°C, the water content can be calculated.

2.3 Weathering Tests

Weathering tests were performed to characterize the accessibility and weathering potentials for the major geological units. After the geochemical characterisation the samples were stored in photo bowls and were exposed to weathering by atmospheric oxygen. For this purpose, the samples were moistened and turned at intervals. The investigated sediments were stored at a constant temperature of about 10°C (mean soil temperature around Leipzig) in the refrigerator.

There was a sampling of the weathering material after approximately 0, 7, 21, 49, 105, 210 and 500 days. In addition to the hydrolytic acidity a geochemical characterisation of the weathered sediments using two different eluates with differing solid-water ratios was carried out. The eluates having a solid water ratio of 1 to 2.5 and 1 to 25 have been shaken by means of the overhead shaker for one hour. Subsequently, the eluates were treated in the centrifuge to separate solid from the liquid. For further analysis the eluate samples were resolved with SCFA-0, 45µm syringe filter. Photometrically the contents of iron, iron (II) and sulphate were measured and the pH-values and conductivities were determined.

2.4 Buffering Tests

With the initial results of the weathering experiments and blending those with the proposed mining technology in the future construction mining fields, the actual procedure has been derived and implemented to the buffering tests. The aim was to investigate the reduction of the acidification by adding the buffering glacial till (carbonate buffer) to Tertiary material. The experiments especially focused on the main acidity-relevant aquifers 2 and 3. Buffering tests were performed after mixing the units in analogy to the weathering tests. Hydrolytic acidity and the parameters pH-value, conductivity and iron release and sulfate release of the eluate 1:2.5 were measured from each single test mixture from the photo bowls. For the recalculation of the values on the dry matter the water content was determined.

For buffering tests of the Aquifer 2 and Aquifer 3 sediments of the mining field Schleenhain and Peres, five individual approaches with differing mixing ratios of tills and acidic sediments were created (0% to 100%, 10% to 90%, 20% to 80%, 40% to 60% and 100% to 0%). The first and last single test represents the pure, unmixed materials. To obtain enough sample volume of glacial till and aquifer material, relevant, preserved (frozen) samples were combined to composite samples.

3 Short Method by Portable XRF-Sensor?

In the context of research studies, a portable Niton XL3t XRF-sensor could be purchased (Fig. 2). This portable X-ray fluorescence sensor system could be used in recycling, quality assurance, material identification, precious metal analysis, environmental, chemical, and many other applications. The analyzers of "XL3 classic series" are versatile, portable XRF instruments for fast element analysis within a few seconds.

Fig. 2 Niton XL3t XRD-sensor, Sample chamber and agate mortar

The main advantages of these devices are program modules for many applications as well as individual calibrations which are possible. In our case the pre-installed "Mining Mode" was used. Elements measured in the "Mining mode" are exposed colorly in Figure 3. All other items are specified as "balance" and mainly contain hydrogen, oxygen and carbon.

The XRF-measurement was performed on dried samples, ground in an agate mortar. Then the measured values were sorted and plotted stratigraphically. The results are shown in Figure 4.

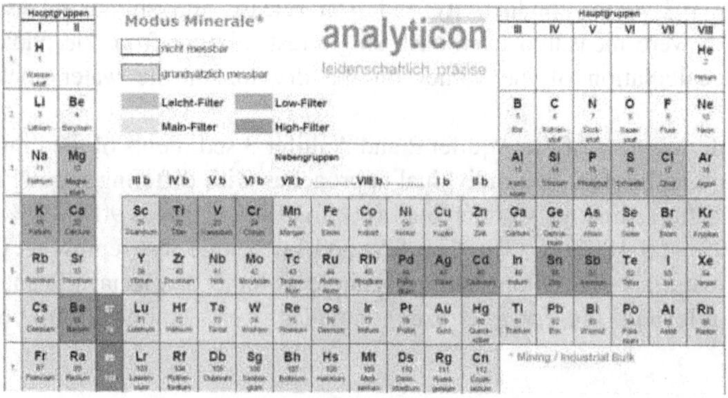

Fig. 3 via XRF un- (gray) and detectable (colored) elements

Looking at the results of the XRF-measurement, geological units can be summed up chemically, due to the distribution of the elements silicon, aluminum, sulfur and calcium, as well as the residual “balance”. In Tab. 1 the results are plotted.

Because carbon is the main component of coal, the "balance" is significantly higher (85 to 95 Mass-%) than in non-cohesive and cohesive sediments. So overburden and raw materials can be distinguished very easily. Even a distinction of different qualities of coal is quite possible by XRF. Especially suitable for distinguishing are the elements calcium, sulfur, iron, titanium and silicon.

All sediments have high aluminum and silicon levels. These characterize the overburden matrix. By means of the calcium amount the glacial till can be detected, because the carbonate is present as $CaCO_3$. In the sediment calcium can also be detected outside the glacial till, but in a much smaller level, for example as gypsum. 0.3 percent by mass have been determined as a distinctive criterion. This way, the buffer material can be determined.

The other sediments need to be differentiated in the terms of acidification tendency. As prior knowledge from previous studies for lignite mining, a division in three sediment classes is sufficient. Due to the sulfur content of the residual, sediments can be divided in non-acidifying, low-acidifying and acidifying sediments. As separation limits 0.3 and 1.5 mass percent were selected. These are no fixed values. By the means of the aluminum amount adjustments are possible,

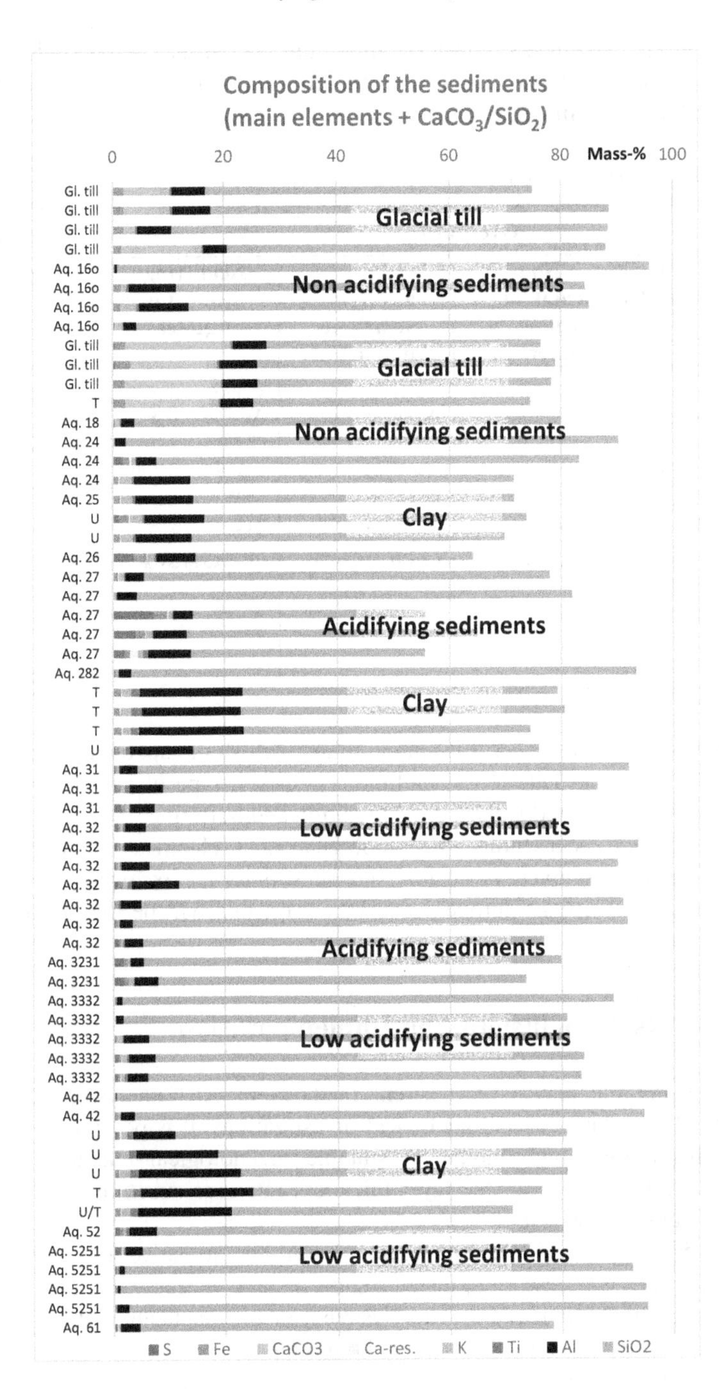

Fig. 4 Elemental composition characterization of dry drilled sediments classified by stratigraphic units

because cohesive sediments with high aluminum and sulfur contents release low acidity. If no subdivision is necessary, the sulfur amount can be associated as a numeric value to a sediment volume.

Table 1 Distinction excavated material

Overburden				Resource
Glacial Till Buffermaterial	**Non acidifying Sediments Aquifer/Clay**	**Low acidifying Sediments Aquifer/Clay**	**Acidifying Sediments Aquifer**	**Lignite**
sediment matrix: Al, SiO_2 + Al-Si-phases Si as SiO_2 = 50-70 Mass-% Al = 5-7 Mass-% Ca as $CaCO_3$ up to 20 Mass-% S < 0,5 Mass-%	sediment matrix: Al, SiO_2 + Al-Si-phases Si as SiO_2 = 70-90 Mass-% Al = 5-7 Mass-% Ca < 0,3 Mass-% S < 0,3 Mass-%	sediment matrix: Al, SiO_2 + Al-Si-phases Si as SiO_2 = 50-80 Mass-% Al = 5-7 Mass-% Ca < 0,3 Mass-% S = 0,3-1,5 Mass-%	sediment matrix: Al, SiO_2 + Al-Si-phases Si as SiO_2 = 40-70 Mass-% Al = 5-7 Mass-% Ca < 0,3 Mass-% S = 1,5-8 Mass-%	high residual phase, balance = 85-95 Mass-% S = 1-5 Mass-%

By means of a XRF sensor it is simple to distinguish the upcoming overburden and lignite in mining. Due to the preliminary tests, a type of calibration can be performed with abstracted values of acidification and buffering.

Perhaps in the near future this geological and chemical examination of sediments can help to invent or manufacture sensors for the excavator or conveyor belt that allow operating improvements. However, this will be the task of other working groups.

4 Results, Conclusions and Implementation

By superposing the results of different methods of XRF-analysis, as well as geotechnical, geochemical and geological investigations, it is possible to identify main problem areas and implement suitable technological counter-measures. Therefore the knowledge, of the layer thickness ratios or cubature of substantial overburden units is essential.

Figure 5 shows the different ways or methods to solve the task. Basic preliminary work is necessary to install a short method successfully. This requires a gauging of the XRF method. This system can be practicable, shown by the activities of RWE AG in the Rhenish lignite mining area.

As another result of the research, a new dump structure was created (Simon 2013). Rather than in conical strips, the overburden is now dumped in fine layers, so that a large contact area between the substratess and the problematic aquifers is created. Figure 6 shows the dumpsite of the open cast Schleenhain. On the right side it's possible to see the old unstructured dump area. On the left is the new layer dump structure casted by the spreader.

Proven method	Short method
Dry drilling	Excavation
Geological examination	Geological examination by geological model
On-site eluate	XRF-measurement
C-S-analysis	
Sieving	
Weathering test	
Buffering test	
Structured tilting	

Calibrati on

Fig. 5 Flowchart of test methods - qualification of the excavation

Fig. 6 Dump structure in mine Schleenhain, left new structure with thin layers, right old unstructured dump

The combination of the results of XRF-analysis, buffering tests and geological model allows efficient mixing of rare buffering material to achieve a good geochemical compositions and neutral pH values in the dump body.

Thus, it is possible to save the cost-intensive admixing of buffer materials such as lime or dolomite. The realization of the technology transition requires no additional major equipment. The implementation can be done with the present excavators, conveyors, conveyor switches and spreaders. However, a qualified technological control of mining, transport and deposit is necessary.

References

Drebenstedt, S.: Overburden Management for Formation of Internal Dumps in Coal mines. In: Fourie (ed.) Rock Dumps. Australian Centre for Geomechanics, Perth (2008) ISBN 978-0-9804185-3-8

Hoth: Modellgestützte Untersuchungen zur Grundwassergüteentwicklung in Braunkohleabraumkippen und deren Abstrom unter Berücksichtigung natürlicher Rückhalt- und Abbauprozesse. Schriftenreihe für Geowissenschaften, Heft 15, 214 p. (2004)

Hoth, et al.: Leitfaden Natürliche Schadstoffminderungs¬prozesse an großräumigen Bergbaukippen/-halden und Flussauensedimenten. Empfehlungen zur Untersuchung und Bewertung der natürlichen Quelltermminimierung. KORA Themenverbund 6: Bergbau und Sedimente. Institut für Bohrtechnik und Fluidbergbau, TU Bergakademie Freiberg (2008)

Pfütze, Drebenstedt.: Sensor based selective mining concepts in coal mining. In: ISCSM, Continuous Surface Mining, pp. 201–208 (2012) ISBN 987-615-5216-09-1

Rascher, et al.: Abschlussbericht zum FuE-Vorhaben: Lithofazielle Modellierung tertiärer Fazieseinheiten in Bergbaufolgelandschaften. - GEO montan GmbH Freiberg, TU Bergakademie Freiberg i. A. Sächs. Landesamt f. Umwelt u. Geologie, Freiberg, pp. 1–102 (2010)

Simon, et al.: Methods for the determination of geogenic acid mine drainage and buffering potentials formed in central German lignite mining. Scientific Reports on Resource Issues, pp. 10–19 (2012) ISSN 2190-555X

Simon, N.: Challenges in modern lignite mining – mass management to protect the environment and the budget. In: Proceedings of the 22nd MPES Conference 2013, vol. 2, pp. 833–842 (2013) ISBN 978-3-319-02677-0

Wisotzky: Untersuchungen zur Pyritoxidation in Sedimenten des Rheinischen Braunkohlerevieres und deren Auswirkungen auf die Chemie des Grundwassers, Besondere Mitteilung zum Deutschen Gewässerkundlichen Jahrbuch, 58, Ed.. LUA NRW, 153 p. (1994)

Market-Oriented, Flexible and Energy-Efficient Operations Management in RWE Power AG's Opencast Mines

Dieter Gärtner[1], Ralf Hempel[2], and Heinrich Rosenberg[3]

[1] RWE Power AG Head of Opencast Mines segment
50129 Bergheim, Germany
dieter.gaertner@rwe.com
[2] RWE Power AG Opencast Mines segment/Mining Technology,
50129 Bergheim, Germany
ralf.hempel@rwe.com
[3] RWE Power AG Opencast Mines segment/Operations
Management Systems 50129 Bergheim, Germany
heinrich.rosenberg@rwe.com

1 Introduction

RWE Power AG operates three large-scale opencast mines in the Rhenish lignite mining area with an output of about 100 million tons per year to supply the Company-owned lignite-fired power plants and refining factories (see Fig. 1). Some 450 million cubic metres of overburden must be moved each year to expose the lignite.

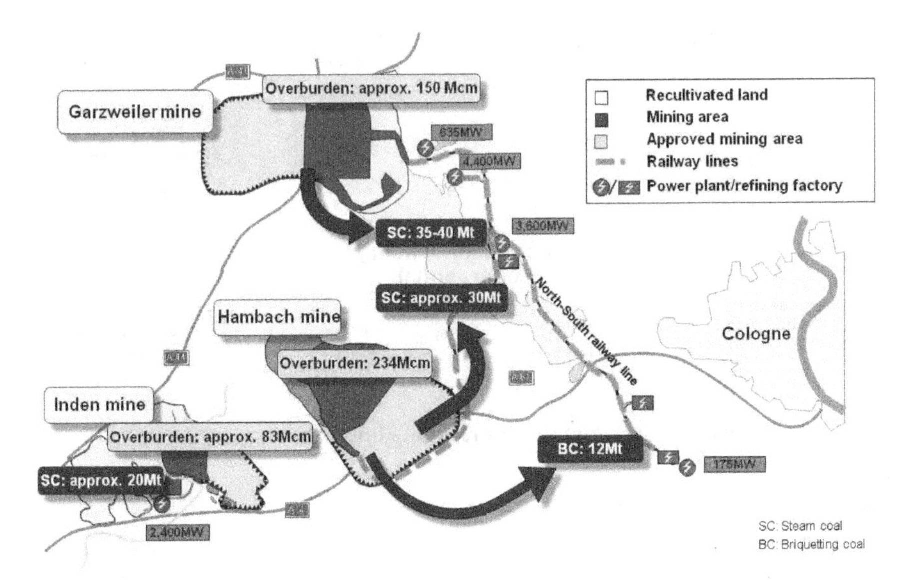

Fig. 1 Overview of the Rhenish lignite mining area and mass movements

C. Niemann-Delius. (ed.), *Proc. of the 12th Int. Symp Continuous Surface Mining - Aachen 2014*,
Lecture Notes in Production Engineering, DOI: 10.1007/978-3-319-12301-1_30

The development and operation of the opencast mines are determined by a multitude of "external" and "internal" factors, which will continue to increase in number both today and in future – not least due to the transformation in energy systems. Market orientation and the resulting demand for flexibility as well as energy-efficient business processes play a significant role in this regard. The complex interplay of the associated processes calls for carefully coordinated planning, scheduling and execution. Here, the operating team is supported by advanced IT and PDP systems, which can be adjusted to match the growing challenges.

2 Challenges Posed to Operations Management in the Opencast Mines

The development and the operation of an opencast mine are determined both by "external" factors such as power plants, factories, the environment, the neighbourhood, policy-makers and society and by "internal" factors such as geology, opencast mine technology, opencast mine planning and bunker management.

Some of these factors are becoming more influential, among other things, due to the transformation in energy systems (see Fig. 2). To meet these challenges, mining processes need to be optimised and must become more flexible, and further cost-cutting potential must be tapped.

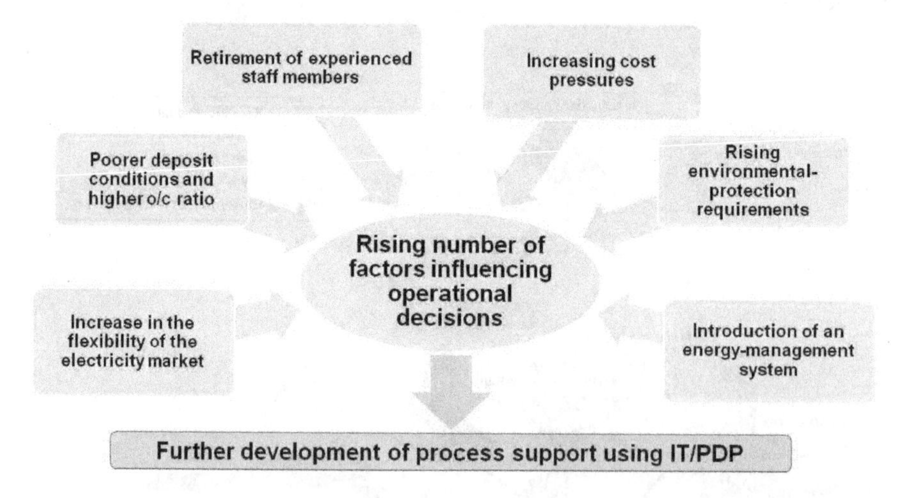

Fig. 2 Challenges posed to operations management

Several examples of the increasing challenges are presented below:

2.1 Increase in Flexibility

Owing to the volatile generation from renewable energy, the demand for lignite is subjected to ever greater fluctuations as lignite-fired power plants are redispatched, e.g. when the feed-in from wind power plants is high and at the same time electricity needs are low.

For operations management, the need for greater flexibility means:

- Short response times
- Assessment of response action with a view to its impact on other processes and on planning horizons
- Inclusion of personnel deployment planning in mining and downtime planning

2.2 Energy Management, Energy Efficiency

Following the introduction and certification of the energy management system according to DIN 50001 for RWE Power AG's opencast mines, the future will be all about implementing measures that reduce energy consumption in the long term and make the success measurable. For mining-related operations management this means that the power demand of large-scale consumers, particularly of the belt conveyors, must be considered as a further criterion when deciding on planning and scheduling options. This criterion will rely on the continuous measurement of energy performance indicators.

2.3 Environmental-Protection Requirements and Coal Quality

From 2014 on, the power stations need to comply with even stricter sulphur limits for flue gas. In the opencast mines, by contrast, the mean sulphur content will rise in the years to come, with significant differences on the individual coal benches. For the operations management of the opencast mines along the North-South railway line it is therefore important that the equipment deployment areas are coordinated in detail and with foresight across all opencast mines, with due regard to the blending rules for the specific power-plant units.

2.4 Mixed Soils

The extraction of overburden is increasingly facing drops in output due to difficult mining materials, so-called mixed soils (M2). They have a high share of cohesive components and are difficult to drain. The fact that only a limited quantity of these unstable mixed soils can be placed results in downtimes and bottlenecks in the placement process on the dumping side. Figure 3 shows one example of the imbalance between M2 supply and placement options. To deal with this imbalance and to ensure safe dump slopes, pinpointed and far-sighted operations management is required.

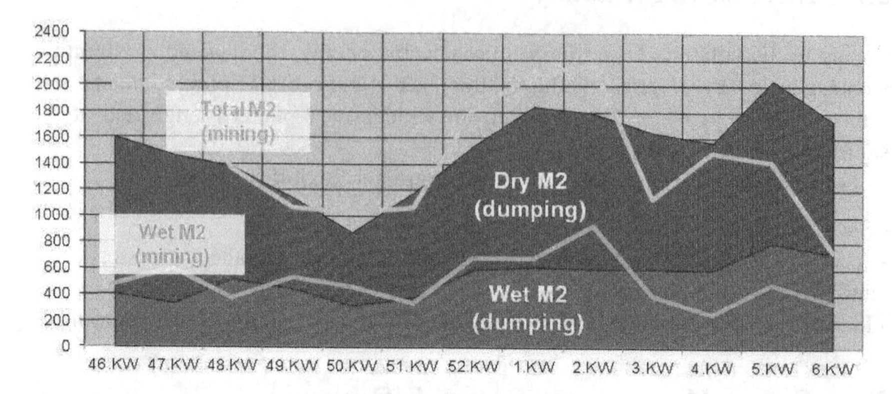

Fig. 3 Comparison of M2 supply and placement options

2.5 *Retirement of Experienced Staff Members*

In the years to come, operations will be more and more forced to do without experienced staff members (on age grounds), with requirements increasing at the same time. On the basis of programmed operational dependencies and rules, operations management systems can assume a supporting role in this context. The challenge faced in the development of this kind of system is to keep operation user-friendly despite significant intricacies in the background.

2.6 *Increasing Cost Pressures*

The further increasing cost pressures call for the inclusion of cost-related criteria – apart from the factors mentioned above – when assessing planning alternatives. One example is the evaluation of transports on the bench. The specific costs of travelling-gear repairs associated with these transports must be included in production planning by means of a key indicator.

3 Strategy for Developing Mining-Related Operations Management Systems

The development of operations management systems in mining looks back on a long history. Depending on the operational requirements and IT-related possibilities, IT and PDP systems were implemented. While in the past the systems tended to support individual processes, today's focus is also on main process chains and their interconnection with adjacent activities in ancillary processes. This makes it possible to achieve the joint overall objective of process optimisation.

An opencast-mine process model was defined to provide a clear basic structure for the development of the systems (see Fig. 4).

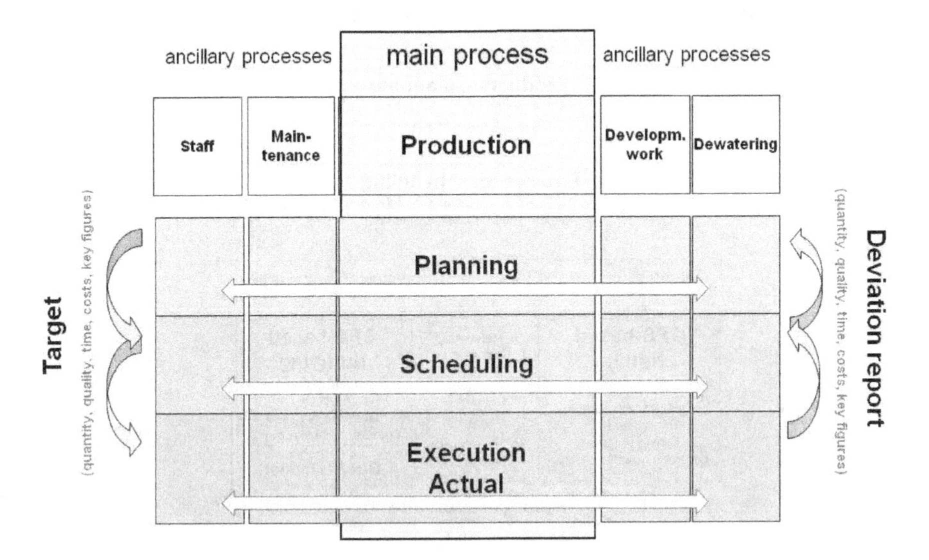

Fig. 4 Integrated opencast-mine process model

In this model, the main processes are understood to comprise the mining, transport and interim storage of lignite as well as overburden movement including the transport and dumping of the material. Personnel deployment, maintenance, the dewatering of the opencast mine and development work as supporting service including the use of auxiliary equipment are summarised as ancillary processes.

In principle, each opencast-mine process has three process levels. The first level involves the planning of processes all the way to equipment-deployment planning. The second level is about scheduling, i.e. the short-term use of resources. The third level deals with the actual execution, i.e. the real process with its operating, controlling and automation functions. At this level the actual situation of the opencast mine is mapped in the form of a digital model, serving as a basis for updating planning and scheduling.

To achieve optimum overall opencast mine operations in economic, safety and technical terms, the dovetailing of the individual processes is essential. In other words: the information systems need to be vertically and horizontally integrated, with the individual process levels being able to exchange information and to develop in line with the specified quantities, qualities, times and costs, while at the same time responding to deviations by way of a feedback process.

4 Status and Development of Operations Management Systems

There is a mining-area-consistent target structure for mining processes, which was developed on the basis of the integrated opencast-mine process model. This target structure provides all information relevant for production and ensures the exchange of information with associated processes (Fig. 5).

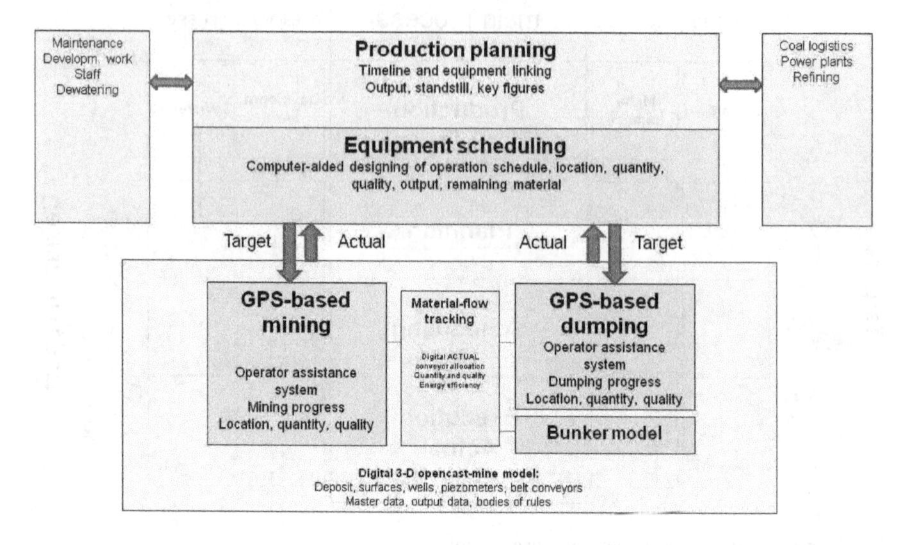

Fig. 5 Overview of operations management systems for mining production

Important individual systems are presented in what follows.

4.1 Digital Opencast-Mine Model

The reliable presentation and continuation of the actual mining and dumping situation in near real-time are of central importance. The actual situation is compiled and made available in the form of a digital 3-D opencast-mine model. Information from different areas is merged. In essence, this information includes the following modules:

- Geological deposit model indicating relevant rock strata, quality parameters, lithographs and tectonic structural elements. The material varieties that were actually encountered during mining and were classified by overburden material identification and coal online analysis are added to the model. This leads to a "learning" deposit model.
- Obstacles in the rock such as wells, piezometers and piping from water-management systems

- Locations of belt conveyors
- Surface of the opencast mine in order to display the mining and dumping progress as an important basis for determining the remaining material quantities. Here, photogrammetric data and the "quantity survey" from the high-capacity equipment's GPS systems are merged.

4.2 GPS Systems and Material-Flow Tracking

GPS-based control systems for excavators (satellite-assisted excavator operation control, *SABAS*) and spreaders (satellite-assisted spreader operation control, *SATAS*) and a material-flow tracking system (*MAFLU*) were developed in several steps in the opencast mines. This system group called *SAMASA* has been developed into a Rhenish mining area standard, which will now be rolled out in all operations.

These systems fulfil three important basic functions:

1. Operator assistance including the display of the area to be excavated or to be dumped at present
2. Precise recording of the mining and dumping progress by volume and material variety for documentation purposes and the determination of remaining material quantities
3. Identification of the masses and material varieties transported on the belt conveyors for the connection of excavators and spreaders and the calculation of energy efficiency

4.2.1 Satellite-Assisted Excavator Operation Control

For the digital mapping of the mining process, several GPS receivers and an evaluation system continuously determine the spatial movement of the so-called bucket-wheel enveloping surface on the basis of the equipment dimensions data. This surface is intersected with the read-in 3-D opencast-mine model (actual surface, deposit model and further information).

The surface, volume and material varieties are analysed at regular intervals between two points in time and are made available for the calculation of the remaining material, for documentation purposes and for the *MAFLU* transport model.

SABAS constantly compares the target surfaces from equipment scheduling with the actual position of the bucket wheel. On this basis, the operator assistance system calculates the current distances to the target surfaces. The remaining slewing range is continuously displayed in the horizontal direction, while in the vertical direction the precise distance to the working floor or slice bottom is shown (see Fig. 6). This enables the operator to cut an accurate working floor and to meet the requirements of the operation schedule, without complex geodetic survey staking being necessary as in the past. Since obstacles in the rock such as wells and piezometers are registered in the read-in opencast-mine model, *SABAS* is able to tell the operator at an early stage if they are approaching such operational installations and inform them of the remaining spatial distance to these installations. This likewise applies to cases in which operators approach tectonic faults (tectonic displacement areas from the deposit model).

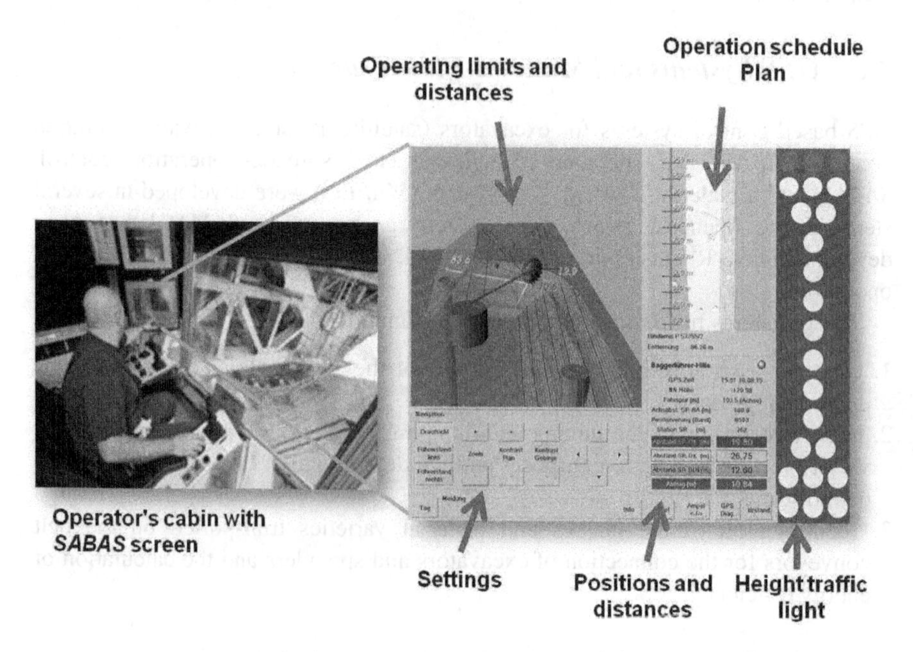

Fig. 6 *SABAS* operator assistance system

4.2.2 Material-Flow Tracking

Material packages from *SABAS* are transmitted to the *MAFLU* transport model which tracks transport via all belt conveyors of the conveying routes to the receivers in high resolution. The conveyor-belt speed, drop times at transfer points as well as interim measurements performed by belt scales and possibly online analysers are included in the calculation. The overall accuracy is basically determined by the measuring accuracy of the individual sensors.

By comparing the material movement with current consumption, *MAFLU* establishes the friction coefficient and consequently the energy efficiency of the individual belt conveyors. This may provide indications of, for example, poorly-aligned belt conveyors. It also provides the measurement basis for the energy management system.

The display of the changing material varieties on the conveying routes is crucial for operations management as it can help, among other things, to avoid the feeding of incorrect materials to the spreader or the coal bunker.

4.2.3 Satellite-Assisted Spreader Operation Control

In the case of the spreader, GPS is used to define the spatial position of the discharge boom. *SATAS* calculates a model of the developing fill on the basis of the material tracked by *MAFLU* and the discharge trajectory. The model provides a rough overview of the material distribution. Scanning lasers on the discharge boom continuously scan the fill surface and give the discharge model a more precise shape. If the scanning lasers are not available, the discharge model can serve as backup for the continuation of the model (Fig. 7).

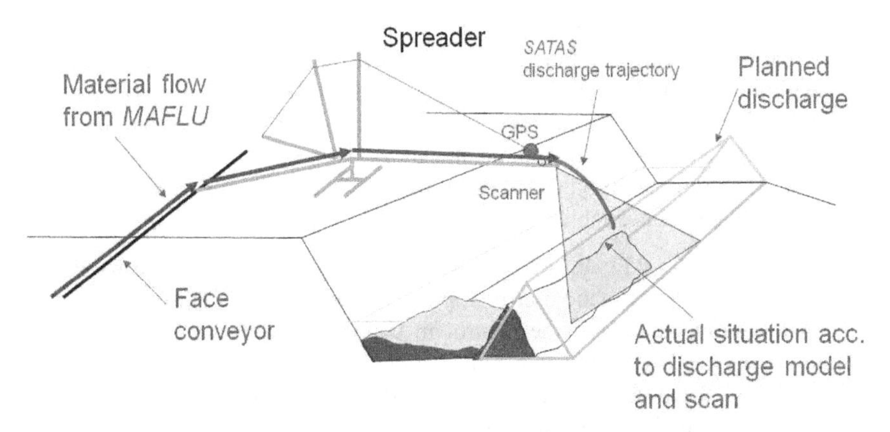

Fig. 7 Operating principle of *SATAS*

The high-capacity equipment operator is constantly informed of the current discharge height and reach in comparison with the operating schedule. In a combined representation of the actual discharge surface and the target surface, discrepancies are stated numerically. The scanned and modelled dump surface is made available for the calculation of the remaining material and for documentation.

4.3 Design Systems Used in Equipment Scheduling

4.3.1 Information Technology

Within the scope of the operation-schedule preparation and based on the current digital opencast-mine model and the specifications of medium-term opencast-mine planning, the specific spatial stipulations for excavator and spreader deployment are elaborated together with the associated material calculation by material varieties in place. For years now, the IT system *LAVA* (deposit management and assessment) has been offering suitable special-purpose applications of various designs (interactive CAD systems, batch-oriented calculation algorithms, assessment and visualisation functions) as well as Excel applications for slice division.

To fulfil the requirements to be met by integrated operations management systems, to cut the maintenance costs of the heterogeneous programme landscape, and to facilitate faster process operations, the slice division and dumping scheme planning modules of the equipment-scheduling software were modernised applying up-to-date IT options and standards. It is planned to use this technology also in other planning areas and in deposit modelling.

This leads to the following IT-related advantages:

- Elimination of system-related redundancies and inconsistencies in data storage by systematic object-oriented storage of the graphical and alphanumeric object elements in the database (RWE standard: ORACLE)
- Creation of a consistent programme-internal information flow in connection with interface reduction
- Use of default development environments and software elements to improve application development and maintenance
- Use of optimised backup processes to reduce the periods in which the IT systems are not available

As the affinity for IT varies among the users of the operations management systems, simple and systematic user prompting was emphasised. This was achieved by workflow-oriented programme control in connection with corresponding user dialogues that visually display complex background information and their individual editing steps to the user in an adequately condensed format. There are only few manually controlled single functions whose sequences are not controlled at system level (CAD-oriented drawing). Proceeding from a consistent data set (graphics and technical data), users are largely guided by the system in their editing steps and are able to work in an object-oriented way.

4.3.2 Slice Planning

As bucket-wheel excavators are predominantly operated in terrace-cutting mode, the individual slice is decisive for materials management. This particularly applies to coal mining, where the face shows widely varying coal quality parameters in the vertical direction. To supply the power plants with the right coal quality, it is necessary to define the slices as precisely as possible, with foresight and in line with needs, and to determine the coal grade.

For a long time, the EXCEL-based software *IKOLA* (interactive determination of lignite qualities based on deposit databases) was used for this purpose. This software, however, no longer meets today's requirements so that it was replaced with the new *IKOLA2* module which incorporates the new technology described above. The new module also offers the option to define slices in the overburden, so that the face can be divided in an optimal way as regards the quantities of mixed soils that are difficult to transport and dump (Fig. 8).

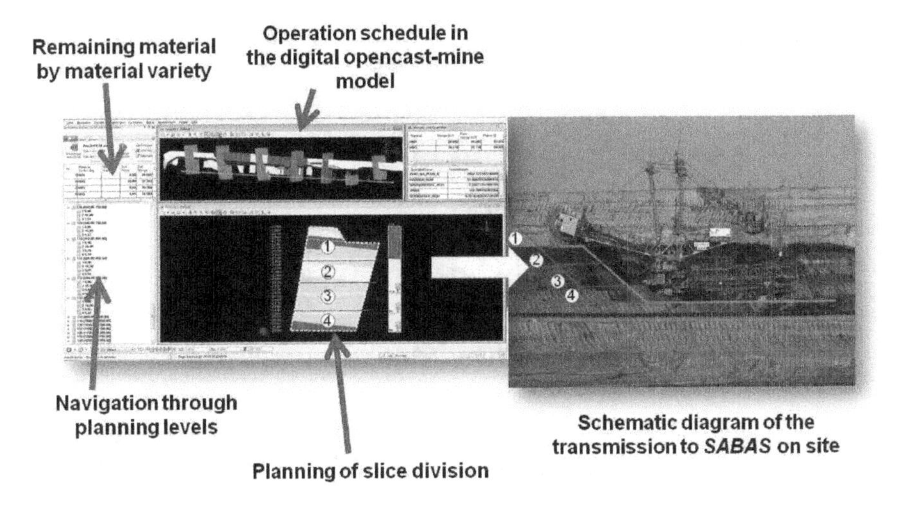

Fig. 8 Slice planning as a guideline for the mining process

IKOLA2 offers the following basic functions for slice definition:

- Interactive, object-oriented working on a 3-D model (3-D or 2-D views)
- Read-in of the current, digital opencast-mine model and general planning specifications
- Automatic slice division using the actually excavated slices of the previous conveyor-belt location
- Editing of the slice limits
- Ad hoc calculation of the slice averages of all quality parameters input to the deposit model
- Option to automatically define deposit horizons (e.g. hanging wall) as slice limit
- Ad hoc calculation of the remaining material either in relation to slices or the deposit structure based on the current surface from *SABAS*
- Transmission of remaining material quantities to the mining schedule
- Transmission of the final operation schedule to *SABAS*

4.3.3 Dumping Scheme Planning

For the placement of unstable materials, geomechanically-tested schemes must be observed which specify the permissible position and sequence for different material varieties. The *KIPMASS3D* module (Fig. 9) facilitates the complex spatial design process with functions which are similar to those offered by *IKOLA2*. Special functions of KIPMASS3D are automatic spatial positioning of the dumping schemes along the dumping level and Calculation of the resulting 3-D bodies per dumping-scheme element for the transmission to *SATAS*.

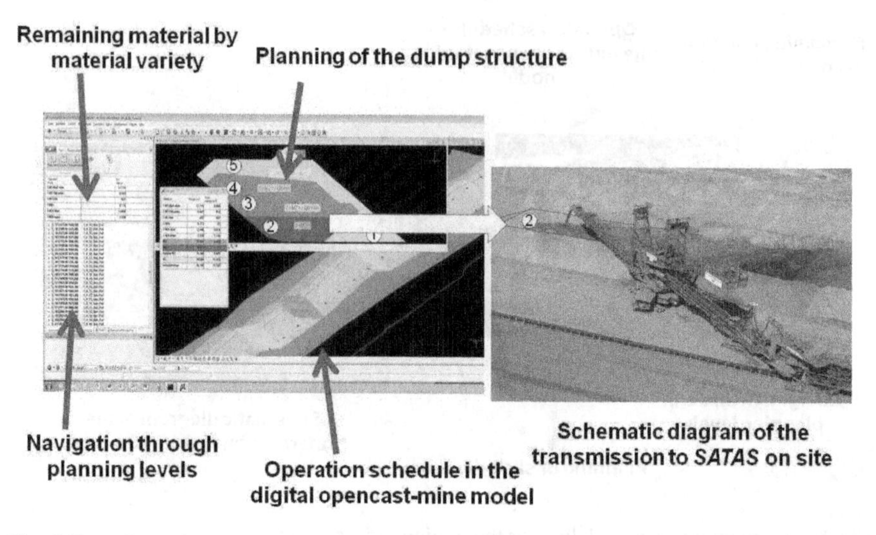

Fig. 9 Dumping scheme planning as a guideline for the dumping process

4.4 Production Planning

The purpose of production planning is to compare the material flow from all bucket-wheel excavators to the spreaders and coal bunkers in the different time horizons and the scheduled outages of equipment, while taking into account the manifold external and internal factors mentioned earlier (Chapter 2, Fig. 2).

The *FÖRDERPLAN* (Production Plan) system assists decision-makers in this complex key task. Typical features of this system are the illustration of the excavator and spreader block sequences and of scheduled outages for mining-related and maintenance purposes in the form of time bars indicating quantities and material varieties, output estimates and descriptive information (Fig. 10).

The level of planning can be adjusted, enabling the user to access even the slice level. The user interface permits interactive modification e.g. of block sequences. As *FÖRDERPLAN* is linked to the calculation of remaining material quantities of *IKOLA2* and *KIPMASS3D*, the current opencast mining progress is incorporated into the system. A wide range of assessment options, e.g. the comparison of mixed-soil supply and possible receivers or the development of the bunker level, opens the way for the immediate evaluation of planning alternatives. The number of key indicators for the evaluation of these alternatives will be stepped up gradually in order to be able to factor in cost and energy-efficiency aspects.

All opencast mines work with the same data set and use standardised software. This facilitates the comparison with other opencast mines, so that bottlenecks or conflicts over the supply of power plants with blended coal from the Hambach and Garzweiler opencast mines can be identified with foresight and in good time. Examples include the cross-mine assessment of the production plans with a view to the future development of the sulphur mean in the supply with blended coal, as this mean may not exceed a certain value. The system points out potentially critical periods, enabling the user to adjust the block sequence of the coal excavators accordingly.

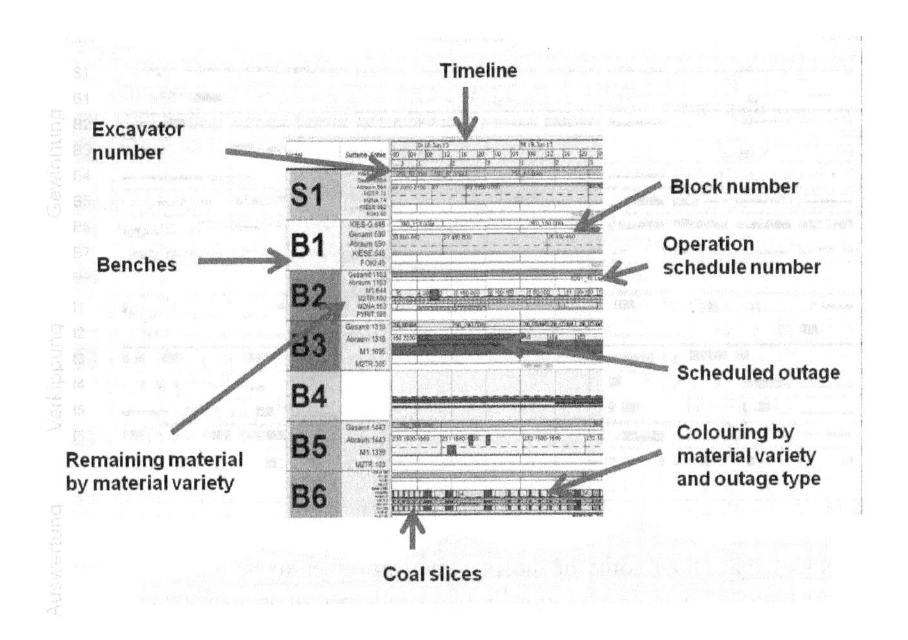

Fig. 10 Production plan with details

Since the timetable of production planning is linked to the spatial development of the blocks, a series of important details must be automatically provided for the purpose of horizontal integration. Examples include:

- Automatic notification of the point in time when wells and piezometers will be shortened
- Geometric opencast-mine position at a future date to provide a planning basis also for other activities (repair sites, pipes, pump stations, etc.)

The *ISSP* (integrated outage planning) system was developed on the same software basis as the production plan. It serves the close, interactive alignment of outages with the maintenance department.

5 Outlook

Further system integration is to the fore in the future development process. The linking of the production plan – being the key element of operations management – to the adjacent systems, e.g. load management and personnel-deployment planning, will be further improved. In addition, the energy management system is to be integrated into the operations management systems (see Fig. 11).

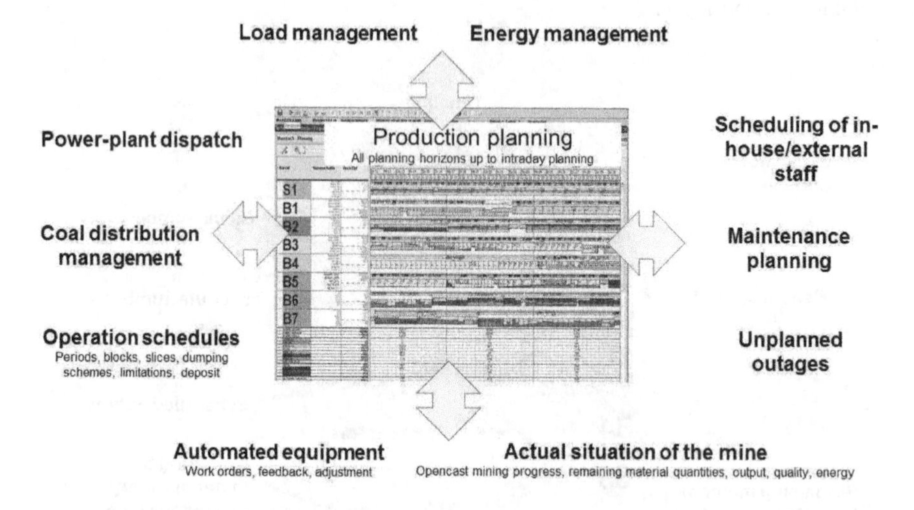

Fig. 11 Production plan with system integration

Modules that allow some of the planning processes to be automated will be integrated into the planning systems. In future, all operations management systems will be joined in a so-called "operations management control centre" and will provide an overview of all information crucial for decisions on the planning and scheduling of equipment deployment.

The simulation and optimisation software initially used for coal logistics in the North-South railway system is to be gradually implemented in other operations-management fields to improve the evaluation of alternatives. In future, (semi-)automatic high-capacity equipment will receive precise work orders from the operations-management systems for model-assisted operation.

6 Summary

The three large-scale opencast mines of RWE Power AG are operated in performance operations to supply the power plants and refining factories in the Rhenish lignite mining area. For some time now, the development and operation of the opencast mines have been determined by a multitude of external and internal factors, which will continue to increase in number both today and in future – not least due to the transformation in energy systems. Among these factors we find market-related flexibility requirements and energy efficiency. The rising number of factors influencing operational decisions calls for the improvement in and further development of processes and supporting IT and PDP systems – so-called operations management systems. This applies to all process levels – from long-term planning to the scheduling of high-capacity equipment – and along the entire coal process chain, starting with mining operations and ending with the coal consumer.

The operations-management systems rely on GPS-based technology which records equipment deployment and delivers a digital image of the opencast-mine progress in near real-time. At the associated scheduling and planning levels, equipment deployment, material distribution and coal quality control are carefully prepared. The result is a set of precise instructions for the operator assistance systems on site. Cutting-edge information technology is used, with user-friendly operational concepts promoting acceptance and being key to success. Upcoming further-development activities focus on continued system integration and the automation of planning processes and equipment.

Options for the Use of Sensor Technologies for a Quality-Controlled Selective Mining in Central German Lignite Mines – Results of the IBI-Project (Innovative Brown-Coal Integration)

Martin Pfütze and Carsten Drebenstedt

Technische Universität Bergakademie Freiberg, Germany

1 Introduction

With over 180 million tons per year, Germany is by far the largest lignite producer in the world. But only a small part of the German lignite is used for material applications. Therefore the Technical University Bergakademie Freiberg worked for 3 years with partners from the industry at the ibi-project (innovative brown-coal integration). The aim of the project was the interdisciplinary research on the material use of lignite in Central Germany. The main development goals were an operating diagram for selective mining of quality lignite as well as the development of mining concept, which uses sensor technology for material and boundary layer detection. The article gives an introduction how different sensors like georadar, XRF, as well as optical and infrared systems were tested and used to achieve these goals. Therefore the main working principles of these sensors will be explained besides the idea how to attach one of these measuring devices directly at the boom of a bucket-wheel excavator to measure and recognize the underneath lying surfaces or guiding layers. In the end the consideration and results how these technologies could work as an aid and the dividing layer recognition and selective mining process should occur is also examined.

2 Actual State of That in Selective Mining

The mining technology in the current situation is designed for a high output, in which the selectivity plays a minor role. But only selective mining guarantees a material application with a maximum product yield. Selective mining yet again is dependent on a precise material and boundary detection, which sets high level of sensory detection ahead. Furthermore in case of the actual production technology used in German opencast mines, like bucket-wheel excavators, bucket-chain excavators, Continuous Surface Miners and Mobile Technology, it is to mention that the machines with a high selective digging height own only a low dividing

C. Niemann-Delius. (ed.), *Proc. of the 12th Int. Symp Continuous Surface Mining - Aachen 2014*,
Lecture Notes in Production Engineering, DOI: 10.1007/978-3-319-12301-1_31

sharpness. Nevertheless primarily for reasons of the existing production technology in the Central German mining region and the economic efficiency as well as the demanded performances, the consideration of selective mining by bucket-wheel excavator was made as a result of suitable consultations. Therefore the bucket-wheel excavator was used as basic production technology in this project on which the operations of different sensor systems were tested [1, 2].

A selective mining of horizontally stored layers can be carried out with the bucket-wheel excavator in the side block cutting effectively. Thereby the excavator will proceed beside the opencast mining-sided embankment of the passageway and can remove so bigger bench parts in slices. Condition for this is that the thickness of the single layers to be separated of each other is largely enough to get still a sufficient production amount with the bucket-wheel. The separation of the layers can be carried out in the terrace cut in best when the bucket-wheel proceed after a slewing process (cut) several times in cutting direction and the height position of the bucket-wheel is only reasonable changed.

With the terrace cut the dividing surface can be laid between the single terraces in such a way that they correspond to the dividing surface between the single layers. However, the uppermost dividing surface may not lie higher than the lower edge of the bucket-wheel in his highest position. This is the selective digging height of the excavator (h_a, figure 1). For the operation of the excavator it is important that the maximum height position of the bucket-wheel can be reached in every position of the excavator, so also under a certain excavator slope [3, 4].

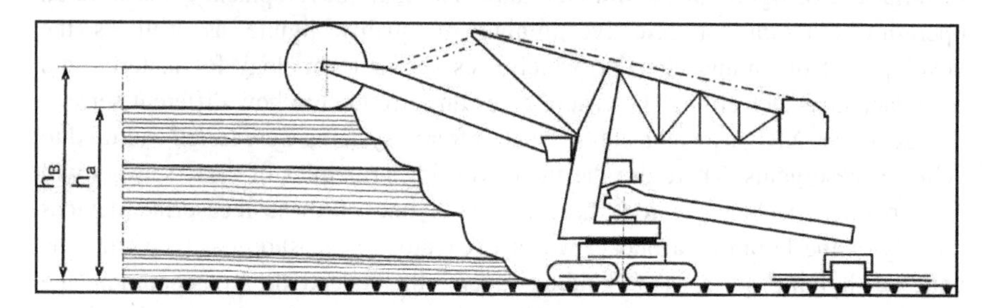

Fig. 1 Selective mining with bucket-wheel excavator (h_B-cutting height, h_a-selective digging height) [3]

3 Optimization Approaches

To optimize selective mining a high level of sensory detection is very important; whereby the fact here is that it needs to be sensors suitable for mining technology. The number of the possible sensor systems was limited after detailed search, considering the requirements for the sensors and on account of the past research results by the numerous procedures with which a dividing surfaces recognition is possible. While looking for potential solution attempts in the area of sensor technology it is a matter of considering that with a choice of a suitable measuring method no direct contact of the sensor with the coal face is aimed to bend forward oversized wear and damages of the sensor technology. Because in addition measuring procedures working with contact are often very sensitive and complicated, therefore as a result the decision was made to investigate measuring procedures working without contact. These offer the advantage that the sensors must be hardly protected against wear by abrasion. An overview of the possible measuring procedures is shown in figure 2. Hereby the technologies of georadar, X-ray fluorescence (XRF) as well as infrared detection or optical sensors have been identified as especially promising. Therefore these sensor technologies and the way they were tested will be explained in the following passage [5].

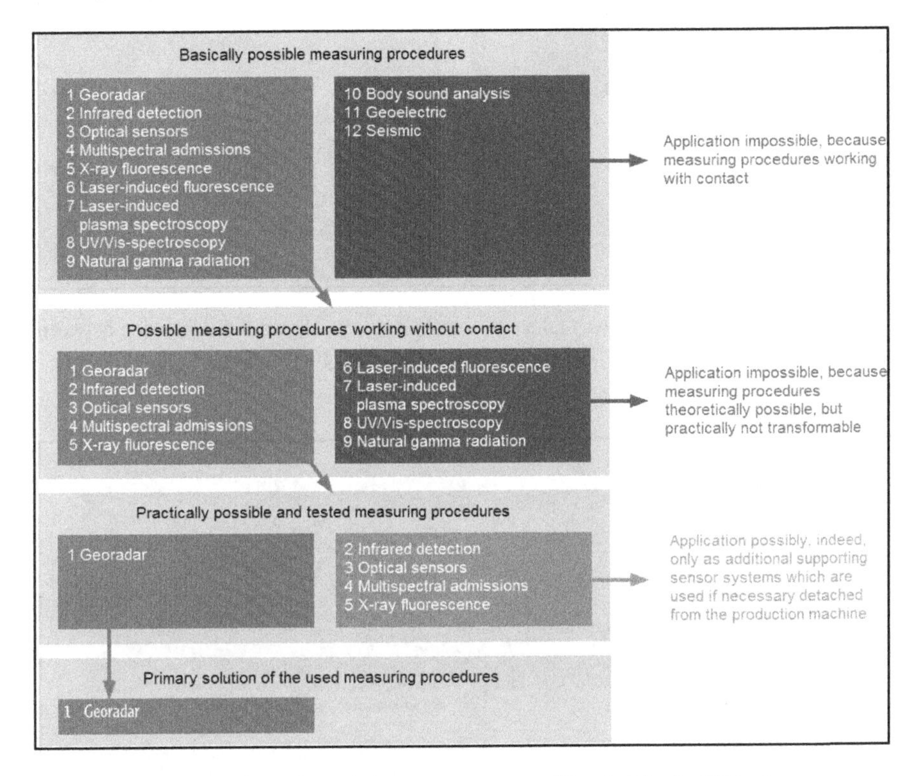

Fig. 2 Selection procedures for suitable sensor systems [Graph: Martin Pfütze]

In the case of the sensor systems suitable for wall face mapping it is basically important to compare the data with the geologic model and to have suitable clues for the part of the layer recognition. In this area the technologies of optical detection, infrared sensors and XRF have given helpful and promising results. For optical sensors or picture processing programs as well as infrared cameras the consideration consisted in using different software systems to analyze pictures and in extracting certain signs. The experiments have proved that layer courses are indeed recognizable; however, certain conditions like the right day light are needed. Figure 3 shows the result of an infrared measurement at a coal seam. In the case of X-ray fluorescence (XRF) it concerns an optical procedure that with the help of typical fluorescence signals recognizes qualitative differences between significant material types. In different field tests carried out 60 point measurements were taken at a coal seam in an open pit mine in Central Germany. In the analysis by which it was concentrated on the elements iron, titanium, calcium, sulphur and silicon it has become clear that the layer sequence of the coal seam can be exactly defined and understood on account of measured values.

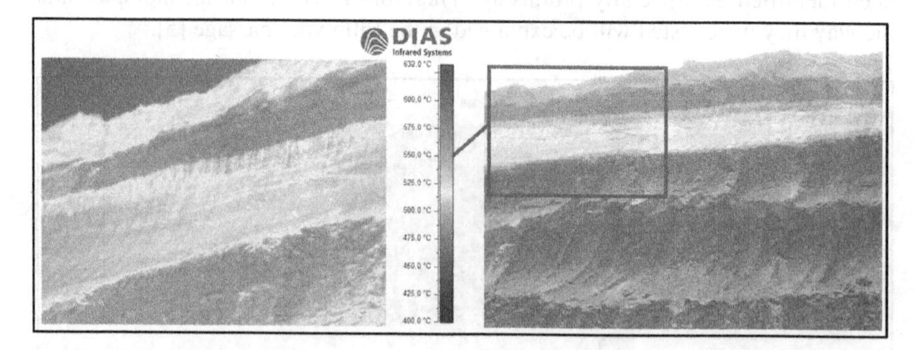

Fig. 3 Infrared picture which was taken at a coal seam and shows the difference between overburden and coal as well as the interburden in the coal seam [Pictures: Martin Pfütze]

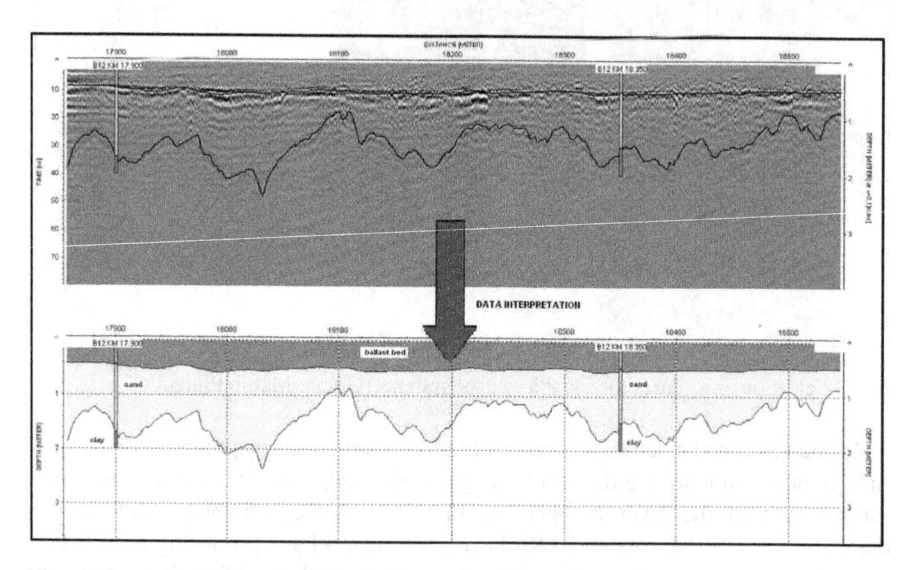

Fig. 4 Example for the operation of georadar (The layer difference between the both boreholes is clearly recognizable) [6].

In the case of the application of georadar, the measuring instrument sends out electromagnetic waves which are reflected of heterogeneities in the ground and in the loose rock (areas in which the electric conductivity and the relative permittivity change by leaps and bounds). Transitions are mostly geologic border layers, changes of the mineralogical composition or the moist salary. The reflected waves will be conceived with an aerial and the term, the phase and the amplitude will be taped (figure 4), whereby the water content of the ground or loose rock influences the measuring result substantially.

Specific experiments which were carried out on freely accessible seam surfaces as well as directly at a bucket-wheel excavator proceeded very promisingly. Thus it was possible to recognize clay layers as well as other distinguishable layers. Furthermore the idea exists to attach the measuring device for the georadar directly at the boom of the excavator parallel to the bucket-wheel to measure and recognize the underneath lying surfaces or layers (figure 5).

Fig. 5 Constructing-ready measuring device for the georadar (a) and idea-image of the position of the georadar at the end of the boom of an excavator parallel to the bucket-wheel (b) [Pictures: Martin Pfütze].

Thus in particular the consideration how the sensor technology with georadar could work as an aid and the dividing layer recognition should occur will be further examined. Besides it was a matter of analyzing also which dividing sharpness and exactness is accessible according to today´s state and which dividing sharpness is necessary now for a selective production and should be transformed by the sensor technology with the georadar. This was also valid for the technical transformation on the machine (figure 6) [5].

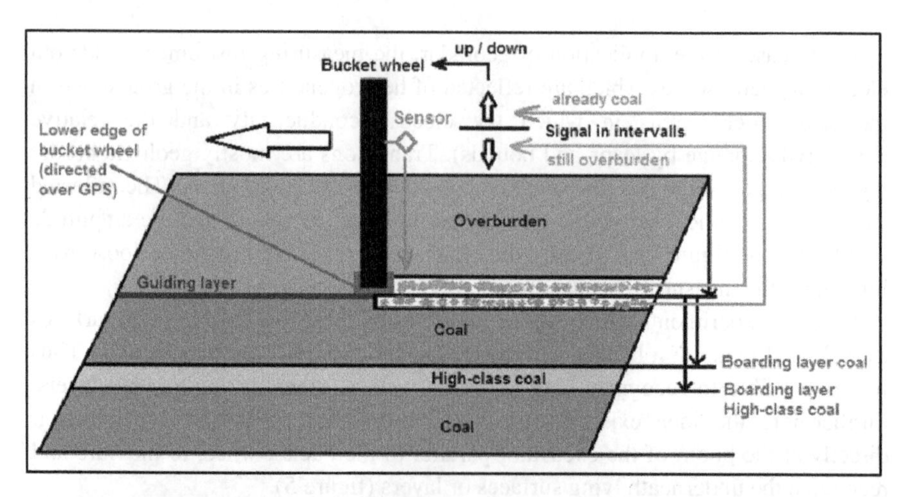

Fig. 6 Control of the bucket-wheel by sensor technology with recognition of the guiding layer [Graph: Martin Pfütze].

4 Summary and Outlook

During appropriate measurement tests in the field work the aim was to characterize different sensor technologies as well as to recognize the relevant border dimensions and the adaptation to the working conditions. The technologies of georadar, X-ray fluorescence (XRF) as well as infrared detection or optical sensors have given first promising results to define different layers in the coal seam. However these technologies need to be divided into systems suitable for wall face mapping (optical procedures, infrared sensors, XRF) and a system which can be used for layer recognition in the direction of advance on the seam surface as well as directly at the excavator (georadar) (table 1).

Table 1 Division of sensor technologies based on their suitability [Source: Martin Pfütze]

sensor technologies suitable for	
wall face mapping (2D)	layer recognition (3D)
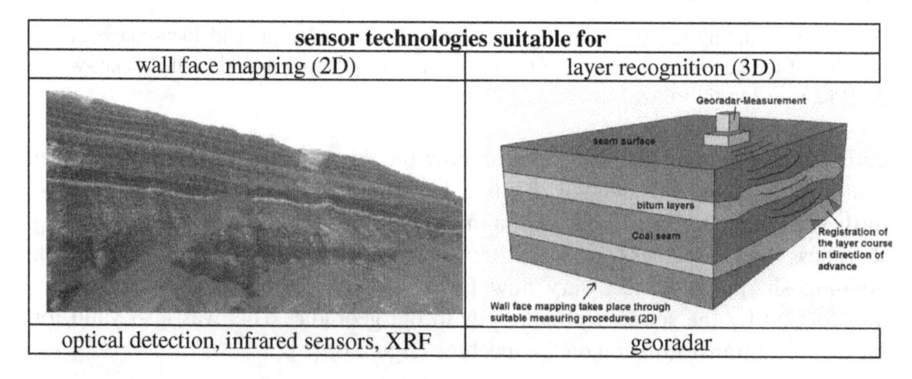	
optical detection, infrared sensors, XRF	georadar

Furthermore as a result of the differentiation a suitable approach for the development of a procedure for sensor recognition of different coal quality layers was defined. In the 1st step suitable data over the respective layer courses are gained by the different procedures which are suited for the wall face mapping. Afterwards the data are transmitted as a 2D-representation of the coal face in the existing geologic model where each single layer can be assigned an x- and z-coordinate (for the direction of working and the dismantling height). A sequence of the representations of the respective coal faces originates from the repetition of the measuring processes for every new dismantling block which leads in their sum to an increasing prediction exactness of the single layer courses. In the 2nd step the seam surface is measured by the georadar and therefore data is gained about how the layer courses develop in direction of advance. Afterwards the data will be transferred into the already revised and refined geologic model to receive a detailed 3D-representation of the upcoming dismantling blocks. Thereby the respective layers become next to the already determined x- and z-coordinate from the wall face mapping one more y-coordinate assigned (for the direction of advance). Finally in the 3rd step additional data of the single layer courses is registered and taken by a georadar which is directly at the boom of a bucket-wheel excavator and works continuously during the dismantling process. Again the gained data should serve to refine the geologic model and to define the authoritative predictions for the right cut division for selective mining. An overview of the approach of the single steps is shown in figure 7.

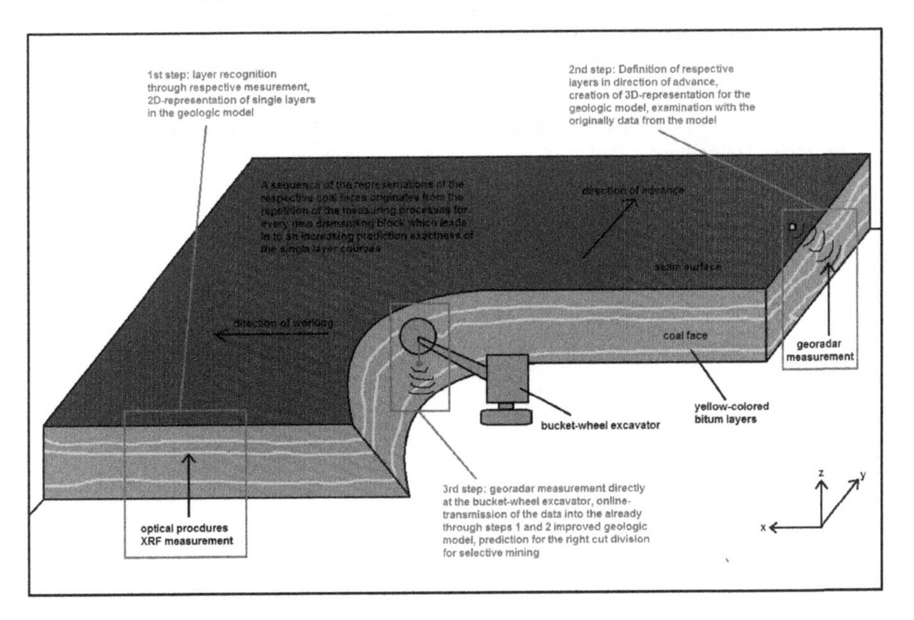

Fig. 7 Overview of gradual approach for the development of a procedure of sensor recognition of different coal quality layers [7].

For the upcoming work mainly the technical implementation on the bucket-wheel excavator will play a great role, because the production equipment with the sensor will be used for tracking the deposit model and the determination of the distribution of the coal quality in the open pit.

References

[1] Martin, P., Carsten, D.: Actual state of technique for selective mining and materials identification. In: Continuous Surface Mining – Latest Developments in Mine Planning, Equipment, and Environmental Protection, TU Bergakademie Freiberg, pp. 158–164 (2010) ISBN 978-3-86012-406-2

[2] P. Steffen, D. Carsten: Berechnungsmethodik für Surface Miner (in Russisch: Методика расчета для горных комбайнов). In: Bergbau-Forum, Dnepropetrovsk, pp. 26–33 (2006) ISBN 966-350-050-6

[3] Stoll, R.D., Niemann-Delius, C., Drebenstedt, C., Müllensiefen, K.: Der Braunkohlentagebau – Bedeutung, Planung, Betrieb, Technik, Umwelt, 605 Seiten. Springer, Heidelberg (2009)

[4] Durst, W., Vogt, W.: Schaufelradbagger, 391 Seiten. Trans Tech Publications, Clausthal-Zellerfeld (1986) ISBN 0-87849-057-4

[5] Martin, P., Carsten, D.: Sensor based selective mining concepts in coal mining. In: Continuous Surface Mining – Latest Developments in Mine Planning, Equipment and Environmental Protection, University of Miskolc, Hungary, pp. 201–208 (2012) ISBN 978-615-5216-09-1

[6] Staccone, G.: High Speed Radar from Ground Control. Company presentation for MIBRAG mbH. Profen, 20 pages (March 30, 2012) (unpublished)

[7] Pfütze, M., Drebenstedt, C.: Approaches for a quality-controlled selective mining in german opencast lignite mines using different sensor technologies. In: Drebenstedt, C., Singhal, R. (eds.) Mine Planning and Equipment Selection, vol. 145, pp. 1277–1284. Springer, Heidelberg (2014)

Highly Selective Lignite Mining and Supply

Stefanie Schultze[1] and Madleine König[2]

[1] MIBRAG mbH – Investment and strategic planning, Germany
[2] MIBRAG mbH – Mine planning, Profen mine, Germany

Abstract. Lignite is still the most important source of energy in Germany. In 2012, a total of 185.4 million tons of lignite was mined here. In addition to electricity generation and district heating, processed products such as briquettes, dust and coke are also produced. Besides this, a very small amount of lignite (~1% of the lignite mined worldwide) is used for non-energetic utilisation. In central Germany, the ibi alliance was formed to further promote this particular sector. The aim is to use lignite as a source of hydrocarbons by further developing technology, plants and processes. An alternative method of producing basic chemical products would reduce Germany's dependence on oil and natural gas imports. The successful implementation of this goal requires not only the development of chemical processes but also a secure supply of lignite.

Since deposit conditions in Germany will become more and more challenging (in deposit depth, seam splitting and - dips), more well developed extraction and conveyance techniques as well as mining technology will become preconditions if deposits are to be mined sustainably and economically.

It is this issue that the companies MIBRAG, FAM, TAKRAF, ABB and Freiberg University of Mining and Technology are addressing in one of a total of five joint research projects within the ibi alliance.

MIBRAG's involvement includes the conception of mining technology where the selective extraction of different coal qualities plays a key role. Typical deposit data were defined and then used to formulate the specific technical requirements in terms of mechanical engineering.

In most cases, the use of large machinery is favoured when planning mining operations. The research focuses on compact bucket wheel excavators that allow improved handling of biaxial inclination (up to 10%) as well as greater mobility.

On the stopes and at the border slope system, continuous conveyor systems are considered for the transportation of overburden and coal. Due to the depth of the deposit, more challenging requirements have to be set for the conveyer system along the border slope system. Moveable transfer systems are to replace the conventional central, stationary mass distributors. Since the intention is to use the conveyer system in the border slope system to transport several qualities of coal before then transferring them to 2 main conveyer belts leading towards the coal pile, another crucial focus of the joint research project is on material tracking, controlling the conveyer routes (including for overburden) and stockpiling.

Further research activities are focusing on the identification of mayor geological layers which shall be done during the mining process on the excavator. The

C. Niemann-Delius. (ed.), *Proc. of the 12th Int. Symp Continuous Surface Mining - Aachen 2014*,
Lecture Notes in Production Engineering, DOI: 10.1007/978-3-319-12301-1_32

main focus here is on contactless systems. Online raw material analysis can be used to write the data back directly into the deposit model.

These innovations in the fields of engineering and sensor technology allow modified mining technologies. Furthermore it opens up the potential for significant improvements by mining geologically challenging deposits. In the ibi process, research is currently focused on lignite products only. Additional focus on “overburden products” in the future will make it possible to carry out accepted, resource-efficient and sustainable surface mining operations.

1 Introduction

Deposit conditions in Central Germany will become more and more difficult, which is especially caused by increasing deposit depths, seam splitting and – inclination. The following figures show the difference between present and future deposits.

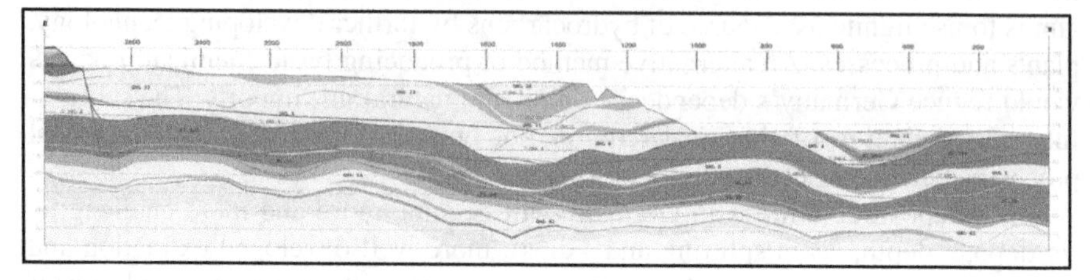

Fig. 1 Present deposit conditions

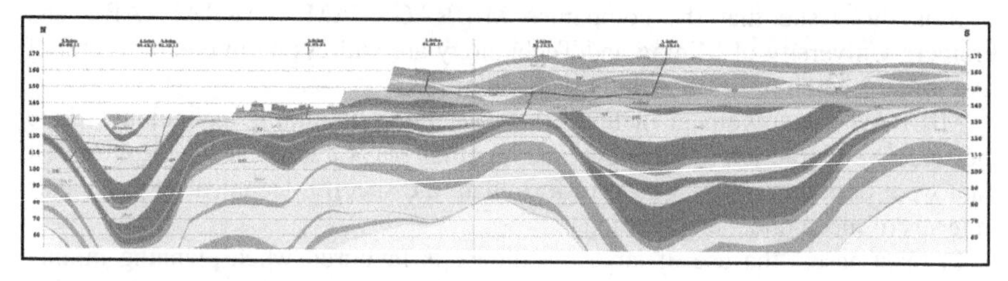

Fig. 2 Future deposit conditions

In spite of this challenging deposit conditions the demand for mining products with high and consistent quality increases. At the same time the requirement for a sustainable mine development becomes more and more important, not only to the industry, but also to the public. Therefore lignite coal should not only be used to generate electricity and heat but also to produce chemical products.

In Central Germany existed a high knowledge about selective extraction of lignite coal and the production of basic chemical products out of lignite. To preserve and develop this knowledge the ibi alliance was founded. Twelve partners belong to the ibi alliance. On the scientific side there are the University of Freiberg and the University of Applied Science Merseburg. The industrial partners are ABB Automation, EPC Engineering Consulting GmbH, FAM GmbH, IHU mbH, InfraLeuna GmbH, isw gGmbH, Linde Group/ Linde Gas, MIBRAG mbH, Romonta GmbH and TAKRAF GmbH.

The ibi project consists of six composite projects (VP1 .... VP6):

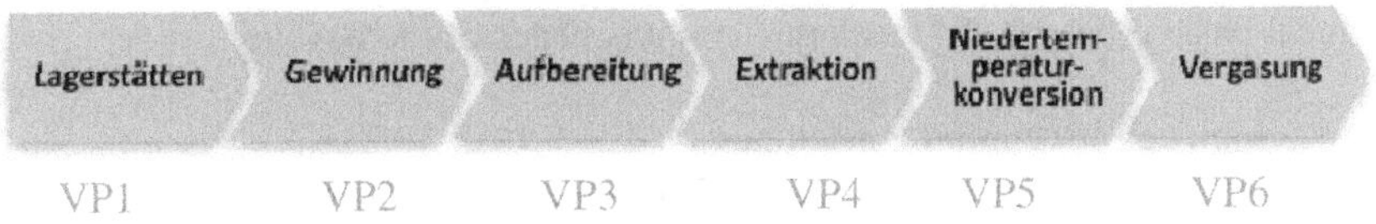

Lagerstätten | Gewinnung | Aufbereitung | Extraktion | Niedertemperatur-konversion | Vergasung

VP1 VP2 VP3 VP4 VP5 VP6

VP1	DEPOSITS
VP2	MINING
VP3	PREPARATION
VP4	EXTRACTION
VP5	LOW-TEMPERATURE CONVERSION
VP6	GASIFICATION

MIBRAG works together with other partners on the first two subprojects (VP1 and VP2).

2 Deposit Management (VP1)

In the first subproject MIBRAG and the University of Freiberg work on improving the prognosis of geological models and increasing its content.

Today's geological models are based on interpolation, in which the space in between the Exploration drillholes is displayed through continuous functions. Because of this smoothing, the variability of deposits remains unconsidered.

Future geological models will be based on stochastically approaches. In this case the variability of deposits is furthermore considered by random elements. These are for example the mapping of the wall face, sensor systems at the mining machine or online raw material analysis.

The Usage of stochastically methods has several advantages, like the inclusion of different data sources and a more realistic display of the deposit.

This improved deposit model is mainly used to create a mine layout, which enables the selective extraction of lignite coal. The main objective of the VP1 (Integrated deposit management) is the development of an optimized mine layout. Besides the stochastically, geological model other criteria are considered, for example: customer requirements, economical influences, safety issues and efficient use of mining machines.

3 Highly Selective Lignite Mining and Supply (VP 2)

Because of the geological conditions it is not only important to develop a new way of geological modelling but also more suitable mining technologies, as well as extraction and conveyance techniques. VP 2 deals with this topic.

The main objective is the continuous production of lignite coal with consistent quality parameters. Up to 4 coal qualities need to be provided to produce different chemical basic products. To reach this aim it is necessary to develop:

- high selective mining machines
- methods, that recognize parting surfaces and track material flows
- sensor and measuring technologies
- new conveyer systems (on the stopes and border slope system)
- material flow and stockpiling models

To reach these aims 6 work packages have been determined.

3.1 *Mining Organization*

The first part is about general mining organization. Therefore MIBRAG defined an example mine. This mine consists of several coal seams, which are variable in depth. The form of the seams is trough shaped.

In the course of the consideration a mine layout has been created by MIBRAG. The stope system changes a lot because the stopes are supposed to follow the seams as good as possible. Consequently the individual stopes often vary in the height of cut and the inclining during the mining process. That is why several face advance technologies have been created. These face advance technologies are mainly focused on coal qualities and geological conditions.

This geometric mine structure leads to requirements for the mechanical engineering of mining machines and conveying systems, especially on the border slope system.

3.2 *Mechanical Requirements*

Due to the mine layout three possible machinery uses are considered. Only two of them are studied in detail. These are on one side the exclusive usage of bucket wheel excavators and on the other side the combination of bucket wheel excavators and mobile technique.

Although mobile technique is able to work efficient in difficult and rapidly changing deposit conditions, the only use of mobile technique is too expensive for a deposit, which has an overburden thickness up to 160 m and a total depth up to 270 m.

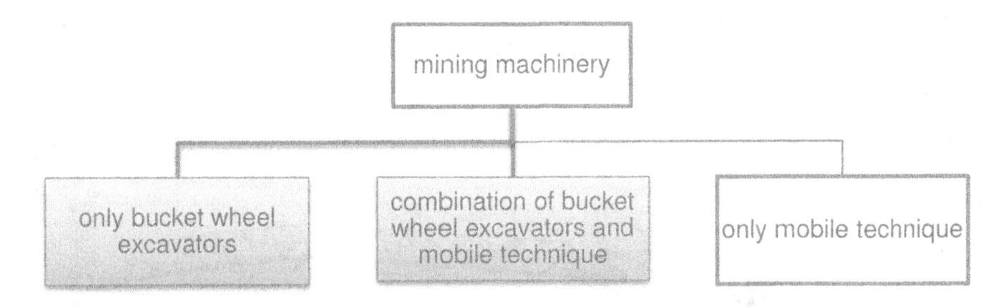

Fig. 3 Possible machinery usage

In detail the planning was focused on bucket wheel excavators. Although bucket chain excavators have a very high separation grade, they are not suitable. Their heavy weight and high wear makes bucket chain excavators more expensive than bucket wheel excavators. Furthermore the mass to performance ratio is not that good.

To reach a maximum of lignite selectivity the use of bucket wheel excavators in combination with mobile technique is the better choice. Especially in slope systems the cutting precision of bucket wheel excavators cannot reach the selectivity of mobile technique. Mobile Equipment should be used as a supporting technique.

In the future bucket wheel excavators have to work on inclining surfaces with better product selectivity. Therefore the ibi - partners determined the following requirements:

- Increase of mobility
- Working on surfaces with biaxial inclination up to 1:10
- Cutting precision up to ± 5 cm
- Working on intermediate levels

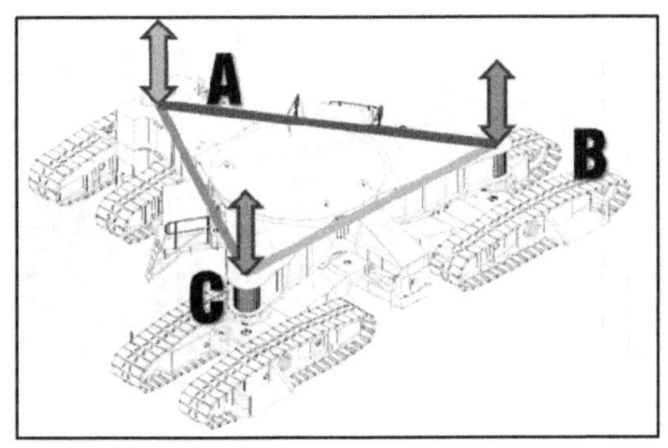

That is why TAKRAF develops a compact bucket wheel excavator, which is able to work efficiently and with a high tonnage during biaxial inclination up to 1:10.

The main attention lies on the development of a new crawler track (undercarriage) with a levelling construction.

Due to the fact, that variable face advance technologies are developed, mobile conveyor (feeder) systems that connect the excavator to the bench belt need to be adjusted.

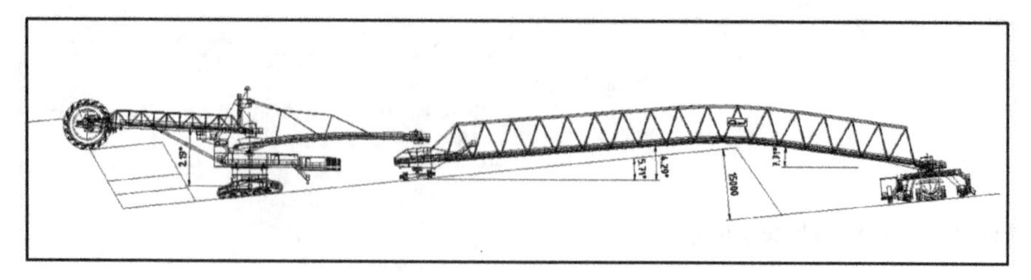

Fig. 5 Combination of compact bucket wheel excavator and conveyor bridge [1]

MIBRAG created a border slope system without an edge tube. This means the coal and the overburden need to be transported along the border slope system. Since the internal spoil tip follows the winning face, the conveyor belts on the border slope system have to be mobile. Generally there are 3 possibilities to integrate the transportation system on the border slope:

- movable conveyor belts (traditional)
- grasshopper
- conveyor bridge on crawlers

FAM focused on the conveyor bridge on crawlers. Therefore they planned a new support and joint system that is able to work on inclined surfaces (figure 6).

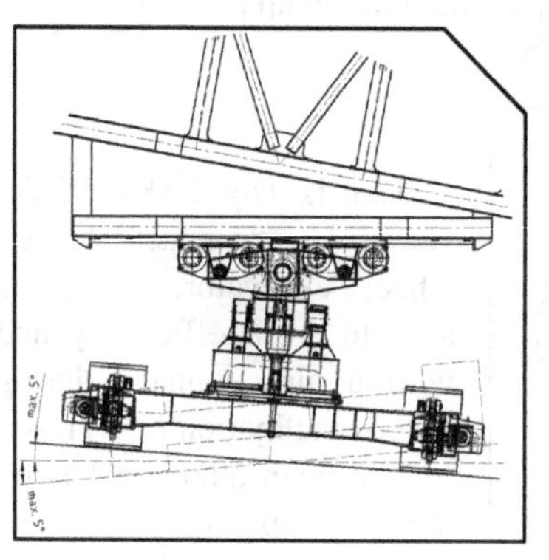

Fig. 6 Possible border slope conveyor system [2]

The biaxial inclination that the support and joint system can handle is 5° at a maximum.

During the mining process the border slope system will change in height and inclination so that the conveyor bridges on crawlers have to be flexible.

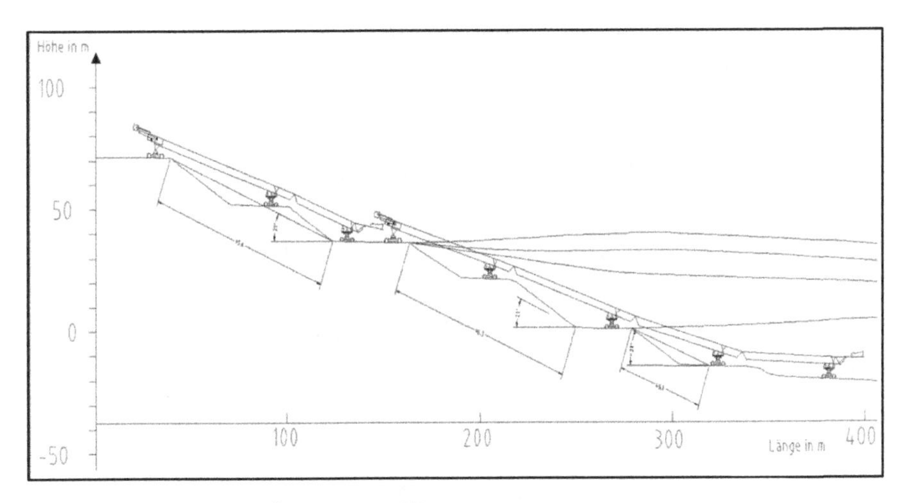

Fig. 7 Schema of a border slope system [2]

On the border slope system the conveyor bridges on crawlers manage inclinations up to 18°. As seen in figure 7 the bridges can have varying lengths. The conveyor bridges consist of 3 to 6 segments and one segment is 30 to 80 m long. As shown in figure 7 to reach the bottom of the mine it is necessary to connect several bridges. The transfer points in between these bridges are essential for a continuous mass flow. Since the mass flow is either coal or overburden, there has to be a distribution on different belts. The coal needs to get to the stock piling system and the overburden has to be distributed on the dump. Therefore a material tracking system needs to be introduced.

3.3 Material Analysis and Tracking

One requirement for a material tracking system is a continuous supply of characteristic raw material data. First of all the system needs to know if it is coal or overburden. This is sometimes hard to recognize because of difficult geological conditions (for example coal and overburden look alike). Secondly it is challenging to identify characteristical coal parameters within one coal seam. These characteristics include for example the calorific value, bitumen, sulphur and water content. In this context the Technical University of Freiberg works on subproject four. The aim of this working package is the development of measurement methods to identify material characteristics and parting surfaces. Therefore characteristic coal parameters have been analysed in a deposit of MIBRAG. The following figure shows a wall face of the open pit mine “Vereinigtes Schleenhain”.

Fig. 8 Reference profile of a coal seam with different horizons [3]

Further the University studied the usage of several analysis methods, for example X-Ray fluorescence analysis (XRF) and ground-penetrating radar (GPR).

After the data registration on the wall face this data is used by ABB to work on the 5th subproject of VP 2.

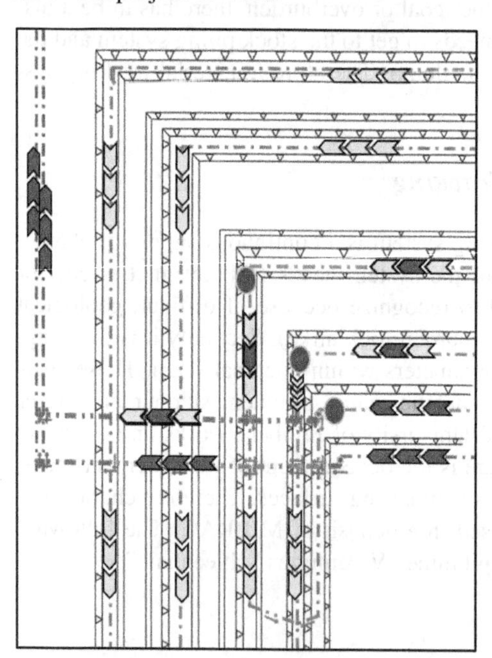

Fig. 9 Possible material flow [4]

The aim of this subproject is to create an upgraded data collection and control system, which makes it possible to follow the material flow and even coal qualities on the conveyor belts. This material tracking system allows the piling of coal in the right quality and quantity on the stock piles. Since there are 4 coal qualities the system has to calculate whether and, if so, then how much raw material out of each working level has to be distributed on the stock pile to reach one quality.

The figure shows a possible material flow in the reference deposit. It represents a snapshot of a typical mining situation with 6 working levels.

In the first 2 working levels there is only overburden. The four following working levels contain coal and overburden in exchange. At each excavator quality parameters are determined.

The material flow system follows that coal over the slope and head conveyor belts as well as the conveyor bridges and the two coal collecting conveyors to the stockpiles outside the mine.

The incoming coal on the stockpile is distributed to defined sections of which each contains one coal quality. At the same time ABB creates a stockpile model, which records the amount and qualities in each stockpile section. So the customers can be supplied by the necessary amount and quality of coal just in time. The following figure represents one possible coal content in a stockpile.

Fig. 10 Graphical schema of a stockpile [5]

The main objective within the VP2 project is high selective lignite mining and the supply of lignite coal with defined quality parameters. This is the basic requirement for the following processes within the ibi alliance.

[1] ibi – Fachsymposium Juni 2013, Poster TAKRAF
[2] ibi – Fachsymposium Juni 2013, Poster FAM
[3] MIBRAG mbH
[4] ibi – Fachsymposium Juni 2013, Poster ABB
[5] ibi – Fachsymposium Juni 2013; Rollup VP 2

Accidents Prevention Strategy in the Surface Coal Mining in Indonesia; Vision Zero

H. Permana[1] and Carsten Drebenstedt[2]

[1] Institut für bergbau und spezialtiefbau, Gustav-Zeuner-Str.1A, 09599 Freiberg, Sachsen, Germany
[2] TU Bergakademie Freiberg, Germany
herry.permana@student.tu-freiberg.de, hpermana71@ymail.com

Abstract. The coal mining industry plays an important role contributed to the national economy and development in Indonesia, however the mine accidents become as a sensitive issue or problem. The main cause of the mining accidents is still dominating the low safety awareness and accountable by workers, and also improperly cost spent in the occupational health and safety (OHS) program. This paper describes the concept behind the return on prevention and workers' perception in the occupational health and safety in relation to prevent accidents or improve safety performance. The proactive action plan related to accident prevention is as an essential step of the risk management process. Participation and intervention for all employees are important and urgent, especially for the frontline workers, which are crucial in achieving good safety performance with the financial support properly. The investment of the people and money properly are an effective strategy to prevent accidents in order to protect people, property, process and profit of the company for a short or long term benefit, and will also give a good image for sustainability in the mining business.

Keywords: coal mining, mine accidents, prevention, return on prevention, workers perception, intervention, occupational health and safety.

1 Introduction

The mining industry in Indonesia is expected to grow rapidly in the 5 to 10 years into the future and will become an increasingly strategic sector for development and economic improvement. It will be encouraging increased domestic and foreign investment in this sector with national and international banks supporting. Indonesia is one of the key players in the to supply the industry needs to the developed countries through the export of minerals and coal, however must also be fulfill the domestic needs. The established of the mining law No. 4 of 2009 related to the minerals and coal mining, which one of the important point is improving of the value added of the minerals as the main priority by the processing or refining until the final product. This all mining process activities

C. Niemann-Delius. (ed.), *Proc. of the 12th Int. Symp Continuous Surface Mining - Aachen 2014*,
Lecture Notes in Production Engineering, DOI: 10.1007/978-3-319-12301-1_33

should be complied with the law in order to increase state revenues to achieve welfare of the Indonesian people equitably in accordance to mandate of the constitution of 1945 in article 33 point 3.

The energy and mineral resources sector has an important role in the development and economic of Indonesia. The economic contribution of these sectors is significantly to the economy of Indonesia, with amount IDR 415.20 trillion[1] or 32.13% of the total state of revenue (minerals & coal mining share 9.56% in 2012), but the tax income is still the largest contributor to the state revenue[1]. Most of mining operations in Indonesia are using surface mine methods of the open pit mine, because the majority of the mineral and coal reserves found in the shallow areas. The main characteristic of mine operations is using truck and shovel system with a large amount of manpowers. Indonesia has an important role player as a supplier of world coal. According to the estimation of the world coal association (WCA)[2] in 2012 has been said that Indonesia is the largest coal exporter country with total coal's share of the world market achieved 383 mio.tons following Australia with 301 mio.tons. The comparison of the coal demands between domestic and export are approximately 21.3% and 78.7% in 2012, and also most influenced to the national economic incomes.

The mine accident is one of the most important issue or problem in the mining industry in Indonesia which can disrupt the production processes. The high number of the mine accidents occurrence in the workplaces is caused by the low safety awareness, low to comply safety regulations and rules, lack of communication and coordination, low leadership, and inadequate cost spent in the occupational health and safety (OHS) program. These problems are going to become as a big issue in the mining industry of Indonesia, due to, it needs the effective solution how to reduce or prevent the accidents at workplace. Several of the mining accidents show in the figure 1.1[3], it can be influenced the mindset or concentration of the workers during work.

Fig. 1 Coal mining accidents in Indonesia, a. hauling road, b. pit area

[1] Bank of Indonesia exchange rates, USD 1.0 = IDR 12,226 (date 30.01.2014).

The important issue relates to the coal mining industry before the Mining Law no. 4 of 2009 established, that most of the coal mining operations are conducted by the mining contractors as the main player for mined excavation.

The criteria of the mine accidents in Indonesia must involve five elements according to the ministry of mining and energy decree of the *Kepmen 555.K/26/M.PE/1995*, in the article no. 39. The mine accident is an unplanned, unexpected, undesired, uncontrolled and undesigned events which occur suddenly and caused injury to the workers in connection with a work relationship of the mining operations. The five elements of the mining accident are the actual accident occurred, resulting injured to the mine worker/s or people who get a permit from the technical mine manager, consequences of the mining business activities, the accident occurred during working hours to injure of mine worker/s or people who get permits, and the accident occurred within the area of the mining activity or operation or in the project area.

Improperly managed of the potential hazard and risk in the mine site activities, it will cause an accident which resulted in losses, such as the people, production, property, and profit (4Ps). Investments in the occupational health and safety (OHS), however, will contribute to the commercial success of the enterprise, and safety performance improvement. The companies are not following the statutory and social responsibility requirements but also their own business concern (e.g. image, performance).

2 Objective and Methodology

2.1 Objective

The main objective of this study is evaluated of the workers' perception (qualitative) and cost spend in the occupational health and safety program (quantitative) at the companies in relation to accident prevention. The accident is preventable if the hazard and risk in the workplace or mine site were managed properly with a good condition, proper compliance to regulations, good budgeting of occupational health and safety (OHS), and competent workers[3].

In general, the objectives are described to evaluate of the mine accident statistics, the validity and reliability tests of the questionnaire, focus group interviews, worker relationships, and the cost effectiveness or cost benefit analysis. The objective is assessed with the return on prevention of the micro economic model to justify the benefit of the prevention work and intervention by the workers' perception in order to prevent accidents for safety improvement to achieve the zero accident vision. The idea framework concept of the strategy to prevent accidents has been modified from the risk management process[4], it is shown in the Figure 2.1.

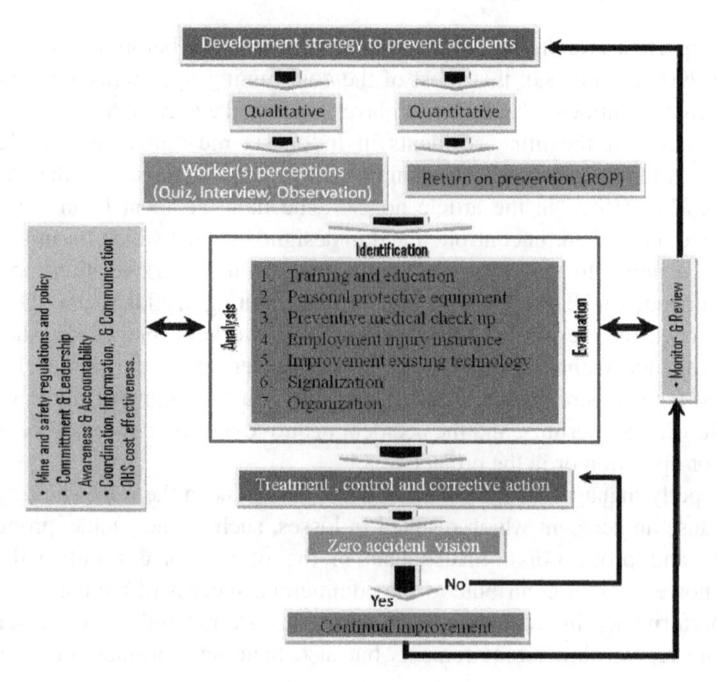

Fig. 2 The idea concept to prevent accidents

The paper is intended to consider of the mine accidents in relating to the surface coal mining operations only. The danger occurs and occupational diseases or work related to illness were not considered for a limitation of this case.

2.2 *Methodology*

The methodology concerns to gather data which is based on the assessment offer by the workers perception through the questionnaire, interview, observation, and money invest in the occupational health and safety. The survey activity of this paper involves amount of 1,600 respondents as a total sampling of the four companies (e.g. total employment of the four companies is 68,519 employees in 2012). The time period of survey is 3 months since August to October in 2012.

The questionnaire, interview, and observation method are a new created by the author according to the mining conditions in Indonesia. The questionnaire determines into two types of question, the Q-35 and respondent suggestions in relation to improve safety performance. The questionnaire will be scored by the Likert scale assessment. The five level scale of the Likerts[5] are strongly disagree to strongly agree (score range from 1 to 5). Calculation the validity and reliability tests to justify the linear correlations between two variables or bivariate (x and y variables). The value of the coefficient of correlation (r) is defined by the *Pearson's* product moment, the *Spearman-Brown*, and *Cronbach Alpha* equations of the validity and reliability tests[6].

The seven basic elements of OHS are as a basic for assessment approaches. It is involving of the training and education (TE), personal protective equipment (PPE), preventive medical check up (PMCU), and employment injury insurance (EII), improvement existing technology (IET), signalization, and organization.

The methodology develops using the seven steps assessment to prevent or reduce accidents in order to improve safety performance of the company, as follows:

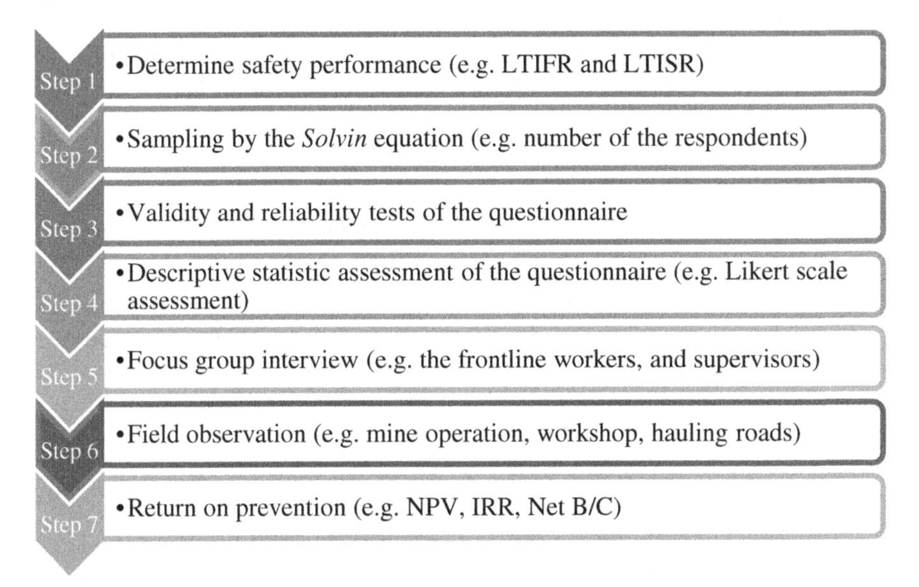

Fig. 3 Seven steps to prevent accidents

The explanation of the seven steps to prevent accidents or improvement safety performance are described as following details:

Step 1: Determining safety performances
The frequency and severity rates are calculating by the equations below:

$$\textbf{LTIFR} = \frac{\textbf{(Number of lost time injuries)}10^6}{\textbf{(Total number of employee hours worked in an accounted period)}} \quad 2.1$$

$$\textbf{LTISR} = \frac{\textbf{(Number of work days lost)}10^6}{\textbf{(Total number of employee hours worked in an accounted period)}} \quad 2.2$$

The standard level of the frequency rates for mineral and coal mining in Indonesia are explained in Table 2.1, but the severity rates are calculated by the author with assumed the employee's number maximum in the company of amount 30,000 people. The fatal injury is counted by 6,000 lost work days.

Table 1 Standard of the safety performance indicator

Indicator	The criteria of accident rates		
	Low	Medium	High
LTIFR	0.00 up to 0.40	> 0.40 up to 0.99	> 0.99
LTISR	0.00 up to 80	> 80.00 up to 240.00	> 240
Safety performances	Good	Fair	Bad

Step 2: Sampling
Before measure the validity and reliability tests, the number of the respondents must be determined by the *Solvin's* equation. The equation shows below:

$$s = \frac{N}{1+(N \cdot e^2)} \quad 2.3$$

Where, s is the number of sample, N is the total population, e is a margin of error.

Step 3: Validity and reliability tests
Testing the validity and reliability of the new questionnaires in order to ensure the coefficient of correlation values (r). The coeficient of correlation (r) values are calculated using the *Pearson's product moment*, *Spearman-Brown*, and *Cronbach's Alpha* equations.

The value of r is defined by *Pearson's product moment* equation:

$$r = \frac{n\sum x.y - \sum x.\sum y}{\sqrt{\{(n\sum x^2 - (\sum x)^2\}\{n\sum y^2 - (\sum y)^2\}}} \quad 2.4$$

Where, r is the coefficient of linear correlation, n = number of the respondents, x = score item of the question (e.g. question number 1 for all respondents), and y = total score of the respondent (e.g. respondent number 1 for all questions).

The next calculation is using the *Spearman-Brown* formula to repeated the calculation according the validity test. The equation is the following:

$$r_{odd\ even} = \frac{2r}{(1+r)} \quad 2.5$$

Where, $r_{\text{odd even}}$ is the coefficient of correlation odd even, r is the coefficient of correlation.

The final step calculation is using *Cronbach's alpha* equation. Alpha is developed by Lee Cronbach in 1951[7] to provide a measure of the internal consistency of a test or scale, it is expressed as a number ranged of scale between 0 and 1. Alpha is an important concept in the evaluation of assessment and questionnaires[8]. The comparison of the calculation resulted values of r will compare with the supporting table data of the *Person's product moment* can see in Table 2.2. The interpretation of alpha score is defined into five classes [9-11], see in Table 2.3.

Table 2 The values of the *Person's product moment*

Number of respondents (n)	Margin of error (5%)
3	0.997
4	0.95
5	0.878
10	0.632
100	0.195
200	0.138
300	0.113
400	**0.098**
600	0.08
1000	0.065

Table 3 Interpretation values of the $r_{\text{cronbach alpha}}$ (Triton, 2005)

$r_{\text{cronbach alpha}}$ values	Interpretation
0.00 up to 0.20	Less reliable
0.21 up to 0.40	Little reliable
0.41 up to 0.60	Quite reliable
0.61 up to 0.80	Reliable
0.81 up to 1.00	Extremely reliable

The coefficient of correlation of the *Cronbach's alpha* is defined as follows:

$$r_{ca} = \left\{\frac{n}{(n-1)}\right\}\{1 - \frac{\Sigma(s_{ev}{}^{2})}{s_v{}^{2}}\} \text{ Where, } \sum(s_{ev}{}^{2}) = \frac{\left\{\Sigma x^2 - \frac{(\Sigma X)^2}{n}\right\}}{n} \text{ and } s_v{}^{2} = \frac{\left\{\Sigma y^2 - \frac{(\Sigma y)^2}{n}\right\}}{n-1} \quad 2.6$$

Where, r_{ca} is the coefficient of correlation of *cronbach alpha*, that use to determine of the consistency of correlations (reliability test equation of *Cronbach alpha*), $\Sigma(s_{ev}^{2})$ is the total variance of the score on each question, $s_v{}^2$ is the total variance of samples.

The valid criteria result is defined if the value of the r or r_{count} is bigger than the r_{table}, otherwise if the r or r_{count} is less than the r_{table} value is not valid. The comparison of the calculation results of the coefficient of correlation (r_{count}) will be compared to the r_{table} from table data of the *Pearson's* product moment and *Cronbach's* alpha. In this paper, the number of respondents (n) is 400, and significant level is 5%, so the r_{table} is 0.098.

Step 4: Descriptive statistic assessment of the questionnaire
The questionnaire is consisting the seven elements of OHS with 35 questions (Q-35). The questionnaires shall be scored by the Likert scale assessment. The five level of the likerts are defined as a strongly disagree to strongly agree with score 1 to 5, where 1 is the lowest score and 5 is the highest score. A mean score over 3.00 is generally considered as a positive result when it is the mean value of slighly higher. The questionnaires are a diagnostic or intervention tools which can be used to evaluate the the organization goals on safety in the company.

Step 5: Interview
The interviews conduct as a focus group interviews to the representative respondents of the supervisors and frontline workers for getting the actual informations related to accidents prevention at work. The summary result of the focus group interview will show in a bar graph that represent of the low, moderate, and high of the workers performance.

Step 6: Observation
Field observations have been conducted in several workplaces location such as the mine pit operation, hauling roads traffic, workshop, and the mine support facilities. The field observations are also based on the seven elements of OHS as a guidelines. The objective of the field observations are finding the real situation or condition of workers during work.

Step 7: Return on prevention assessment (financial)
The return on prevention assessment is calculating using the net present value (NPV), internal rate of return (IRR), and net benefit-cost ratio (Net B/C ratio) in order to find the cost effectiveness or the break even point. This assessment is using to determine the minimum standard of the money invest in an occupational health and safety programs at the companies. The assessments are defined using several equations as follows:

The net present value is defined by the equation as follows:

$$\mathbf{NPV} = \sum_{t=0}^{T} \{(\frac{\mathbf{PV_{t\,inflows}}}{(1+i)^t} + \frac{\mathbf{PV_{t\,outflows}}}{(1+i)^t})\} \quad 2.7$$

$$NPV = \sum_{t=1}^{T} \frac{C_t}{(1+r)^t} - C_o \quad 2.8$$

Where, NPV is net present value, T is the number of time periods of investment, t is the time accounting period, *i* is the interest or discounted rate, $PV_{t\ inflows}$ is the present value incomes in the time accounting period, $PV_{t\ outflows}$ is the present value expenses in the time accounting period, C_t is the cash of income or receive, and C_o is cost of capital investment in the initial year of zero.

The NPV criterias can see in the following Table 2.4.

Table 4 The NPV values criteria

NPV criteria of investment	Explanation
NPV > 0	The project is profit or acceptable
NPV < 0	The project is not profit or not acceptable or rejected
NPV = 0	The project is in balance condition or the project is break event point (BEP)

The internal rate of return (IRR) is defined as follows:

$$IRR\ is\ the\ NPV = \sum_{t=1}^{T} \frac{C_t}{(1+r)^t} - C_o = 0 \quad 2.9$$

Where, IRR is internal rate of return.

The Internal rate of return (IRR) is the discount rate at which the net present value of an investment becomes zero.

The Net B/C discounted rate equation is defined as follows:

$$Net\frac{B}{C}\,or\,ROP = \sum_{t=1}^{T}\left|\frac{\frac{\text{Benefits}_t}{(1+r)^t}}{\frac{\text{Costs}_t}{(1+r)^t}}\right| \quad \text{or} = \sum_{t=1}^{T}\left|\frac{\left(\frac{\text{Benefits}_t}{(1+r)^t}-\frac{\text{Costs}_t}{(1+r)^t}\right)}{\frac{\text{Costs}_t}{(1+r)^t}}\right| \text{x } 100\% \qquad 2.10$$

Where, Benefits_t are refresented of benefits of the seven elements of OHS, and Costs_t are also refresented of costs of the seven elements of OHS, and I I absolute symbol.

3 Case Study

In this paper, the author has been chosen for case study of the representative of the four mining companies are Adaro Indonesia, Berau Coal, Trubaindo Coal Mining, and Kaltim Prima Coal. These companies were significantly contributed to the national coal output (e.g. 275 mio.tons) about 37.06% from 2006 to 2012. The cumulative of the Indonesian mineral and coal of mine accidents that reported by the CCOWs from 2006 up to 2012 are accounted of minor injury for 918, serious injury of 618, and fatality accidents of 172. Comparison of the fatal accidents between mineral and coal mines are 33.14% and 66.86%. Top four open pit coal mine companies are contributed amount 17.33% of the total mine accidents of Indonesia, but especially for coal mine fatal accidents contributed about 30.86%. The comparison between coal production and mining accident shows in the Figure 3.1[12].

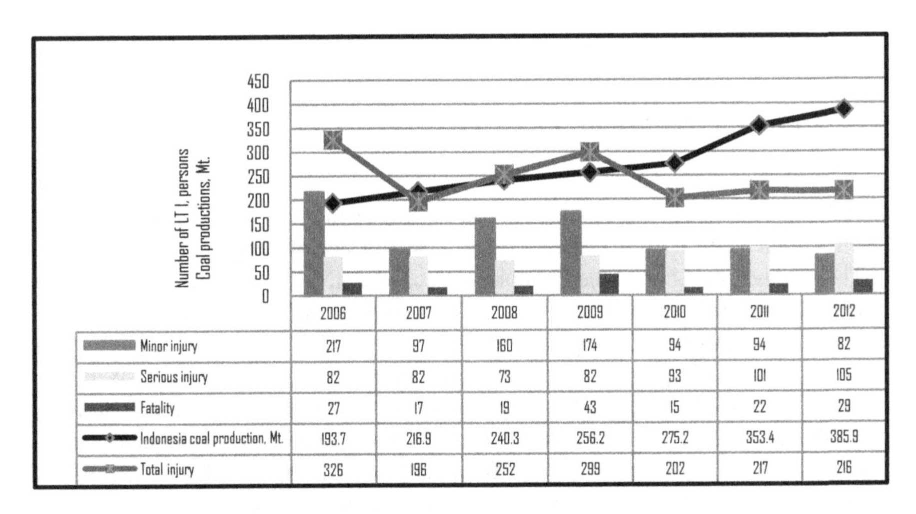

	2006	2007	2008	2009	2010	2011	2012
Minor injury	217	97	160	174	94	94	82
Serious injury	82	82	73	82	93	101	105
Fatality	27	17	19	43	15	22	29
Indonesia coal production, Mt.	193.7	216.9	240.3	256.2	275.2	353.4	385.9
Total injury	326	196	252	299	202	217	216

Fig. 4 Comparison between coal production and mining accident

The case study concerns into the four surface coal mining companies as a mention in Chapter 2. The four companies are located in Kalimantan island. The total number of respondents are 1,600 persons. The respondents are divided into two groups, the management level contribution is about of 9% and the frontline workers contribution is about of 91%. The management level is involving the supervisor, superintendent, and manager positions. The frontline workers are involving the heavy equipment operator, light vehicle driver, and mechanic positions. The process of survey has been done in 2012 by delivery of the questionnaire, focus group interview, and field observation directly to the workers during work and working conditions, it can see in a Figure 3.2.

a. Questionnaire b. Focus group interview c. Field observation

Fig. 5 The survey or investigation processes in the surface coal mine companies

4 Result and Discussion

4.1 Frequency and Severity Rates

Lost-time injury is defined as an occurrence that resulted in a fatality, permanent disability or time lost from work of one day/shift or more. The LTIFR and LTISR are important to justified the safety performance of the company.

The figure 4.1 shows the comparison of the frequency rates of the four companies and Indonesia. Berau Coal (BC) and Kaltim Prima Coal (KPC) show the average values of the frequency rates of 0.19 and 0.28, it rates smaller than 0.40 as the threshold point within a good safety performance conditions. With the exception for KPC in year 2009 with value of 0.54. However, Adaro Indonesia (AI), Trubaindo Coal Mining (TCM) and the Indonesian average indicate values of 0.53, 0.45 and 0.52 which is larger than 0.40 and smaller than 1.00. This means the safety performance conditions are fair. For example, frequency rate average of Kaltim Prima Coal is 0.19, it means in one million manhours worked are involving 0.19 person injured, or in one person injured within 5,263,158 manhours worked.

The figure 4.2 shows the comparison of the LTISR between four companies and Indonesia. The average of the severity rates of the four companies and Indonesia are larger than the threshold point of 80. It means that the four companies is contributed fatal injury to national statistic as well. BC and KPC show the average values of the severity rates of 157.73 and 80.08 that it rates smaller than 240 as the threshold point of the fair safety performance conditions. However, the AI, TCM and Indonesia average values of 264.44, 386.27, and 324.84. It means that the severity rates are larger than 240 as an threshold point of the bad safety performance condition. For example, in 2012, the severity rate of Trubaindo Coal Mining is 488.59. It means in every one million manhours worked of the company will lost 488.59 work days, or in every one lost work day within 2,047 manhours worked.

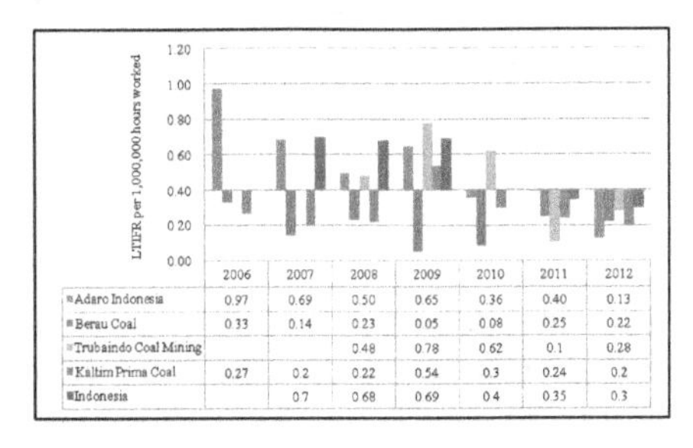

	2006	2007	2008	2009	2010	2011	2012
Adaro Indonesia	0.97	0.69	0.50	0.65	0.36	0.40	0.13
Berau Coal	0.33	0.14	0.23	0.05	0.08	0.25	0.22
Trubaindo Coal Mining			0.48	0.78	0.62	0.1	0.28
Kaltim Prima Coal	0.27	0.2	0.22	0.54	0.3	0.24	0.2
Indonesia		0.7	0.68	0.69	0.4	0.35	0.3

Fig. 6 Comparison of the frequency rates

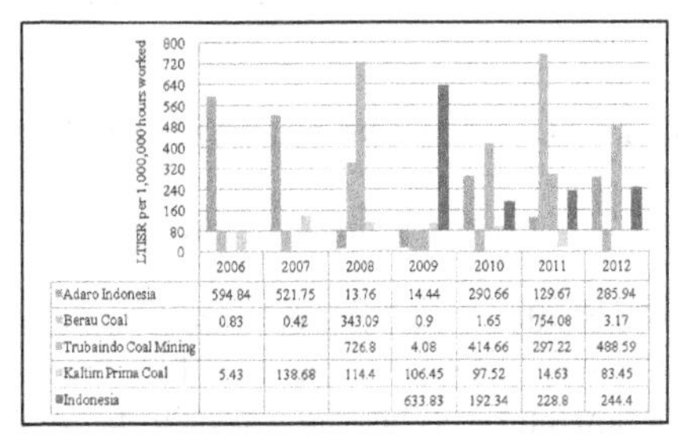

	2006	2007	2008	2009	2010	2011	2012
Adaro Indonesia	594.84	521.75	13.76	14.44	290.66	129.67	285.94
Berau Coal	0.83	0.42	343.09	0.9	1.65	754.08	3.17
Trubaindo Coal Mining			726.8	4.08	414.66	297.22	488.59
Kaltim Prima Coal	5.43	138.68	114.4	106.45	97.52	14.63	83.45
Indonesia				633.83	192.34	228.8	244.4

Fig. 7 Comparison of the severity rates

4.2 *Sampling*

Slovin's equation will find out the sample size of a population, for example the number of employees of Adaro Indonesia in 2012 is 23,555. The calculation result is defined with the confidence level of 95%, and then the margin of error of 5%, as follows:

$$s = \frac{23,555}{1 + (23,555 \times 0.05^2)} = \frac{23,555}{59.89} = 393.32$$

The number 393.32 is equal 394 employees, in this paper will use approximately of 400 respondents in the survey of the questionnaire within each company.

4.3 *Validity and Reliability Tests*

The vaildity and reliability tests of the questionnaires are complied to the requirements, and it can be used as an instrument to measure the working conditions of the company in order to accidents prevention. The calculation steps are determined of the coefficient of correlation (r or r_{count}) by the Person's product moment, *Spearman Brown, and Cronbach alpha* equations, then the result compare to the r_{table} values. The result of the coefficient of correlation (r) values of the questionnaire are valid and reliable, show in Table 4.1. It means two variables are in good linear correlations.

Table 5 The calculation results of the coefficient of correlation (r) values

No.	Mine company	The coefficient of correlation (r) values			Remarks
		r (e.g. quiz no.1)	$r_{odd\ even}$	r_{ca}	
1	Adaro Indonesia	0.52	0.75	0.72	valid & reliable
2	Berau Coal	0.38	0.86	0.77	valid & reliable
3	Trubaindo Coal Mining	0.44	0.86	0.81	valid & extrem reliable
4	Kaltim Prima Coal	0.37	0.79	0.71	valid & reliable

4.4 Descriptive statistic analysis

4.4.1 Likert Scale Assessment

The result of the Figure 4.3 relates to the four coal mine companies. Perception of the respondents from the four coal mine companies think that the OHS organization is the main priority to reviewed first in order to prevent accidents at work (the highest average score of 4.18). Followed by the PMCU, TE, IET, PPE, Signalization, and the last priority is the employment injury insurance but should be considered too (lowest average score of 3.87, and the total average score is 4.04). In the figure shows that the respondent's perception of the four companies almost the same of the tendencies in the radar plot graph, it means no matter small as well as big scale of the mining companies, the perceptions show almost the same results.

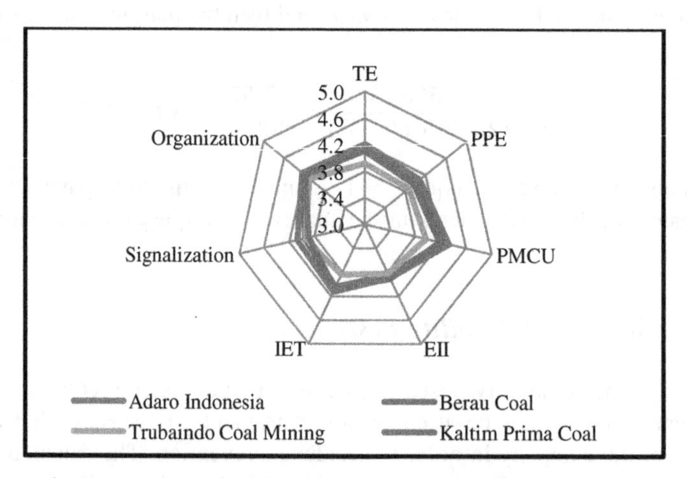

Fig. 8 Respondents perception of the four mine companies

4.4.2 Suggestion Summary

The OHS organization is important things to consider in order to prevent accidents. The OHS organization is defined into several sub element improvement efforts, it can see in Figure 4.4 and 4.5.

The result of the Figure 4.4 relates to respondent suggestions from the four coal mine companies. Every respondent gives one or more suggestions to the company in order to improve safety performance or the accidents prevention. In this radar plot graph, the respondents' suggestion from the four coal mine companies have the same options which the OHS organization should be reviewed as the main priority. All respondents think the organization is the main tools to coordinate all personnel in order to prevent accidents. This radar plot indicates that small as well as big scale mining companies have the same target to be improved.

The following result of the Figure 4.5 relates to respondents' suggestion for organization improvement in the companies. In this radar plot graph is elaborated improvement efforts of the OHS organization as a mention in Fig. 4.4. Suggestion of the respondents are involved the five sub elements priority in order to prevent accidents at work such as improve the safety accountable and awareness as the main priority for all workers, followed by comply and obey of the OHS rules, including follow the safety work instructions, hazards identification, inspection and supervisory towards all type of the risky jobs, improvement the commitment to safety and personnel communication, and the last but not least priority is improved the support facilities (e.g. comfortable rest building area at mine site operations). All these sub elements should be considered and applied by the OHS organization for a good business in mining.

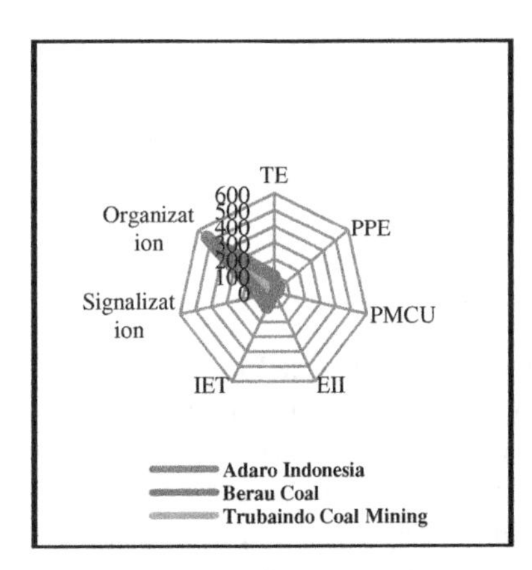

Fig. 9 Respondents suggestion to prevent accidents

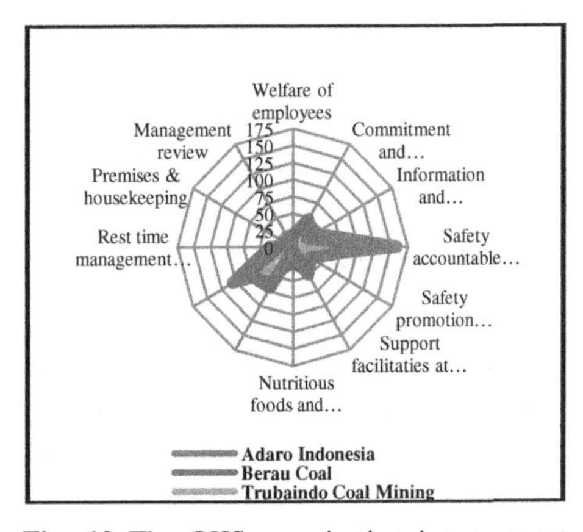

Fig. 10 The OHS organization improvement efforts

4.5 Interview

The summary results of the interview shall be described in terms of the rating points in the Figure 4.6. The rating points are determined into three types of a low performance is one, moderate is two, and high performance is three. The summary results of the supervisor level show above the average rating point of two, with the highest rating point is 2.35 (Adaro Indonesia/AI) and the lowest rating point is 2.11 (Trubaindo Coal Mining/TCM). However, the frontline workers show result lower than the average rating point, with the highest rating point is 1.59 (Adaro Indonesia) and the lowest rating point is 1.30 (Trubaindo Coal Mining). Based on the rating points as a mention above, the frontline workers should be an important concern in order to improve the knowledge and behavior through the OHS training and education, accurate information given, also improve and intensive personal communication constructively. The supervisor level is concerned to improve two ways communication to safety as an intense and constructive.

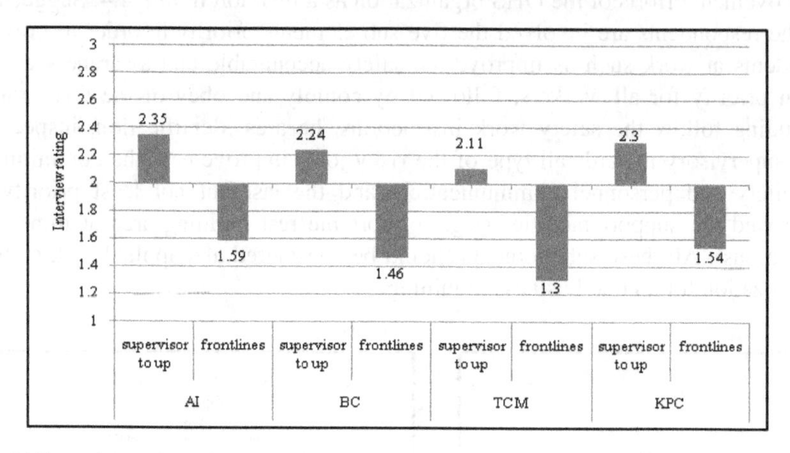

Fig. 11 Interview rating of the respondents

4.6 Observation

The field observations have been conducted in several workplace locations such as mine pit operations, hauling road traffics, workshop, and the mine supports facilities. The field observations are also based on the seven elements of OHS as a guideline. The worker violations still occur during work at the workplace or at the mine site activities, such as violation of traffic and safety signs, improper usage of the personal protective equipments. Also found the several working conditions or support facility at mine sites still need to improve and maintain properly, such as lack of the lighting during working at night at disposal areas or mine pits, restroom facility in a bad condition. The standard level of the training, PMCU, and EII systems are not equal between mine operator and contractors. The personal communication gap between the frontline workers and supervisors are still happening. the The personal safety promotion and enforcement are not run in well yet, it means still focus on the production target goals.

Fig. 12 Field observation in the day and night

4.7 Return on Prevention

The investments consist of the seven elements of the occupational health and safety (OHS) on each company. The cost calculation is based on the work and budget plan data and information from the companies.

The Figure 4.8 shows the Bar graph related to the OHS cost expenses of the companies from 2008 to 2012. The comparison of the required, invested, and calculated of the OHS costs of the companies show in different average of the values. The budget of OHS costs should be reviewed and improved to become a progressive values of 1.68 times (ratio). It means, the company should pay the OHS cost as much as US $ 0.12 per ton of coal produced (US$ 0.07 per ton (as a standard by the government) x 1.68) or US $ 207.17 per employee in a year.

Figure 4.9 shows the net benefit-cost ratio of the four companies with and without accident costs. The comparison results show that AI is 1.30 and 2.31 (ratio 1.78, highest), BC is 1.19 and 1.38 (ratio 1.16, lowest), TCM is 1.36 and 2.02 (ratio 1.49), and KPC is 1.47 and 2.01 (ratio 1.36). For an example of AI, the Net B/C ratio with accident costs is 1.30, and without accident costs is 2.31. It means, every one dollar invested will give the benefit 1.3 dollars (the accident occurred), and otherwise, if without any accident occurred with one dollar invested will give the benefit 2.31 dollars. For an example, the ratio 1.78 (2.31/1.3=1.78) of AI, it means without accident occurred in the mining operations will give 1.78 times within every dollar invested in OHS. The total average value of the Net B/C ratio of the seven elements of OHS is 1.63, and the total average value of the Net B/C ratio of the OHS program only (i.e. 4 elements;TE,PPE,signalization, and organization) for the companies is 1.72.

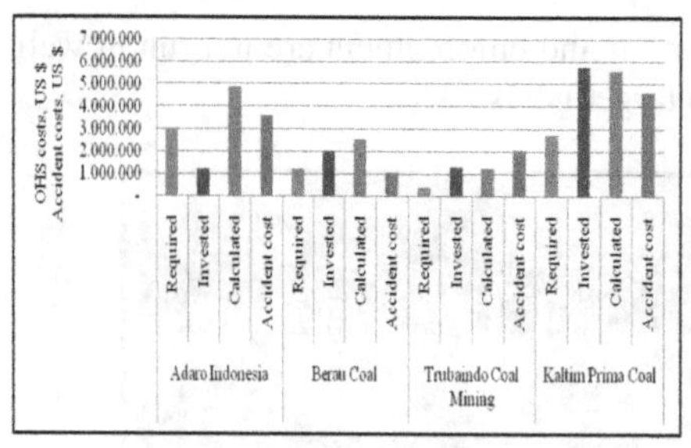

Fig. 13 The OHS costs required, invested, and calculated

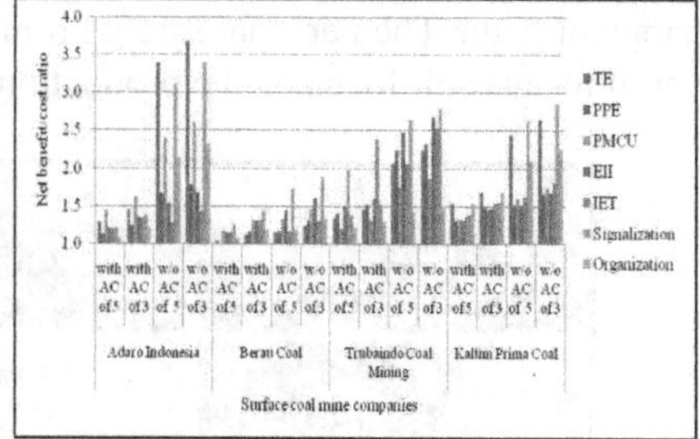

Fig. 14 Comparison of the Net benefit-cost ratios

Notes: With AC of 5 is with accident cost (AC) and benefit start in the fifth of the year, with an AC of 3 is with accident cost and benefit start in the third of the year, w/o AC of 5 is without accident cost and benefit start in the fifth of the year, w/o AC of 3 is without accident cost and benefit start in the third of the year.

5 Conclusion

According to the explanation of the result and discussion in Chapter 4 about the preventing accidents at the workplace, especially in the surface coal mining companies which represent the coal contract of works of the mining industries in Indonesia. The survey or investigation in this paper the mine accident rates, workers perception and the return on prevention. The assessment approaches use to determine the strategy to prevent accidents at work in order to achieve the zero accident vision. The conclusion of the survey or investigation results show in various expectations:

- The frequency and severity rates of the mining accidents are important to justified the company performance on safety, but the severity rate is a better choice for interpreting on safety performances of the company
- The validity and reliability tests of the questionnaires have complied with the requirements, with the coefficient of correlation *cronbach alpha* is larger than 0.6 (reliable threshold value), and it can be used as an instrument to measure the safety performances of the company
- The perception of the respondents is concerned that the OHS organization should be reviewed (e.g. management intervention) as the main priority in order to prevent accidents at workplace. The main sub element priority of the OHS organization reviewed is improving the safety accountable and awareness of the workers
- Based on the rating points of the focus group interview, the frontline workers should improve the knowledge and behavior through the training and education, accurate information given, and also improve the personal communication intense and constructive

- The worker violations still occur during work at the workplace, and there is not equal safety standard or guideline between the mine operator and contractor in relation to give a training and education process, personal protective equipment, preventive medical check up, and the employment injury insurance. Personal communication gap between the frontline workers and supervisor, and also production target oriented.
- Some companies invested money in the occupational health and safety (OHS) program are lower than the cost calculation results (78.35%), and the OHS costs required are also lower than the OHS cost calculation results for all companies surveyed (47.80%). The money invested in OHS program is almost the same with the accident costs spent (1:1.02).
- The net present value (NPV), internal rate of return (IRR), and net benefit-cost ratio (Net B/C ratio) are important financial calculation approaches of the return on prevention in order to find the minimum standard of the money invest in the occupational health and safety program at the company or the OHS cost effectiveness. The break even point (BEP) of the financial calculation of the seven elements of OHS is US $788.45 per employee per year, or the company should be invested US $ 207.17 per employee per year for the OHS cost program only (4 elements).
- The OHS cost standard amount of US $ 0.07 per ton of coal produced should be reviewed, increasing become US $ 0.12 per ton of coal produced according the calculation results of the Net benefit-cost ratio (Net B/C ratio) is 1.72 for OHS only (the ratio of the seven elements is 1.63).

The investment of the workers intervention and money into the occupational health and safety properly will give more benefits in relating to protect people, property and profit of the company for a short and long term of period. The competence people and money invested in the OHS properly will contribute significantly to the company image and performance as well. It can be reduced number of accidents (will be *zero*), reduced cost of accidents, reduced employee turnover, increase production and profits, and finally will be improved the satisfaction and pride of the workers in the company. In general, invested for every one dollar in prevention work, the expectation may return of almost two dollars or more. The accidents are preventable with a good condition and safety practices complied effective and can let the economic and social benefits.

References

[1] MEMRI, Ministry of Energy and Mineral Resources of Indonesia (MEMRI), Work performances of MEMRI, State revenues from energy and mineral resources sector, Jakarta (2012), `http://www.esdm.go.id/siaran-pers/55-siaran-pers/6127-kinerja-sektor-esdm-tahun-2012-.html` (cited December 05, 2013)

[2] WCA, World Coal Association, Coal Statistics 2012 estimation, http://www.worldcoal.org/resources/coal-statistics/ (cited December 04, 2013)

[3] Permana, H., Drebenstedt, C.: Development strategy to prevent accidents in Indonesian surface coal mines. In: Proceeding of the 22nd International Mining Conference, Mine Planning and Equipment Selection, Dresden, Germany, October 15-17 (2013)

[4] Head, G.L., Horn, S.: Essentials of the risk management process, 1st edn. Insurance Institute of America, Malvern (1985)

[5] Wuensch, K.L.: What is a Likert Scale and How Do You Pronounce 'Likert' (April 30, 2009)

[6] Cronbach, L.J.: Essentials of psychological testing, 5th edn. Harper & Row, New York (1990)

[7] Cronbach, L.: Coefficient alpha and internal structure of tests. Psychometrika, 297–334 (April 30, 1951)

[8] Tavakol, M.: Making sense of Cronbach's alpha. International Journal of Medical Education 2, 53–55 (2011), http://www.ijme.net/archive/2/cronbachs-alpha.pdf (cited September 11, 2013)

[9] Cohen, R., Swerdlik, M.: Psychological testing and assessment. McGraw-Hill Higher Education, Boston (2010)

[10] DeVellis, R.: Scale development: Theory and Applications. Sage, Thousand Okas (2003)

[11] Graham, J.: Congeric and (Essentially) Tau-Equivalent estimates of score reliability: what they are and how to use them. Educational Psychological Measurement 66, 930 (2006)

[12] DGMC, Directorate General of Minerals and Coal, Ministry of Energy and Mineral Resources of Indonesia, Annual report 2012, Jakarta, http://www.minerba.esdm.go.id/ (cited December 01, 2013)

Procedures for Decision Thresholds Finding in Maintenance Management of Belt Conveyor System – Statistical Modeling of Diagnostic Data

Paweł K. Stefaniak[1], Agnieszka Wyłomańska[2], Jakub Obuchowski[1], and Radoslaw Zimroz[1,3]

[1] Diagnostics and Vibro-Acoustics Science Laboratory,
Wroclaw University of Technology, Na Grobli 15, 50-421 Wroclaw, Poland
{pawel.stefaniak,jakub.obuchowski,radoslaw.zimroz}@pwr.wroc.pl
[2] Hugo Steinhaus Center, Institute of Mathematics and Computer Science,
Wroclaw University of Technology, Janiszewskiego 14a, 50-370 Wroclaw, Poland
agnieszka.wylomanska@pwr.wroc.pl
[3] System Analysis and Process Management Department,
KGHM Cuprum Ltd., Research & Development Centre,
Sikorskiego 2-8, 53-659 Wroclaw, Poland

Abstract. Belt conveyors are a key component in material transportation system in both opencast lignite mining and underground copper mines in Poland. Regardless of the structure of the mine, the problem of maintenance of belt conveyors is important (from the entire mining process point of view) for many reasons, such as: (a) conveyors are spatially distributed over a large area, (b) they create logically structured form of complex and heavy components, (c) they are operating in harsh mining environmental conditions, (d) failure of any belt conveyor might result in downtime of the entire production line or its major part. The paper discusses the issue of maintenance of gearboxes used in the conveyor drive systems. The authors have developed a CMMS-class system using GIS technology to support management of conveyors' network. Its fundamental role is to make right decisions for the exchange of components of the drive systems or allow them to continue their work. Such defined problem requires determination of complex decision rules and the definition of appropriate thresholds of diagnostic parameters. The article presents the procedures for determining decision thresholds, based on statistical modeling of diagnostic data and multidimensional data clustering. By selection of suitable distribution of the data and appropriate statistical parameters,

C. Niemann-Delius. (ed.), *Proc. of the 12th Int. Symp Continuous Surface Mining - Aachen 2014*,
Lecture Notes in Production Engineering, DOI: 10.1007/978-3-319-12301-1_34

multidimensional data analysis has been performed to determine threshold values for the effective identification of the condition of machines and their components.

Keywords: belt conveyor, maintenance management, diagnostics, data modeling.

1 Introduction

Belt conveyor system is a key component in the whole transportation system of Polish copper ore mines. Thus, it highly affects both volume and costs of production and safety of a company as well. For this reason the issue of proper maintenance of belt conveyor system is very important. Managing the process of conveyor operation is a singularly complicated case. Nearly one hundred of conveyors linked together form a very long transport network having a complex series-parallel structure. Thus, unexpected failure of a single element might cause a downtime of the whole system, which might result in decrease in productivity of the mine [2,13,15-16]. Such downtime are often associated with large financial losses. Moreover they might endanger the safety of workers. There exist several ways of maintaining machinery, i.e. (a) operating until failure, (b) planned-preventive maintenance and, the most effective one – (c) condition based maintenance. Operation of the mining network of belt conveyors is a complex issue, thus it needs a multifaceted approach. It is caused by a large number of driving units (in various conditions) scattered on a broad area and various operating load. Additionally, it must be taken into account that the driving units are complex and huge machines which operates in various conditions e.g. high humidity, temperature, dust and salinity. Such complicated environmental and operating conditions make the use of classical diagnostic methods non-effective. However, availability of current solutions and technology makes it is possible to develop an intelligent system oriented to data fusion from many different sources [5,11-12,15]. A computer system allows for extraction of knowledge about factors affecting occurrence of unexpected events and phenomena describing behavior of both a single object and a set of items including all interactions among them. This is carried out through initial validation, processing and systematization of data in appropriate database structures and adaptation of advanced data mining methods (inspired by artificial intelligence). Integration of the data describing various phenomena, physical quantities and processes in the machines and proper interpretation of the knowledge contained in their multidimensional space allow for detection of unknown patterns and dependencies, thus, it lead to model the operation process, i.e. determining the current and forecast future condition of machinery for a given time period. The authors are convinced that such a multi-faceted approach might provide a wide range of opportunities for appropriate policies to maintain the machinery oriented on improving reliability, reducing failures, increase of safety, production efficiency, reduction in repair costs, etc.) [13,15-16].

2 Data Acquisition System and Input Data Description

The process of monitoring the technical condition of the drive units is based on the diagnostic data acquisition using vibroacoustic methods. From the point of view of spatial dispersion of the drive units and their amount, it was necessary to build a portable module for periodic data acquisition and processing of the vibration data in simultaneously with the measurement of the operating load variability. A scheme of the data acquisition module is consisted of two parts: (a) laptop acting as the control and measurement manager and (b) sensors' layer, i.e. accelerometers and tachometers. Measurements are performed on the drive unit operating under external load. The conveyor load is expressed via input shaft rotation speed determined by the tachometric signal. Vibration data measurement takes into account technical configuration of the analyzed object [13,15-16]. Fig. 1 presents a general scheme of an exemplary arrangement of sensors.

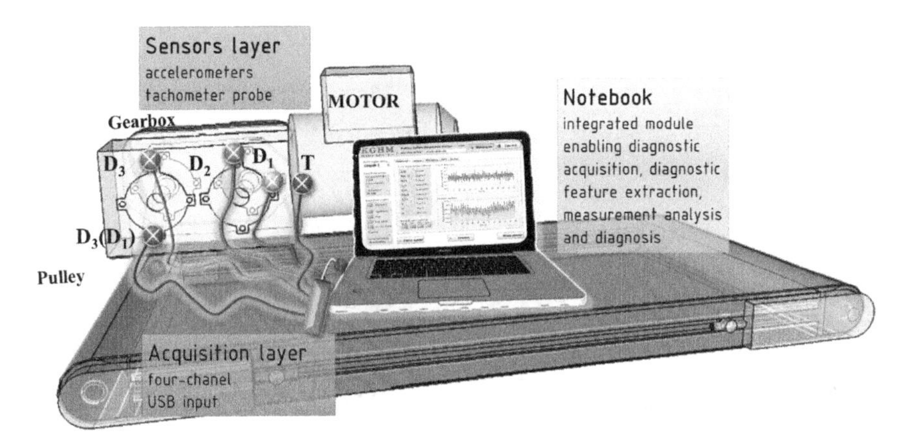

Fig. 1 Diagram of the vibration data acquisition module with an exemplary arrangement of the measurement points. The measuring module consists of a portable computer and sensors.

Raw signals acquired this way are segmented and spectral analysis of them are performed to determine the diagnostic features. Such processing leads to data of average rotational speeds and diagnostic features DF1, DF2 and DF3 related to condition of shafts, gearwheels and bearings. Frequency sampling of measurements of a single diagnostic feature is 1 Hz. Length of the measurement is 60 s. 3 channels of data acquisition are used for measuring each single diagnostic feature. Also a tachometric signal is measured simultaneously. Fig. 2 presents an exemplary measurement of operational load and the diagnostic feature. As mentioned above, we acquire diagnostic data from all drive units consisting the machinery park with different technical configuration. Therefore, it is necessary to categorize the data with respect to the type of a gearbox to determine technical condition of their components, i.e. gears, bearings and shafts. Moreover, one gearbox type usually operates with several types of couplings. Depending on the type of the coupling, gearboxes work in different ranges of rotational speed. Therefore, operational load values measured as the inverse of the rotational speed of gearbox input shaft considering all types of couplings is inconsistent. For this reason, additional categorize data in relation to type of coupling is reasonable (Fig 3). In further analysis diagnostic data from only one gearbox type operating only with fluid coupling have been used (gearbox KA-134). Fig. 4 presents input dataset obtained from 119 60-second long measurements performed on the same gearbox type.

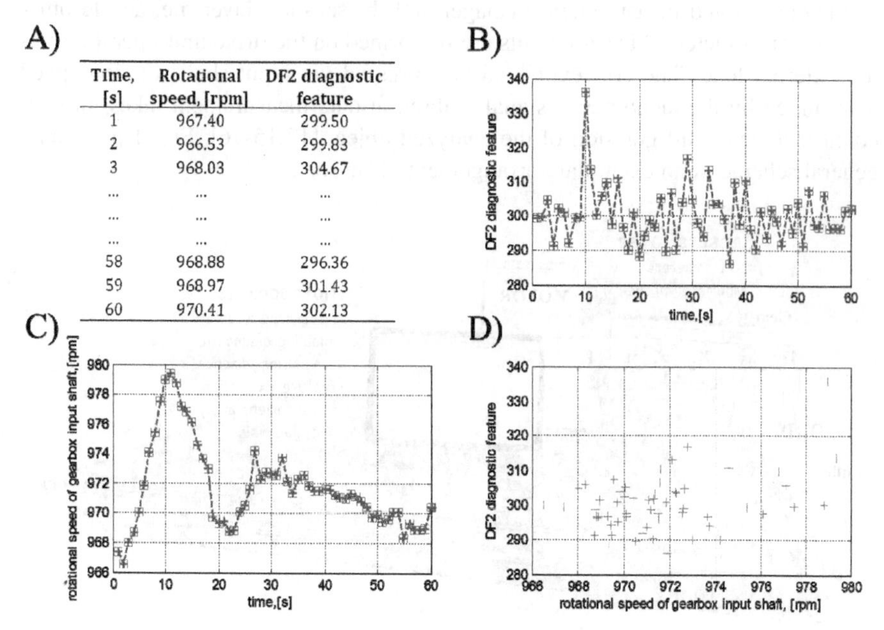

Time, [s]	Rotational speed, [rpm]	DF2 diagnostic feature
1	967.40	299.50
2	966.53	299.83
3	968.03	304.67
...	...	...
...	...	...
...	...	...
58	968.88	296.36
59	968.97	301.43
60	970.41	302.13

Fig. 2 A) The fragment of a tachometric measurement and diagnostics feature DF2, B) DF2 variation in time function, C) Operational load in time function, D) DF2 characteristics chart as a function of the operational load.

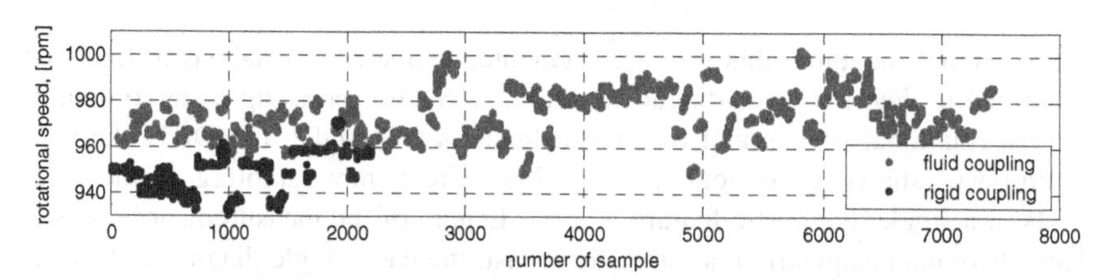

Fig. 3 Dependence between coupling type and rotational speed. Note that rigid-type coupling operates only in a limited spectrum of rotational speed.

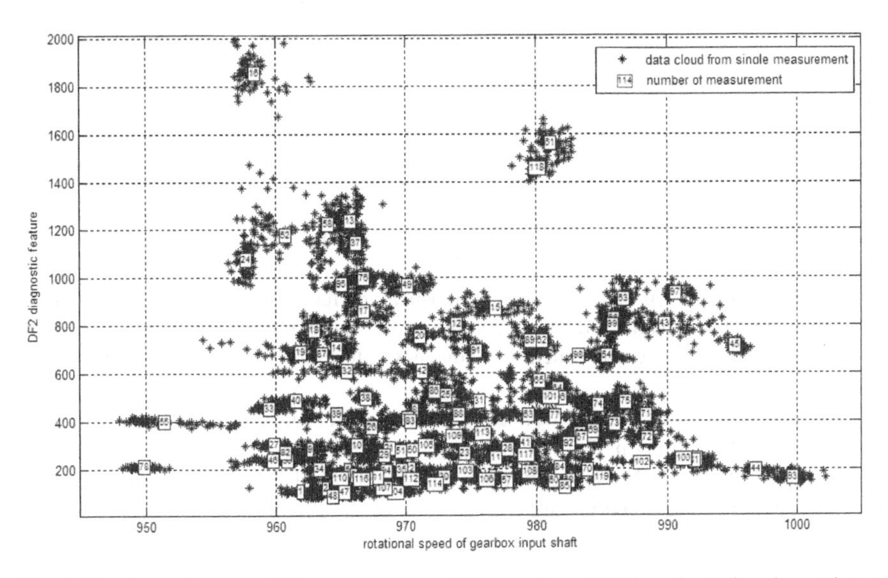

Fig. 4 The input data set: 119 measurements (60 samples of DF2 and rotational speed per single measurement)

3 Methodology of Data Modeling

In the case of mining machinery such as belt conveyors diagnostic decision making based on the spectrum of a single signal might lead to uncertainties due to presence of numerous disturbances in these complex systems. Thus, it might be useful to take advantage of a large set of measurements and statistical analysis performed for determining basic parameters of the set [1].

Statistical approach for bistate pattern recognition (normal state – abnormal state) is based on the probability density function estimating for a particular diagnostic feature for both states. Then, the alarm threshold is determined, see Fig. 5a. Unfortunately, in particular cases it is impossible to distinguish between two states because they are overlapping, Fig. 5b. The problem of overlapping of the probability density functions is often related to a strong dependency between spectrum-based diagnostic features and operational load of the machine. For this reason it is proposed to take into account operational conditions in which the measurements has been performed [3-4,9].

In the considered case it is needed to determine two thresholds, for warning and alarm level of the diagnostic feature (Fig. 5c) which makes the decision making process even more complicated.

In this paper, determining of the thresholds is based on modeling of the distribution of diagnostic features and analysis of volatility of their statistical parameters. The final thresholds are the result of multidimensional clustering of the data, i.e. parameters of the distributions and volatility of operating conditions.

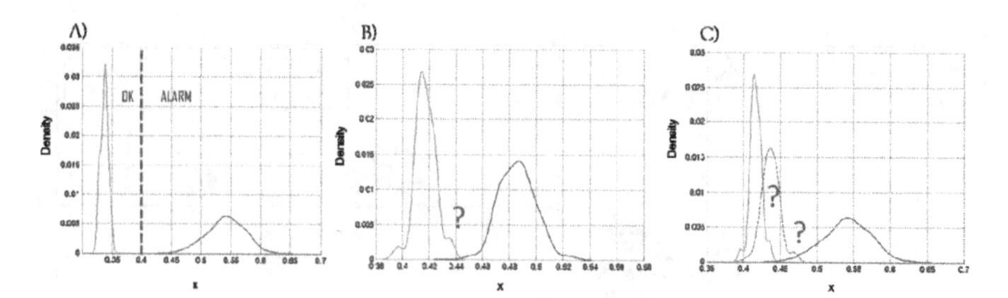

Fig. 5 The concept of setting thresholds for statistical inference: A) case of two-state classification of properly separable, B) not separable sets, C) classification into three states with overlapping probability density functions of diagnostic features (considered in the paper).

Fig. 6 presents empirical probability density functions of every measurement. Empirical PDF has been estimated by using the kernel density estimator with Gaussian kernel [14]. As it was expected, indication of 3 separated group is impossible, because the densities are overlapping each other.

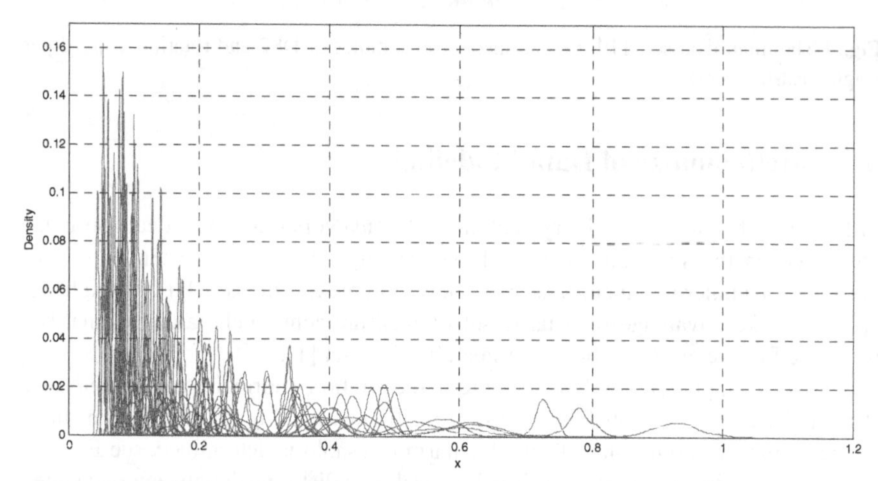

Fig. 6 Empirical probability density functions of DF2 feature for 119 measurements performed on the gears type KA-134. The presence of overlapping effect of empirical density makes indication of three separated states impossible.

Firstly, each one measurement is assumed to be normally distributed, thus empirical mean and standard deviation fully describe the distribution of the diagnostic feature DF2. Moreover, empirical mean and standard deviation of input shaft rotation speed are calculated. In this case we also assume Gaussian distribution.

Finally, we obtain a dataset which might be represented as a matrix with 4 rows and 119 columns, related to 4 statistical parameters and 119 measurements, respectively, Fig. 7. This dataset was subjected to clustering.

We use k-means algorithm for data clustering because of observed overlapping effects of density functions and ability to use a larger number of parameters. It starts with dividing the whole dataset into a given number of classes. Then the partition is being changed to obtain minimal within-class variance. This method is used for preliminary data analysis. The method allows for separation of groups which are subjected for further statistical analysis. K-means algorithm might be also used for data mining, where grouping is used for separation of the analyzed factors into subsets. In general, k-means is used for searching of information and its main task is to provide simple access to the information [6-7,10].

Performed analysis resulted in constituting constant, load-independent warning threshold of DF2 more than 610 and less than 1100 and alarm threshold more than 1100. Results of the method applied to the measurements are presented in Fig. 8.

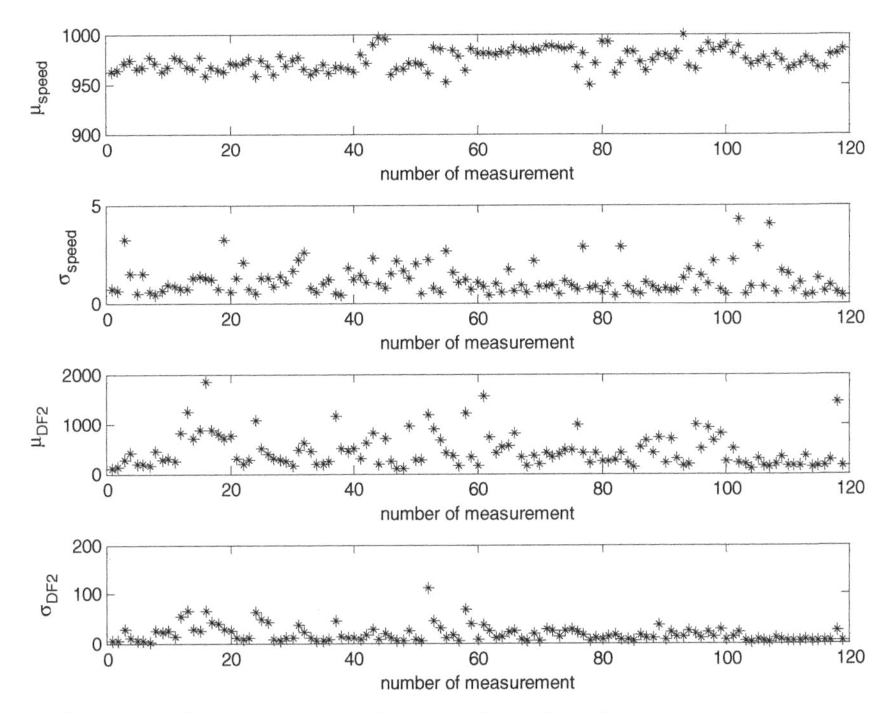

Fig. 7 The input data used in order to determine the decision thresholds

Recall that in the model Gaussian distribution of features is assumed. This assumption is a common one in the literature. In the next step we verify the assumption and try to fit a better distribution to the dataset. After that, the new set of parameters is subjected to clustering. After insightful analysis the assumption of Gaussian distribution of diagnostic features has been rejected. The reason for that is a power law of tails. It means that the complementary cumulative distribution function (tail distribution), i.e. $1-F(x)=P(X>x)$ behaves like a power function x^{α} for a certain parameter α. Fig. 9 presents the empirical tail of the measurement of the diagnostic feature DF2 with a power function with parameter α fitted by using the least squares method.

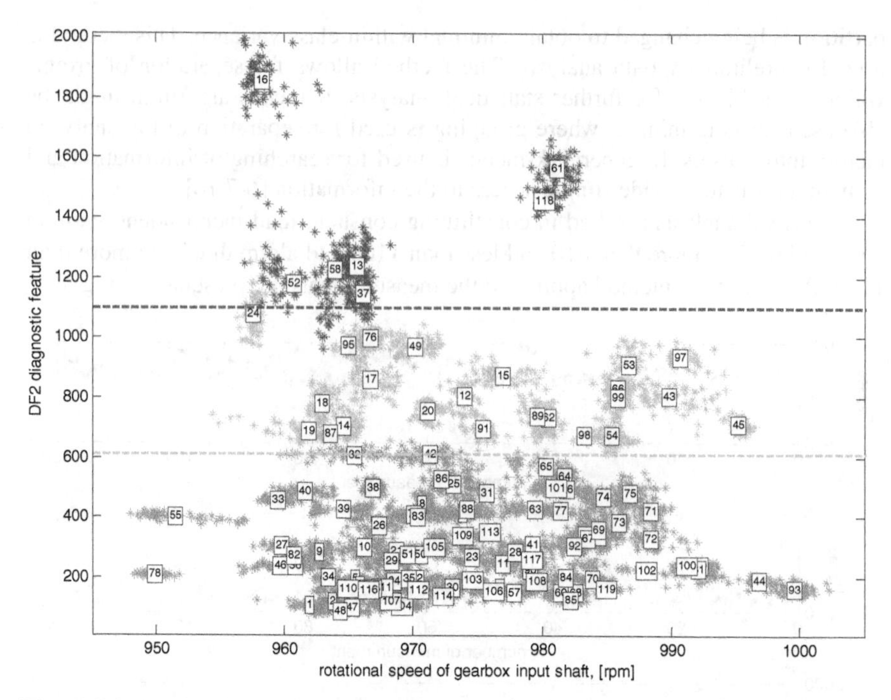

Fig. 8 Diagnostics feature DF2 in rotational speed function. Presented thresholds are determined on the basis of k-means clustering by using μ_{speed}, σ_{speed}, α_{DF2} and λ_{DF2}.

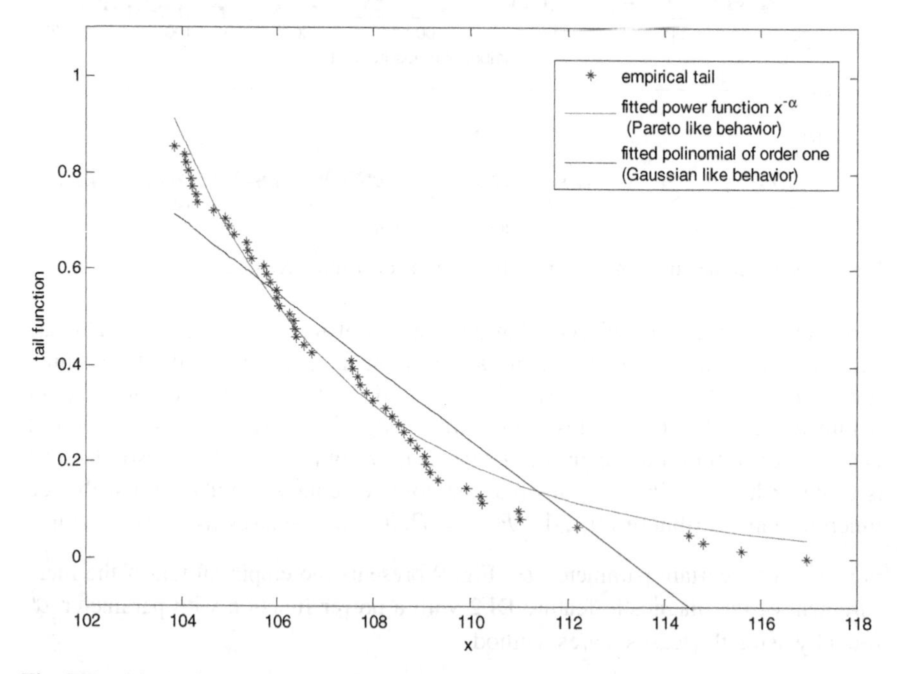

Fig. 9 Empirical tail for the sample data and fitted exponential function, α=29.3

Because the power law in tail is observed we propose to fit a theoretical distribution with tail which also follows the power law. Moreover, values of the measurement are nonnegative, thus we propose to fit the Pareto distribution to the data. The probability density function of the Pareto distribution is:

$$f(x) = \frac{\alpha\lambda^{\alpha}}{x^{\alpha+1}}, \qquad x \geq \lambda. \tag{1}$$

Parameters $\alpha > 0$ and $\lambda > 0$ are shape and scale parameters, respectively. The Pareto distribution might be applied to many phenomena, e.g. insurance mathematics and manufacturing process management [8].

For each of 119 measurements we estimate parameters α and λ the using maximum likelihood method (MLE). Similarly to the previous case, a new set of parameters has been determined and subjected to clustering. Mean and standard deviation of the diagnostic feature measurements have been substituted by α and λ. μ and σ of rotational speed remained unchanged. It is worth mentioning that distribution of the rotational speed measurements do not follow a power law, thus there is no need to characterize them by parameters other than mean and standard deviation.

Size of the new dataset is 4x119 and is consisted of 4 parameters, i.e. μ_{speed}, σ_{speed}, α_{DF2} and λ_{DF2} for 119 measurements, see Fig. 10.

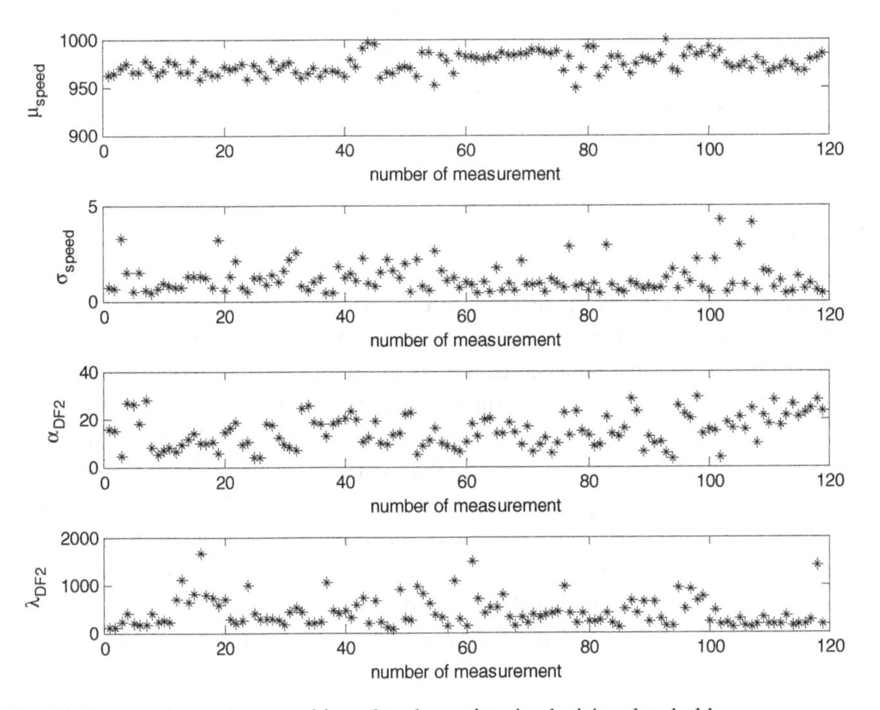

Fig. 10 The new input dataset subjected to determine the decision thresholds.

In the next step, a multidimensional clustering has been performed in order to group the dataset by using the k-means algorithm. In 47 of 119 cases the results are different from those obtained in the previous analysis, where mean and standard deviation of the diagnostic feature DF2 are used for clustering. The current result, where μ_{speed}, σ_{speed}, α_{DF2} and λ_{DF2} are used for clustering, rejects the concept of constant DF2 values thresholds through the whole spectrum of rotational speed. The k-means clustering procedure output is consisted of 5 clusters. Next, according to [1], we fit a regression lines to each cluster and merge similar ones to determine three condition states e.g. good condition, warning and bad condition, Fig. 11.

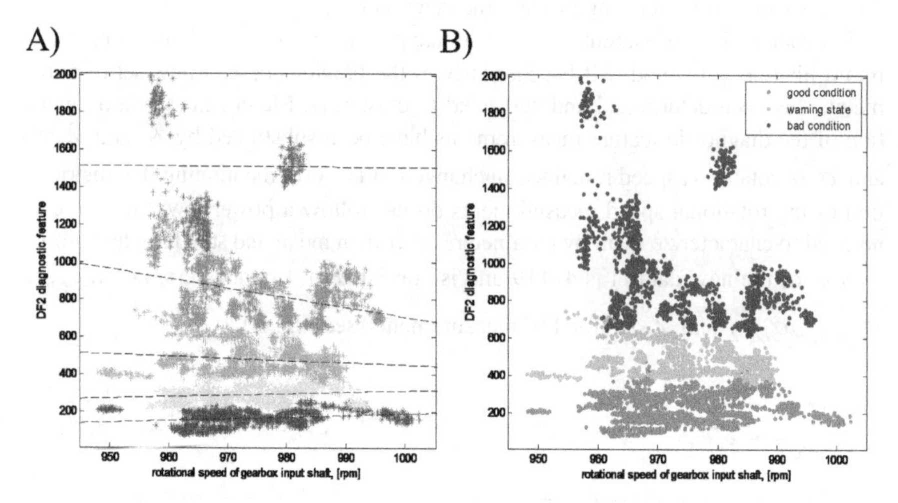

Fig. 11 Results of k-means clustering using μ_{speed}, σ_{speed}, α_{DF2} and λ_{DF2}

4 Conclusions

In this paper the problem of decision thresholds for belt conveyor driving units maintenance is analyzed. Statistical modeling of the diagnostic data has been performed. Appropriate maintenance of the transportation system requires determining of decision making rules – thresholds of diagnostic features. We extended the previous approach which relies on constant thresholds through the whole rotational speed spectrum. Results of our methodology are consistent with those obtained in [1], where the level of diagnostic features depends on rotational speed. This dependency results in overlapping of density functions for every measurement, thus we considered another description of probability distribution. This description is based on the remark that the distribution of diagnostic feature DF2 is not Gaussian – it follows a power law and the Pareto distribution is much better in this case.

Parameters of the Pareto distribution for diagnostic feature DF2 and Gaussian parameters for rotational speed were estimated and their thresholds were determined using the clustering algorithm. Moreover, the clustering algorithm (k-means) does not depend on the distribution of the clustered data, thus one can benefit from it even in cases where data does not follow the Gaussian distribution.

Acknowledgements. This work is partially supported by the statutory grant No. S300973 (J. Obuchowski) and statutory grant No. B300 43 (P. Stefaniak).

References

1. Bartelmus, W., Zimroz, R.: A new feature for monitoring the condition of gearboxes in non-stationary operating conditions. Mechanical Systems and Signal Processing 23, 1528–1534 (2009)
2. Blazej, R., Zimroz, R., Jurdziak, L., Hardygora, M., Kawalec, W.: Conveyor belt condition evaluation vianondestructive testing techniques Mine planning and equipment selection. In: Drebenstedt, C., Singhal, R. (eds.) Proceedings of the 22nd MPES Conference, Dresden, Germany, October 14-19, vol. 2, pp. 1119–1126. Springer, Cham (2014)
3. Brooks, R., Thorpe, R., Wilson, J.: A new method for defining and managing process alarms and for correcting process operation when an alarm occurs. Journal of Hazardous Materials 115 (2004)
4. Cempel, C.: Limit value in practice of vibration diagnosis. Mechanical Systems and Signal Processing 4/6 (1990)
5. Galar, D., Gustafson, A., Tormos, B., Berges, L.: Maintenance Decision Making based on Different types of Data fusion. Eksploatacja i Niezawodnosc – Maintenance and Reliability 14(2), 135–144 (2012)
6. Gordon, A.D.: Classification. Chapman & Hall, London (1999)
7. Hartigan, J.A.: Clustering algorithms. John Wiley & Sons, Inc. (1975)
8. Hazewinkel, M.: Pareto distribution. Encyclopedia of Mathematics. Springer (2001)
9. Jablonski, A., Barszcz, T., Bielecka, M., Breuhaus, P.: Modeling of probability distribution functions forautomatic threshold calculation in condition monitoring systems. Measurement 46(1), 727–738 (2013)
10. Jain, A.K., Dubes, R.C.: Algorithms for Clustering Data. Prentice Hall, Englewood Cliffs (1988)
11. Kacprzak, M., Kulinowski, P., Wedrychowicz, D.: Computerized information system used for management of mining belt conveyors operation. Eksploatacja i Niezawodnosc – Maintenance and Reliability 50(2), 81–93 (2011)
12. Lodewijks, G.: Strategies for Automated Maintenance of Belt Conveyor Systems. BulkSolids Handling 24(1), 16–22 (2004)
13. Sawicki, M., Stefaniak, P.K., Krol, R., Zimroz, R., Hardygora, M.: The integration of thermography data and DiagManager system for diagnostic management of technological system. Transport & Logistics (Belgrade), pp. 263–26 (2012)
14. Silverman, B.W.: Density Estimation for Statistics and Data Analysis. Chapman and Hall, London (1986)

15. Stefaniak, K., Zimroz, R., Krol, R., Gorniak-Zimroz, J., Bartelmus, W., Hardygora, M.: Some remarks on using condition monitoring for spatially distributed mechanical system belt conveyor network in underground mine - a case study. In: Fakhfakh, T. (ed.) Proceedings of the Second International Conference on Condition Monitoring of Machinery in Non-Stationary Operations, CMMNO 2012, pp. 497–507. Springer (2012)
16. Zimroz, R., Krol, R., Hardygora, M., Gorniak-Zimroz, J., Bartelmus, W., Gladysiewicz, L., Biernat, S.: A maintenance strategy for drive units used in belt conveyors network. In: Eskikaya, Ş. (ed.) 22nd World Mining Congress & Expo, Istanbul, September 11-16, vol. 1, pp. 433–440. Aydoğdu Ofset, Ankara (2011)

Low Rank Coal: Future Energy Source in Indonesia

Tri Winarno and Carsten Drebenstedt

Technische Universität Bergakademie Freiberg, Germany

Abstract. World energy grow from year to years. Indonesia, with a population of 240 million, GDP growth rate of 6.23% (2012), estimated final energy demand in 2025 will be reach 2,043 Million BOE (Barrel Oil Equivalent). This requirement when calculated using assumptions on Master Plan of the Acceleration and Expansion (Development) of Indonesia (in Indonesia abbreviated as MP3EI) 2011-2025, the energy demand will reach 2,772 Million BOE.

Oil production decline every year, and Indonesia became oil importer country. In gas sector, prediction from 2019 Indonesia will become importer of gas, and in 2030 the deficit of gas will be reach approximate 640 Million BOE (BPPT, 2013).

Based on data from the Geological Agency of Indonesia (2011), Indonesia total reserves of coal approximate 28 Billion tons of very high (> 7100 kcal/kg) 231 Mt (Million tons), high (6100-7100 kcal/kg) 1,655 Mt, medium (5100-6100 kcal/kg) 16,128 Mt and lower (<5100 kcal/kg) 10,002 Mt.

Coal production reached 426 Mt (2013), dominated by thermal coal, with 74.5% of coal for export. Utilization of LRC (low rank coal) is still lacking, due to the limited market. Besides that, also due to the location is relatively far from the market and have high moisture content (30-55%), so transportation makes LRC not economically.

Domestic coal demand in the coming years is quite significant. In accordance with Presidential Decree no. 5/2006, the role of coal in the energy mix in 2025 become 33%, with 2% addition to a liquid coal. LRC utilization in Indonesia is still limited, and even many LRC companies are delaying production with market reasons. Upgrade (upgrading/drying, coal gasification and coal liquefaction) also enables a reduction in emissions from coal combustion. Considering all the potential for the development of LRC very open. With some economic calculation on investment, making LRC economically profitable and open opportunities for domestic use and for export.

Keywords: energy, low rank coal, upgrading.

1 Energy Needs

In 2000 total energy consumption in the world were 69,110 Million BOE energy mix were oil 38.75%, natural gas 23.31%, coal 24.55%, nuclear 6.25%, hydroelectricity 6.4% and renewable 0.74%. Then, 2012 energy consumption

C. Niemann-Delius. (ed.), *Proc. of the 12th Int. Symp Continuous Surface Mining - Aachen 2014*,
Lecture Notes in Production Engineering, DOI: 10.1007/978-3-319-12301-1_35

became 92,322 Million BOE with mix oil 33.1%, natural gas 23.9%, coal 29.9%, nuclear 4.5%, hydroelectricity 6.7% and renewable 1.9% (BP, 2013)

In Indonesia energy needs increased with the historical average growth of 3.09% per year from 2000-2010, from 737 million BOE (2000) to 1,012 million BOE (2010). In 2011 the final energy to 1,044 million BOE of energy, with energy mix : oil 36%, coal 23%, gas 18%, biomass 16% and others 7% (BPPT, 2013).

Oil production in Indonesia from year to year show a decline. Starting in 2003, Indonesia has become a net importer of oil. In 2012, oil production in Indonesia amounted to 974,000 BOPD (barrel oil perday) with consumption of 1,384,000 BOPD. In 2008 gas consumption 218 Million BOE became 242 Million BOE in 2011. It is estimated that this use will continue to increase in the next years, and prediction, start from 2019 Indonesia will become importer of gas, and in 2030 the deficit of gas reach 640 Million BOE (BPPT, 2013). The production of coal has increased each year, in 2013, 22% were used for the domestic demand and the rest for export. Most coal production is thermal coal. Export coal is still needed as part of the state revenue.

In 2025 final energy demand estimated will be reach 2,043 Million BOE with business as usual scenario (BAU). When calculate using assumptions Master Plan of the Acceleration and Expansion (Development) of Indonesia (MP3EI) 2011-2025, the energy demand will reach 2,772 Million BOE (BPPT, 2012)

In accordance with Presidential Decree no. 5/2006, the role of coal in the energy mix in 2025 33%, with 2% addition of liquid coal (in energy mix 2012, coal 26%).

2 Coal Potential

Indonesia has enough coal potential. Based on data from the Geological Agency of Indonesia (2011), the coal resources approximately 120 Billion tons. Total coal reserves are approximately 28 Billion tons. The coal reserves are classified based on quality; consisting of a very high quality 231 Mt (million tons) (0.8%), high quality 1,655 Mt (5.9%), medium quality 16,128 Mt (57.5%) and low rank quality coal 10,002 Mt (35.8%) (see Figure 1). This quality classification based on Government Regulation No. 45/2003 on Tariffs Applicable for Non Tax Revenue in Energy and Mineral Resources Sector.

In reserves, Indonesia held 3% of world reserves (28 Billion of 948 Billion tons) (EIA, 2008). In terms of world reserves, Indonesia has a 5% low rank coal reserves. Indonesian LRC production in 2013 amounted to 35 Mt (of 391 Mt total coal production), used for coal blending.

Although Indonesia has 53% reserves in South Sumatera, but production is only 8% total coal production, because of LRC. Otherwise, Kalimantan with reserves of 47%, produced 92% (Figure 3).

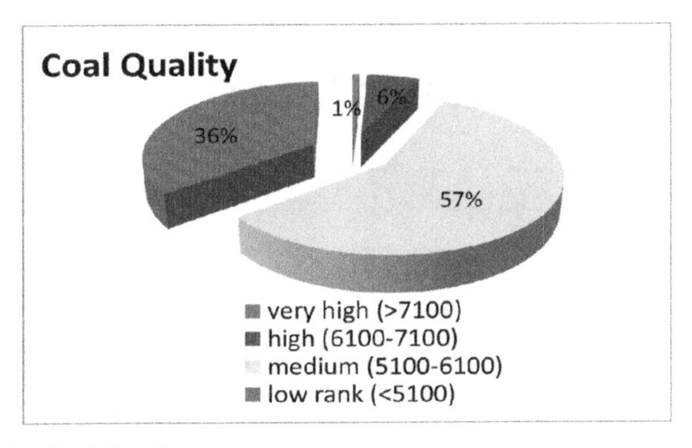

Fig. 1 Indonesian Coal Quality

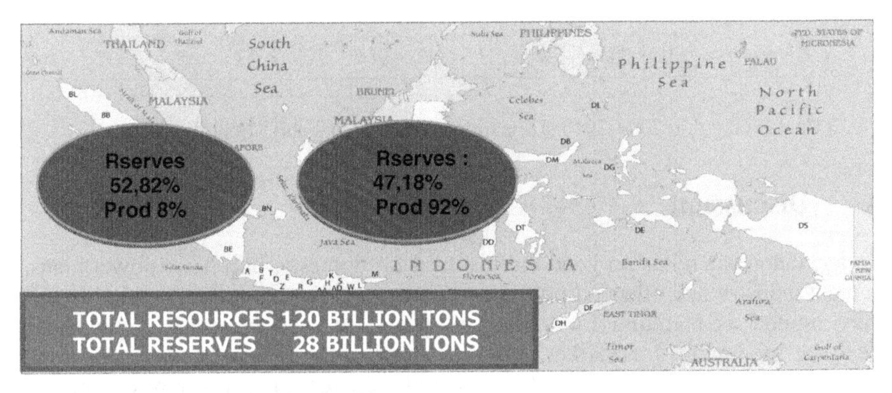

Fig. 2 Indonesian Coal Distribution Map

3 Current Condition

Coal production in the last 10 years showed increasing. Coal production grew by 3 times from 131 Mt in 2004 (95 Mt export and 36 Mt domestic) to 426 Mt in 2013 (317 Mt of export and 97 Mt domestic) (Figure 3). Most of coal production in Indonesia were thermal coal. With production in 2012 and 2013, put the Indonesia as the largest exporter of thermal coal in the world. Coal export in Indonesia is a part of state revenue.

Utilization of coal in Indonesia for domestic market still limited and mostly used in power plant, cement and others industries. Export market for Indonesian coal is Asia. The main coal export destination countries Japan, China and India.

In Indonesia, most of LRC companies stopped their production because of market reason. LRC have some limitation, especially in moisture content. Coal with a high moisture content, will have a low calorific value.

In the world, coal is widely used. Total coal world production in 2011 were 8,440 Mt, with share of LRC (lignite) 1,130 Mt (EIA, 2014).

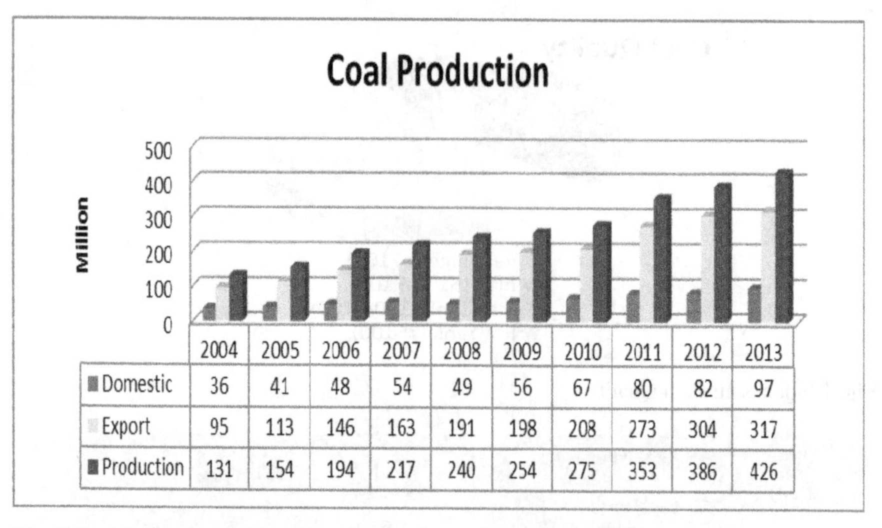

	2004	2005	2006	2007	2008	2009	2010	2011	2012	2013
Domestic	36	41	48	54	49	56	67	80	82	97
Export	95	113	146	163	191	198	208	273	304	317
Production	131	154	194	217	240	254	275	353	386	426

Fig. 3 Coal Production 2004-2013 (Directorate General Coal and Mineral, 2013)

4 Discussion

So far, Indonesia relies on good thermal coal for domestic purposes (power plans, cement industry and other purposes) as well as an export commodity. LRC is still very limited as blending for various purposes. In fact, most LRC companies delaying production for market reasons. Actually, the reason is not very strong because the use of the LRC in the world also increased, e.q. for power plants, gasification, liquefaction and other purposes.

In Act Number 4 year 2009 about Coal and Mineral and Government Regulation Number 23 year 2010 about Coal and Mineral Enterprise, the need to increase the added value of mining commodities including coal. With 10 Billion tons of reserves and the government regulation in the energy mix in 2025, the LRC utilization can be optimized.

Thermal coal could still be maintained as an export commodity, whereas LRC can be used for domestic and export. The main problems of LRC is moisture content. This problems can be reduced by Upgrading/ Drying of Coal. Upgraded can be used as export commodity, besides domestic consumption. With some assumption 1 ton of LRC (lignite) = 2.053 BOE, in 2025 Indonesia need between 324 - 445 million ton of LRC for domestic demand.

Prediction of Indonesia energy needs in 2025 on Table 1.

More advanced utilization in Indonesia from 2008, when the Government announced to change the utilization of kerosene for household gas. By 2015, all the gas need to meet domestic demand supplied by domestic production, while gas needs in 2030 supplied by domestic gas production, gas imports, as well as CBM. Gas imports will reach 640 million BOE, or 25% of the total supply of natural gas. Non-conventional gas resources that can be expected other than natural gas is CBM (with of 10-11%) and coal gasification (BPPT, 2013). Synthetic gas from coal has the potential to meet the demand in industry sector and power plant.

Table 1 Indonesian Energy 2025

	BAU**	MP3EI
Coal	674.19*	914.76*
Gas	612.9*	831.6*
Oil	408.6*	554.4*
Renewable energy	306.45*	415.8*
Coal Liquefaction	40.86*	55.44*

*in million BOE

**Bisnis as Usual (BAU)

In 2030, deficit of gas can be with coal gasification. Gasification can be used to convert solid coal into a gaseous feedstock, which can be used in a range of other products. Gasification produces synthesis gas - syngas, a mixture of mostly carbon monoxide and hydrogen. The process can also help with separation and sequestration of carbon dioxide (Energy and Earth Resources, 2014). Several LRC gasification project have been done in several countries like Germany, America, Australia, etc.

Oil consumption in Indonesia has increased which is not comparable with the production, is predicted to continue to rise (see Figure 4). It is estimated that by 2025 petroleum consumption 408.6 Million BOE (BAU) or 554.4 Million BOE (MP3EI).

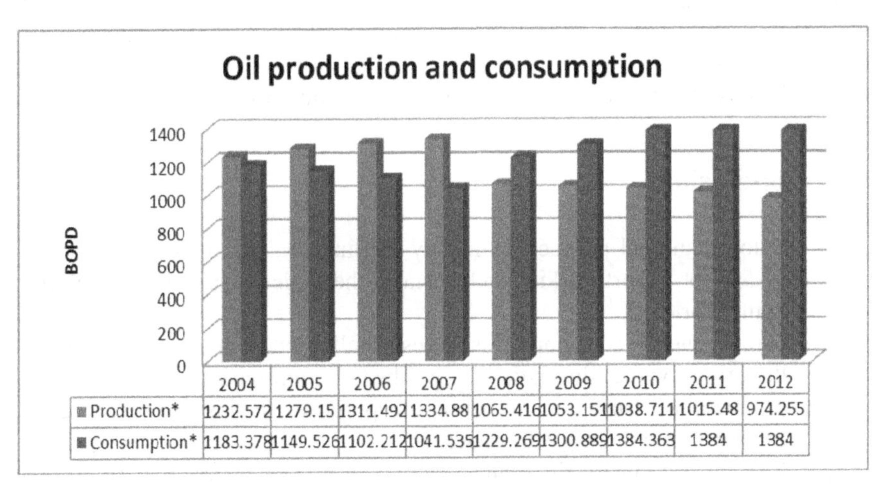

Fig. 4 Indonesia Oil Production and Consumption (EIA, 2014)

High oil consumption can be substituted by coal liquefaction. The development of synthetic fuel production based on coal was first performed in Germany in the 1900s using the Fischer-Tropsch synthesis process developed by Franz Fisher and Hans Tropsch. In 1930, in addition to using the Fischer-Tropsch synthesis process, was developed also Bergius process for producing synthetic fuel.

Coal liquefaction process can be carried out by means of direct and indirect. In indirect coal liquefaction process is converted into a gas and then gas formed is further processed into synthetic oil. Indirect liquefaction process has been carried out commercially in South Africa since 1956 by SASOL (*South African Synthetic Oil Ltd*). Currently direct coal liquefaction technology developed by the Japanese (Brown Coal Liquefaction, BCL) and the United States (Headwaters Technology Innovation, HTI). Japan has successfully tested a pilot plant coal liquefaction capacity of 50 tons / day in Victoria, Australia. While the United States has built a pilot plant capacity of 600 tons / day in Catlettsburg, Kentucky, USA.

Referring to the President Decree No. plan. 5 years in 2006, and also assuming Indonesia's energy consumption in 2025, then there is enough space to utilize the LRC as the future of energy in Indonesia. Concern that the use of the LRC would require high investment. Construction costs for Upgrading Brown Coal (UBC) capacity of 5,000 tons/day (feed coal 7,500 tons/day) requires an investment of over US $ 100 million. Investment cost of a commercial coal liquefaction plant, quite expensive too, at US $ 1.5 billion for the refinery of 13,500 BOPD and could reach US $ 2.1 billion for the refinery with a capacity of 27,000 BOPD.

To assess the investment in these activities, can be used several criteria, such as Net Present Value (NPV), Internal Rate of Return (IRR), Benefit Cost Ratio (BC Ratio) and Payback Period. This analysis can be done separately or together. Additional analysis can be performed include analysis of Break Even Point (BEP) and the Rate of Return on Investment (ROI). In addition is necessary to determine the result of changes in production parameters to changes in the performance of production to make profit by sensitivity analysis. By sensitivity analyzing the possible consequences of such changes can be known and anticipate.The purpose of everything is to assess the feasibility of a project / investment.

5 Conclusions

1. Indonesian coal production, mostly thermal coal continues to rise, with 74.5% for export, making Indonesia as the largest thermal coal exporter in the world.
2. On the energy consumption side, especially petroleum, imports grew at this time while domestic oil production continues to decline. Gas production is sufficient for domestic purposes, but is estimated 2019 Indonesia becomes gas importer.
3. LRC utilization is still limited, with a reserve of more than 10 Billion tons can be optimized by upgrading / drying, gasification and liquefaction to meet the ever-increasing energy consumption.
4. Investment and high risk on upgrading / drying, gasification and liquefaction require economic calculation including calculation of sensitivity analysis. into account a range of possible changes that will occur expected to result in optimal benefits for these activities.

References

[1] Andrew, H.: Matthew Trivett, Indonesian Coal Review. Petersons Securities Limited, Austalia (2012)
[2] Geologi, B.: Sumberdaya dan Cadangan Batubara Indonesia, Badan Geologi, Bandung (2011)
[3] Lucarelli, B.: Benefits of “Drying” Indonesian Low Rank Coals. In: Cleaner Coal Workshop, Ha Long City, Vietnam (August 2008)
[4] Beychok, M.R.: Process and environmentals technology for producing SNG and liquid fuels. U.S, EPA report EPA-660/2-2-75-011 (May 1975)
[5] BPPT, Indonesia Energy Outlook 2012, Agency for the Assessment & Application of Technology, Jakarta (2012)
[6] BPPT, Indonesia Energy Outlook 2013, Agency for the Assessment & Application of Technology, Jakarta (2013)
[7] Directorate General of Mineral and Coal, Mineral and Coal 2013, MEMR, Jakarta (2013)
[8] EIA, International Energy Statistic, http://www.eia.gov/cfapps/ipdbproject
[9] http://www.netl.doe.gov/technology/coalpower/gasification/index.htm
[10] Tanaka, K.: Development of Highly Efficient Coal Gasification Technology and Its Application to Victorian Brown Coal, BCBRA (Brown Coal Business Research Australia (February 2012)
[11] Tekmira, Low Rank Coal Quality Improvement Process With UBC Phase II (2012), http://www.tekmira.esdm.go.id/kp/Batubara/peningktbbprkatrendah.asp
[12] Willson, W.G., Young, B.C., Irwin, W.: Low Rank Coal Drying Advances. Coal 97 (August 1992)
[13] MEMR, Indonesia Energy Outlook 2012, Centre of Data and Information MEMR, Jakarta (2012)
[14] World Energy Council, World Energy Resources 2013 Survey, World Energy Council, London (2013)
[15] http://www.worldcoal.org/resources/coal-statistics
[16] British Petroleum, BP Statistical Review of World Energy June 2013 (2013), http://bp.com/statisticalreview

Impact of Surface Cost on Lignite Mining Project

Michał W. Dudek[1], Leszek Jurdziak[1], Witold Kawalec[1], and Zbigniew Jagodziński[2]

[1] Wroclaw University of Technology, Wybrzeze Wyspianskiego 27, 50-370 Wroclaw, Poland
{michal.dudek,leszek.jurdziak,witold.kawalec}@pwr.wroc.pl
[2] KWB "Konin", Kleczew SA

Abstract. The paper deals with impact of surface cost on lignite mining project. Not only profitability is discussed but also relations between land owners and a Mine Company. Land acquisition is analyzed and statistics are presented. Surface cost as term is introduced. Investigations are conducted on real data obtained from an already closed pit. Geological data is processed and economic block model is built. Lignite quality parameters are estimated with the use of kriging. Surface cost map of deposit area is generated as average measures for purchase/sale of real estate transactions. Ultimate pits with the use of Lerchs-Grossmann algorithm are generated and results are discussed. Additional spatial relation as increase in required land area for operation due to ultimate pits cases is shown. Time factor in land buyout is discussed. Over or underestimating surface cost may have significant impact on viability of lignite mining project.

1 Introduction

Mineral rights, the legal rights to exploit and enjoy the benefits of extraction of minerals located below the surface as country specific issue are subject to government control around the world. Following the Polish law, mineral rights are automatically granted to land owner if only these minerals are not listed in Polish Geology and Mining Act (since 2011) like sand and gravel deposits. Otherwise a concession must be granted. Due to the Polish Geology and Mining Act granting concession for extractive activities to proponent in case of lignite deposits does not require land ownership essential for commencement of operation. There are no straight acquisition regulations so in the event of any legal dispute between land owners and mining or energy company, parties should go to court.

Surface costs, considered as costs of land acquisition and management, mine/milling site clearing, payment of ownership, taxes, costs of reclamation and giving or selling land back to previous owners or government, in developed countries became an important topic because of rising costs of land acquisition

C. Niemann-Delius. (ed.), *Proc. of the 12th Int. Symp Continuous Surface Mining - Aachen 2014*,
Lecture Notes in Production Engineering, DOI: 10.1007/978-3-319-12301-1_36

due to urban area density growth, increasing value of real properties, rising taxes and costs of fulfillment of environmental regulations.

In mineral assets valuation three main approaches are used. The Market Approach, the Income Approach and the Cost Approach. Within each approach there are various methods that have evolved and that are used in different markets. In 2005 the IVSC (International Valuation Standards Council) issued Guidance Note 14, Valuation of Properties in the Extractive Industries but it was withdrawn in 2010 by the IVSB [1] (International Valuation Standard Board) but there are attempts for new regulation to be developed. Mineral assets valuations are subjected to national valuation codes such as: VALMIN Code, CIMVal Code, SAMCODE. In 2008, POLVAL Code - the Polish Code for Mineral Assets Valuation based upon world's best practice was developed in which Fair Market Value of mineral assets is introduced. The Income Approach uses indirect measure of the market value taking into account future income that can be derived from the mineral assets over its entire economic life. Integrated General Approach for mineral assets valuation suggests application of Discounted Cash Flow analysis for projects that are expected to be economically viable. However still there is a problem regarding time factor in commencement of operation. If the project is not expected to be economically viable or time factor is included, IGA (Integrated General Approach) recommends using ROA - Real Option Analysis with e.g. an Option to Delay. Because of ability to choose proper time for commencement NPV of mining project is increased by decision elasticity bonus [2]. ROA approach will be a subject of separate studies.

The aim of research presented include impact of surface costs on profitability of mining opencast lignite project treated as the investment of vertically integrated tandems of a lignite mine and a power plant. To illustrate surface costs scale it is worth to mention that Mine Company from the beginning of its operation has bought about 133 Mm^2 of lands and sold or gave back about 81 Mm^2 what gives about 52 Mm^2 in current disposition. With high fixed costs optimal surface costs management became very important issue to sustain profitability of lignite based energy projects.

2 Surface Cost Identification through Mine Project Valuation

Mineral asset valuation is often a team effort. This paper focuses on isolating surface cost from mining opencast lignite project and measures its impact. Checklist of required data for valuation purposes depends upon approach chosen, nevertheless in sales comparison approach this dataset may look as follows [3]:

Agreement/transaction date, Buyer/acquirer's name, Effective date of appraisal, Price paid per unit, (average) product price at agreement/transaction date, Long-term product price expected at agreement date, (Long-term) product price at effective date of appraisal, Minority interest, Project development status, Deposit grade, Deposit/project size, Property control and security of tenure,

Capital investment requirement, Operating cost/net operating income, Production loss/recovery/metallurgical complexity, Product quality, Product market stability, Discovery and expansion potential, Location and access, Infrastructure, Permitting issues, Reclamation, Country risk, Project risk, Taxes, royalties, levies.

There are other approaches that may be used for valuation purposes but careful cash flow calculation for individual project still remains a key. For cash flow analysis following steps should be considered [4].

a) Calculation of daily production rate for ore, waste and production schedule.
b) Main mine equipment selection and production rate for each type of equipment estimation.
c) Supplementary equipment selection.
d) Number of production and support employees required.
e) Calculation of owning and operating cost for the equipment.
f) Calculation of other costs
g) Overall cost per tonne calculation.
h) Capital cost and the ownership cost for the equipment determination.
i) Other capital expenditures (mine).
j) Mineral processing cost calculation.
k) Expression of the mining costs
l) Mine ownership costs per ton.
m) Total mining cost.
n) Profitability estimation.

While developing geological reserves one may give into consideration many possible opencast mining scenarios from which all have impact on the surface. The influence of lignite base price on the size of ultimate pit (lignite supply) and decrease of lignite quality (decrease of calorific value and increase of sulphur and ash content) has already been analyzed [5].

Additional analysis regarding surface cost and ultimate pit size will be conducted.

Considering overlaying parcels as mineral property implicates fact that it refers to a property that contains a mineral deposit of such quantity and quality that it is profitable for extraction as highest and best land use [6].

For the cost estimation purposes in mining projects land acquisition may be significant as well. Without legal acquisition rules all transactions should be taken under free market conditions. By law, mineral deposit (like cooper, lignite and other) listed in Polish Geology and Mining Act (from 2011) are not a part of real estate in terms of ownership and should not have influence on its value, which is arguable. Land parcels that include sand and gravel mineral rights were investigated with Comparable Sales Approach use in Wroclaw County considered as local market in Poland. Estimated price per 1m^2 of those mineral rights (as possible mining land use) was compared to agricultural lands and vacant lots for residential use. In first case sand and gravel mineral property price was approx.

200% of agricultural lands price, whereas it was only 23% of vacant lots for residential use price [7].

Some mineral property valuation used in real estate appraisals in Poland reflects methodology used in mining projects [8].

With the difference that Cost Approach is not well developed in comparison to mining project valuation where Cost Approach may be applicable to Mineral Assets at prospecting stage or at scaling down and closure [9].

Surface cost is not limited only to acquisition. Additional surface cost might be mine site clearing and property taxes. Mine site clearing (clearing land from trees, plants, topsoil) and access roads costs may be expressed as a function of T_p – tons of ore and waste mined per day (by O`Hara & Suboleski, 1992, found in [4]).

$$A_p = 0.0173T_p^{0.9} \tag{1}$$

Where:

A_p – required area for the pit in acres (approx. 1ac = 4046.9m²).

Total clearing cost is given by following formula:

$$TCC = \begin{cases} \$1600\,A_p^{0.9} & \text{for 20\% slopes with light tree growth} \\ \$300\,A_p^{0.9} & \text{for flat land with shrubs and no trees} \\ \$2000\,A_p^{0.9} & \text{for 30\% slopes with heavy trees} \end{cases} \tag{2}$$

Additional soil and/or waste rock stripping to expose an amount of deposit to sustain four to six months deposit production, then the estimated cost of waste stripping will be

Soil stripping cost $SSC = \$3.20\ T_s^{0.8}$ for soil not more than 20 ft deep (3)

Waste stripping cost $WSC = \$340\ T_{ws}^{0.6}$ for soil not more than 20 ft deep (4)

Where:

T_s – is tons of soil,

T_s – is tons of waste rock.

As for Polish market, average price of soil removal with stockpiling based upon current civil engineering quotations is estimated to be approx. 7.5 PLN/m³. With average weight density 2.65 Mg/m³ and currency exchange 1USD = 3.08PLN soil stripping cost (with stockpiling) in Poland may be expressed simply as follows.

$$SSC_{PL} = \$0.92\ T_s \tag{5}$$

Where:

T_s – is tons of soil.

Applied to mining industry this number is only rough estimate. It does not include trees removal and other services though some general conclusions can be made.

3 Closure and Reclamation Cost

It is believed that legal condition in the scope of mine closure and post-mining land reclamation are unsatisfactory [10].

Closure and reclamation fund must be created and not less than 10% of the required exploitation charge shall be allocated. Entrepreneurs in Poland are obligated to pay operating fees. Royalty rate for lignite is 1.66 PLN/Mg [11].

Average measures for purchase/sale transactions of lands and buildings in Mine Company area (Province) are expressed in following tables as two time frames 2005 and 2012 year.

Table 1 Polish Real Estate Turnover in 2005 (Province of Mine area) [12]

Description	Developed land (land with buildings)			Agricultural land		
	Average price PLN/m^2	Of which: Urban area PLN/m^2	Of which: Rural area PLN/m^2	Average price PLN/m^2	Of which: Urban area PLN/m^2	Of which: Rural area PLN/m^2
1	2	3	4	5	6	7
Mine area	23.70	64.10	7.90	1.20	10.90	1.00

Table 2 Purchase/sale transactions of properties built up with residential buildings in 2012 [13]

Specification A– *number of transactions* B–*value, thous. PLN* C–*area, m^2* D– *average price PLN/m^2*	Total	Location – Province of Mine area				
		Urban areas				*Rural areas*
		Total	*Of which cities with County status*			
			Total	*More than 200 thous. Inhabitants*	*Less than 200 thous. Inhabitants*	
1	2	3	4	5	6	7
A	3486	1600	629	467	162	1886
B	1160990	584716	342902	283837	59065	576274
C	4428415	1729701	690059	523573	166486	2698714
D	262.17	338.04	496.92	542.12	354.77	213.54

Real Estate Tax as surface cost is also a subject of optimization. Lignite deposit may have area of 7 630 000 m^2. Assuming Land Tax for business purposes approx. 0.79-0.85 PLN/m^2 may give annually 6.4 million PLN.

Other surface cost may vary worldwide. Building and development may be treated as sunk cost. In United States special tax consideration for depletion of deposit is given [4].

Table 3 Purchase/sale transactions of residential buildings in 2012 [13]

Specification A– *number of transactions* B–*value, thous. PLN* C– *area, m²* D– *average price PLN/m²*	Total	Location – Province of Mine area				
		Urban areas				*Rural areas*
		Total	*Of which cities with County status*			
			Total	*More than 200 thous. Inhabitants*	*Less than 200 thous. Inhabitants*	
1	2	3	4	5	6	7
A	32	22	-	-	-	10
B	8645	7006	-	-	-	1639
C	4363	3320	-	-	-	1043
D	1981.43	2110.24	-	-	-	1571.43

Table 4 Purchase/sale transactions of built-up agricultural land in 2012 [13]

Specification A – *number of transactions* B–*value, thous. PLN* C –*area, m²* D – *average price PLN/m²*	Total	Location – Province of Mine area				
		Urban areas				*Rural areas*
		Total	*Of which cities with County status*			
			Total	*More than 200 thous. Inhabitants*	*Less than 200 thous. Inhabitants*	
1	2	3	4	5	6	7
A	664	55	3	-	-	589
B	224761	6933	1040	-	-	217828
C	33583597	326390	37939	-	-	33257207
D	6.69	21.24	27.41	-	-	6.55

Table 5 Average measures for purchase/sale transactions of agricultural land in Province of Mine in 2012 [13]

Specification A – *total* B –*urban areas* C –*rural areas*	*Average area sold in single transaction, m²*	*Average value of single transaction in thous. PLN*	*Average transaction Price, PLN/m²*
1	2	3	4
A	17484.4	210.60	12.05
B	3960.7	165.40	41.76
C	19190.4	216.30	11.27

While developing mineral project valuation useful cost information may be derived from many costs reference guides. For an instance acquisition as surface cost – (Canadian Exploration Expense) may be fully deducted in the year incurred to the extent of income from any source for the corporation. Any balance not currently deductible may be carried forward indefinitely in a pool called the Cumulative Canadian Exploration Expense (CCEE) pool. Alternatively, exploration expenses may be transferred through "flow-through-shares" that are a means of obtaining financing for mining exploration and development in Canada by sale of the right to claim the specific expenditures as a tax deduction. Money received from sale of "flow-through-shares" is considered as capital gain from half of which is taxable as income [14].

4 Opencast Lignite Mine Case

Investigation of impact of surface cost on mining opencast lignite project was conducted on existing data collected from already closed pit. For the multi-pit Mine Company it is crucial to choose best possible scenario based upon variant analysis for cost optimization.

Collected data can be divided into three parts: geology, financial information and land management. Cost structure from 17 years of operation (1995 - 2011) was analyzed. Several changes of lignite production unit cost, total production cost, revenue and production rate were observed.

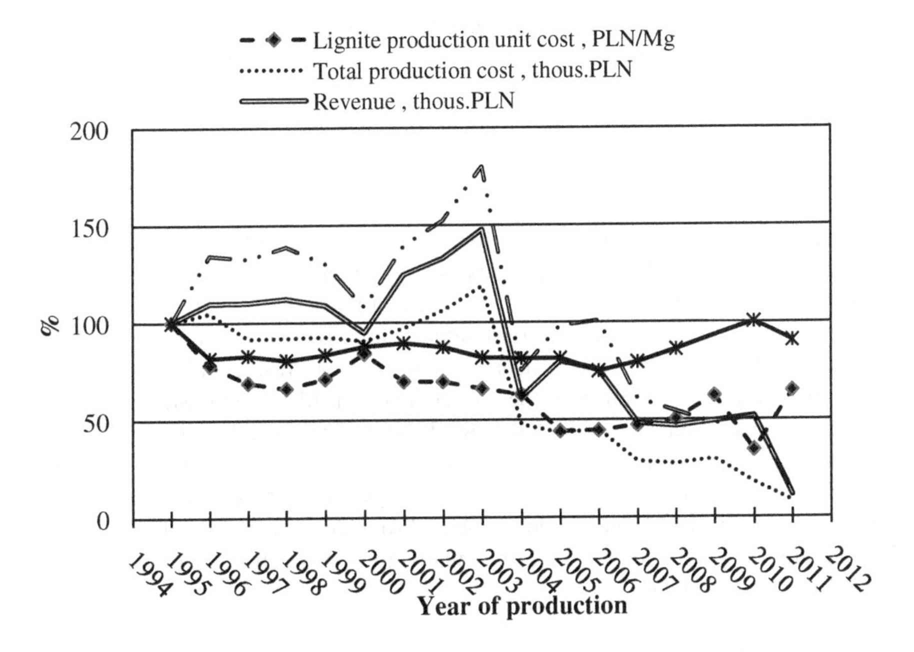

Fig. 1 Changes of: lignite production unit cost, total production cost, revenue, production rate, lignite unit price from last 17 years of operation – to closure, real values. Year 1995 = 100% (due to confidential data).

5 Block Model Development

While building block model David (1977) claims that the size of blocks should not be less than ¼ of average drilling interval. Currently lack of computing power is hardly an issue. One may consider blocks no smaller then the selected mining unit [4]. The smaller the blocks the error may increase [15].

Based on 2558 drillholes, deposit parameters were estimated with the use of ordinary kriging (*ISATIS* by Geovariances) and interpolated into the structural block model (*CAE Mining Studio*). Investigated lignite deposit has stripping ratio 6.9. Deposit quality parameters such as ash and sulphur content and calorific value were integrated into one Quality Index that differentiates lignite quality while the base price for the standard quality lignite is set to 57.65 PLN/Mg (later 80 PLN/Mg from which previous base price 57.65 PLN/Mg is equal approx. to 70% of new base price). The block model consists of overburden and lignite blocks with the prototype block size 50x50x5 m, whereas the deposit block size was 25x25x1 m. For further processing of the quality block model *NPV Scheduler* software was used. Mining cost 4 PLN/m^3, discount rate 8% (without inflation) and the general slope angle 15° as well as the surface costs map as real estate price layer were applied. Basing upon the created economic block model, four ultimate pits (phases) were generated with the use of Lerchs - Grossmann algorithm with 10% of lignite base price incremental change (70-100%).

6 Land Acquisition

Land acquisitions follow established Mine Company regulations. Buyout and land price negotiation usually starts 3 years before pit advancing. Actual land use regarding agricultural activities does not necessarily follow Land and Buildings Registration. To focus on surface cost impact, area of interest was divided into three main forms of land use nearby deposit area: residential, agricultural-residential and agricultural and then price map of Mine area was estimated. Estimation was conducted with Polish Real Estate Turnover In 2012, Statistical Information and Elaborations use. It is an objective statistical data based upon Notarial Acts and includes average measures for purchase/sale transactions of lands and buildings in Mine Company area (Province) selected for this research. This model is quite simple but misses many over estimations and subjectivity made by other researchers.

7 Land Buyout

For this research 23 cases of land buyout were investigated between years 1993 – 2004. Each negotiation was followed by competent person (appraiser) valuation report to estimate market value and then Mine Company made an offer to land owner.

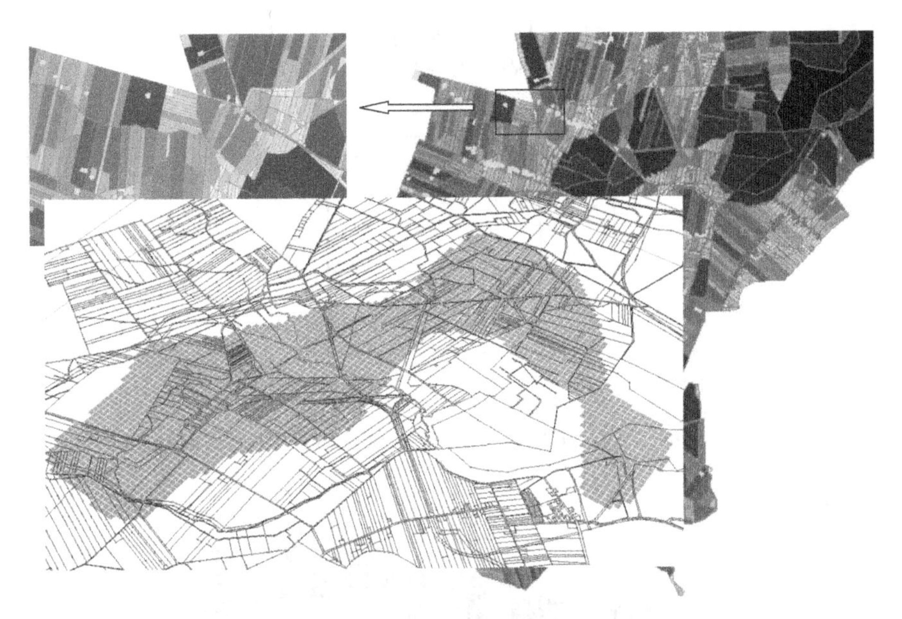

Fig. 2 Map of Mine area with estimated price of land use. Estimation based upon Real Estate Turnover 2012.

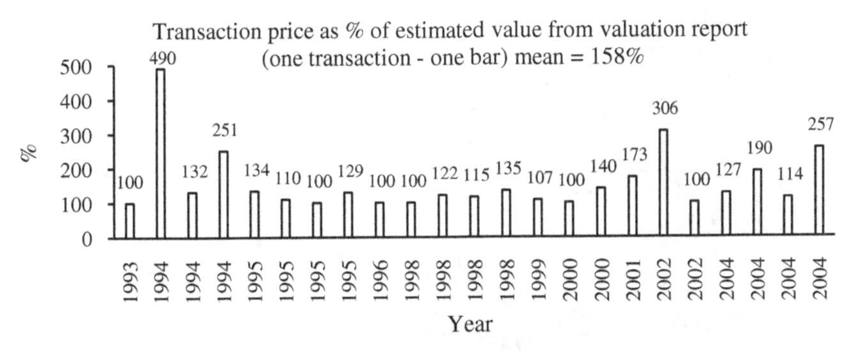

Fig. 3 Ratio (%) of transaction price to estimated value from valuation report of each property (one transaction - one bar) mean = 158%.

Sometimes owners agreed to close to market value price but in few cases there were court cases and counter valuation reports. For few transactions it took longer than 3 years to accomplish. As a rule that could be seen Mine Company and owners started to negotiate from different positions and tend to meet in the middle. In many cases nearly 150% of first valuation report contract was agreed (Fig. 3).

Lignite price formula used in estimations is calculated as multiplication of lignite base price and the Quality Index (QI). Values of the standard coal are set usually as average values for which the parameters of power plant burners have been adjusted [16]. Price formula used by Mining Company is expressed by following equation.

$$C = C_B \times \left[\frac{Q_R}{Q_B} - \frac{A_R - A_B}{200} - \frac{S_R - S_B}{10}\right] \tag{6}$$

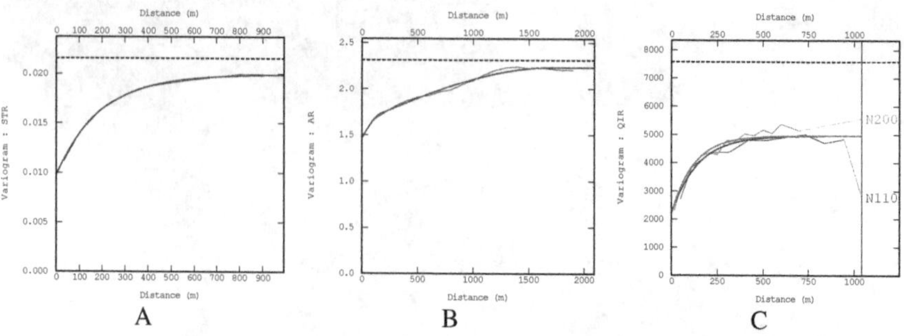

Fig. 4 Fitted variograms. (A) –STR sulphur, (B) – AR Ash, (C) – QIR lignite calorific value

	QIR, kJ/kg	STR, %	AR, %	QI, -
Max	10086	1.54	12.21	1.130
Min	7533	0.49	3.81	0.813
Mean	9339	1.07	5.68	1.039
Std. Dev.	324	0.12	1.30	0.045
Variance	104682	0.01	1.70	0.002

Fig. 5 Deposit basic parameters. From top: (QIR) calorific value, (STR) sulphur, (AR) ash. ('light – dark' scale)

Where:

C_B – is base lignite price, PLN,
Q_B , Q_R – is calorific value of a standard, actual lignite ($Q_B = 8850$), kJ/kg,
A_B , A_R – is ash content in standard, actual lignite ($A_B = 12$), %,
S_B , S_R – is sulphur content in standard, actual lignite ($S_B = 0.6$), %.

Two scenarios were investigated. First case - market value estimated of surface cost map and second case - surface cost map model multiplied by 2.5 as for extreme cases. In the second case pits were not generated due to high surface cost. In the first case four ultimate pits with 10% increment of the base lignite price were generated. To meet market value based surface cost, lignite base price was set to 80 PLN/Mg although ultimate pit 1 with base price of 70% represents original lignite base price (57.65 PLN/Mg).

8 Results

Table 6 Results

% of base price	Rock, Incremental, thous. Mg	Rock Cumulative thous. Mg	Lignite averaged quality, Incremental, thous. Mg	Lignite averaged quality, Cumulative thous. Mg	Revenue, Incremental, thous. PLN	Revenue Cumulative thous. PLN
Pit 1 70%	414 200	414 200	33 800	33 800	2 704 400	2 704 400
Pit 2 80%	37 000	451 200	2 600	36 400	211 100	2 915 400
Pit 3 90%	56 600	507 800	3 800	40 200	303 500	3 219 000
Pit 4 100%	30 000	537 800	1 800	42 000	141 600	3 360 600
Total	537 800		42 000		3 360 600	
% of base price	Mining Cost, Incremental, thous. PLN	Mining Cost Cumulative thous. PLN	Capital Costs, Incremental, thous. PLN	Capital Costs Cumulative thous. PLN	NPV, thous. Incremental, PLN	NPV Cumulative thous. PLN
Pit 1 70%	1 635 200	1 635 200	103 800	103 800	48 200	48 200
Pit 2 80%	145 900	1 781 100	8 700	112 500	13 900	62 000
Pit 3 90%	224 100	2 005 200	24 700	137 200	10 100	72 100
Pit 4 100%	118 700	2 123 900	15 400	152 600	1 100	73 200
Total	2 123 900		152 600		73 200	
% of base price	Pit Area Cumulative thous. m2	Pit area, Incremental, thous. m2	Land Tax, Incremental, thous. PLN	land area, acres	Total Clearing Cost Incremental, TCC, M PLN	Total Clearing Cost Cumulative, TCC, M MLN
Pit 1 70%	5 800	5 800	4 930	1433	3.41	3.41
Pit 2 80%	6 300	500	425	1557	0.38	3.68
Pit 3 90%	7 100	800	680	1754	0.57	4.10
Pit 4 100%	7 500	400	340	1853	0.31	4.30
Total		7 500	6 375			

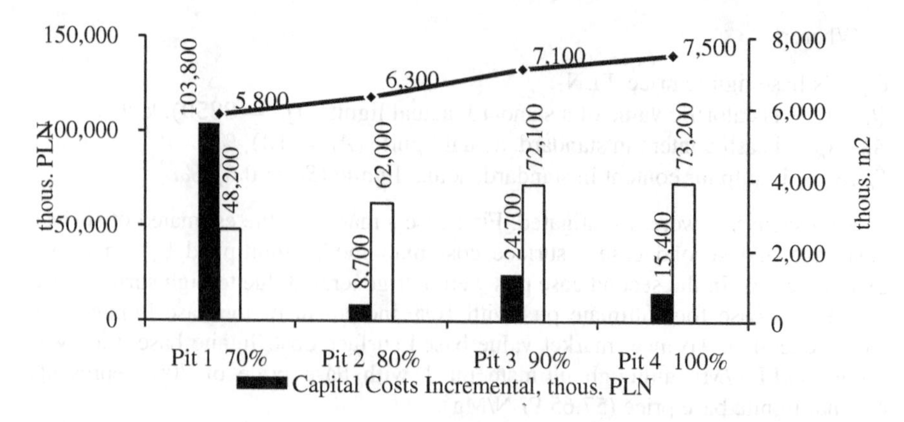

Fig. 6 Ultimate pits. Left axis: Capital costs (surface costs), NPV Cumulative. Right axis: Pit Area Cumulative.

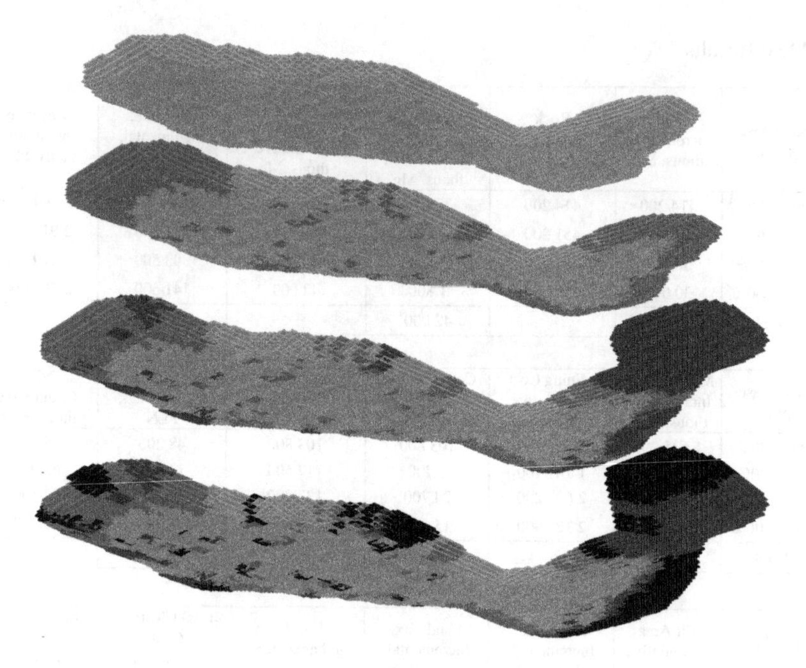

Fig. 7 Ultimate pits, 10% cumulative lignite base price change, bottom side view (from top: ultimate pit 1 4)

9 Conclusions

Assuming the 8% discount rate few conclusions can be made. Required pit area for the mining operation depending on ultimate pit case (1-4) is estimated to be approx. 5 800, 6 300, 7 100 and 7 500 thous. m^2.

One may consider if it is better to purchase all mining lands at once or with pit advancing. Without any special Tax consideration it is better to proceed with progressive land buyout. Two scenarios compared indicated that buying lands at the beginning of operation may lead to extra expenditures regarding Taxes. Ratio of surface cost paid at the beginning to progressive buyout is estimated to be approx. 1.57.

As expected, overestimation of surface costs (more than 2.5 times than estimated market value) did not generate ultimate pits. Surface cost may be decreased by granting shares of Mine Company to land owners in exchange for land/mineral rights (even if deposit below the surface is not a legal part of real estate owner but postponing may undermine mining project).

Transaction data compared to Mine Company negotiation reports showed the advantage that land owners have over the Mine Company. Without any clear buyout regulation such situation will not change. The average free market transaction price was nearly 150% of first valuation report. It is important to notice that only 23 transaction data from 1993-2004 were analyzed. It can be observed (fig. 3.) that distribution in time is quite even and may suggest that progressive buyout did not generate progressive land prices (though 150% is high ratio itself), however more research will be conducted to support or discard this statement. Granting lands to Mine Company (then government owned) at low price from Country Agricultural Office as not market transactions were not taken into account.

As shown in fig. 6. Achieving one million NPV increase in pit 4 compared to pit 3 will require investment of 15 million PLN in land and also 400 000 m^2 to proceed with operation. Such low NPV should raise the question whether the lignite project is economically viable, not to mention unexpected stops that generates additional costs. Despite of the economic issue this example shows that careful surface cost analysis regarding land may lead to social, environmental protection and sustainable land development consideration.

Implementation of the Lerch-Grossmann pit optimization algorithm for the surface mining projects allows to evaluate optional scenarios of the pit size with regard to both mining and surface costs.

Acknowledgment. This work has been financed from the subsidy B40112 for scientific research contributed to the development of young researchers and PhD students by Polish Ministry of Science and Higher Education executed at the Faculty of Mining, Geoengineering and Geology of the Wroclaw University of Technology.

References

[1] Ellis, T.R.: A Review of the Many Cost Approach Methods for Minerals Valuation, presented at SME/CMA/AIMA, Annual Meeting, Denver, CO, 28.02 – 02.03, SME Paper 151–156 (2011)

[2] Saługa, P.: Elastyczność decyzyjna w procesach wyceny projektów geologiczno-górniczych, IGSMiE PAN Kraków (2011)

[3] Ellis, T.R.: Sales Comparison Valuation of Development and Operating Stage Mineral Properties. Mining Engineering 63(4), 89–104 (2011)
[4] Hustrulid, W., Kuchta, M.: Open pit mine planning & design. Fundamentals, vol. 1 (2006)
[5] Jurdziak, L., Kawalec, W.: Sensitivity analysis of lignite ultimate pit size and its parameters on change of lignite base price. Prace Naukowe Instytutu Górnictwa Politechniki Wrocławskiej. Górnictwo i Geologia, vol. VII (2004) ISSN 0370-0798 (now Mining Science ISSN 2300-9586)
[6] Cartwright, M.: An Overview of Mineral Property Appraising, 1–16 (1994)
[7] Dudek, M.: Attempt to estimate the value of sand and gravel mineral rights using the transaction prices analysis of mineral properties in Wroclaw County. In: 10th Students' Science Conference, Wałbrzych, July 12-15, pp. 413–418. Oficyna Wydawnicza Politechniki Wrocławskiej (2012)
[8] Jasiński, J.: Specyfika i metodologia wyceny nieruchomości gruntowych położonych na złożach kopalin. Część II. Biuletyn Stowarzyszenia Rzeczoznawców Majątkowych Województwa Wielkopolskiego (2013) ISSN 1731-1829
[9] Polish Code for the Valuation of Mineral Assets (The POLVAL Code), Cracow (2008)
[10] Uberman, R.: Analysis and assessment of amendments of the provisions concerning mine closure and post-mining areas reclamation in the draft Geological and Mining Law, Prace Naukowe Instytutu Górnictwa Politechniki Wrocławskiej. Górnictwo i Geologia, vol. XV (2011) ISSN 0370-0798 (now Mining Science ISSN 2300-9586)
[11] Polish ACT of 9 June 2011, Geological and Mining Law
[12] Real Estate Turnover In 2005, Statistical Information and Elaborations, Warsaw (2006)
[13] Real Estate Turnover In 2012, Statistical Information and Elaborations, Warsaw (2013)
[14] CostMine, Coal cost guide: A subscription cost data service. InfoMine USA Inc., Spokane Valley (2009)
[15] Rudenno, V.: The Mining Valuation Handbook: Mining and Energy Valuation for Investors and Management, 3rd edn. (2009) ISBN: 9780731409839
[16] Jurdziak, L., Kawalec, W.: Influence of power station efficiency and carbon costs on lignite resources and energy consumption in Poland. In: Proceedings of the 11th International Symposium of Continuous Surface Mining, June 25-27, pp. 125–135 (2012)

Project Management Model for Opening of the Opencast Mine Radljevo in the Kolubara Coal Basin

Vladimir Ivos[1], Slobodan Mitrović[1], and Aleksandar Vučetić[2]

[1] Electric Power Industry of Serbia, Belgrade
[2] Electric Power Industry of Serbia, PD Kolubara, Lazarevac

Abstract. Selection of the optimal variant for opening and development of the opencast mine is based on the organized, systematized and fundamental research of different natural, technical and economic mining project determinants. This is necessary but not sufficient condition for successful implementation of the mining projects. Mining capital projects, such as are the opening of new opencast mines, are very complex due to the influence of a number of internal and external technical, technological, economic and natural factors and the limitations resulting from the natural deposit conditions as well as from the socio - economic environment. Environment of the project for the opening of the opencast coal mine Radljevo is very complex and involves a series of deterministic and stochastic processes. In order to manage efficiently and effectively the project, it is necessary to synthesize a large number of skills, tools and techniques in the project management development plan, which is ensured by a good project management, achieving a better project results.

Keywords: project management, mining project, optimal variant, environment of the project.

1 Introduction

As a rule, mining companies by investment in capital mining projects implements previously defined growth targets, development policies, and business strategies. Investment projects in mining are capital since it spend large financial resources, significant resources and are long lasting. Investing in the projects the company undertakes to affirm the new business environment, adapt to anticipated changes, overcome handicaps of current position, or adjust its production capabilities with the opportunities offered by the market. All this shows that the investment in capital projects is condition for existence, sustainable growth and the acquisition of mining companies competitive advantages.

C. Niemann-Delius. (ed.), *Proc. of the 12th Int. Symp Continuous Surface Mining - Aachen 2014*,
Lecture Notes in Production Engineering, DOI: 10.1007/978-3-319-12301-1_37

However, investment in capital mining projects is a very complex process consisting of multivariate investigation activities of all relevant determinants for future conditions and changes that the project entails. Bearing in mind that projects for opening of new coalmines, operating within the Electric Power Industry of Serbia, are connected with its strategic development goals, in the implementation of such projects it has to be taken into account all internal and external factors of development and a number of constraints arising from the socio-economic environment.

Selection of the optimal variants for the opening and development of the opencast mine is based on an organized, systematized and fundamental studies of different natural, technical and economic determinants of the mining project. This is a necessary but not sufficient condition for successful implementation of the mining project. High quality investment projects can be successfully implemented only if the project implementation process is managed and the process is by design methodology organized. Using the methodology for project management is a business strategy that allows maximizing the value of the project by the organization [1].

A good methodology contains all the important processes of project management, and subject involved in the process continue to spread further [3]. Characteristics of a good methodology are necessary level of details, the use of quality basic data and standardized techniques for planning, time determination, cost control and reporting forms. In addition, one should be flexible and reasonable in the application and based on standardized phases of the project life cycle.

In contrast to the accepted methodology that represent the best practice and is basically a framework for the model implementation building management of the mining project, the model itself should be developed specifically for a particular project. This is very important considering the specifics that these projects have relating to the location predisposition, the impact on the population and the environment, a long time frame and large investments.

2 Analysis of the Opencast Mine Radljevo Mining Project in the Function of the Proces Model Developmen for the Project Management

The strategic and development plans of the Electric Power Industry of Serbia envisage the construction of new thermal capacities, which requires the provision of a new quantity of coal. The Pre-Feasibility Study: Study on the Selection of Limitation and Opening of Opencast Mines South Field and Radljevo with Comparative Overview of Technical and Economic Aspects of Coal Mining for the Selection of Priority Coal Supplier of CHP Kolubara B [9], it was concluded that the opencast Radljevo with production of 13 million tons per year is a priority for the further development of new capacities in coal mining in Kolubara coal basin.

Considering the findings of preliminary feasibility study for the opencast mine Radljevo it was developed a Feasibility Study with Preliminary Coal Mining Design at opencast mine Radljevo [9], where are determined geological, hydrogeological, technological, environmental and economic parameters of coal mining.

This document stipulates that the coal mining and overburden removal is to be performed by continuous systems with large bucket wheel excavators. The total weights to be mined from the opencast mine Radljevo amounts to 1.7688 bcm, from which 1.385 m is solid cubic meters of waste consisting of overburden, interburden, interlayers, and 449 m tons of regular and low-quality coal. Figure 1 shows a general overview map of the Kolubara coal basin and the location of Radljevo deposit.

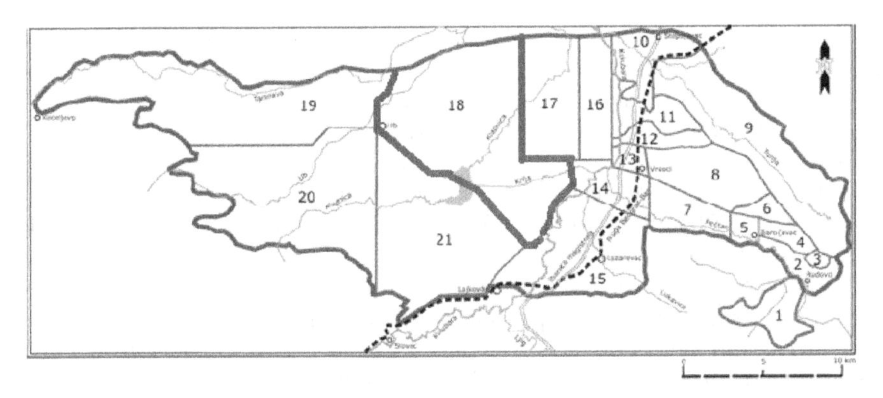

Fig. 1 Overview map of the Kolubara coal basin with the location of Radljevo deposit (18)

Annual production capacity of the opencast mine Radljevo is designed at 13 m tons of coal. According to the ToR annual capacity of the mine in the first phase will be limited to 7 million tons. In relation to the Feasibility Study, opencast mine opening is dynamically moved for two years, so that instead of in the year 2014 the opening starts in the year 2016, and coal production in the second year of operation (in the 2017) should be 3 million tons. The production capacity of 7 million is to be made in the year 2019. The second stage of the designed annual coal production with capacity of 13 million tons will be achieved in the year 2022. Opencast mine opening begins along the western border of the Tamnava West Field. In the first phase of mining the opencast mine is developed in parallel to the west, and then in the radial direction. Such opencast mine development provides long-term stable conditions of the system shifting devices, and initial parallel opencast mine opening provides gradual relocation of the village Radljevo parts.

In the initial period overburden and interlayers material will be dumped on the inside dumping sites of the opencast mines Tamnava-East and Tamnava West Field. Since the year 2022/23 onwards, all the material will be dumped on the outside dumping site of the opencast mine Radljevo.

For overburden removal will be used two bucket wheel excavators (type 6800 lcm/h) with a belt bench conveyor system and belt width of 2 m. Selective mining of coal and interburden is planned to be carried out with up to five bucket wheel excavators (type 4500 lcm/h) with a separate belt conveyors of 1.6 m width. By the year 2022 in operation are to be used two bucket wheel excavators 4500 on the first and second coal bench, from the year 2022 onwards three units, from the year 2029 four, and from the year 2031 five excavator units.

Requirements for selective mining have been analyzed, too. Calculation of bucket wheel excavators' capacity on coal is carried out for different blocks and conditions of the site face during selective mining, which resulted when sizing equipment with reduced equipment utilization coefficient and increased capacity reserve.

For the hydro geological conditions are carried out detailed explorations and designed dewatering system comprising a drainage wells with submersible pumps.

It has been identified and preliminarily designed required infrastructure facilities in the opencast mine.

The impact of the planned mining activities on the environment has been analysed and planned measures to reduce negative impacts.

Special attention is given to the issue of population and public infrastructure d resettlement. Total investments for the resettlement of people and public infrastructure are €170 million over the life of the project, of which €85 million are till the year 2020.

On the opencast mine Radljevo is to be directly engaged up to 1,800 employees, which has a positive impact on the employment situation in the region.

Based on the opencast mine preliminary design it have been identified and listed required investments per years. Total investment cost for the main and auxiliary equipment to be procured until the year 2022 and reaching its full production capacity are approximately €360 million. The investment costs for infrastructure and facilities in the opencast mine, for the same period are estimated at around €127 million. Total investment cost until the year 2022 was estimated at €487 million.

Regarding the implementation of this project, it is indisputably a major challenge in terms of meeting short-term plan by the beginning of mining operations.

By this analysis it is evident all the complexity and multidisciplinary nature of the Radljevo project, ranging from continued intensive geological, hydrogeological, geotechnical, mining, technological, environmental and other researches, elaboration of studies and technical documentation in the fields of geology, mining, mechanical engineering, electrical engineering, civil engineering, IT technology, resettlement, procurement of different types of services and main and auxiliary equipment for the mine up to the performance of works at the opening and development of the opencast mine. Despite the delay in relation to the timeline of the project implementation, the fact is that according to the time schedule provided by the Feasibility Study, time for opening and opencast mine development up to achieving the full capacity is limited to seven years, and that foreseen budget for the project in this period is €487 million.

3 Project Management Model for Opening and Development of the Opencast Mine RADLJEVO

Project management for opening and development of the opencast mine Radljevo is characterized by a strategic approach, particularly with regard to the project director who assumes a key role as a system integrator, integrating all the required processes and knowledge to successful project implementation. In addition, an important role is played by the integration of communication and IT services as the support of the management process. Thanks to these technologies, project management obtains distribution dimension, better coordination and communication between project teams, remote locations, interested parties and all relevant factors within the project.

Based on the complexity of the project management definition, as well as by its very nature, project management methodology involves division to the process groups, which are easier to be managed. Process groups of project management represents essential, basic processes and can be practically identified with the project phases being in reciprocal interaction.

On the analysis of the project opening and the open pit mine Radljevo development up to achieving designed annual capacity of 13 million tons of coal are defined process groups (phases) for this project process model management determining:

- Process group for project initiating;
- Process groups for project planning;
- Process groups for project implementation;
- Process group for project control;
- Process group for project closure.

Each of the above-mentioned project phases is unique and the iterative cycle during which is undertaken integration in order to set each project process in an appropriate level and connected with other process to establish coordination.

Figure 2 shows a model for the project management on the context-level used further for development of the process model for the opencast mine Radljevo project management.

Starting phase and corresponding group of basic processes is the identification of the need for the implementation of a business idea. Clear and unambiguous guidance on the project at this early stage of the project, are formally discuss by the feasibility and viability studies, which provide a picture of whether the project is feasible and whether it should proceed with its execution. Speaking about the project of the opencast mine Radljevo, activities of this phase is virtually complete, and the output from this phase is a verified Feasibility Study with the Conceptual Design of coal mining in the opencast mine Radljevo [9].

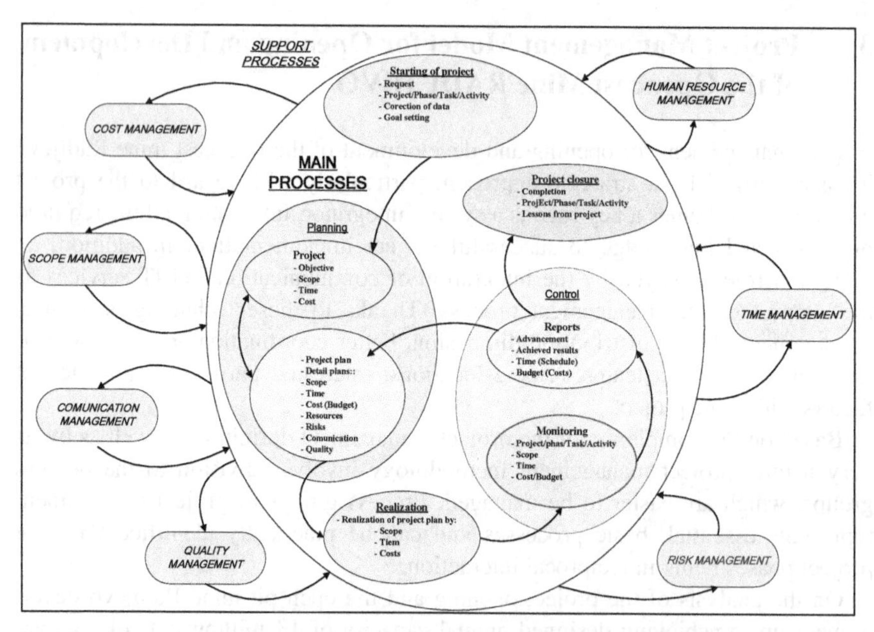

Fig. 2 Project management model for the opencast mine Radljevo on the context-level

From the perspective of the defined model, the opencast mine Radljevo project is in the phase of planning, where over the process groups for project planning is defined the project scope, are identified all activities relevant to the project implementation and for which is carried out detailed planning in terms of the order execution, budget projections, and time required for implementation. In addition in this phase is to be done allocation of resources, both human as well as material for the execution of each planned project work. During this phase are discussed potential risks that may jeopardize the execution of the project, and even lead to total failure of the project as well as preventive measures in a function of the risk occurrence, then plan communications plan and project quality plan.

When in question is the status of the project in this stage are define the scope, duration and budget of the project. Regarding time, the project began in the second half of the year 2014, and ends in the year 2023. As from the diagram in Figure 3 can be seen, it has been defined the key items on the project as the starting of the opencast mine opening in the year 2016, then reaching the full capacity of 13 million tons of coal in the year 2022, and the project end in the year 2023. In the last year of the project generally it should be held administrative activities related to the closure of the project. Regarding the scope of the project, it has been defined key activities related to the preparation of urbanism planning documents, resettlement, additional geological, hydro-geological and geotechnical exploration works, the preparation of the main mining project, procurement of basic and auxiliary equipment, mining operations of the opencast mine opening up to achieving the full capacity. The project budget is practically defined by the Feasibility Study with the Conceptual design for coal mining in the opencast mine Radljévo.

Figure 3 provides the extent of the activities of process groups during the project duration. Considering that the extent of the project implementation budget is in direct function of the volume of activity in time, the illustration in Figure 3 can be regarded as a presentation review of the budget scope per process groups during the project duration.

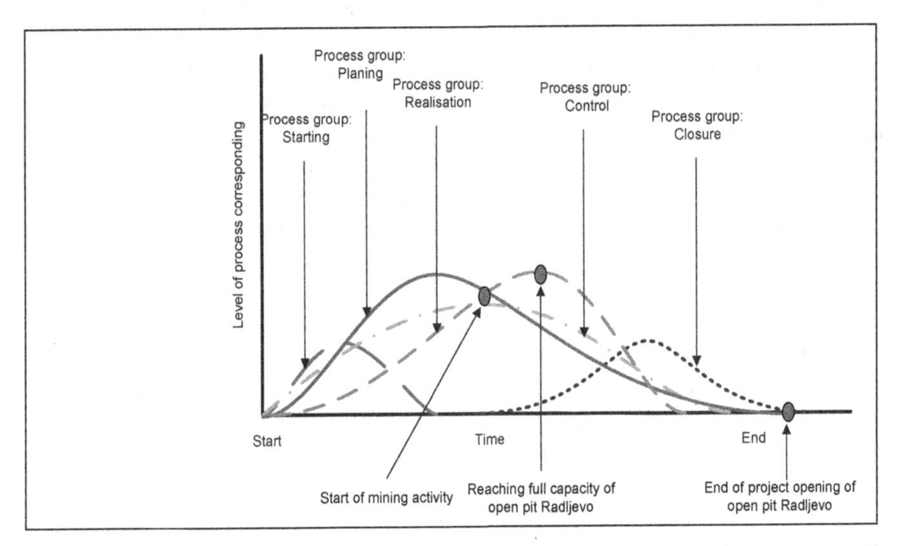

Fig. 3 The scope of activities and the extent of budget implementation per process groups in time and the key points of the opening project for the opencast mine

During the third phase, the implementation phase, over the corresponding process group are implemented the planned activities from the previous phase, with regular monitoring and control of performance. Implementation phase requires maximum focus and commitment of the project team in the implementation of all planned activities from the previous phase, in order to achieve the desired target - the successful completion of the project.

In the control phase through a defined process groups, is carried out continuous monitoring of the planned work execution within the project. During the monitoring is performed monitoring progress, comparing with the time schedule, planned material resources, and are perceived the possible deviations from the plan and accordingly corrective actions are taken.

The final phase or the conclusion phase and the associated process group aim the formal statement that a result of the project is in line with expectations, in terms of quantity and quality, due to which the implementation of the project started. Formally, the project at this stage completes its life cycle and it changes to the project results exploitation, which is actually a post-project cycle. As a work on the project is formally ended, after accepting the results of the project there is no more required for the project team.

Strategic, or integrative, project management is made out of the process aimed to identify and define of all the processes and activities required to be integrated within the project management, as well as their mutual correlation and impact on the project itself to be managed (Figure 2). This implies the definition and integration of process groups of essential, basic processes and auxiliary processes, as well, in terms of the project management process model. In this sense, are defined out from the described basic the auxiliary project management processes for the opening of the opencast mine Radljevo: human resource management, cost management, quality management, risk management, scope management, and time management (Figure 2).

Time management is aimed to define the time frame of the project, as well as the duration of all sub-processes, tasks and works that make the project itself.

Management by project scope integrates all required activities and tasks needed to be made in order to achieve comprehensive implementation of the project as successful. Scope of work planning, defining the mutual structure on works, verification of scope and scope volume are integral processes of the scope management.

Quality management is a series of processes that ensure that the project will result in the required goals both by volume and by the quality. Some of these processes are quality planning, providing quality assurance and quality control achievement by individual processes, tasks and the project as a whole.

Human resource management aims to organize and manage the project team as a whole. Integral parts are planning of human resources, the creation of a project team and management by the project team during project activities. The success of the project is, among other things, based on the ability of project managers to explain to the project team their importance for the achieving the project results.

Cost management synthesize processes aimed the implementation of the project within the foreseen projected budget. The processes in this group are: cost estimates, the types of costs determination and cost control during the implementation of project tasks.

Communications management integrates the processes aimed to generate, select and distribute information and data about the project. These processes are communications planning, distribution of information and use of information.

Risk Management includes the processes aimed to identify, systematize and response to risks that may arise in the project and the project environment and adversely affect to the project. Quantitative and qualitative risk analysis, response planning to risks, risk consequences, monitoring and control are of some of the processes in this group.

Procurement management - logistics encompasses all the processes of procurement of goods and services required for the project. Procurement planning, tenders, selection of suppliers, contracting and immediate implementation of the procurement are processes of this group.

For the successful implementation of the above mentioned project management process model for the opening and development of the opencast mine Radljevo, it is required by the project manager and project team to possesses a number of specialized knowledge. In addition to specialized knowledge in the field of mining and geology, this project requires specific skills, such as:

- Technical skills that are specialized for particular industries such as civil engineering, electrical engineering, mechanical engineering, economics, ecology, etc.
- Knowledge of operational processes management, such as logistics, commercial jobs, etc.
- Knowledge related to state regulations, such as laws, regulations, customs regulations, standards, etc.

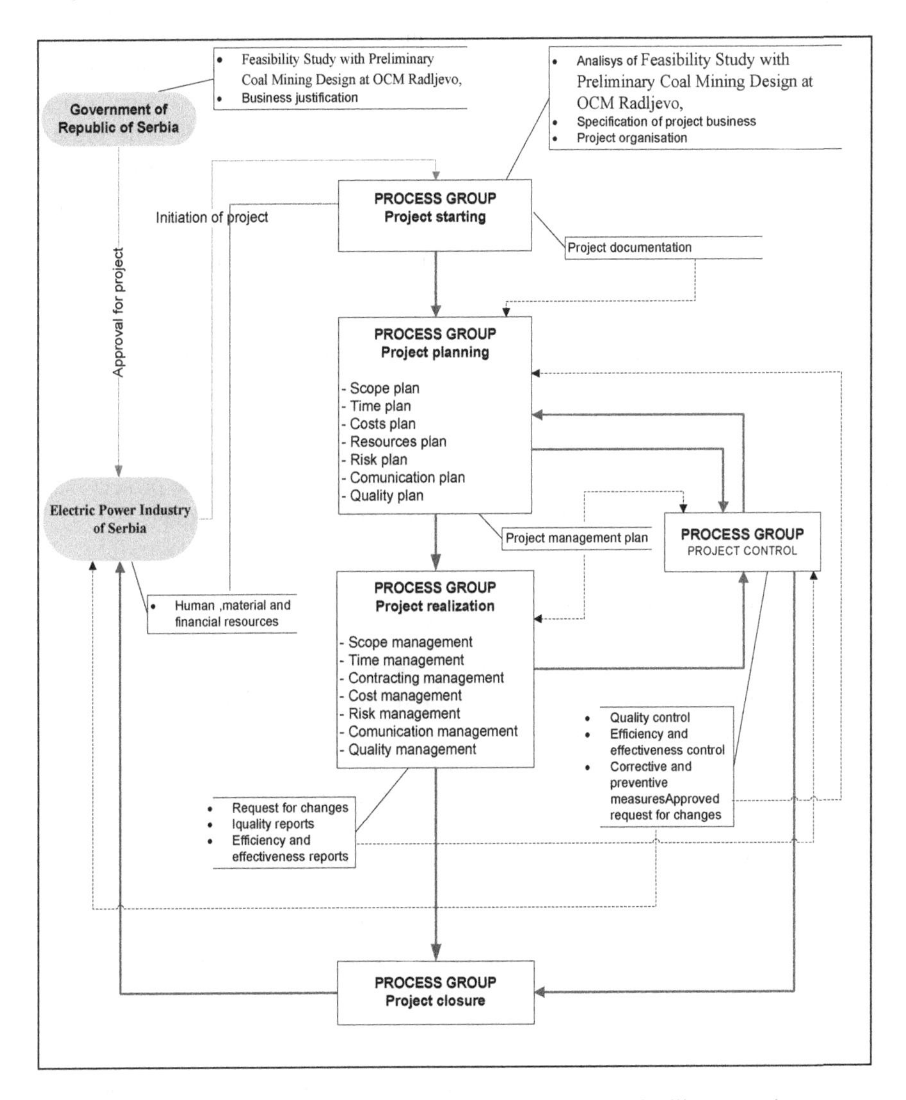

Fig. 4 Project Management Process Model for the opencast mine Radljevo opening

As each mining project is implemented in almost a specific and unique environment for the success of the project is of vital importance the knowledge of the environment. Project Manager with the project team during the planning and implementation of the project must necessarily take into account all the specifics of the environment, to gather the necessary information in order to integrate the positive aspects of the project environment, and minimize or eliminate the impact of the negative aspects of the environment. A very important aspect of the project environment is a cultural and social environment. As the opencast mine Radljevo project involves direct interaction with the environment, then it is really important how this project affect the environment and vice versa, how the environment affects the project. Aspects of the environment such as economic, environmental, demographic, cultural and ethical may have significant implications at the project.

Based on the preceding analysis is determined the process project management model for the opening of the opencast mine Radljevo (Figure 4). The present model, in terms of this paper, is sufficient to fully examine the multidisciplinary nature and complexity of this mining project implementation. This model as per scope and structure is on the context-level and allows defining for each process group a complete process model with defined process activities. An additional benefit of the developed model is that on its base can be defined procedures for the management of individual processes and activities of the project. Integrated models and procedures make a system for project management, which to the significant extent should to increase the reliability of implementation for the each of the planned activities.

Considering that the project is in the early planning stage, when only are defined scope, time and budget of the project, presented model for the project management process in Figure 4 is to be the basis for defining of other plans.

4 Final Considerations

Opening and development project for the opencast mine Radljevo is multidisciplinary and highly complex project that requires the implementation of complex research as well as development of study and technical documentation in the field of spatial planning, resettlement, environment, geology, mining engineering, mechanical engineering, electrical engineering, civil engineering, IT technologies, then the procurement of various types of services as well as the main and auxiliary equipment at the mine and in the end the performance of works at the opening and development of the opencast mine in order to achieve full capacity. In addition, when in question is this project it should not be ignored the value of investments till the year 2022 amounting to around half a billion of Euros and the environment in which the project will be implemented.

Such complex procedure can be successfully implemented only by applying the best practices in the project management. The described project management process model for opening and development of the opencast mine Radljevo in order to achieve design capacity is based on the latest knowledge in the field of project management and is consistent with the best practices in the field of mining projects management. Process model is developed in the in the early planning stages of this project and it is required but is not sufficient condition for the successful implementation of this project and it is the basis for the development of the opening and development of the opencast mine Radljevo project management. The developed system to manage the project should include a complete process model on the required level of details for the each process of developed procedures according to the model. Considering that investment in capital mining projects is a very complex process composed of multidimensional activities of investigation all relevant determinants of future conditions and changes that the project carries this project management model with process procedures are the basis for the development of project management systems, and makes a reliable basis for the efficient and effective opening of the opencast mine Radljevo project implementation.

References

[1] Charvat, J.: Project Management Methodologies - Selecting, Implementing, and Supporting Methodologies and Processes for Projects. John Wiley & Sons Inc. (2003)
[2] Harvard Business School: Project management manual (1997)
[3] Kerzner, H.: Project Management: A Systems Approach to Planning, 8th edn. John Wiley & Sons, Inc., Hoboken (2003)
[4] Kerzner, H.: Strategic Planning for Project Management using a Project Management Maturity. John Wiley & Sons, Inc. (2001)
[5] Maylor, H.: Project management, 3rd edn. Prentice Hall (2003)
[6] Guide, P.A.: to the Project Management Body of Knowledge (PMBOK Guide), 4th edn. Project management Institute (2010)
[7] Smith, N.J.: Engineering Project Management. Blackwell Publishing (2009)
[8] Zambruski, M.: A Standard for enterprise project management. CRC Press (2009)
[9] Feasibility Study with Preliminary Coal Mining Design at Opencast Mine Radljevo, Vattenfall Europe Mining AG and University in Belgrade, Mining and Geology Faculty (2010)

Problems and Prospects of Cyclic-and-Continuous Technology in Development of Large Ore-and Coalfields

Yuri Agafonov, Valeri Suprun, Denis Pastikhin, and Sergei Radchenko

Moscow State Mining University, Russia

An effective way to reduce unit costs of stripping work is to improve the transport scheme and a pit overburden removal method. This is especially important for motor transport. Transport expenses dominate in the cost structure of overburden removal production processes and can reach up to 60% of the total costs.

Resolution of this problem by increasing carrying capacity of dump trucks is problematic due to the following reasons. Cost of cargo transportation as a result of increase in dump trucks capacity becomes lower. However, intense reduction takes place for trucks capacity from 40 to 110-130 t (area A, see Fig. 1).

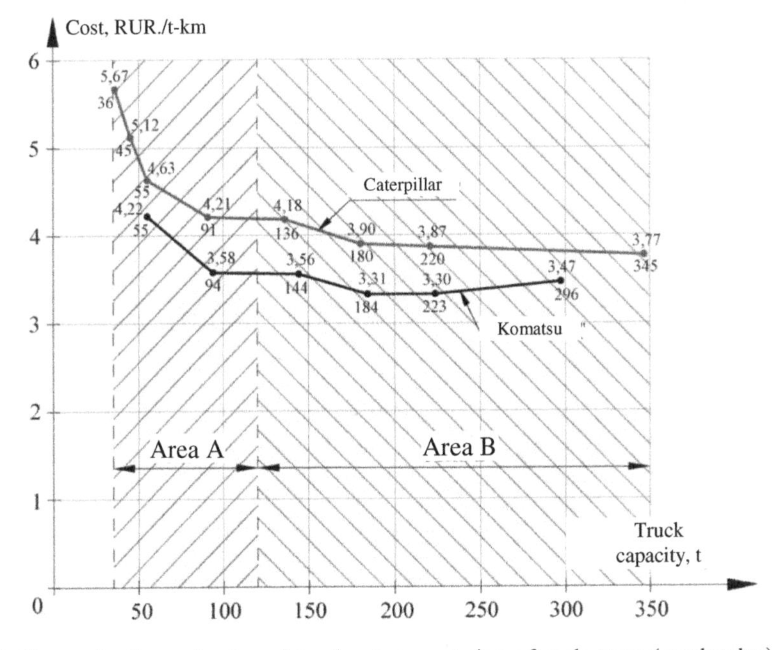

Fig. 1 Change in the cost price of ton-km transportation of rock mass (overburden) by dump trucks with different capacity with account of operating costs and owning costs (prepared on request by the «Caterpillar» and «Komatsu» companies)

C. Niemann-Delius. (ed.), *Proc. of the 12th Int. Symp Continuous Surface Mining - Aachen 2014*, Lecture Notes in Production Engineering, DOI: 10.1007/978-3-319-12301-1_38

Within a 120-350 t carrying capacity range, rates of reduction in ton-km cost sharply decrease (area B, see Fig. 1).

Conveyer and railway (mains) mode of transport do not react to increase in the length of haul so dramatically as the motor transport does. That is why in order to reduce costs, associated with transportation, for the purposes of development of mines with dipping and steep mineral deposits it is advisable to switch over to a combination of a few modes of transport.

For development of modern quarries, up to three types of transport may be used having different cost price for ton-km transportation of rock mass (overburden).

The biggest prospects for the development of large quarries lie in cyclical–and-continuous technology (CCT), where motor transport is used as a gathering transport, while conveyer is used as the main transport, carrying out the main work for lifting of crushed material to the surface. Use of the CCT has the biggest efficiency for transportation of ore. This is explained by the fact that in the total ore dressing complex, ore reduction is the initial stage of ore treatment, preceding subsequent enrichment and conversion.

As of today, CCT complexes for transportation of overburden rocks are used to a lesser extent, which is explained by additional costs for crushing of starting material for its subsequent transportation by conveyor transport.

Despite the fact that capital expenditure for organization of CCT method is higher, overall costs are 25-35% lower. The reason for this is that operating costs in CCT schemes are twice as small because of reduced need in trucks, drivers and technical staff (Fig. 2) [1].

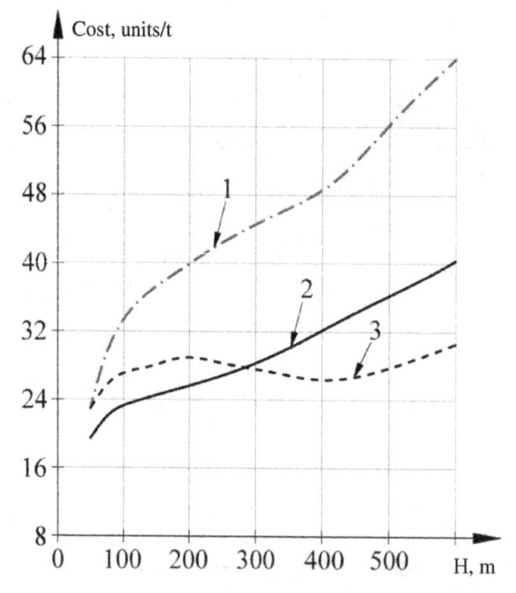

Fig. 2 Dependence of economic indicators of the transport from the opencast depth (according to M.V.Vasilyev). 1 – autotruck transport; 2 – railway transport; 3 – autotruck - conveyer transport; C – cost price of ton of material, transported from opencast by various modes of transport; H – depth of opencast

In ore-crushing traffic flows, transfer points (crushing stations) of CCT complexes are equipped by primary gyratory cone crushers and Blake sledgers.

At the first phases of development of CCT complexes (1980-1995), the main drawback of crushing stations was that they were "strictly fixed" transfer points at stationary edges of open-pit mines. Their transfer (as a result of deepening of mining operations) represented a complex engineering and technical problem and was economically a very costly process.

Given an average rate of progress in depth of 10-20 m/year, opencast depth increased rapidly, which due to "strictly fixed" points of crushing stations did not enable to effectively stabilize the volume of open-cut transport operations for a longer period. An undoubted progress in the development of CCT complexes became a creation of semi-permanent crushing and transfer units, equipped with single-roll, two-roll, screw and hybrid crushers.

Leaders in creation of screw and toothed-roll, roll and hybrid crushers are «MMD», «ThyssenKrupp Fördertechnik», «Joy Global», «Sandvik», «FAM» and «FLSmidth». A prototype model of two-roll crusher was manufactured by «Tenova TAKRAF».

Semi-permanent crushing stations, equipped with roll and screw crushers are used currently at many quarries of the world.

Use of single-roll, two-roll and screw (roll-sizer) crushers in CCT complexes is a progressive technical solution, ensuring compactness of crushing unit with lower energy and metal consumption.

Screw two-roll crushers process material without use of prescreening (Fig. 3).

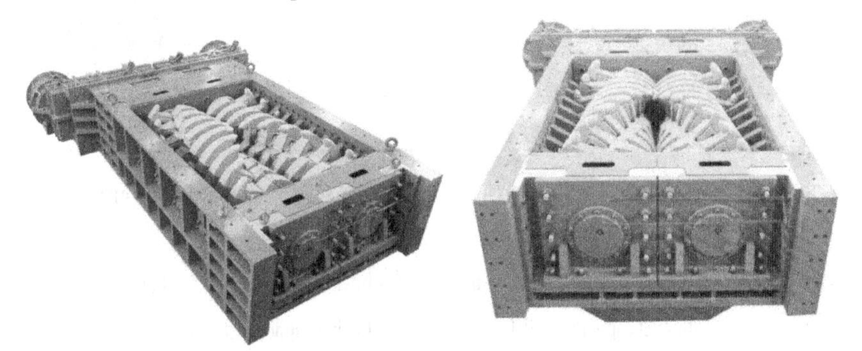

Fig. 3 Screw and toothed-roll crusher, manufactured by «MMD»

Configuration of teeth is such that it enables small pieces of rock to spill without additional recrushing (overgrinding). Crushing principle is based on enhancement of structural defects in material by the crusher's teeth through tensile stresses thereby allowing to control the size of the output piece.

«Joy Global» manufactures BF-38 and BF-43 single-roll crushers, successfully used for primary crushing of rocks.

For fitting out of semi-permanent crushing units, «Sandvik» is using CR810 series hybrid crushers. Hybrid crushers combine working principles of traditional crushers – sizers (classifiers) and two-roll crushers, which made them suitable for primary crushing.

Improvement of excavator-dump truck complexes moves towards increase in the capacity of excavator buckets and trucks capacity. Excavators with bucket capacity of 50-54 m^3 are capable in a stable mode to load up to 2-2,5 mln. m^3 and with buckets of 28-32 m^3 – up to 1-1,5 mln. m^3 of rocks per month respectively.

Use of powerful excavator-truck complexes reduces concentration of mining equipment in the open-pit mine and increases the intensity of development of area of mining operations. Their use further results in reduction of costs on excavation and transfer of overburden rocks. Experience of operation of PC-4000, PC-5500, EX-5500, and P&H-2800 machines with bucket capacity of 24-32 m^3 show that unit costs of excavation of 1 m^3 of overburden are reduced by 20-25% compared with performance indicators of lower class mining power shovels (Fig.4).

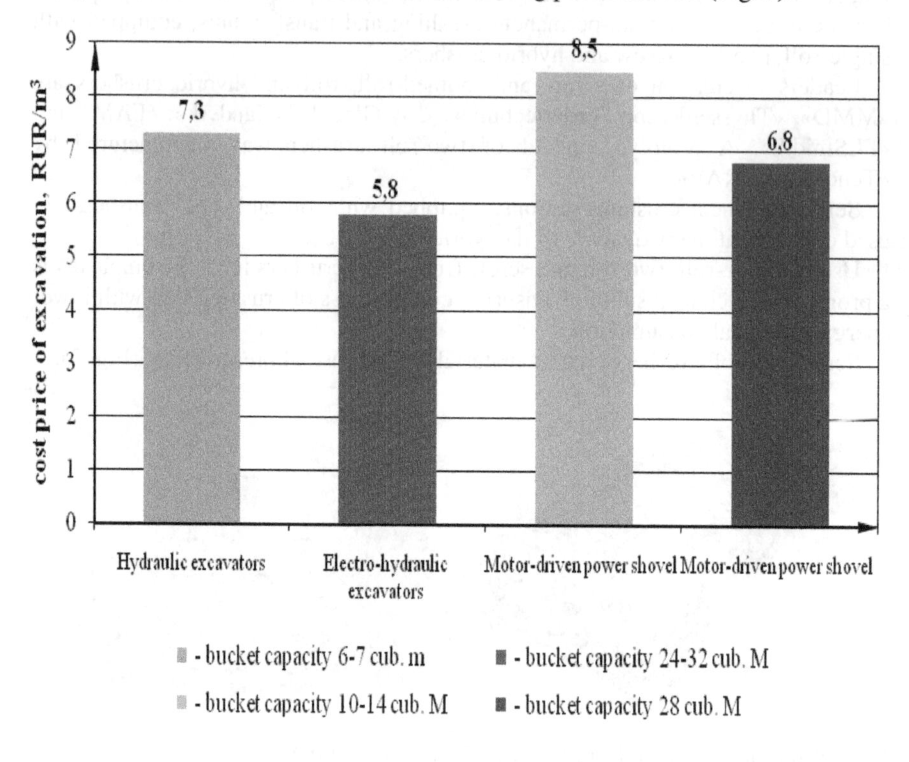

Fig. 4 Production costs of excavation of 1 m^3 of overburden by different types excavators

A factor, constraining a switch over to mining power shovels with bucket capacity of 54-70 m^3, operating as part of excavator-dump truck complexes, is the motor transport.

The main problems of working with 360-450 t load class trucks are complicacy of design of motor tyres, difficulty in transportation of large parts by railroad and need to increase width of auto roads. The latter circumstance leads to an increase in the extent of overburden and mining capital works when constructing open-pit motor ways in nonworking and working flanks of an opencast.

That is why «ThyssenKrupp Fџrdertechnik», «Tenova TAKRAF», «Joy Global», PJSC «Novokramatorsk machine-building plant», «Metso» and «Kleemann» have developed a concept of cyclical–and-continuous technology (CCT), using fully mobile crushing stations. Innovation lies in the possibility of crushing stations to move during excavation, which enables to fully eliminate previously necessary transportation by dump trucks.

Use of continuously operating machinery compared with cyclically operating dump trucks gives an economic effect not only as regards increase in output and decrease in transport costs, but also as regards saving of energy costs and protection of the environment.

CCT complexes, using fully mobile crushing stations, include a power shovel, a mobile crushing station, an overloader, and a conveyer system [2].

However, CCT complexes, using fully mobile crushing stations with big capacity are still largely experimental equipment.

The most promising for use in removing overburden at large coal and ore mines in the near 5-7 years CCT complexes, built on the basis of use of semi-mobile crushing stations and gathering motor transport.

A key risk in selecting these CCT complexes is a type of crusher, used in the crushing station. Faultless in this respect are jaw and cone crushers, however these types of crushers have several-fold higher cost of crushing process compared with screw-toothed and toothed roll crushers (table. 1).

Table 1 Estimated values of cost of crushing by crushers of various types (designs)

№ i/i	Quantity and parameters of crushers within a crushing station	Cost of crushing by crushers of various designs (types)
1.	Three crushers C-200 (ЩДС 2000Ч1500) by «Метсо». The total annual capacity of 3-jaw crushers – 5,2 mln. m^3 for solid block (7,3 mln. m^3 in loose condition)	5,24 RUR/t
2.	Two gyratory crushers Superior MK-II. The total annual capacity – 11,5 mln. m^3 for solid block (16,3 mln. m^3 in loose condition)	3,34 RUR/t
3.	One screw crusher MMD 1500. Annual capacity for solid block - 12,8 mln. m^3 (in loose material - 18,1 mln. m^3)	1,7 RUR/t
4.	One two-roll crusher by «ThyssenKrupp Fџrdertechnik». Annual capacity for solid block - 10,4 mln. m^3 (in loose material - 14,6 mln. m^3)	1,5 RUR/t

An efficient way controlling performance of a crusher is decrease in the size of starting material coming for crushing. This condition can be realized through efficient management of preparation of rocks, delivered to crushing stations of CCT complexes.

Gradation of material entering the CCT complex feed for crushing, is determined, on the one hand, by the geological structure of the rock mass (in the first place by fracture pattern and mechanical properties), on the other hand, by the way the rock is prepared for excavation.

With standard methods of blasting preparation of sedimentary rocks (sandstones, claystones, siltstones, limestones) of coal deposits and of overburden rocks of many ore deposits (of different geological origin), up to 70% of material (in shotpile) has a fractional composition of 0 - 400 mm. Rocks with this grain composition are potentially suitable for transportation by conveyer transport.

By improving blasting preparation of original rock, one can achieve a fraction with size of 0-400 mm (in shotpile) to the extent of 80-85% (Fig. 5). This circumstance at the turn of 1980-1990 became the reason for creation of experimental and pilot commercial crushing stations of CCN complexes based on the primary screening of the original rock [3].

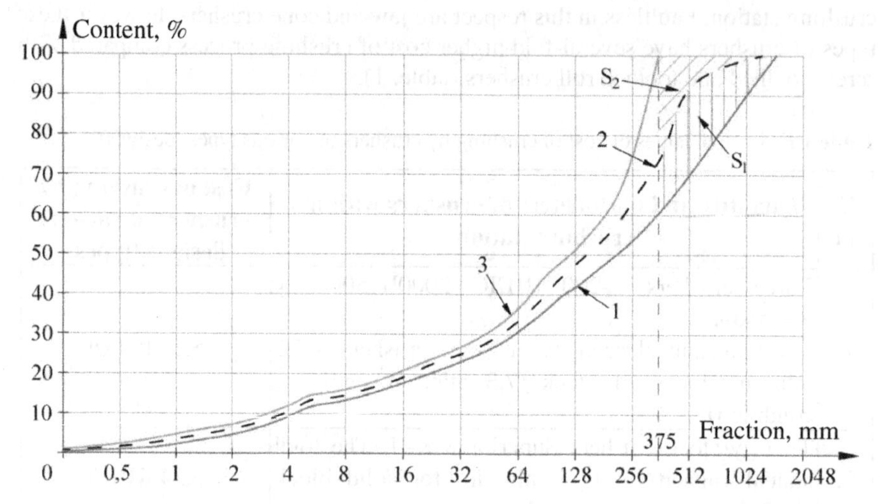

Fig. 5 Change in the size of granulometric composition of rocks at various ways of their preparation for transportation. 1 – change in fractional composition of the original rock for crusher feed when standard technology of blasting in used; 2 – change in fractional composition of the original rock for crusher feed in optimization of parameters of blasting preparation of original rock; 3 – change in fractional composition of rocks when mining of rock mass is made with cutting and loading machine; S_1+S_2 –original rock crushing , characterized by granulometric curve 1; S_2 –crushing of original material, characterized by curve 2

Because of imperfection of grizzly screens, the idea of creation of crushing stations, where up to 70-80% of 0-350 mm material fractions (suitable for transportation by conveyer transport) is separated by screening process has not been fully implemented.

At the same time, the use of single-roll, two-roll, screw-toothed and hybrid crushers is a reasonable compromise in the process of preparation of material for transportation by conveyor. These crushers effectively pass the original rock of fine and medium fractions and, in fact, combine the functions of screening and crusher (Fig.6) For the purposes of selecting CCT, it is important to determine capacity of crushers, operated as part of crushing stations.

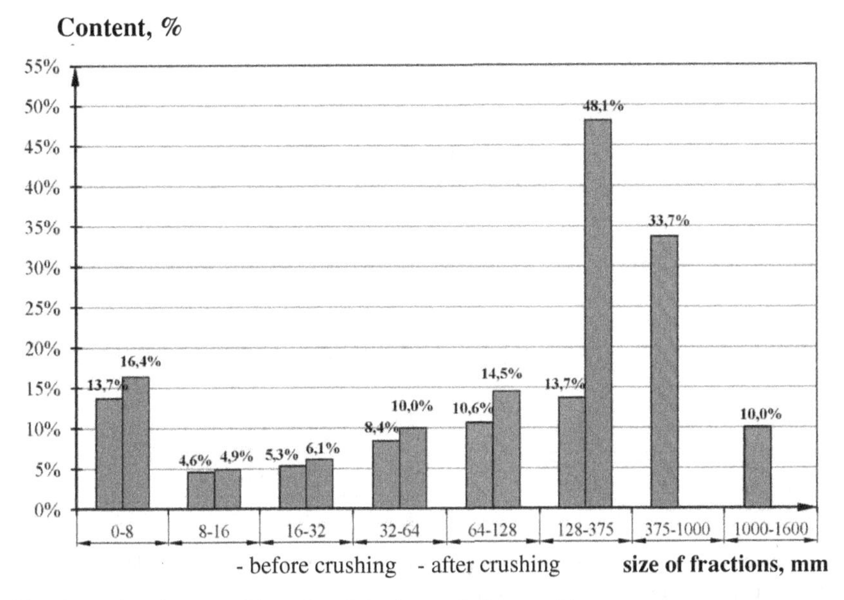

Fig. 6 Fractional composition of rock before and after crushing

When designing crushing stations of CCT complex for overburden rocks it is required to be guided by the principle of "do not crush anything extra» [3]. According to this principle, the cost of formation of granulometric composition of the initial rock and the actual cost of crushing need to be optimized.

One of the problems of CCT complexes operating is that during rock blasting in a majority of cases it is impossible to reliably eliminate output of coarse fractions (>1000 mm) of material delivered for crushing.

For liquidation of separate oversize pieces of initial rock (by conditions of crusher feeding), CCT complexes incorporate hydraulic breakers, capable to destroy oversize pieces in the crusher hopper or in the feed hopper.

In case of systematic getting into composition of original rock of 1200-1500 mm fractions, it is required to select a roll crusher or a screw-toothed roll crusher of a larger class, capable to cope with crushing of material of large fractions. In the latter case crushing capacity is underused.

An advanced engineering solution in improving crushing stations seems to be the use of feeder-crushers and a scheme of two-phase crushing.

In the first phase, a feeder-crusher is used, which brings original material fractions (0-2000 mm in size) to 0-600 mm fraction (Fig. 7).

In the second phase, a less powerful crusher processes a stabilized flow of original material (0-600 mm) to 0-350 mm fraction, suitable for conveyer transporting (see Fig.7).

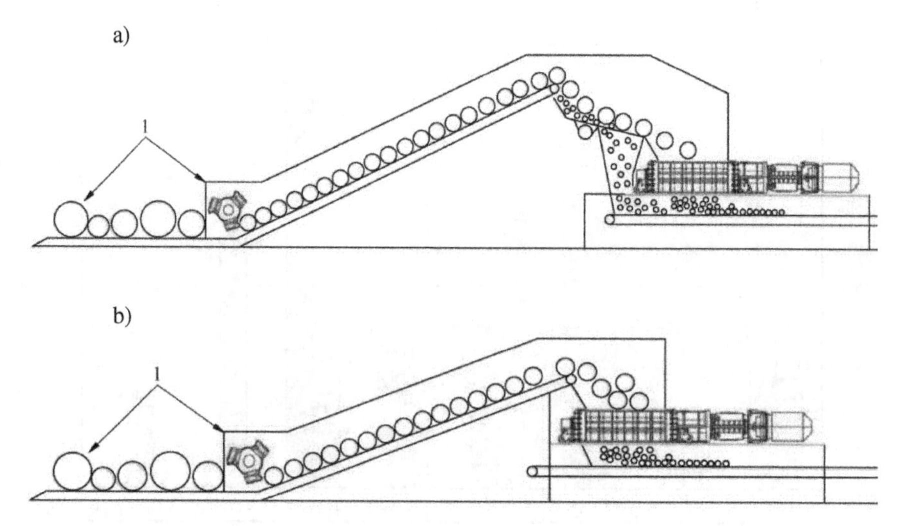

Fig. 7 Schemes of two-phase crushing using a feeder-crusher. a) scheme includes a grizzley screen; b) scheme without a grizzley screen.1 – a feeder-crusher

Use of two-stage crushing schemes enables to remove part of problems, associated with operation of crushing stations with standard design, namely:

- to reduce capital and operating expenses, associated with operation of large crushers;
- to ensure loading into crusher of original material 1600-2000 mm in size;
- to ensure synchronous start of the main drive motors with a total capacity of over 1000 kW.

Power consumption of secondary crusher will be several-fold less through guaranteed bringing of feed size up to 0-500 mm and reducing share of fractions, greater 300 mm.

An advantage of this technical solution using feeder-crushers is also a substantial simplification in the design of the crushing unit.

The problem in that the civil part of crushing units constructions for dump trucks with capacity over 200 t becomes more complicated, since heavy duty bearing metalwork is needed in order to cope with dynamic load, arising in the course of discharge of dump trucks. The latter circumstance results in significant increase in costs for their construction.

If it is required to transfer crushing stations, then a crusher, a feeder and part of metalwork can be easily enough transferred to a new place. However, transfer of the building structures (concrete foundations under supporting structures of hopper, feeder, crusher, retaining wall etc) is virtually an impossible task.

References

1. Vasilyev, M.V.: Transport in deep pits. – M.: Nedra, 295 pages (1983)
2. Mentges, Y.: Fully mobile crawler-tracked crushing complex for large open pits and strip mines. Ugol. (4), 28–31 (2009)
3. Olyunin, V.V.: Processing of non-metallic construction materials. – M.: Nedra, p. 232 (1988)

Analysis-Specific Standardization of Quarries to Determine the Potential for the Application of Belt Conveyor Systems

Christian Niemann-Delius and Tobias Braun

Institute for Surface Mining and Drilling, RTWH Aachen, Germany

Abstract. With a yearly power consumption of about 1,700,000 MWh, the quarry industry appears to be one of the most energy-intensive sectors among German industries. In-pit transport operations require approx. 40% of the entire power demand, generating some 50% of the total mining costs. To date haulage is mostly done by mine trucks, driven by diesel engines with a total annual fuel consumption of 68 million liters, emitting an equivalent of 208,000 tons of carbon dioxide. Substitution of the mine trucks by establishing belt conveyors will on the one hand significantly reduce the emissions, and on the other increase energy efficiency in in-pit transportation. As every open pit requires a specific approach in order to find a suitable conveying system, a standardized analysis is not possible. In order to achieve a differentiated statement on the suitability of continuous conveyors according to various structural properties, a classification and standardization of quarries with similar transport parameters has been established. An individual scheduling of conveying and machine usage in the different types of quarries will allow a comparison of optimized constellations of the different haulage systems within the respective field of application, in order to quantify and compare ecological and economical aspects among the haulage systems.

The following paper deals with a standardization of the German quarries. First the general quarry information and parameters are to be capture. Secondly the restrictions on the application of conveying systems will be argued. Referring on the different open-pit types and their characteristics, the results of the standardization will be presented. Results will include the prediction of the ecological and economical potential of applying continuous haulage systems for the different open-pit types in Germany.

1 Introduction

In Germany, hard rock mining is almost exclusively done in open-pit mines. Regarding an adequate haulage method there is a choice between continuous and non-continuous transport techniques. Currently, apart from few exceptions, discontinuous methods are being used. This is due to the given transport conditions and the general framework regarding the optimal use of transport technologies.

C. Niemann-Delius. (ed.), *Proc. of the 12th Int. Symp Continuous Surface Mining - Aachen 2014*
Lecture Notes in Production Engineering, DOI: 10.1007/978-3-319-12301-1_39

The bulk of open-pit mines where hard rock extraction is done are of comparatively small size. This is also true of the yearly tonnage. Such operational conditions basically speak against the use of continuous methods transport because these are primarily apt for transport of large masses over long distances. What is more, the material to be transported has an improper great size after being loosened from the ground mass and has to be reduced or at least size controlled before transportation. The economic and technical operational restrictions for continuous handling techniques that limit the operation of belt conveyors almost exclusively to the use in loose rock are based on data and experience from the early application of such technologies. Technological progress in general and especially the technological advance of mobile crusher units in the last two decades have however from an engineering point of view expanded the operational range of continuous transport techniques and thus created potentials for the use in hard rock mining. The question remains whether it is basically economically feasible and if so under which operational conditions to use continuous transport techniques in open-pit mines. The testing of the hitherto largely unexplored application in the most cost-intensive processing step in hard rock mining however carries an high economic risk since the decision in favour of a transport variant not only involves big investments but also has repercussions on the design of the open-pit layout, the personnel employed and the dimensioning of the corresponding equipment.

The in-pit transport accounts approx. 50% of total costs and some 40% of the total energy consumption in hard rock mining. The total energy consumption in hard rock mining amounts to approximately 1.7 MWh/a of which ca. 680.000 MWh/a go into in-pit haulage [1,2]. This is currently almost exclusively caused by diesel-engine dumper trucks, which leads to a yearly consumption of 68 million litres of diesel fuel with a fuel value of 9.9972 kWh per litre of diesel fuel, by which in turn 208.000 t of CO_2 equivalents are emitted. Belt conveyor operations however are driven by electrical drives and only have a fifth of the specific energy consumption per mined ton compared with dumper trucks [3]. So, CO_2 emissions in in-pit transport may be reduced by 49.5% by alternatively using belt conveyors and can even be completely avoided locally [4]. If the required power is entirely generated from renewable energy sources the percentage of emission reduction rises to 96% in terms of the current initial value [5].

The economic and ecologic potential resulting from the application of continuous transport techniques primarily depends on the characteristics of the open-pit mine and the specific requirements to material extraction [6]. The quantification of this potential in terms of the considered area of Germany therefore implies knowledge of the underlying structure of the local hard rock mining. The present paper the classification and standardization of the German hard rock mines is first described in order to approximately quantify the described potential.

2 Standardization of Quarries with Regard to the Use of Different Transport Techniques

The characteristics of an open-pit mine are influenced by a series of factors. The layout is based on e.g. the form and the shape of the deposit and local circumstances such as protection of nature or residential areas. Furthermore, in choosing an appropriate mining design more factors such as the quality of the rock and the requirements of the material to be extracted come into play. The differences between open-pit mines are partly significant. Regarding the used area there are differences of a few hundred thousand square meters. In Germany alone, there are more than 1.000 quarries which more or less differ from each other.

In order to come to a detailed conclusion regarding the fitness for the use of continuous transport techniques in these open-pit mines, different classes are defined in terms of the characteristics of certain features. These are subsequently intentionally defined by the representational properties so that a well-arranged amount of standardized open-pit mines of different properties can be used as reference.

2.1 Classification

In order to reduce the large number of quarries, the analysis of a subset or control sample is effected. The research is based on a control sample resulting in a control sample error of 9% referred to 107 surveyed open-pit mines based on a population of 1.500. Since the observation refers to the open-pit mines situated in Germany quarries of each region of the country are considered. The distribution is oriented towards the federal states concerned where the open-pit mines are located. Thus in order to obtain a realistic distribution an amount of quarries is chosen which approximately corresponds to the percentage share of the open-pit mine distribution per federal state based on the basic population (see Figure 1).

Of the quarries 30% are located in Bavaria. Rhineland-Palatinate, Hessia, North-Rhine-Westphalia and Baden Württemberg each have a share of 10%, while Thuringia as well as Saxony represents 5%. The other quarries spread over the other federal states are in low percentage rates [7].

To obtain a pertinent classification of the quarries classification features are examined which create a relevant grading for the research question and which on the other hand are equivalent regarding the contents and the characteristic significance [8]. Probably, the most decisive influence on the use of different transport methods in open-pit mines are their areal size and extension and the dependent length and layout of the transport routes. As a consequence, the classification is done by the areal size of the pit opening and another sub-division is effected by the type of mining and the resulting pit design.

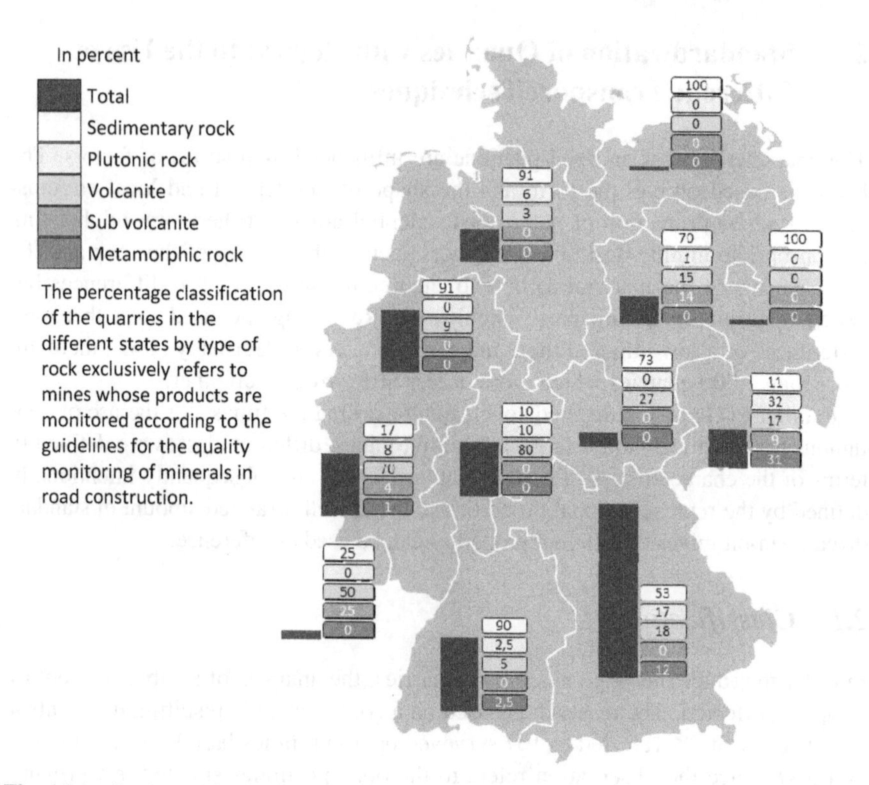

Fig. 1 Distribution of German hard rock mines by federal state

For the present research question an amount of 10 classes is considered appropriate. As the final number of classes results from the multiplication of the number of the variable „type of mining", whose maximum number with three different types of mining is already certain, with the not yet defined number of the variable „size levels" a scale of four size levels is considered sensible. The alternative of a scale of 3 size levels is refused because in this way only an insufficient structuration especially regarding the similarity and dissimilarity structures is obtained. From a first classification in terms of the similarity/dissimilarity structure the following levels result from the size of the open-pit mines:

- open-pit mines smaller than 200.000 m² (K 200)
- open-pit mines smaller than 500.000 m² (K 500)
- open-pit mines smaller than 700.000 m² (K 700)
- open-pit mines greater than 700.000 m² (G 700)

The characterization of the classes is effected by the arithmetic means of the corresponding property value. In Table 1 the property values of the determined size levels are shown. The quality of the found values can be determined by the variation coefficient, which indicates the relative size in terms of the mean. A sufficient characterization of the property values by the mean prerequisites a variation coefficient smaller than 50% [1].

Table 1 Distribution of German hard rock pit mines according to size levels

size level	number of observations	interval in 1000m²	Ø x in 1000m²	s in 1000m²	vc in %
K 200	50	30 - 180	126	45	36
K 500	36	202 - 495	289	77	27
K 700	12	500 - 680	574	64	11
G 700	6	720 – 1.600	1.098	323	29

x: mean s: standard deviaton vc: variation coefficient

For a final grouping an amount-dependent distribution of the classes is aimed at. The percentage distribution on the different size levels, established by simply counting the observed quarries is further subdivided by the classification of the open-pit mines according to the type of mining used in each case. These are arranged as follows, according to an survey of the Institute for Surface Mining and Drilling of the RWTH Aachen from 2008, where 110 hard-rock mines in the Federal Republic of Germany were considered: in 44% of the open-pit mines slope and hill mining is done, in 28% areal mining and in the other 28% mining to depth.

From the classification according to the properties size level and type of mining results the percentage distribution of the open-pit mines in question over twelve classes presented in Figure 2.

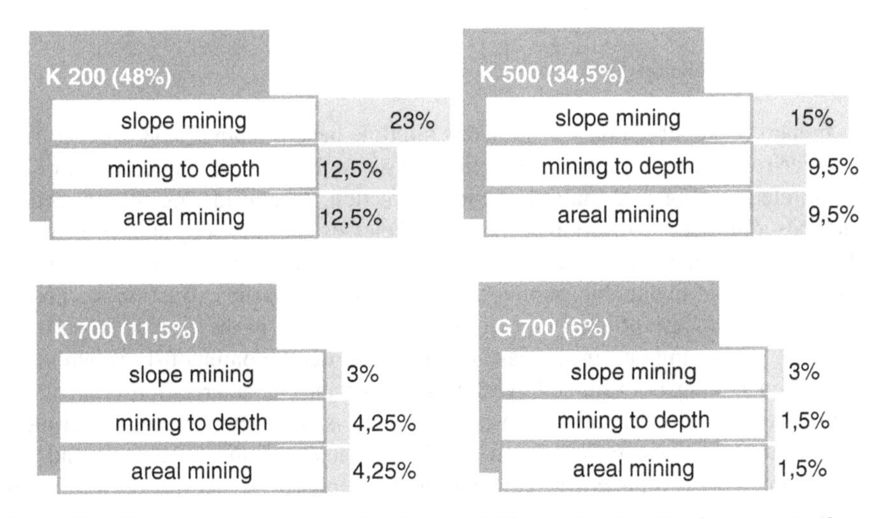

Fig. 2 Classification and percentage distribution of German hard rock mines over twelve classes

2.2 *Standardization*

In order to allow a comparison of the different transport methods regarding their fitness for use in hard rock mines with regard to economic feasibility and ecological aspects more descriptive features are assigned to the open-pit mine classes.

Next to the **size of the mine opening** and the **type of mining** the **gradient** of the transport ramp is an important factor for the accessibility of the open-pit mine. The average difference in level to overcome results from the **maximum pit depth** and the **number of bench levels** in the open-pit mine. The relevant distance in application considerations is the **transport distance** i.e. the length of the distance between the mining operation and the transport reference point. As this constantly changes over the lifetime of an open-pit mine the mean distance will be considered. The required belt length for the use of continuous hauling means is also determined by the applied mining method. Since the belt conveyor cannot be easily moved to another bench level the necessary belt length increases with every additional **active bench.** The different extraction methods are used in German hard rock mines in the following frequency distribution:

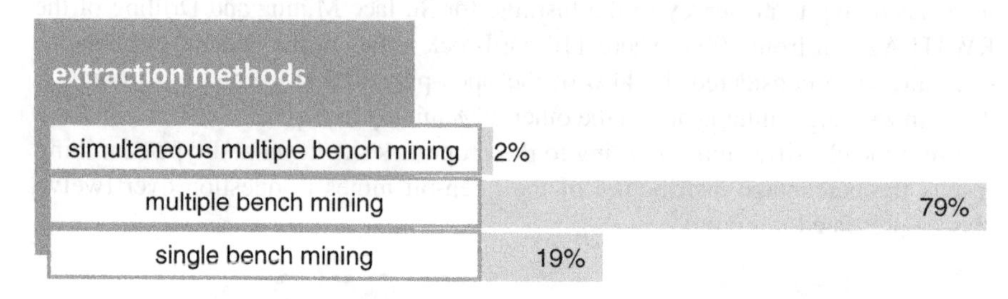

Fig. 3 Mining operation distribution in German hard rock mines [1]

In hard rock extraction, simultaneous multiple bench mining plays more or less a minor role and is applied in exceptional cases only. That is why considerations are limited to single or multiple bench extraction. In considering the extraction methods a further sub division of the formed classes is effected by standardization. As the standardization primarily refers to the use of continuous transport techniques the most favourable mining design depending on the extraction method is selected. In the case of areal extraction, swivel mining, in the case of slope mining, parallel mining and in the case of mining to depth continuously progressing expansion mining is examined. The most important characterization point to be considered next to the mean transport distance is the **yearly haulage tonnage** which has a decisive influence on the dimensioning of the equipment and the development of the mine. For calculation an average bulk density of 1.7 t/m^3 is considered.

The calculated feature characteristics are identical for the sub classes and types of a size level. Table 2 summarizes the relevant properties and the characteristic properties of the individual classes..

Table 2 Characteristic attributes of the individual open-pit mine classes

	K 200	K 500	K 700	G 700
pit depth* [m]	61	76	81	100
ramp gradient [°]	7,3	6,1	4,9	4,7
ramp length [m]	480	715	948	1.220
Ø average transport distance [m]	441	658	885	1.187
conveyance tonnage* [1000t/a]	223	632	679	1.600
number of horizons* [#]	3	4	4	5

* The data stem from a survey of companies of the industry sector. Between four and six companies per size level were questioned

The sub-division by mining method is reflected in the consideration of the slope direction of the ramps, which influences the performance and the energy consumption of the means of haulage. In the case of the extraction by drilling, e.g. in the loaded state up-hill move and in the unloaded state down-hill move is implied. This mining method is only interesting for the transport by belt because it influences the required belt length as mentioned earlier. For the mobile discontinuous haulage changing loading points play a minor role.

Briefly summarized, from the combination of different extraction methods (slope mining (HA), areal mining (FA) and mining to depth (AT)) and different extraction procedures (multi-bench extraction (MA) and single floor extraction (EA)) result 24 open-pit mine types, which are presented in Table 3.

Table 3 Sub-divided presentation and naming of the 24 types of open-pit mine

mining method / size level	AT		FA		HA	
K200	K200 AT MA	K200 AT EA	K200 FA MA	K200 FA EA	K200 HA MA	K200 HA EA
K500	K500 AT MA	K500 AT EA	K500 FA MA	K500 FA EA	K500 HA MA	K500 HA EA
K700	K700 AT MA	K700 AT EA	K700 FA MA	K700 FA EA	K700 HA MA	K700 HA EA
G700	G700 AT MA	G700 AT EA	G700 FA MA	G700 FA EA	G700 HA MA	G700 HA EA

The external shape of the open-pit mines is largely dependent on the mining method and is only little influenced by the direction of the mine development. The size levels of the standardization have an impact by the bigger or smaller dimensioning of the presented forms and/or a differing length-width relation. The basic shape will not be changed by the mining method either because only the number of the actively used levels changes. As in here, the mining direction is considered as conditioned by the mining method there are three basic shapes of open-pit mines, which are presented by screenshots from aerial photo analysis in Figure 4.

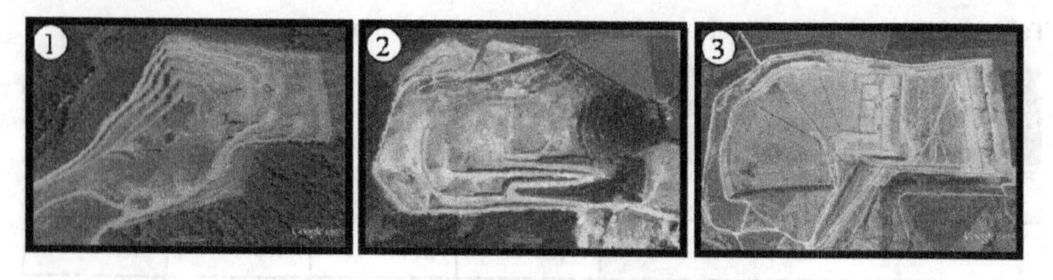

Fig. 4 Shape of open-pit mines: 1) slope mining in parallel extraction 2) depth mining in surface bench extraction 3) areal mining in swivel extraction [1]

As the open-pit mine models should be accessible for discontinuous and continuous transport techniques for reasons of comparison the safety requirements of the more demanding transport technique in that respect are taken as a basis. The required width of the roads and ramps results from the threefold width of the dumper truck so that it can be estimated at ca. 20m. The minimum bench width is determined by the blasting-engineering safety clearance of the belt conveyor system to the blast location. For high-strength rocks and 20m as the height of the mining slope it is 50m. The bench gradient of the slope can be set at 75° for unweathered rock. [12].

3 Comparison of Transport Techniques

To determine the economically most advantageous choice of equipment for a transport a comparison is made of the cost of any possible equipment grouping based on a specific application. For the comparison of the transport techniques among one another only the most advantageous equipment grouping are subsequently taken into consideration. In the process, the transport techniques listed below regarding the yearly operational costs and energy consumption:

Discontinuous: hydraulic excavator – dump trucker – stationary primary crusher

Continuous: hydraulic excavator – mobile primary crusher – mobile belt conveyor – stationary belt conveyor system

Combined: wheel loader in load & carry operation – semi mobile primary crusher – stationary belt conveyor system

The comparison of the transport methods based on the data of the open-pit mine type is at the same time the starting point and main basis for the problem defined at the beginning, if and if so, under which application conditions cost efficiency potentials result by using continuous transport techniques in hard rock mining, especially aggregates. Here, departing from a constant hauling tonnage and a haulage distance averaged over the operation period it is assumed that the result calculated for one year is also representative for the remaining years of operation. The formulas used, characteristics such equipment specifications and their conditions of application stem from the relevant literature, producers' brochures and the questioning of manufacturers. The underlying data base for equipment information such as durability, availability, original price and repair costs for different equipment types and sizes is provided by the "Construction equipment list 2007" published by the Main Association of the German Construction Industry Registered Association. Data on the more exact specification of the equipment in question such as speeds, energy consumptions, dimensions, cycle times and behavior under various operating conditions are taken off „Specifications & Application Handbook" of „Komatsu" company from 2007 as well as the manufacturer's leaflets of the same company and „Caterpillar", „Metso Minerals" und „Kleemann" companies. The formulas required to compute performance and cost effectiveness also arise from these brochures and the usual relevant literature.

The calculations the methods comparison is based upon do not include the entire costs and energy consumptions caused by the extraction, but only the part that varies as different transport techniques are applied. Thus, the results are not apt for the analysis of a certain transport but enable a comparison of the transport techniques among each other. Table 4 shows an example of the open pit specific analysis.

Table 4 Comparison of results for the open-pit mine types K 500 EA

conveyance techniques	K 500 AT EA		K 500 HA EA		K 500 FA EA	
	€/a	kWh	€/a	kWh	€/a	kWh
discontinuous	115%	118%	100%	101%	100%	100%
continuous	100%	100%	107%	100%	112%	125%
combined	191%	209%	207%	243%	216%	281%

For this reason, the presentation of the results is not in absolute values but on the basis of the percentage deviations of the results among each other under the condition that the most advantageous value is set at 100% costs and energy consumption in each case.

Analysis of the results shows that a continuous transport proves to be the most cost effective method in 8 of the 24 mines making up ca. 19% of all German quarries. However, in other cases, too, only marginal cost differences are found compared with the individually most advantageous transport. However, in other cases, too, there are marginal cost differences as opposed to the most cost effective means of haulage in each case. Presentation of the quarry-specific most cost effective means of haulage follows in Table 5.

Table 5 Presentation of the quarry-specific most cost effective means of haulage

size level \ mining method	AT		FA		HA	
K200	K200 AT MA	K200 AT EA	K200 FA MA	K200 FA EA	K200 HA MA	K200 HA EA
	DK	**DK**	**DK**	**DK**	**DK**	**DK**
K500	K500 AT MA	K500 AT EA	K500 FA MA	K500 FA EA	K500 HA MA	K500 HA EA
	K	**K**	**DK**	**DK**	**DK**	**DK**
K700	K700 AT MA	K700 AT EA	K700 FA MA	K700 FA EA	K700 HA MA	K700 HA EA
	K	**K**	**DK**	**DK**	**DK**	**DK**
G700	G700 AT MA	G700 AT EA	G700 FA MA	G700 FA EA	G700 HA MA	G700 HA EA
	K	**K**	**DK**	**DK**	**K**	**K**

DK: discontinuous K: continuous KM: combined

The analysis provides some factors that influence the cost effectiveness of the application of continuous haulage. So, e.g., the costs for the pay mineral extraction decrease as the yearly haulage tonnage increases. What is more, it becomes evident that the difference between the operational costs for continuous transport in single and multi-bench extraction is significantly higher than in the other transport techniques. Regarding the gradient of the transport distance to be covered, the situation is reverse: the steeper it becomes, especially when loaded, the lower the specific cost for the pay mineral transport compared with the other methods of transport.

Regarding energy consumption, it becomes evident that continuous transport in the majority of cases represents the most cost effective haulage alternative. However areal mining favours the use of dumper trucks operating discontinuously (see Table 6).

Table 6 Presentation of the quarry-specific most cost effective means of haulage

size level \ mining method	AT		FA		HA	
K200	K200 AT MA	K200 AT EA	K200 FA MA	K200 FA EA	K200 HA MA	K200 HA EA
	DK	**K**	**DK**	**DK**	**DK**	**K**
K500	K500 AT MA	K500 AT EA	K500 FA MA	K500 FA EA	K500 HA MA	K500 HA EA
	K	**K**	**DK**	**DK**	**DK**	**K**
K700	K700 AT MA	K700 AT EA	K700 FA MA	K700 FA EA	K700 HA MA	K700 HA EA
	K	**K**	**DK**	**DK**	**K**	**K**
G700	G700 AT MA	G700 AT EA	G700 FA MA	G700 FA EA	G700 HA MA	G700 HA EA
	K	**K**	**K**	**K**	**K**	**K**

DK: discontinuous K: continuous KM: combined

A sensitivity analysis made in the wake of the result evaluation underlines that in some open-pit mine types where the comparison of techniques showed that only little cost differences in relation to the most advantageous haulage alternative already little changes or deviations of the pit parameters “ramp gradient”, “transport distance” and “yearly transport tonnage” may lead to cost reductions by the use of continuous haulage equipment. If under this aspect open-pits are considered, which show a cost difference of less than 10% between the expenses for continuous transport and the expenses for the most advantageous alternative technique in that case a potential of cost reductions by continuous transport may arise in 10% of the German hard rock mines. If only open-pit mines operating in single bench extraction mode are regarded the share even rises to 60% of all single cases.

4 Summary

The paper shows that the use of continuous transport techniques in Germany’s hard rock open-pits is economically feasible and in some cases is even economically advantageous. Based on the results of the comparison of techniques and sensitivity analysis it is also assumed that in some cases by artful modifications of some pit parameters (especially mining method and yearly transport tonnage) and the use of continuous transport techniques possible savings can be elaborated and realized. Provided that application requirements and deposit characteristics do not speak against the use of continuous transport techniques right from the outset a potential quantification specific to the application is recommended.

References

[1] Deutsche Rohstoffagentur (DERA): DERA Rohstoffinformationen. Deutschland Rohstoffsituation, Hannover. Bundesanstalt für Geowissenschaften und Rohstoffe. Online verfügbar unter (2011), `http://www.bgr.bund.de/DE/Gemeinsames/Produkte/Downloads/DERA_Rohstoffinformationen/rohstoffinformationen-07.pdf;jsessionid=C8D610D3E337403445D0F66D67AC7E54.1_cid297?__blob=publicationFile&v=9` (abgerufen am July 09, 2007)

[2] Statistisches Bundesamt: Energieverwendung. Energieverbrauch des Verarbeitenden Gewerbes nach ausgewählten Wirtschaftszweigen 2011. destatis.de. Online verfügbar unter (2011) `https://www.destatis.de/DE/ZahlenFakten/Wirtschaftsbereiche/Energie/Verwendung/Tabellen/EnergieverwendungBeschaeftigte11.html` (abgerufen am July 11, 2013)

[3] Zimmermann, E., Kruse, W.: Mobile crushing and conveying in quarries - a chance for better and cheaper production. In: Aachen (ed.) Institut für Bergbaukunde III der RWTH Aachen: 8th International Symposium Continuous Surface Mining, pp. S.481–S.487. Druck und Verlagshaus Mainz GmbH, Aachen (2006)

[4] Bundesministerium für Wirtschaft und Technologie (BMWI): Zahlen und Fakten Energiedaten. Nationale und Internationale Entwicklung. Berlin, Bonn. Online verfügbar unter (2013), http://www.bmwi.de/DE/Themen/Energie/Energiedaten/gesamtausgabe.html (abgerufen am July 09, 2013)

[5] Fritsche, U., Greß H.: Der nichterneuerbare Primärenergieverbrauch des nationalen Strommix in Deutschland im Jahr 2011. Darmstadt. Internationale Institut für Nachhaltigkeitsanalysen und -strategien (IINAS). Online verfügbar unter (2012) http://www.iinas.org/tl_files/iinas/downloads/IINAS_2012_KEV-Strom-2011_%28HEA%29.pdf (abgerufen am July 11, 2013)

[6] Korak, J.: Technisch-wirtschaftliche Untersuchung der Transportbetriebsmittel unter besonderer Berücksichtigung der Transportmittelkombination fahrbare Brecheranlage - Gurtbandanlage für den Transport der Haufwerke im engeren Festgesteinstagebau. Aachen. Rheinisch-Westfälische Technische Hochschule Aachen (1978)

[7] Drozdzewski, G.: Gewinnungsstätten von Festgesteinen in Deutschland. In: 2. überarbeitete und ergänzte Auflage. Geologisches Landesamt Nordrhein-Westfalen, Krefeld (1999)

[8] Vogel, F.: Probleme und Verfahren der numerischen Klassifikation. Unter besonderer Berücksichtigung von Alternativmerkmalen. Vandenhoeck & Ruprecht, Göttingen (1975)

[9] Meißner, J.: Statistik verstehen und sinnvoll nutzen. Anwendungsorientierte Einführung für Wirtschaftler. Oldenbourg Wissenschaftsverlag, München (2004)

[10] Niemann-Delius, C.: Grundbegriffe. Vorlesungsreihe: Allgemeine Tagebautechnik I. Aachen: Institut für Rohstoffgewinnung über Tage und Bohrtechnik der RWTH-Aachen (2008)

[11] Google Earth: 1) Diabassteinbruch Altenkirchen; 2) Granulitsteinbruch Elzing; 3) Kalksteinbruch bei Karsdorf

[12] Dachroth, W.R.: Handbuch der Baugeologie und Geotechnik. 3., erweiterte und überarbeitete Auflage. Springer, Berlin (2002)

Regarding the Selection of Dumping Station Construction and Parameters of Concentration Horizon

Bayan R. Rakishev and Serik K. Moldabaev

Kazakh National Technical University,
Almaty, Republic of Kazakhstan
{b.rakishev,moldabaev_s_k}@mail.ru

Abstract. As a result, finding reserves to improve efficiency of excavator-truck systems (ETS) in crushing and conveying overburden complex and implementation of their power for the period of commissioning developed:

- schemes of dumping station, that is suitable for the period of the introduction of a combined truck- conveyor transport at the stripping works. Temporary concentration horizon is fixed up at the boundary of using excavator-railway and excavator-truck systems;
- method of justification of block-panels width, which are excavated by cross panels with varying levels of work sites with using ETS. It conjuncts with intensity of development of the working sites and commensurate mining of stripping zone is kept up. Step of transferring of dumping station is rational.

During the transition period of cyclic-stream stripping complex (CSSC) implementation the upper stripping zone is worked out by ERC, and the lower part of it – by ETS. Carrying out commissioning works at the CSSC No. 1 and No. 2 and the achievement of their designed horse-power on the border of ERC and ETC application predetermines the leaving of concentration horizon. A part of the upper stripping of overburden benches by straight drives is rapidly worked out through ERC. Preliminary calculations show that according to the condition of required pace of advancing the scope of work it is advisable to leave up to 5 highwalls in a given period of time under ERC. Therefore, there is a chance for a part of the working platform on the fifth rock horizon at mark +125/0 m to be used to overload the auto stripping to the railway transport. Increasing its width to accommodate the bulk of auto stripping in the transition period is compensated by more intensive working off the bottom of the stripping zone EAC.

Productivity single bucket excavators is determined in accordance with applicable regulations (see Table 1 According to academician K.N. Trubetskoy, the automobile and railroad complexes represent a series connection of excavator-automobile and excavator – railroad complexes with the unloading of dump trucks at the dumping stations inside the open-cut mining, followed by excavator loading

C. Niemann-Delius. (ed.), *Proc. of the 12th Int. Symp Continuous Surface Mining - Aachen 2014*
Lecture Notes in Production Engineering, DOI: 10.1007/978-3-319-12301-1_40

into dump cars. Due to this, distance for transporting by vehicles is reduced. The use of bottom-hole railway tracks at the lower difficult to access horizons is excluded. The dumping station (DS) with accumulating warehouse ensures the independence of works of automobile and railroad sub-systems; the distance of conveyances by transport vehicles does not exceed 1.2-1.5 km.

Table 1 Aggregated indicators of stripping excavators productivity

Name of the excavator and type of the transport	Hourly, m^3	Shiftable thousand m^3	Daily, thousand m^3	Annual, mln. m^3
EKG-12.5 to the railroad.	984.4	4.2	8.4	2.3
ESH-13/50 to the bulk	885.0	7.6	15.2	4.3
EG-18 from the bulk to the railroad.	1704.0	8.4	16.8	4.6
EKG-12US from the bulk to the railroad	1098	6.4	12.8	3.5
EKG-12.5 from the bulk to the railroad	1144.0	6.7	13.4	3.7
EKG -12,5 to the auto	984.4	4.7	9.4	2.6
EG -18 to the auto	1417.2	7.1	14.2	3.9
EKG -15 to the auto	1188.0	5.6	11.2	3.1

Location of the warehouse on the one horizon for a long period of time, as well as the proximity of the exchange station to the warehouse creates conditions of high productivity work of excavators in transloading. An annual production of excavators at the dumping stations per 1 m^3 bucket capacity is 1.4 times higher than in the faces.

Storage capacity is expedient to take within 10 to 15 daily transportation volume; it ranges from 100 to 500 thousand m^3.

Dumping station consists of two bings. To select parameters of bings it is required to take into account ensuring productive safe operation of a 90-ton dumptrucks HD-785-5 Comatsu Company. In this regard, the minimum width of bing on top is assumed to be 28 m, while its maximum height does not exceed 8 m. These parameters of the bing will limit the width in the bottom up to 49 m.

A two-lane highway is left with two DS from the side of worked off interspace between bings of DS and the upper edge of automobile stripping zone at a safe distance from the bottom edge of bulks. Its width without an auxiliary panel for the ETS is 36 meters. With one DS on the concentration horizon, the width of this area is reduced to at least up to 11 meters, Then the minimum width of the working area of the concentration horizon without consideration of a block panel ERC from 187 meters with two DS will be decreased up to 161 meters with one DS.

As an extraction-loading machine with such parameters of the bing it is advisable to use an excavator of EKG- 12US type. This will allow to produce an overload in the bulk volume of auto stripping into railway transport means with the almost complete cross-section of dass with 30 m in width while leaving the minimum volume of non-dispatched portion of rock at the warehouse.

Then the "passive" volume will be located by the ridge along the front works at a height of about 7 meters. The maximum productivity from a single dumping station on the concentration horizon will be 3.5 million m^3 per year.

Setting the parameters of the dumping station, depending on the coupling with the adjacent horizons in terms and traffic schemes of dump trucks within the concentration horizon are specified in Fig. 1.

Reducing the width of the bing from below, and accordingly its height will decrease the productivity of dump trucks and the excavator at the DS, and will require to use an abutment to increase the length of flexible cable of the excavator. It would be possible to accept productivity of one DS while planning mining operations from 2.5 million m^3 to 3.5 million m^3 per year, and with two DS on the concentration horizon - from 5.0 million m^3 to 7.0 million m^3 per year.

The proposed design of DS will require changing of the excavator with the switching substation to the high-voltage feeder lines only after the formation of bings in the new position on the advance increment F . Therefore, based on the length of flexible cable of the excavator, the capacity of one bing will be limited to a 5- daily transportation volume of auto stripping.

The length of one bing along the front works $l_{нш}$ consists of the length of the bulk of auto stripping $l_{ш}$ and the length of adjoining cross-over l_{c}. The length of the bulk of auto stripping includes the length of front work along the bulk of the excavator of DS within one bing $l_{фн}$ and the length of retained portion of bulk $l_{но}$, after the shipment from one of the bings of volume warehouse on dass.

The length of front works on bulk of the excavator within one bing is determined by the expression:

$$l_{фн} = \frac{t_{рг} P_{э.сут}}{A_{о} h_{он}} + h_{он} ctg\beta_{н},$$

where $t_{рг}$ - a reserve capacity of the bing of DS per day, depending on the transportation volume of auto stripping, daily; $P_{э.сут}$ - daily productivity of the excavator at the DS, m^3; $A_{о}$ - the width of dass on the bing bulk, m; $h_{он}$ - the height of dumped bing, m; $\beta_{н}$ - an angle of slope of auto stripping bulk, degree.

According to the calculation, with $t_{рг}$ = 5 days $\Rightarrow$ $l_{фн}$ = 278 m. Then the length of the bing along the front works will be 406 m. In this case, the length of the flexible power cable of the excavator will not exceed 430 m. The length of the dumping station along the front works $l_{пп}$, taking into account the distance between the bings $l_{сп}$ = 30 m, will not exceed 842 m, i.e. almost two times less than half of the length of the front works on the ledge of the concentration horizon.

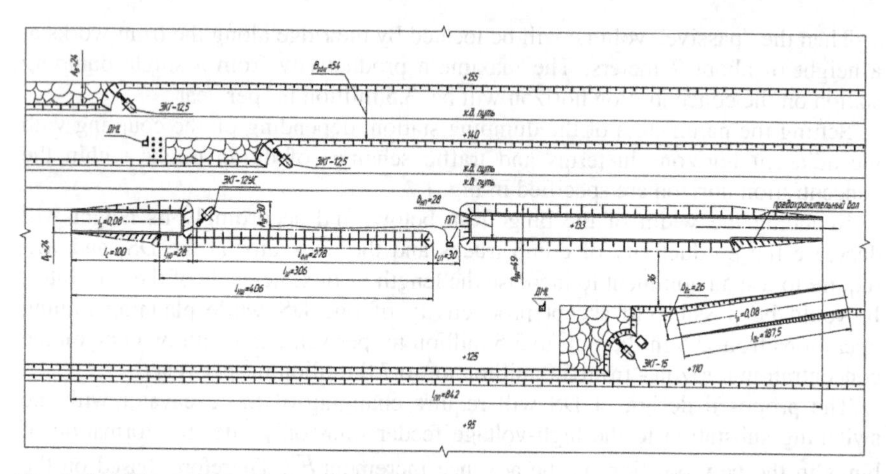

Fig. 1 The Scheme of the Dumping Station

Elements and design values of DS are shown in Table 2.

While developing the ETS block panels from flanks to the center of the open-pit field, this will allow to freely place in one line up to two DS on the concentration horizon and apply a simplified scheme of lay-out of tracks.

All rock horizons are transport. One or two railways are located in the upper stripping zone on the highwalls, worked off onto the railway transport. While working out the highwall onto the vehicles – a transport lane for the two-lane strips of road traffic. According to the design, the working out of stripping blocks panels onto the railway transport is envisaged both with the breakdown of the rock mass and two excavator passages, and blasting operation in the clamped environment to a buffer and one excavator passage.

Currently, the development of coal and stripping is carried out by using blast hole drilling. Drilling of wells is conducted by boring rigs of rotary drilling.

The design provides the use of boring rigs of rotary drilling, respectively DM-45 and DML with a diameter of wells, respectively 175 and 220 mm in the mining and stripping works.

For the estimated year, a large portion of volume of stripping rocks from the horizon +95,0 ÷ +110,0 m is delivered by dump trucks to the concentration horizon followed by the railway transportation and shipment to the external dumps. The design envisages the creation of dumping station from automobile to the railway transport on the horizon + 125.0 m.

Dumping station will be located on the working side of the open-cast. Stripping delivered by automobile transport is stored in a temporary bulk, then it is overloaded by the shovel-excavator EKG-12US onto the railway transport means.

Placement of DS and freight transportation link with the concentration horizon of cargo traffic of automobile and railway stripping are shown in Figures 2a (with two DS) and 4b (with single DS).

Table 2 Elements and design values of the Dumping Station

№№ Ser. No.	Name of the elements and parameters	Description	Unit of measurement	Value
1	Number of bings: - dumped; - shipped	-	pieces	2 1 1
2	The number of flanking cross-over adjacent to the front works	-	pieces	2
3	Type of the excavator	-	-	EKG-12US
4	Excavator-type bulldozer	-	-	D3-94S
5	The length of the dumping station along the front works	$l_{пп}$	m	842
6	The height of dumped bing	$h_{он}$	m	8
7	The length of bing along the front works	$l_{нш}$	m	406
8	The length of auto stripping bulk	$l_{ш}$	m	306
9	The width across the bottom of bing	$b_{нн}$	m	49
10	The width across the top of bing	$b_{нп}$	m	28
11	The distance between bings	$l_{сп}$	m	30
12	The length of the front works on the excavator bulk EKG-12US within one bing	$l_{фн}$	m	278
13	The width of dass across the bulk of bing	$A_{о}$	m	30
14	Area of dass across the bulk of bing	$S_{о}$	m^2	240
15	The length of cross-over of bing	$l_{с}$	m	100
16	The slope of cross-over of bing	$i_{р}$	‰	80
17	The width of cross-over of bing	$b_{с}$	m	24
18	The length of retained portion of bulk on top	$l_{но}$	m	28
19	The angle of slope of auto stripping bulk	$\beta_{н}$	degree	36
20	Excavator's shifting productivity at the DS	$P_{э.см}$	thous. m^3	6,4
21	Excavator's daily productivity at the DS ПП	$P_{э.сут}$	thous. m^3	12,8
22	The maximum annual productivity of the excavator at the DS	$P_{э.г.мах}$	mln. m^3	3,5
23	Accepted annual productivity of the excavator at the DS	$P_{э.г}$	mln. m^3	2,5

With two dumping stations, each of them is located along the bench rail-track with orientation in terms of center of one of the two halves of the concentration horizon. The midst of free platform left between the bings (bulks) of each of the two DS (warehouses) is taken as such center. It is placed in the center of the concentration horizon with a single DS. (Figure 3b).

Due to the need of transfer of a part of the equipment of a reloading point (RP) on a new place in process of deepening of pits and development of mining works in practice of design of ore fields the corresponding methodical base is applied to justification of questions of transfer and placement of such points [1, 2].

Fig. 2 Scheme of Stripping Zone with two DS (a) and single DS (b) on the concentration horizon

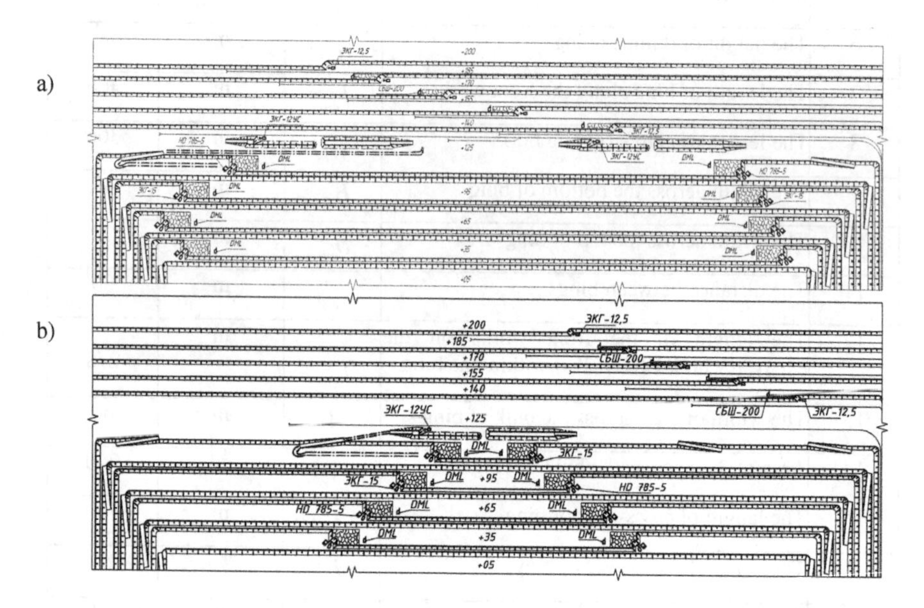

In work [1] the technique of a choice of the horizons of a pit and the placement period is given in them the reloading crushing points, allowing in process of career development to optimize distribution of traffics between automobile and conveyor links of transport system.

It also provides definition of border of transition to automobile and conveyor transport and the solution of problems of formation of the concentration horizons taking into account influence of location of RP and conveyors in career on development of mining works.

On coal mines the cyclic and line technology (CLT) only takes root. On Vostochny coal mine five top overburden benches are fulfilled by excavator and railway complexes (ERC), the others – EAC. The creation of the concentration horizon with realization of the combined automobile railway transport is provided on border of their application. The main volumes of automobile overburden move on an external dump, with combined automobile and conveyor transport. Feature of working off of overburden benches of ETS is serial dredging of two subbenches from both flanks of a career field cross-section cut with changing level of a working platform. The construction of temporary auto descents for moving of breeds from top subbench is provided for purpose in end faces of a cut.

The main direction of researches is ensuring compliance of parameters of the concentration horizon and ETS blocks panels of design capacity of a cut on coal. Design of development of an overburden zone with realization of an intensive way of working off of ETS blocks panels consists in ensuring demanded rate of development of all overburden zone with various complexes of the equipment. It depends directly on productivity of a cut on coal for planned year.

In turn it is required to observe proportionality of an advance of parts of an overburden zone of rather temporary concentration horizon. Speed of an advance of a scope of work of benches in the top part of an overburden zone, in which ERC are maintained (v_{ERC}), should allow to transfer RP to new situation in due time. The demanded advancing of the top part of an overburden zone should be created by the time of completion of working off excavator-automobile complexes of the blocks on both flanks of a career field.

Below the concentration horizon where ETS are maintained, it is necessary to allocate consistently from top to down average and bottom parts of an overburden zone. In an average part of an overburden zone working off of overburden benches of ETS is made on offered double-subbench scheme with changing level of a working platform.

The cutting new overburden subbench and cleaning of a roof of the top coal layer is made in the lower part of an overburden zone. During certain moments of operation of a cut the lower part of an overburden zone can be absent. The accepted width of a working platform of the concentration horizon (B_{wchj}), without width of blocks of ERC and ETS panels at certain stages of working off, makes 117 m at two RP and 91 m at one RP.

The key adjustable parameter in an average part of an overburden zone is the width of the block panel fulfilled by ETS (B_b). Especially the long period of time reloading points will be maintained without transfer in new situation. In turn the width of the block panel should provide timely opening of demanded stocks of layers for performance of planned productivity of a cut on coal at the minimum demanded volumes of dredging of overburden breeds. Criterion function then should answer a minimum of the current overburden ratio (k_{corij}) in planned year at a certain stage of working off under condition of proportional development of various parts of an overburden zone with the greatest step of transfer of RP:

$$k_{corij} = \frac{V_{oij}}{Q_{kij}} \rightarrow \min \tag{1}$$

at the following restrictions:

$$V_{oij} = \frac{V_{oj} Q_{kc.ij}}{Z_j} \quad \text{at} \quad V_{oij} \leq V_{o(i+1)j},\; V_{oij} \geq V_{o(i-1)(j-1)}; \tag{2}$$

$$Q_{kc.ij} = \frac{\sum_{i=1}^{f=\text{int}(t_{yj})} Q_{cij}}{\text{int}(t_{yj})} + (Z_j - \sum_{i=1}^{f=\text{int}(t_{yj})} Q_{kij})\left[t_{yj} - \text{int}(t_{yj})\right]$$

at

$$(Z_{j-1} - \sum_{i=1}^{f=\text{int}(t_{y(j-1)})} Q_{ki(j-1)})\left[t_{y(j-1)} - \text{int}(t_{y(j-1)})\right] = 0 \tag{3}$$

$$Q_{kc.ij} = (Z_{j-1} - \sum_{i=1}^{f=\text{int}(t_{y(j-1)})} Q_{ki(j-1)}\left[t_{y(j-1)} - \text{int}(t_{(yj-1)})\right] +$$

$$+ \frac{\sum_{i=1}^{f=\text{int}(t_{yj})} Q_{kij}}{\text{int}(t_{yj})} + (Z_j - \sum_{i=1}^{f=\text{int}(t_{yj})} Q_{kij})\left[t_{yj} - \text{int}(t_{yj})\right];$$

$$\text{at} \quad (Z_{j-1} - \sum_{i=1}^{f=\text{int}(t_{y(j-1)})} Q_{ki(j-1)})\left[t_{y(j-1)} - \text{int}(t_{y(j-1)})\right] > 0$$

$$t_{yj} = \frac{V_{oj}}{Z_j};$$

$$v_{ERCi} \geq v_{EACi}; \tag{4}$$

$$v_{EACi} \geq \frac{h_e Q_{kij}}{Z_{yj}} (ctg\gamma_{oij} + ctg\beta_{kij}); \tag{5}$$

$$\gamma_{oij} = arctg\left\{(\sum_{i=1}^{n_{orij}} h_{or} + \sum_{i=1}^{n_{oanan}} h_{oan}) / \left[ctg\alpha_o(\sum_{i=1}^{n_{orij}} h_{or} + \sum_{i=1}^{n_{oanan}} h_{oan}) + (n_{oanan} - 1) + \right.\right.$$

$$\left.\left. + N_{bij} B_b + (n_{orij} - 1) B_{wor} + B_{uoij} + B_{wchij}\right]\right\}; \tag{6}$$

$$n_{oanan} = \frac{H_{eij} - H_{ezij} - \sum_{i=1}^{n_{orij}} h_{or}}{h_{oan}} \quad \text{at} \quad n_{oanan} := \text{int}(n_{oanan}), \tag{7}$$

$$\text{if } n_{oanan} > \text{int}(n_{oanan}), \quad n_{oanan} = \text{int}(n_{oanan}) + 1; \tag{8}$$

$$H_{eij} = h_c + h_{eb} + \sum_{i=1}^{n_j} (H_{ezij} - h_{eb}); \tag{9}$$

$$H_{ezij} = h_{eb} + \sum_{i=1}^{n_{eij}} h_e; \tag{10}$$

$$n_{eij} = \frac{M_j - в_{ic} - h_{eb}(ctg\alpha_e + ctg\alpha_{ey}) - B_{tbij}}{B_{we} + h_e(ctg\alpha_e + ctg\beta_j)}; \tag{11}$$

$$N_{bij} = 0{,}5 n_{oanan} \quad \text{at } n_{oanan} = 2{,}4{,}6{,}8\ldots, \tag{12}$$

$$N_{bij} = 0{,}5 n_{oanan} + 0{,}5 \quad \text{at } n_{oanan} = 1{,}3{,}5{,}7\ldots;$$

$$B_{tb} \le B_{coij} < B_b; \tag{13}$$

$$v_{EACi} =: \frac{V_{oij} - V_{ori}}{n_{oanan} h_{oan} L_{s.a.a.i}} \quad \text{at } v_{EAAC} \ge \frac{h_e Q_{kij}}{Z_{yj}} (ctg\gamma_{oij} + ctg\beta_{kij}), \tag{14}$$

$$V_{ori} = n_{erij} P_{aer};$$

$$v_{ERCi} = \frac{n_{erij} P_{aer}}{n_{orij} h_{or} L_{s.r.a.i}} \quad \text{at } v_{EACi} n_{orij} L_{s.r.a.i} \le n_{erij} P_{aer}; \tag{15}$$

$$F \ge B_b \quad \text{at } 0{,}5 B_b h_{oa} L_{s.a.a.i} \le P_{aea}; \tag{16}$$

$$N_{crych} = \text{int}\left(\frac{n_{erych} P_{aer} t_{rp}}{h_{or} L_{s.r.ych} A_{ERC}}\right) \quad \text{at } t_{rp} = \frac{B_b}{v_{EAC}}; \tag{17}$$

$$F := A_{ERC}\, \text{int}(N_{crych}); \tag{18}$$

$$\text{if } F > B_b, \quad B_b := B_b + \Delta B_b; \tag{19}$$

$$\text{if } F < B_b, \quad B_b := B_b - \Delta B_b; \tag{20}$$

$$\text{if } F = B_b \pm \Delta B_b, \quad B_b := B_b \tag{21}$$

$$\text{at } n_{cri} A_{ERC} \le v_{ERCi}, \quad n_{cri} A_{ERC} h_{or} n_{orij} L_{s.r.a.i} \le n_{erij} P_{aer}, \tag{22}$$

where V_{oij} - annual production rate of a coal mine on overburden in the i-th year within the j-th of a stage of working off, m^3; Q_{kij} - annual production rate of a coal mine on coal in the i-th year within the j-th of a stage of working off, t; V_{oj}

- volume of overburden rocks within the j-th of a stage of working off, m^3; $Q_{kc.ij}$ - the average annual production rate on coal within the j-th of a stage of working off, t; Z_j - coal stocks within the j-th of a stage of working off, t; t_{yj} - time of working off of the j-th of a stage of working off, year; v_{ERCi} - speed of an advance of a scope of work of the benches fulfilled by ERC, in the i-th year, m; v_{EACi} - speed of an advance of a scope of work of the benches fulfilled by EAC, in the i-th year, m; h_e - height of an extraction bench, м; γ_{oij} - corner of a slope of an overburden zone in the i-th year within the j-th of a stage of working off, degree; $\beta_{\kappa ij}$ - hade of a roof of the top coal layer in borders the i-th years within the j-th of a stage of working off, degree; h_{or} - height of an overburden benches at their working off of ERC, m; n_{orij} - quantity of overburden benches at their working off of ERK in the i-th year on the j-th a working off stage, pie.; h_{oan} - height of an overburden subbench at their working off of EAC, m; n_{oanan}- quantity of an overburden subbench at their working off of ETS in the i-th year on the j-th working off stage , pie.; α_o - working corner of a slope of an overburden bench or subbench, degree; B_{sb} – width of safety berm, m; N_{bij} - number of the created ETS block panels in the i-th year on the j-th a working off stage, pie.; B_b - width of the block panel, fulfilled by EAC, m; B_{wor} – width of a working platform of the overburden ledge fulfilled by ERC, m; B_{coij} - width of a cutting trench on a bottom on overburden in the i-th year on the j-th working off stage, m; B_{wchij} – width of a working platform of the concentration horizon (without width of the blocks panels fulfilled by ERC and EAC) in the i-th year on the j-th working off stage, m; H_{eij} - depth of coal mine in the i-th year on the j-th working off stage, m; H_{ezij} - height of a extraction zone in the i-th year on the j-th stage of working off, m; h_c – capacity of deposits equal to height of the advanced ledge, m; h_{ec} - height of a cuting extraction bench, m; n_j – quantity of the extinguished stages of working off since the beginning of development of a field of a cut, piece; n_{eij} – number of extraction benches in simultaneous work in the i-th year on the j-th working off stage at line technology of their development, piece M_j – horizontal capacity of coal layers on the j-th stage of working off, m; α_e – working corner of a slope of an extraction bench, degree; α_{ey} – steady corner of a slope of an extraction bench, degree; B_{tbij} – transport berm on a non-working board in the i-th year on the j-th stage of working off, m; $B_{p\partial}$ – minimum width of a working platform of an extraction bench, m; β_j – a hade on a sole of a fulfilled bottom coal layer in borders of the j-th stage of working off, degree; $\hat{A}_{ta}$ - width of transport berm at automobile transport, m; V_{ori} - annual capacity of a cut on railway overburden in the i-th year, m^3; $L_{s.a.a.i}$ - the average length of a scope of work of an overburden zone where EAC, are maintained by, m; n_{erij} - number of at the same time working excavators in a complex with railway transport in the i-th year on the j-th stage of working off, piece; P_{aer} - annual production operational capacity of the excavator to loading in means of railway transport, m^3; $L_{s.r.a.i}$ - the average length of a scope of work of an overburden zone where ERC, are maintained by m; F - step of transfer of RP, m; h_{oa} - height of an overburden bench when their working off by EAC, m; P_{aea} - annual production operational capacity of the excavator to loading in means of automobile transport, m^3; N_{crych} - quantity of cuts, fulfilled by ERC on the concentration horizon in borders of reloading points, piece; n_{erych} - number of at the same time working excavators in a complex with railway transport on the concentration horizon at working off of an overburden bench, piece; t_{rp} - time of transfer of RP, year; $L_{s.r.ych}$ - length of a scope of work of an overburden bench on the concentration horizon, fulfilled by ERC, m; A_{ERC} - width of the ERC block panel, m; ΔB_b - step of an increment or reduction of width of the block panel fulfilled by EAC, м; n_{cri} - quantity of cuts, fulfilled by ERC in a year, piece.

The given mathematical model allows to prove a step of transfer of reloading points to interrelation with intensity of development of a working zone with observance of a proportional advance of parts of an overburden zone of rather temporary concentration horizon at the rational width of blocks panels fulfilled by excavator and automobile complexes by cross-section cuts with changing level of a working platform.

Conditions of a formula (2) provide the most rational mode of mining works during a certain period of operation of a cut, and restriction (3) are necessary for a complete sample of stocks upon transition from one stage of working off to another.

The proportional development of various parts of the overburden zone fulfilled with application of railway and automobile transport is reached at performance of condition (4), and for the productivity of a cut on coal, the opening of demanded stocks is reached at condition (5).

The change of a corner of a slope of an overburden zone is considered in a formula (6) when element's parameters of development system, including width of a working platform of the concentration horizon and the blocks panels fulfilled by ETS are regulated. Compliance of a design of an overburden and extraction zones with realization of the offered way of working off of ETS blocks panels is provided with expressions (7-12) and restriction (13).

The greatest step of transfer of RP is reached on conditions of formulas and restrictions (14-22) with providing rational width of blocks panels for EAC, depending on intensity of development of overburden zone parts and with application of settlement quantity of the equipment complexes of concentration horizon.

For the accepted productivity of a cut on coal, the technique of justification of width of the block panel for ETS depending on a step of transfer of reloading points and intensity of development of a working zone is expedient for using at justification and choice of the mining transport equipment on overburden works with CLT introduction in design. It will provide observance of conditions of a proportional advance of parts of an overburden zone by various complexes of the equipment of rather temporary concentration horizon.

In work the integrated calculations for justification of width of blocks panels for ETS depending on intensity of development of an overburden zone in borders of application of the accepted complexes of the equipment are executed at capacity of a cut of 25 million t/year.

The analysis of calculations shows that at exhaustion of capacity of two CLOC lines it is necessary to maintain originally with a maximum load excavators on RP, and after exhaustion of this possibility to transfer the concentration horizon on a bench below and to execute reconstruction of opening developments of a flank external capital trench for an institution of tracks on this horizon [3].

Results of calculations show that for increase in a step of transfer of RP it is necessary to increase speed of an advance of a scope of work of a zone of work of ERC. It will allow to increase or width of a working platform of the concentration horizon, or size of width of blocks panels in a zone of work of EAC. However the size of the last is in direct dependence on existence of a demanded reserve of the prepared stocks on coal.

Technique approbation at the choice of width of blocks panels for ETS proves expediency of effective application cross-section cut according to the offered double-subbench scheme of working off of average and bottom parts of an overburden zone excavator and automobile complexes from both flanks of a career field.

In building of capacity of two CLOC lines it is necessary to aspire to use reloading points of the concentration horizon with the maximum loading for part transfer automobile overburden in means of railway transport.

Novelty of the offered technique on justification of a step of transfer of reloading points is that for the first time this task is solved in interrelation with intensity of development of a working zone for the accepted productivity of a cut on coal. Taking into account a proportional advance of parts of an overburden zone of rather temporary concentration horizon she allows to define rational width of the blocks panels fulfilled by excavator and automobile complexes by cross-section cut with changing level of a working platform.

At introduction of cyclic and line technology on overburden works of coal mines with line technology of coal mining the best results are reached at performance of the following conditions:

- providing the most rational mode of mining works during a certain period of operation of a cut with a complete sample of stocks upon transition from one stage of working off to another;
- achievement of proportional development of various parts of the overburden zone fulfilled with application of railway and motor transport for opening of demanded stocks with set productivity of a cut on coal;
- influence on a corner of a slope of an overburden zone of change of width of working platforms of the concentration horizon and ETS blocks panels at regulation in parameters of elements of system of development;
- observance of compliance of a design of an overburden and extraction zones with realization of the offered way of working off of ETS blocks panels;
- achievement of the greatest step of transfer of RP with rational width of ETS blocks panels depending on intensity of development of parts of an overburden zone with application of settlement quantity of complexes of the equipment of rather concentration horizon.

References

1. Stolyarov, V.F.: The problem of cyclic-flow technology of deep pits, 232 p. UrRAS, Ekaterinburg (2004)
2. Drizhenko, A.Y., et al.: Open-cast mining of iron ore in Ukraine: Status and Improvement: Monograph. In: Drizhenko, A.Y., Kozenko, G.V., Rykus, A.A. (eds.), p. 452. Poltavian Litterateur, Poltava (2009)
3. Rakishev, B.R., Moldabayev, S.K., Samenov, G.K., Aben, Y., Anafin, K.M.: Innovative technology of expanding the boundaries of efficient application of the open cast mining method. In: Proceedings of the 23rd World Mining Congress, Montreal, Canada (2013), `http://file:wmc2013/pdfs/wmc2013Paper872.pdf`

Investigations to Apply Continuous Mining Equipment in a Shovel and Truck Coal Operation in Australia

Arie-Johann Heiertz

RWE Power International, RE GmbH, Germany
www.rwepi.com

Abstract. Investigations to apply continuous mining equipment in a shovel and truck coal operation in Australia

Beside bucket wheel excavators (BWE) several high performance In-Pit Crushing and Conveying Systems (IPCC), which allow the adoption of continuous mining systems, even for typical Shovel and Truck (S&T) operations, are available today. RWE Power International assists operators technically and investigative considering a change from the S&T- to IPCC-technology, both, in the planning process involving large scale belt conveyor systems and later with operational know-how transfer. In a number of cases the challenges associated with this system change have been addressed and solved successfully. One key challenge is managing the change in mindset and development of skills for operating the belt conveyor systems efficiently.

OPEX drivers as high personnel costs and diesel fuel cost together with a deteriorating coal price make Australian coal mines less competitive in the international market. Australian mining companies are looking for alternatives to optimize the mining process. RWE Power International, RE GmbH, has been requested by a major coal miner to undertake an IPCC scoping study with a view on reducing cost by implementing continuous mining equipment.

As the reviewed company plans to increase its coal production to 7 Mt per annum, RE´s task is to optimize the mine production in regard to reduce production costs per tonne of coal. As the operation has a certain amount of very soft overburden material, the special focus of this paper lies on the application of fully continuous mining systems in these soft overburden horizons. For the underling harder zones with multiple coal seams the application of semi-mobile IPCC systems is further investigated.

This paper compares the effectiveness of S&T to continuous mining systems for an up to 300 m deep mine with an average annual output of 106 Mbcm. The paper starts with an explanation of the conceptional mine design from a classical shovel and truck operation to favor the application of conveyor belts. Later on, the selection of the mining method will be further explained at the hand of the advantages and disadvantage of BWE and IPCC equipment in regard to the

C. Niemann-Delius. (ed.), *Proc. of the 12th Int. Symp Continuous Surface Mining - Aachen 2014*
Lecture Notes in Production Engineering, DOI: 10.1007/978-3-319-12301-1_41

classical S&T pit. Annual effective operating hours, the quantity of staffing required and maintenance factors are explained in detail.

1 Introduction and Project Location

The client is currently mining about 2 Mt per annum of hard coal in its operations in the Collie Basin in Western Australia, located approximately 230 Km southeast of Perth. The project site is located about 12 Km east of the city of Collie, WA. The client plans to increase its coal production up to 7 Mt per annum in the next years. Currently the operation is split into two single pits: Ewington I which is currently in production, mining steam coal for the local power plant and Ewington II which is currently ceased due to cost reasons.

Fig. 1 Arial photography on both Ewington (EW 1 and EW 2) operations in the North, Permier Coal deposit (P)in the middle and Muja operation in the South (Google Maps)

As the client plans to increase the production to 7 Mt per annum for overseas export, RE´s task is to optimize the mine production in regard to reduce production costs per tonne of coal and to extend the lifetime of the operation. The client is interested to optimise the operation with continuous mining systems. As the operation has a certain amount of very soft overburden material, the special focus of this advisory mandate lies on the application of fully continuous mining systems in these soft horizons. For the underling harder zones with multiple coal seams the application of Inpit Crushing and Conveying (IPCC) technology is further reviewed.

2 Ewington Mine: Geology and Mining Conditions

The Collie Basin coal deposit is an isolated Permian fault bounded sedimentary basin in the Archean basement rocks of Yilgarn Craton. The basin has a length of 27 km

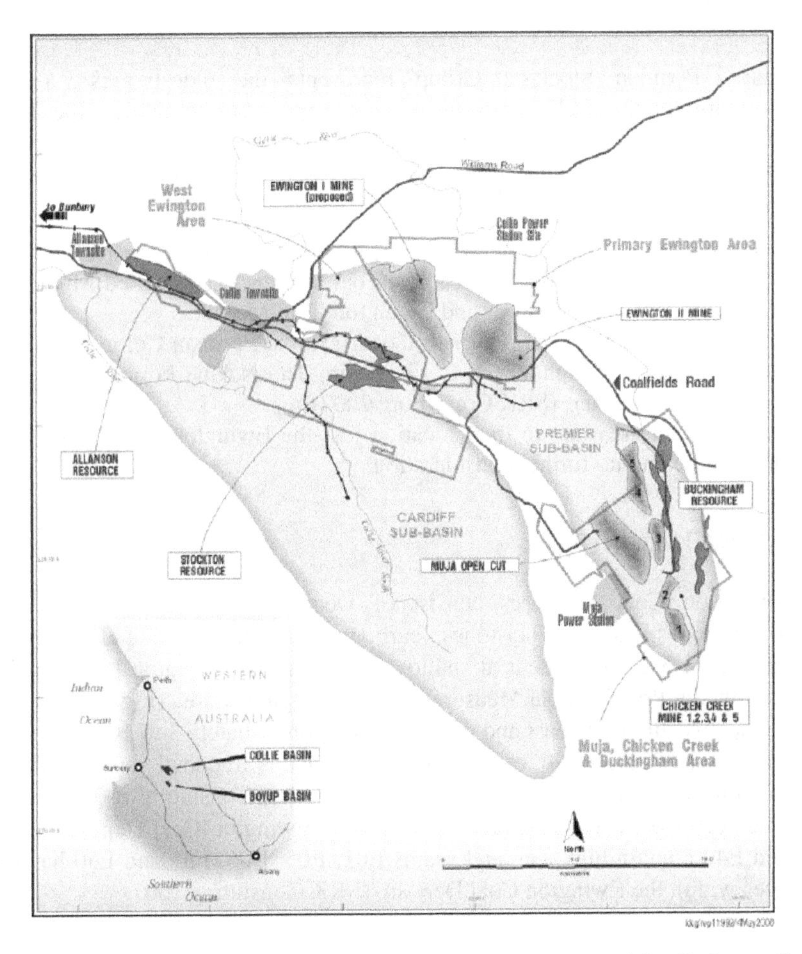

Fig. 2 Location Map of the Collie Basin showing the two areas comprising Ewington Coal Deposit (SRK Consulting, 2007)

and a width of 13 km, covering an area of approximately 224 km² (Figure 2.1). It is stretched in a NW-SE direction and has a sub-parallel orientation to the major faults of the surrounding basement rocks. (SRK Consulting 2007)

The basin is divided into two unequal lobes – the easterly lobe is known as the Premier Sub-Basin, the westerly lobe is known as the Cardiff Sub-Basin. Both are partly separated by a fault controlled basement, called Stockton Ridge. The shape and boundaries of both lobes are mainly determined by the north-west trending faults. These faults also limit the general horst and graben configuration of the Collie Basin. The coal-bearing sedimentary rock inside the basin has a maximum thickness of 1,500 m and range in age from Sakmarian to Late Permian.

The stratigraphy of the Collie Basin features three main sedimentary formations:

- Early Permian Stockton Group
- Permian Collie Group
- Nakina Formation

The Early Permian Stockton Group represents the oldest rocks, resting unconformably on the Archean basement. This group is present all over the Basin, but with a considerable variation in its thickness. The Stockton Group is created by glacial or fluvioglacial processes and consist of sandstones, pebble conglomerates, mudstone and tillites.

The Permian Collie Group is the coal-bearing formation and is overlying the Early Permian Stockton Group. The three coal sequences are called Muja Coal Measures, Premier Coal Measures and Ewington Coal Measures.

The Nakina Formation unconformably overlays the Permian Collie Group and is generally present throughout the Collie Basin. The Nakina Formation itself is covered by a laterite layer. (SRK Consulting 2007)

For the project area due to their location just the Ewington and Premier coal measures are taken into further consideration.

2.1 Coal Measures

The Ewington Coal Measures consist of weakly cemented sandstones and conglomerates, siltstones, claystones, carbonaceous mudstones and coal. The Ewington Coal Measures occur at shallow depths in the Ewington I area, and at depth below the Premier Coal Measures in the Ewington II area (**Figure 2.2**). The strata generally dip to the east and north-east at approximately 3 to 10 degrees but can steepen to 15 to 20 degrees near faults. The seams frequently split, coalesce, and sometimes display washouts (seam thinning). (SRK Consulting 2007).

Sixteen (16) seams have Resources within the Ewington Coal Deposit. Seams E26 and E40 contain high ash, and seams E01, E05, E06, E35 and E50 have no resources within the Ewington Coal Deposit. (SRK Consulting 2007)

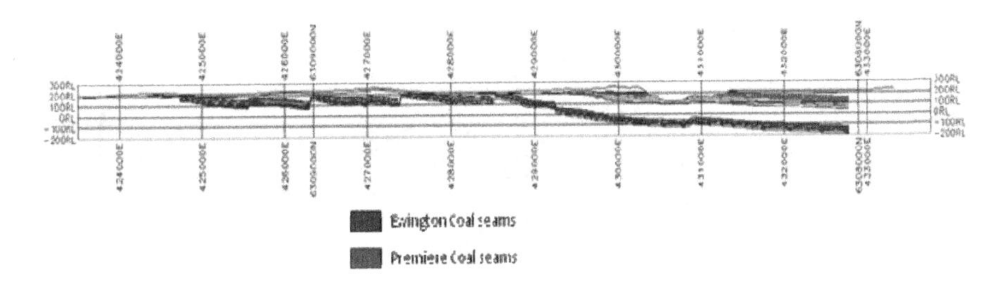

Fig. 3 Crosssection northwest – southeast through both Ewington operations, showing the both seam groups (RWE Power Int.)

The Premier Coal Measures overlie the Ewington Coal Measures, and exist only in the eastern part of the Ewington Coal Deposit area (Ewington II). The Premier Coal Measures consist of a sequence of thick and thin coal seams, with interburdens comprising sandstones, siltstones and claystones, which generally dip to the east and south-east.

3 Conceptual Mine Design in Favor of Continuous Mining Technology and in Consequence of Geology

3.1 General Strategy

The complete Ewington operation can be splitted in two different major overburden layers. The "soft" material down to the base of weathering (BOW) with an UCS strength below 4 MPa and the harder layer including the coal seams down to the E45SF (lowest viable seam of E-seam group) with sandstones up to 90 MPa.

As to be seen in Figure 3.1 the thickness of the soft material above the Base of Weathering (BOW) is shown. The coulours in the figure show that most of the weathered material is below 20m thickness. In certain areas the thickness is above 20m, shown as red in the graphic. The dumps of the existing Ewington I and Ewington II operation are clearly visible with a thickness of up to 80m above the BOW. (Heiertz 2013)

The waste dump volume of Ewington II needs to be excavated in advance of the mining face. Here it makes sense to utilize a contractor. As there is not sufficient data available about the volumes of the outside dump, an interpolation of the natural weatherd heights in the area and the dump volume has been done.

The soft layers seem to be very favourable for the application of cost effective Bucket Wheel Excavator (BWE) technology in combination with conveyor belts and spreader on the dumping side as these horizons allow free digging no blasting of the weathered material is required. A high clay content and cohesive material bring the bucket wheel excavator in favour to the discontinuous haulage with shovel and truck as they are less prone for slippery surface and ground conditions. Based on the experiences of the client the operation of the mine must be stopped for several days in the year as precipitation makes the benches too slippery and thus too dangerous for truck traffic. (Heiertz 2013)

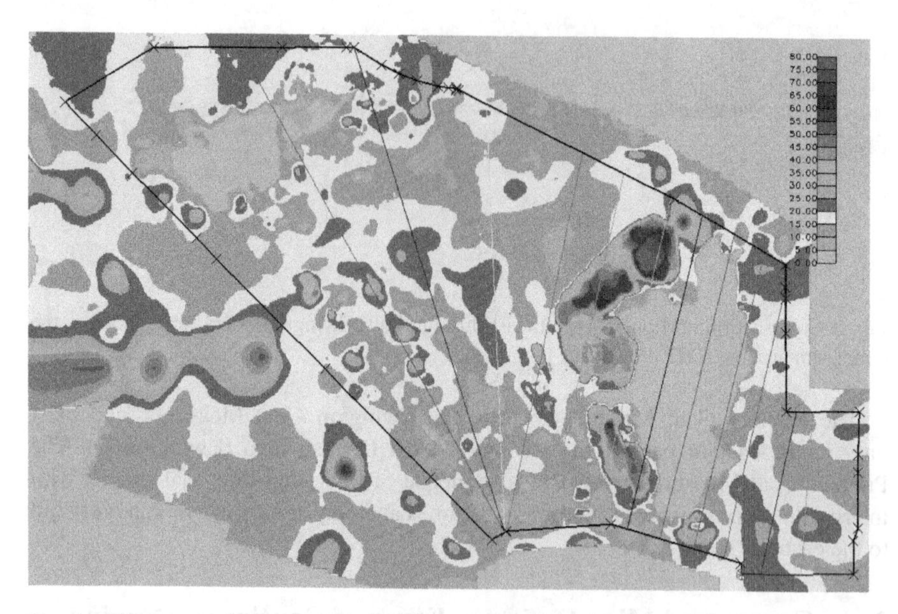

Fig. 4 Thickness of Weathered Layer above BOW in m (critical height above 20 meters marked in red)

The underlying layers below the BOW of the operation need to be blasted as most of the layers are to hard for free digging. The coal seams bigger than 1.3 m, which can´t be ripped, also needs to be blasted. As RE has bad experience with the combination of BWE and blasting operation due to the lump sizes and productivity of the BWE, which is designed for softer material, we recommend the utilisation of conventional shovels in combination with trucks in these horizons. In this case to lower the production and operational costs we recommend the combination of discontinuous loading with high effective continuous haulage and dumping. For the Ewington operation this would be Semi Mobile Crushing (SMC). Due to the multiple seam structure and faulting below the BOW a high flexibility of the mining system is required. Therefore the utilization of Fully Mobile Crushing (FMC) applications is not recommended by the consultant.

A current mining lease area defined in the Jorc Report 2010 has been given by the Client. As to be seen in Figure 3.2 the mining lease area contains the current operations Ewington I, Ewington II, the coal handling plant with rail loop and the outside dumps of the mines.

The current mine in Ewington I is operated in a long strip from NW to SE (Figure 3.3). In sense of applying continuous mining systems this long strip operation is already very favourable for the application of conveyor belts and continuous mining equipment. With a view on applying this equipment it has been discussed to operate the mine in the first years in a pivoting operation and later on in a parallel movement, a very favourable operation mode for conveyor belts and also in regard to the shape of the whole mining lease area. The pivoting point as a center of the operation will be in the south of the mining lease area (Figure 3.4), thus the deposit can be fully mined.

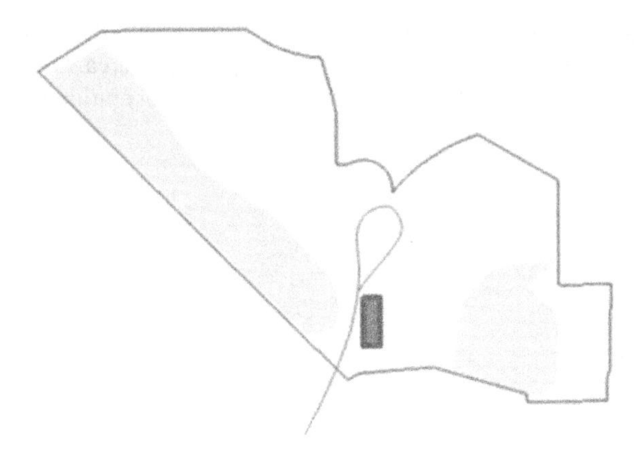

Fig. 5 Boundary defined in Jorc Report 2010 showing the existing operations EW 1 and EW2 (marked green) and the coal handling facilities with rail infrastructure (blue)

Fig. 6 View from Griffin I outside dump on the running operation. The mine is operated in a long strip NW-SE

As discussed with the client, the mining lease area has been modified in order to optimise the operation for continuous mining equipment (CME) and to maximise the coal recovery (Figure 3.4). The optimisation of the mining lease area requires additional land purchase and governmental permission.

With respect to the given mine planning parameters, the optimized boundary and the geological model, the coal excavation at the Ewington deposit has been modelled down to the E45 Seam Floor in the whole project area.

In case of following the E45SF the final depth of the operation would be about 300m below surface level as to be seen in Figure 3.5.

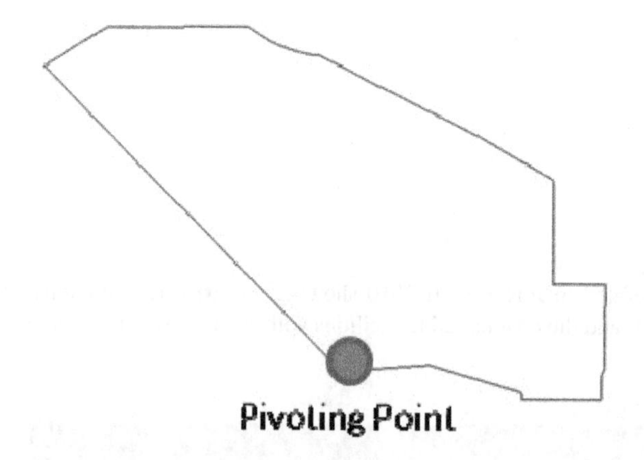

Fig. 7 Optimised Boundary by RWE to apply CME

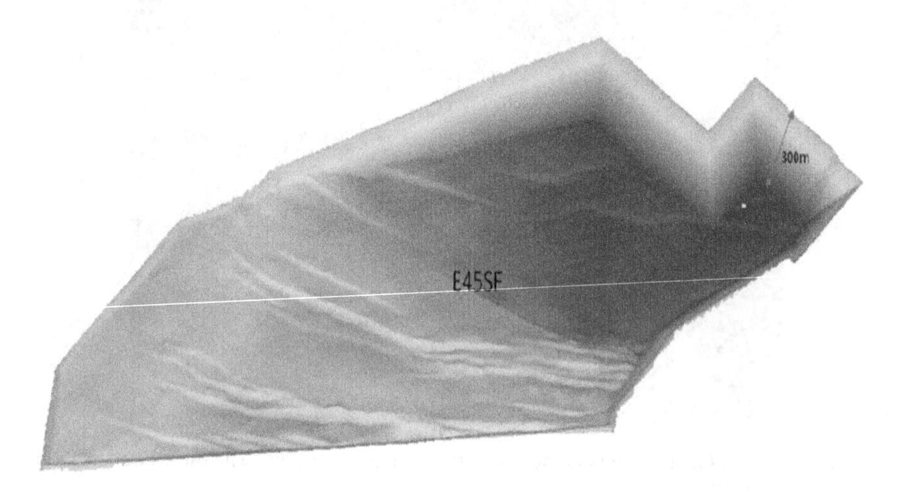

Fig. 8 Ewington ultimate pit looking east - mining down to E45SF - final depth of about 300 m below surface

In order to get a first approach and a feeling for the volumes to be excavated in the different layers the deposit has been split into 13 strips. The first strip is the existing Ewington I operation to be seen in Figure 3.6.

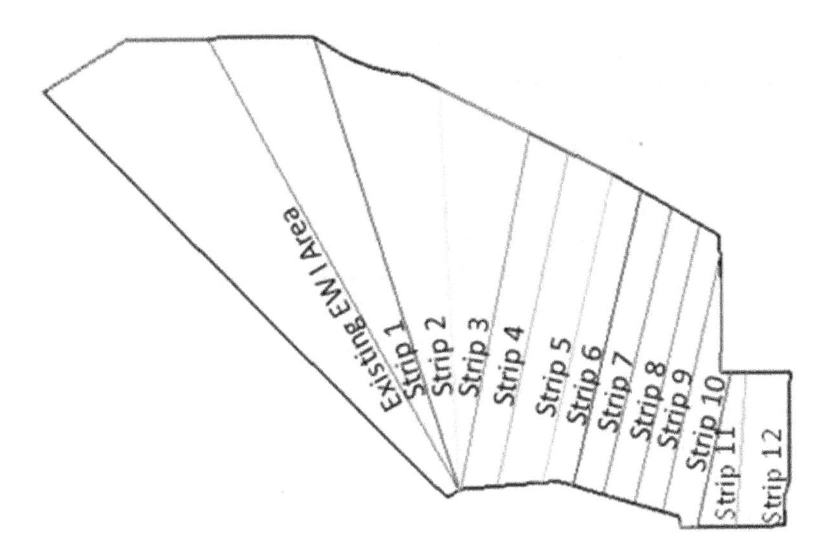

Fig. 9 Ewington deposit splitted into strips for volume scheduling in pivoting operation

The following tonnages can be mined over the deposit. As the final coal recovery is 90% the final Run of Mine will be less then the numbers shown here:

Table 1 Insitu volume and tonnages created by stripping of the Ewington deposit

Strip	Volume Weathered	Volume Hard Waste	Volume Coal	Ratio
NR	[Mbcm]	[Mbcm]	[Mt]	1:X
Strip 1	40.9	149.8	24.3	7.8
Strip 2	41.1	202.3	28.4	8.6
Strip 3	27.6	259.1	18.5	15.5
Strip 4	26.7	240.0	14.5	18.3
Strip 5	44.7	351.4	18.5	21.4
Strip 6	19.7	232.6	14.1	17.9
Strip 7	5.9	225.7	15.8	14.6
Strip 8	5.4	216.8	19.1	11.6
Strip 9	12.6	166.9	17.6	10.2
Strip 10	12.7	126.0	13.5	10.3
Strip 11	7.2	92.2	9.3	10.7
Strip 12	10.7	84.6	8.4	11.3
Total	255	2,347	202	12.9

4 Selection of Mining Method and Technology in Regard to Optimization

4.1 Mining Methods

Various mining methods can be imagined to mine the Ewington Coal Deposit. To plan and select the equipment of an open pit mine, the following important basic facts and conditions in the specific area and the specific mine must be considered:

- meteorological, geological, hydrological conditions.
- required annual capacities of overburden and coal
- material properties
- mine planning
- operation planning
- mine depth and logistical constraints
- outside dump height
- transport distances
- and other local conditions

The coal production aim is high (a 7 Mt/y). Thus the volume of overburden to be handled is also quite high due to the average stripping ratio of approx 13:1 (bcm:t).

The following two mining methods are further reviewed:

1. Bucket Wheel Excavator for weathered overburden material,
2. the IPCC method (shovel & truck – semi-mobile In-pit crushing unit – belt Conveyor – spreader) for harder waste material and coal below the BOW

4.1.1 Bucket Wheel Excavator

Free diggable waste is mined by the bucket wheel excavator (BWE), moved via a belt wagon or mobile transfer conveyor to the bench conveyor and transported to the transfer conveyor and further to the dump conveyor, from where it is picked up by a tripper car and dumped by a spreader.

Compact sized bucket wheel excavators (Figure 4.1) do not achieve the performance of conventional C-frame machines, but are more robust, can dig harder material and are more flexible in their deployment. Also capital expenditure / m³ capacity is in RE's experience below the conventional C-frame machines. For this reason matching with the projects conditions the use of compact BWE´s is considered. The annual average volume of weathered material is about 10 Mbcm which is a good diggable volume for a compact BWE.

Fig. 10 Sandvik–manufactured Compact BWE (6,700 Lm³/h) at RWE coal mining operation in Hungary (RWE Power Int.)

Advantages

- No need for diesel-burning equipment (except for auxiliary equipment, usually one dozer)
- Using belt conveyors, higher gradients can be covered compared to truck transport.
- Extensive haul road maintenance is avoided.
- Reduced number of trucks in the mine.
- Dust emission control requires less effort.
- Nearly no bad weather downtimes.
- High noise reduction compared to S&T operation.

Disadvantages

- The initial investment is slightly higher if compared to a pure shovel & truck operation.
- The use of bucket wheel excavators requires a more detailed long-term planning and equipment specific training of mine personnel.
- More difficult to adapt to changing mining conditions.

4.1.2 SMCC Method (Shovel & Truck – Semi-Mobile In-pit Crusher – Belt Conveyors – Spreaders)

All the waste and coal is excavated by shovels and transported by trucks to semi-mobile in-pit crusher located near the benches at the low wall side, close to the pivot point of the mine. The location of the crusher close to the pivot point has the advantage of reducing the shifting intervals to a minimum amount. Via belt conveyors the waste material is transported to a spreader or the coal to a stockyard. Semi-mobile crushers as lump reducer and belt conveyors are an option to reduce the number of trucks in operations based on pure shovel & truck technology when long transport distances and/or significant lifting heights must be covered.

Advantages

- Lower diesel fuel demand compared to a pure shovel & truck operation can result in lower operation costs.
- Using belt conveyors, higher gradients can be covered compared to truck transport.
- Extensive haul road maintenance can be limited to bench transport.
- Reduced number of trucks in the mine,
- Dust emission control requires less effort,
- Fewer operators (truck drivers) compared to pure S&T operation.
- Trucks don't have to climb steep slopes and the operation is less effected to bad weather downtimes

Disadvantages

- The initial investment is higher compared to a pure shovel & truck operation.
- Limited flexibility with respect to the operation of the dumps.

Fig. 11 Semi-mobile Crushing Plant (4,500 tph) (Thyssen Krupp Fördertechnik)

A more flexible alternative to the conventional semi-mobile crushers could be the Dual Truck Mobile Sizer (DTMS) developed by FLSmith.

Fig. 12 Dual Truck Mobile Sizer (FLSMITH)

The advantage of a DMTS compared to semi-mobile crushers is the lack of need for a loading pocket, the low loading height (can be located on same level as conveyors) and higher mobility (can be built on crawlers or moved by transport crawler) to be located close to the loading points. Despite not being 'proven technology', due to these advantages DTMS should be considered at least for coal loading. However, this paper is based on conventional semi-mobile crushers.

5 Two Alternatives Pure Shovel and Truck and IPCC with BWE

5.1 Operational Data

The estimated lifetime of the operation with an annual coal production of 7 Mt/a will be 26 years. To mine this amount of coal an average annual required waste movement of 100 Mbcm/a will be required. This results in an average stripping ratio of 1:13 incorporating a coal recovery of 90 %. The total overburden volume for the whole mining operation results in a volume of 2.6 Bbcm. Due to this large scale mining operation for further planning and comparison economies of scale are considered where applicable.

For a better comparison of the different mining variants and further calculations a pivoting operation in the beginning and a parallel movement of the mine in the later years will be assumed (Figure 3.5). In regard to meet the advantages of S&T mining truck bridges are incorporated in the mine planning to minimize the haulage costs of the equipment. The mining operation requires from the machines a high selectability of different interburden materials due to the multiple coal seams.

A shift regime of two working shifts per day of maximum twelve hours (hot seat change) each must be applied due to the Australian regulations. For having 2 crews on duty there will be always one crew off on holiday. Additionally taking unforeseen illness of employees and travelling into consideration for the further calculations a shift factor of 4 will be applied.

The mine is operated in two major horizons. This means on the first bench in majority soft laterites and clay due to the weathering of the material. The underlying benches are a variation of coal seams, stiff clays and medium hard sandstones with a maximum strength of 90 MPa. So blasting will be required below the base of weathering to allow high production rates. Due to the high amount of cohesive waste material and in some periods heavy rainfalls a very good and personnel intensive haul road preparation and maintenance for the application of heavy mine trucks will be required to avoid that major areas of the mine will come to a full stop.

As described above, the different overburden layers need to bc excavated separately. The coal must be cleaned from overburden or interburden material. Therefore sometimes excavation tools must be used that can excavate from a face at any height dictated by the material boundaries. For this, hydraulic excavators have a clear advantage over rope shovels.

Maintenance is one of activities that require a large number of personnel. Wear due to the abrasiveness of the ground materials is high. For example maintenance cost of the crawler systems are linearly related to the distance travelled. Even short advances during working at the face add up to reach significant travel amount. Therefore the proposed layout was designed such that the moving distances for loading equipment are minimised.

5.2 *Operation of the Mine by Pure Shovel and Truck (S&T)*

Assuming the Ewington mine as one single operation is run as a S&T mine due to the size of the operation and the high overburden volumes the company would require economies of scale to run the mine most effectively.

But as the overburden, especially the weathered material is very soft and also the underlying benches are due to their clay content prone for slippery surfaces, the largest machines on the market are not the optimum solution here. They would be far to heavy for the local ground conditions. Therefore for the further calculations the medium sized equipment as rear dump trucks with a payload of 200 tonnes and excavators in a range of 29 m³ buckets will be applied used to run this operation in S&T mode.

The total depth of the mine will be up to 300 m below surface with an overall stripping ratio up to 1:13. To meet this high overburden stripping requirement the benches will be maximised in area and depth. Also when the coal seam is dipping down to south east the overburden benches will be kept near horizontal to maximise output and to generate optimum working conditions for the big shovels. The horizontal layout of the benches also has the advantage of better haulage conditions due to a nearly complete reduction of lifts in the haulage ways. This will effectively reduce the energy costs and the number of trucks on the benches. The material will be hauled via the shortest route across the benches of the low- and high-wall side to the dumping bench on the same level. Where applicable for the lower benches, truck bridges will used for very short haulage to the dump side (Figure 5.1).

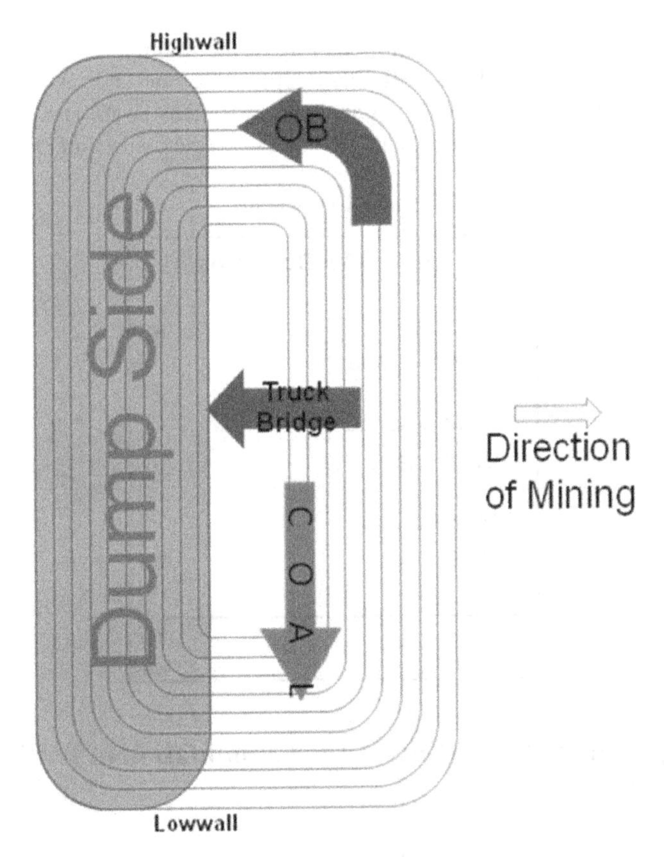

Fig. 13 Overall mine development strategy along strike

Depending on the height of the coal seams they will be mined by backhoe excavators to allow a good cleaning of the seams and to allow a high selectability of the material. Two backhoe configuration excavators with 10m³ buckets are planned. For production purpose they are matched with 77t dump trucks to haul the coal to the stockpile. The single haulage distance would be about 5 Km.

5.3 Personnel Structure Pure S&T

The mine developed fully by S&T will require 84 Trucks as maximum amount per year for the complete operation. 20 digging units in a range of 29m³ and 10m³ will be used. The production numbers here are based on an annual effective operating time of 5600 hours. A large number of trucks are required to transport the coal out of the pit to the stockpile at the power plant, a haulage distances of up to 5km. Also, when operating trucks in soft rock conditions high road maintenance standards are essential. Therefore a significant amount of auxiliary equipment, especially grader and articulated trucks are required to prepare and maintain the mine roads and infrastructure. Total manning of the mining services is 245. The mechanical maintenance fleet will require personnel of 260. As most of the equipment is Diesel driven the personnel of electricians is about 87. The factor "others" incorporates the white collar employees and the management.

Table 2 -1 Number of employees in pure S&T operation

Pure S&T Operation	
Mining-production	868
Mining-service	245
Electricians	87
Mechanics	260
Others	64
Sum	1524

6 Operation of the Mine by the Combination of BWE and IPCC

For the IPCC scenario a combination of fully continuous BWE for the weathered material and a very flexible shovel and truck operation with cost effective crushing (SMC) and conveying by conveyor belt (IPCC) is considered. As the annual effective operating of continuous mining systems is less than a pure shovel and operation an calculation based on the proposed system configuration has been done and for the further review an annual effective operating time of 4900 hours is assumed.

BWE on Bench 1, Weathered Material

This means in general an application of compact type BWE attached to a conveyor system to transport the material to a spreader on the dumping side of the mine. The conveyor will be routed on the low wall side of the operations as a belt distribution point for material distribution will be required.

SMC on Overburden Benches below Weathered Bench

To reduce truck numbers and therefore the personnel requirements the continuous conveying of waste around the pit to the dumping side is a good way for optimizing the operational cost of the operation.

As described above the different interburden layers need to be excavated separately. The lignite must be cleaned from overburden or interburden material. Therefore excavation tools must be flexible enough to operate efficiently at variable bench heights. For this hydraulic excavators have a clear advantage over rope shovels. Also in regard to flexible material movement the truck is the favoured choice here.

SMC or FMC on Coal Benches

The underlying coal benches are planned to be mined by two $10m^3$ hydraulic shovels due to the need for blending and material distribution. Trucks will be hauling the coal in a discontinuous operation to two semi mobile crusher systems which are connected to a coal conveyor. The two coal conveyor are continuously hauling the produced coal out of the mine.

From year 8 on the waste volumes of the operation are continuously increasing up to 110 Mbcm. To mine this material additional crushing and conveying capacity is required. Five waste crushers are serving five spreaders in the same range (Figure 6.1). The system combination above the BOW stays the same. The coal crushing systems need to be adapted to the height differences and the climb on the low wall.

Direct Dumping on the Lower Benches by Truck Bridge

Due to the short haulage distance to the dump side the trucks are directly moving the waste material by truck bridges to the dump side. This can be just done on the lowest benches.

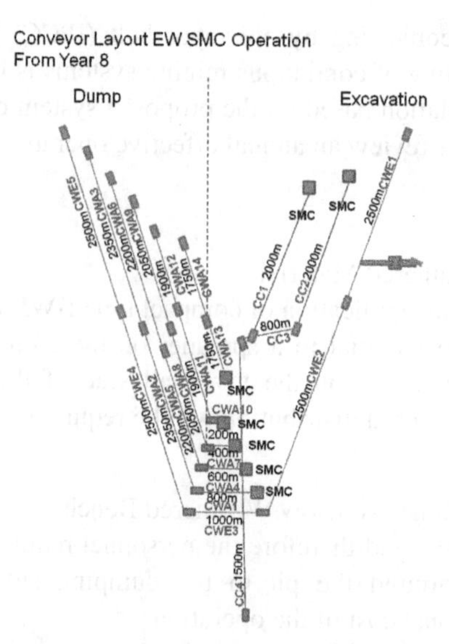

Fig. 14 Schematic view of conveyor layout from year 8 until the end of the operation in Ewington

6.1 Personnel Structure by the Combination of FMC and S&T

Due to the implementation of IPCC Systems the number of trucks can be drastically reduced. Especially for the long coal haulage distance the conveyor belts replace nearly all the trucks on the way out of the mine. The weathered bench normally operated with ultra-class trucks in a range of up to 200t payload will now run truck-less as just one BWE is operated here. The utilization of more continuous transport systems has also the advantage of a reduction in auxiliary equipment. Especially by avoiding the operation of heavy trucks in very soft material conditions the number of road building equipment is much less.

Overall the change in the haulage method for the main benches will cause a reduction of down to 1177 employees (Table 6-1) in contrast to the pure S&T method (Table 5-1).

Table 3 Number of employees in FMC / S&T operation

BWE Operation	
Mining-production	452
Mining-service	240
Electricians	136
Mechanics	316
Others	33
Sum	1177
100%	

7 Conclusion and Comparison of the Two Variants

OPEX drivers as high personnel costs and diesel fuel cost together with a deteriorating coal price make Australian coal mines less competitive in the international market. Australian mining companies are looking for alternatives to optimize the mining process.

This paper compares the effectiveness of S&T with a continuous mining operation in two different mining variants for an Australian mine (up to 300m deep) with a total average output of 106 Mbcm. The mining operation with BWE´s and IPCC with supplementary truck transport is compared with a full S&T operation. The comparison between these two variants demonstrates the reduction of manning requirements with the implementation of continuous conveying, IPCC, systems.

The calculations show that continuous mining systems in long-term mining operations have advantages in sense of reduction in staffing in material movement. Especially in normal S&T operations the haulage of the material is personnel and energy intensive. For large and deep operations with high stripping ratios the differences increase with increasing stripping ratios and depth. In the direct comparison between pure S&T and IPCC in the same mine, by the production with minor truck transport the number of employees is reduced. A reduction in truck transport is also followed by a reduction of ancillary equipment, hence a reduction of employees.

The following Table 7-1 finally shows the comparison in employee numbers for the same deposit mined by two different variants:

Table 4 Comparison in employee numbers for the same deposit mined by different variants

Number of Employees			
BWE-IPCC Operation		**Pure S&T Operation**	
Mining-production	452	Mining-production	868
Mining-service	240	Mining-service	245
Electricians	136	Electricians	87
Mechanics	316	Mechanics	260
Others	33	Others	64
Sum	1177	Sum	1524
100%		129%	

References

SRK Consulting. JORC Resources Report – Ewington Coal Deposit. Brisbane (July 2007)

Heiertz, A.-J.: IPCC Scoping Study Ewington Coal Mine. Cologne (January 2013)

A Prototype Dynamics Model for Finding an Optimum Truck and Shovel of a New Surface Lignite Mining in Thailand

Phongpat Sontamino[1] and Carsten Drebenstedt[2]

[1] Faculty of Engineering, Prince of Songkla University, Thailand
phongpat.s@psu.ac.th
[2] Institut für Bergbau und Spezialtiefbau, TU Bergakademie Freiberg, Germany
drebenst@mabb.tu-freiberg.de

Abstract. At present, a surface lignite mining has been operating in the north of Thailand, but there are other reserves of lignite in the south of Thailand that not yet operating. All of lignite production in Thailand used to produce electricity and the demand of electric is increasing every year. So, in the future, the lignite reserve in the south of Thailand may need to operate in serving electricity demand of Thailand.

One of the main investment cost (capital cost) when the mining company wants to operate a mine is equipments cost, and the main equipment cost is the transportation system cost. While, in Thailand, truck and shovel were used in the transportation system of the lignite mining because of the flexibility of the movement and selecting the zone of lignite and waste easily, especially in the geological condition of the steeply bed of ore. So, in the new mining operation, the question of the number of truck and shovel and which capacity needs to buy for achieving the target of production planning, can be found by using this prototype model.

A prototype system dynamics model (SDM) for finding an optimal truck and shovel of a new surface lignite mining in Thailand is a prototype dynamic decision making tool by using the system dynamics theory and computer software modeling (Vensim DSS software). The SDM was decided an optimum condition of the truck and shovel using an optimum cycle time theory and it's also a fast and flexible tool to change the mining production rate and other variables. Thus, the optimum cost of buying trucks and shovels which shown the number and capacity of truck and shovel were found with this dynamic decision making tool.

Keywords: System Dynamics Model, Optimum Shovel and Truck Selection, Decision Making Tool

1 Introduction

Coal is an energy resource from hydrocarbon which is currently a major source of electricity and energy in industries. From past to present, there are many coal

C. Niemann-Delius. (ed.), *Proc. of the 12th Int. Symp Continuous Surface Mining - Aachen 2014*,
Lecture Notes in Production Engineering, DOI: 10.1007/978-3-319-12301-1_42

reserves around the world and the large coal mining are operating in diverse countries such as USA, Russia, China, Australia, Germany, India, and Indonesia, etc. Thailand coal resources found in many areas, and the most of coal in Thailand are lignite. A lignite mining in the north of Thailand has been operating, that is the Mae Moh Lignite Mining. Furthermore, it has some reserves in the south of Thailand (Krabi and Songkhla) which have not been operating yet.

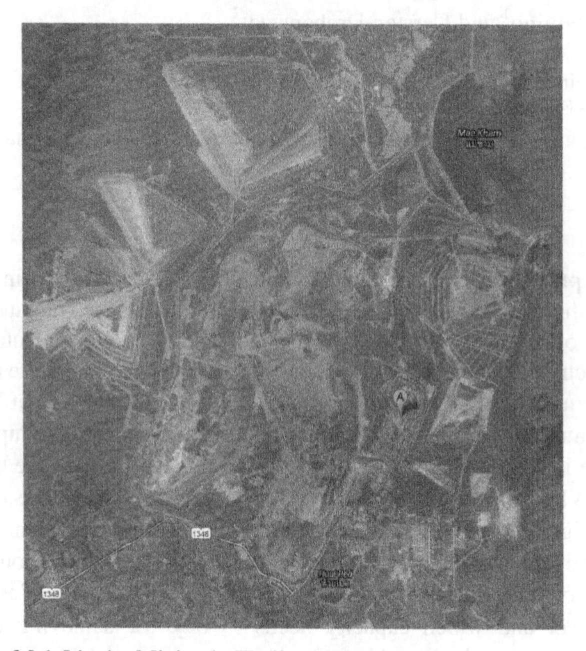

Fig. 1 The Mae Moh Lignite Mining in Thailand [1]

1.1 Objective

To develop a prototype system dynamics model for finding an optimal truck and shovel of a new surface lignite mining in Thailand.

2 Status of Coal in the World and Coal Mining in Thailand

Coal is one of the World's most plentiful energy resources, it was estimated about 861 billion tones reserves around the world (2008, 2012) [2-3] which cab be an average used in 109 years, compared with oil and gas are 52.9 years and 55.7 years respectively [3]. It is mainly source use to generate electricity. Thus, it is today and will be in the future the most important global source of electricity, and also many industries use for making the merchandise. Nowadays, there still have many coal resources around the world and large coal mining is also operating in several countries.

Fig. 2 Top 10 world coal reserves (2008) [2]

The energy consumption of the world is increasing each year. At 2012, the second of the world energy source is coal, that consumed about 3,730 million tonnes of oil equivalent, compared with oil and gas that consumed 4,130 and 3,314 million tonnes of oil equivalent respectively [3].

Coal resources in Thailand found disrupted, from the north to the south of Thailand, most of them are lignite and mainly in small scale resources. However, some coal resources are more than 100 million tons, and economically to be operated such as, (1) Mae Moh reserves, (2) Krabi reserves, and (3) Songkhla reserves [4]. Estimated at the end of 2012, Thailand has proved reserves about 1,239 million tonnes which can be used for 68 years [3], and the Mae Moh Lignite Mining is the biggest coal reserves and still operating. It has economical reserves around 825 million tons, which used 364 million tons and remain 461 million tons (2011) [5]. All of the production of lignite in a Mae Moh Lignite Mining used for produce electricity, which is produced around 15-17 million tons/year. It is used for 13 units of electric generators that have the total capacity around 2,625 MW, which supported around 20% of the electricity demand of Thailand [6]. Moreover, The Ministry of Energy of Thailand has plans to build a coal power plant in Krabi province, which capacity 800 MW, and plan to start the coal power plant at 2019. So, in this research the Krabi coal reserves estimated 111 million tons [4] is an alternatively sources to serve the Krabi coal power plant.

3 Concept of Truck and Shovel Selection

The cycle time of trucks and shovels is one of the popular theories to optimize trucks and shovels. Truck cycle time can separate into 4 parts; (1) a loading time, (2) a hauling time, (3) a dumping time, and (4) a return time, and also shovel cycle time can separate into 4 parts; (1) a digging time, (2) a swing time to truck, (3) a dumping time, and (4) a swing back to dig. But, in actual work, it might have a waiting time in the cycle time of both equipments.

However, in the ideal of optimum condition of trucks and shovels, it should have a continuous working without a waiting time. Moreover, the optimum trucks and shovels in term of size and amount also related to coal production rate, working time, truck driving speed, etc.

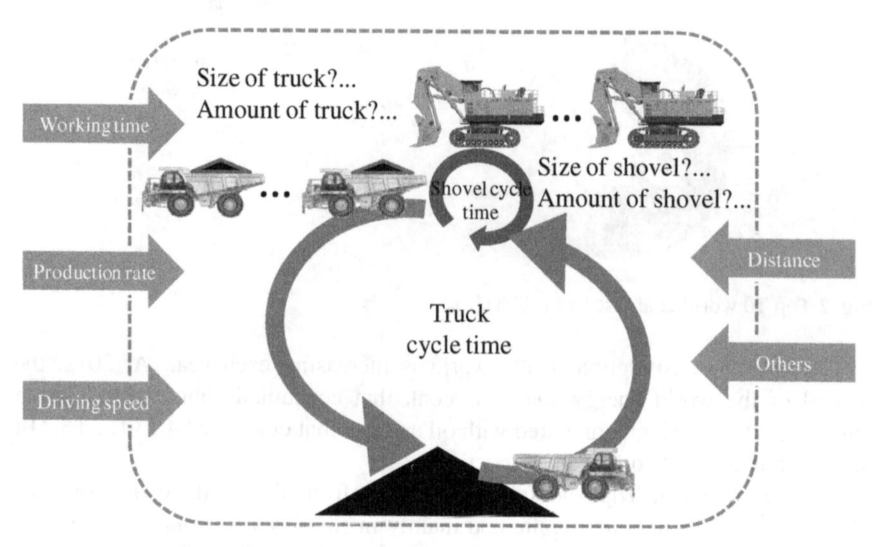

Fig. 3 Influence of trucks and shovels selection

4 Research Methodology

4.1 Background Theory and the Research Working Process

System dynamics theory [7] was used to support the logic and thinking of this research, and the system dynamics model (SDM) was developed by using Vensim DSS Software [8], it helps to connect each variable and calculating the results, which cover the theory of cycle time of truck and shovel selection. The simple research working process is shown in Fig. 4 below.

In this research, the case study simulation result of the prototype SDM focuses only optimum trucks and shovels for the coal production rate. Mae Moh Lignite Mining produced 16 million tons/year [5], and Krabi reserves estimated the production rate around 5.5 million tons/year, which excluded the activity of the topsoil and overburden transportation.

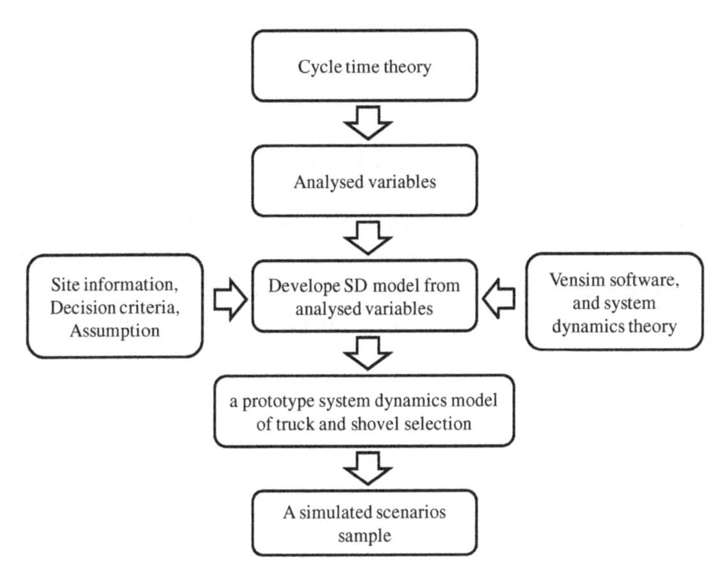

Fig. 4 Simple working process of the SDM development

4.2 Optimization Criteria

The optimization criteria of this research, focusing on the minimum cost of buying the truck and shovel. The list of price data supported by Mae Moh Lignite Mining show in the table 1 and 2 below. This price list also applies to the new coal mining that may operate in Krabi.

Table 1 Price list of shovel and loader [9]

No.	Capacity (m^3)	Price (M.Baht)	Price* (M.Euro)
1	9.2	19	0.475
2	5.4	7	0.175
3	2.5	3	0.075
4	1.4	1.5	0.038

* calculated at currency 40 Baht/Euro

Table 2 Price list of off-highway truck [9]

No.	Capacity (tons)	Price (M.Baht)	Price* (M.Euro)
1	90	27	0.675
2	85	16	0.400
3	77	12	0.300
4	40	8	0.200

* calculated at currency 40 Baht/Euro

5 Results

The succeed of development a prototype SDM of truck and shovel selection shows the usefulness of the model to answer the question of finding the optimum truck and shovel for a new coal mining in Krabi reserves, Thailand. It is also flexible to change the value of the input variable and condition for other mining. The result of this research can separate into 4 parts:

5.1 Model Variables Analysis Result

The variables of the model are separated into 2 types, (1) input variables which have 22 variables such as, initial coal reserves, coal production rate, distance between mining site and dumping site, and truck driving speed, etc., and (2) output variables which have 30 variables such as, mining stocks, total cost of buying trucks and shovels, optimum shovel and optimum truck, etc. The connected of each variable can see in the model structure in Fig. 5 below.

5.2 Model Structure and User Interface Result

The prototype system dynamics model structure shows the connection of all variables and behind of each variable is a value or equation, that means if one or many variables change, it affects to others automatically.

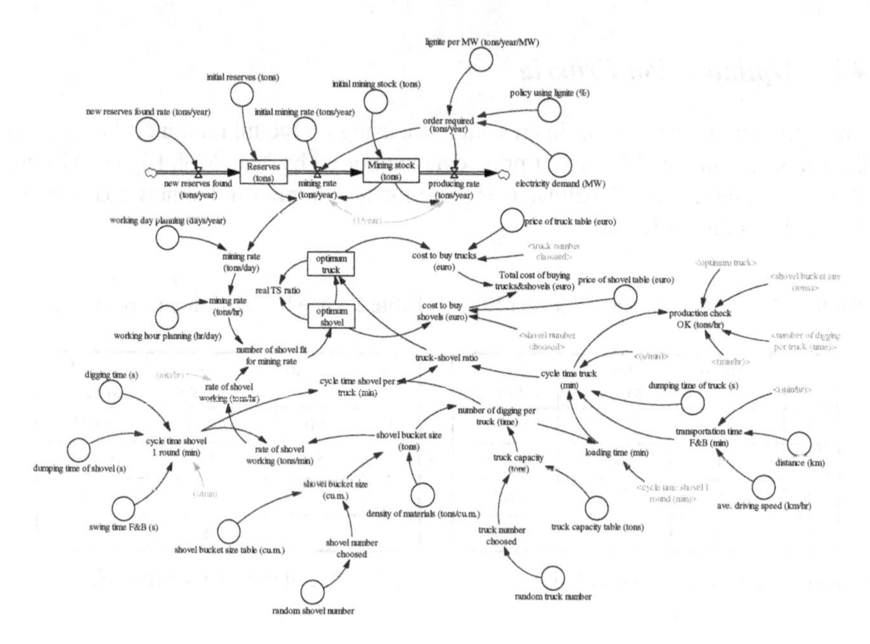

Fig. 5 The prototype SDM structure of truck and shovel selection

Because of the model structure is hardly used for a general user, so the user interface was made. The simple model user interface is shown in Fig. 6 below.

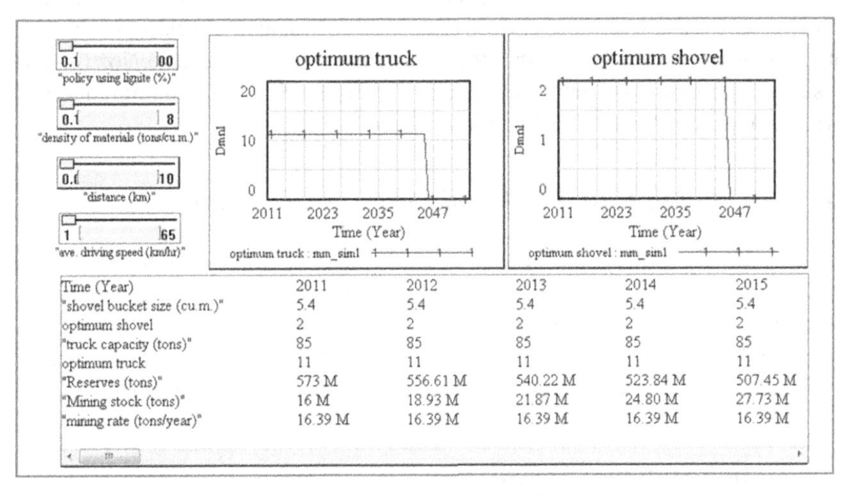

Time (Year)	2011	2012	2013	2014	2015
"shovel bucket size (cu.m.)"	5.4	5.4	5.4	5.4	5.4
optimum shovel	2	2	2	2	2
"truck capacity (tons)"	85	85	85	85	85
optimum truck	11	11	11	11	11
"Reserves (tons)"	573 M	556.61 M	540.22 M	523.84 M	507.45 M
"Mining stock (tons)"	16 M	18.93 M	21.87 M	24.80 M	27.73 M
"mining rate (tons/year)"	16.39 M	16.39 M	16.39 M	16.39 M	16.39 M

Fig. 6 The example of a simple user interface of the model

5.3 Simulation Result of Mae Moh Lignite Mining Condition

The minimum cost of buying trucks and shovels in Mae Moh Lignite Mining condition is shovel type 2nd (bucket size 5.4 m^3), truck type 3rd (capacity 77 tons). Numbers of shovel are 3 shovels, the numbers of trucks are 8 trucks, and the cost of buying trucks and shovels at optimum point is around 2.92 million euro.

Table 3 Manual simulation truck and shovel buying cost of Mae Moh Lignite Mining

No.	Scenarios*	Costs (euro)	No.	Scenarios*	Costs (euro)
1	: mm_s4_t4	3,742,000	9	: mm_s2_t4	3,325,000
2	: mm_s4_t3	4,242,000	10	: mm_s2_t3	2,925,000
3	: mm_s4_t2	5,142,000	11	: mm_s2_t2	3,725,000
4	: mm_s4_t1	8,442,000	12	: mm_s2_t1	5,925,000
5	: mm_s3_t4	2,975,000	13	: mm_s1_t4	4,350,000
6	: mm_s3_t3	3,075,000	14	: mm_s1_t3	3,650,000
7	: mm_s3_t2	3,575,000	15	: mm_s1_t2	4,150,000
8	: mm_s3_t1	5,775,000	16	: mm_s1_t1	6,350,000

* The meaning of scenario label, for example mm_s2_t3 is the mm= Mae Moh Lignite Mining case, s2= shovel type 2nd, and t3=truck type 3rd which can check the detail of each type in table 1 and 2.

The optimum point result of a minimum cost of buying trucks and shovels supported by the optimization function in Vensim DSS Software. So, the optimum point was re-checked by manually changed the input variable, that make 16 scenarios to choose truck and shovel. The result of manual checking is shown in Table 3 below. It is confirmed that the optimum point is in scenario No. 10, gives the same as using an optimum function in Vensim DSS, that the minimum cost of buying trucks and shovels is 2.925 million euro.

5.4 Simulation Result of Krabi Reserves Condition

In the Krabi reserves condition, it was used only the Vensim DSS optimization function. So, the minimum cost of buying trucks and shovels in the condition of Krabi reserves is shovel type 3rd (bucket size 2.5 m^3), truck type 3rd (capacity 77 tons). Numbers of shovels are 2 shovels, numbers of trucks are 4 trucks, and cost of buying trucks and shovels at optimum point is around 1.35 million euro.

6 Conclusion and Discussion

The prototype dynamic model of optimum truck and shovel selection can help to decide the suitable number of trucks and shovels automatically and rapidly. It's an alternative tool which fast and flexible to change multiple value of input variables for creating many alternative scenarios. We can use Vensim DSS optimization function to find the optimum point for supporting decision making. In the case study condition, the objective of optimum point is to search the minimum cost of buying trucks and shovels, and the prototype SDM can show the calculation result in a different optimum point in the different condition.

In case of Mae Moh Lignite Mining, the optimum number of truck is 8 trucks with capacity 77 tons, and the optimum number of shovel is 3 shovels which have a bucket size 5.4 m^3 and cost of buying trucks and shovels at optimum point is around 2.92 million euro. Furthermore, in case of Krabi lignite reserves, the optimum number of trucks is 4 trucks with capacity 77 tons, and the optimum number of shovel is 2 shovels which have a bucket size 2.5 m^3 and cost of buying trucks and shovels at optimum point is around 1.35 million euro.

For the further development, the prototype SDM can add more criteria in term of cost such as maintenance cost, operating cost, and mine closure cost, etc., and summarize in a total cost to be a better decision criteria. The prototype SDM also should verify with more case study for example, in Songkhla reserves, and include more detail of variables in the future.

Acknowledgement. I acknowledge for the primary data and other information supported by the Mae Moh Lignite Mining and specially Mr. Ampon Kittichotkul, the senior mining engineer of Mae Moh Lignite Mining.

References

[1] Sontamino, P., Drebenstedt, C.: A Development of System Dynamics Model of Optimum Shovel and Truck Selection: Power Point Presentation, ed: TU Bergakademie Freiberg, p. 29 (2013)

[2] Heinberg, R., Fridley, D.: The End of Cheap Coal (2011), http://www.postcarbon.org/article/406162-the-end-of-cheap-coal

[3] BP, "BP Statistical Review of World Energy" (2013)

[4] Chumrum, P., et al.: Coal Mining Technology: Electricity Generating Authority of Thailand: EGAT (1995)

[5] Kittichotkul, A.: "Mae Moh Lignite Mining 2011: PowerPoint Presentation," Electricity Generating Authority of Thailand (EGAT) (2011)

[6] EGAT, About us (2013), http://maemohmine.egat.co.th/aboutus/index.html

[7] Wikipedia.org, System dynamics (February 15, 2011), http://en.wikipedia.org/wiki/System_dynamics

[8] Ventana System Inc. Vensim Software. (May 28, 2011), http://www.vensim.com/software.html

[9] Kittichotkul, A.: "Mae Moh Machine List 2011: Excel,"Electricity Generating Authority of Thailand (EGAT) (2011)

Research on Energy Consumption in Open Pits of the German Quarry Industry

Thorsten Skrypzak, Alexander Hennig, and Christian Niemann-Delius

RWTH Aachen University, Aachen, Germany

Abstract. Quarrying of natural stone is an energy-intensive process. The utilization of this type of raw materials will remain essential in the future. Thus, a systematic improvement of the energy efficiency in this mining branch will gain sustainable benefit.

A research project at the Department for Surface Mining and Drilling at RWTH Aachen University focuses on the development of a systematic approach for the identification and utilization of possible energy savings in quarrying of natural stone. This research project is promoted by the German Federal Environmental Foundation (Deutsche Bundesstiftung Umwelt, DBU). Within this project, energy consumption data was collected in several quarries. Also key aspects and quantity of energy data collection were evaluated regarding their benefit for the companies.

1 Introduction – Relevance of the Topic for the Quarry Industry

In the Federal Republic of Germany 500 million tons of industrial rocks and minerals, mainly sand and gravel, limestone and other hard rocks, are annually excavated in open pit mines. [MIRO 2013] The production and processing techniques include various sub-processes, which differ in type and number depending on the characteristics of the raw material and the mineral deposit. In general, the extraction of raw materials is an energy-intensive production process. This applies especially to the quarrying of natural stone, which usually requires multiple crushing stages in addition to the loosening, loading and transport of the mineral. The energy intensity of mining can be illustrated in relation to other fields of industry. With an energy consumption having a share of 9 % of the gross production value, the industrial rocks and minerals industry is one of the most energy-intensive manufacturing industries in Germany. [STATBUNDESAMT 2011]

The importance of industrial rocks and minerals, which are mainly used in the construction industry, will remain essential in the future. Thus, a systematic increase of the energy efficiency in this mining branch will gain sustainable benefit. In addition, the economic advantages, which energy savings generate for the companies, have to be considered. The topicality and importance of these aspects increased in recent years. This is not only due to energy prices, which were more or less continuously rising in the recent past. Also current political trends and

C. Niemann-Delius. (ed.), *Proc. of the 12th Int. Symp Continuous Surface Mining - Aachen 2014*,
Lecture Notes in Production Engineering, DOI: 10.1007/978-3-319-12301-1_43

resulting legal regulations reflect the importance of the topic of energy efficiency for the German industrial rocks and minerals industry.

Based on the energy concept the Federal Government passed in September 2010, measures were initialized by the legislator to encourage the utilization of energy efficiency potentials. For example, this concerns tax benefits and especially the energy tax cap for companies of the manufacturing industry. From 2013 onwards, only companies who are actively contributing to energy savings can apply for this tax privilege. The contribution can be evidenced by the implementation of a certified energy management system referring to DIN 50001 or alternatively by an equivalent system of different kind. [BMWI 2010]

Due to this legal regulation, which was enacted in November 2012, many companies of the quarry industry have started with the implementation of an energy management system. But the systematic collection of energy data can be way more than an evidence to be eligible for the energy tax cap. Though it is linked with financial and organizational effort, it is the basic requirement for the identification and elimination of possible weaknesses in energy use. Irrespective of the type of energy management system, the methodical collection of energy data can be an option to achieve economic advantages. This applies especially to an energy-intensive field of industry like the quarrying of stone.

2 Mining Methods and Equipment Used in Quarries

A research project at the Department for Surface Mining and Drilling at RWTH Aachen University focuses on the development of a systematic approach for the identification and utilization of possible energy savings in quarrying of natural stone. Of particular interest are optimization potentials in the fields of the equipment available on site, equipment combinations and the machinery operation in the quarrying of stone.

Nine quarries were integrated in the project. In these open pits, which produce crushed rock, discontinuous mining methods and discontinuous machinery are commonly operated. The equipment is mainly diesel-electrically powered. This applies to the loosening and loading process as well as to the haulage operation. In the participating quarries the loosening of the rock material is carried out exclusively by drilling and blasting. This represents the most common loosening technique in hard rock mining. Diesel-powered drilling equipment is used to create the blast holes. As an alternative hydraulic excavators are operated in some quarries to loosen the rock material by ripping, if this is possible due to the rock properties and the characteristics of the mineral deposit. Other methods as for example cutting techniques are almost exclusively used for the production of dimension stones.

Hydraulic excavators, more precise front shovels and crawler-based backhoes are the main loading equipment in quarries, though the use of wheel loaders has increased in the recent past. Hydraulic shovels are able to generate higher digging forces due to their way of construction. Hence they are more suitable to handle blasted rock, which is difficult to load. [EYMER *et al.* 2006a] In addition they can be used for the fragmentation of boulders. Wheel loaders provide a higher level of mobility. That allows a faster relocation between different extraction points. Furthermore, wheel loaders can be operated as a combination of loading and haulage equipment, if the transport distance does not exceed approximately 250 m. [EYMER *et al.* 2006b] This so called load-ad-carry method used in one of the quarries that participate in the project. In this case, a wheel loader hauls the blasted rock to a semi-mobile crusher. After primary crushing the material is transferred to a belt conveyor and transported to the processing plant.

The use of belt conveyors in hard rock mines of the German quarry industry rather represents an exception so far. The haulage process from the production site to the processing plant is typically carried out by diesel-powered vehicles. Mainly rigid frame dump trucks are used in hard rock operations, whereas articulated trucks are less common in quarries. Usually a stationary primary crushing plant is the point of destination for the haulage process. When compared to belt conveyors, the main advantages given for dumpers are mobility and the greater flexibility for the pit design. In addition, issues regarding the required investment costs are mentioned, because the average production rate in the field of the German quarry industry is only 250,000 tonnes per year. [MIRO 2013]

The methods and processes described illustrate the importance of diesel fuel in quarry mining. The price of diesel fuel for large-scale consumers averaged 112 Cent per litre plus purchase tax between January and November 2013. [STATBUNDESAMT 2013] Compared to the previous year, that means a slight decrease in the fuel price. By analyzing thé development over a period of several years, a consistent increase in prices is obvious. As equipment with diesel engines is mainly used in quarries, significant research on fuel efficient combustion engines and drive concepts has already been conducted for several years. In contrast to this, research on possible savings due to the optimization of the interaction between different sub-processes and equipment has rarely been done.

3 Capture of Energy Consumption Data in Production Sites

Part of the research project was the capture of data concerning the energy consumption and the production process in the participating quarries. The intention was to document and analyze the energy consumption in order to derive key figures of the energy demand of quarrying. In this context, it should be determined how much effort has to be spend on data recording and how detailed the information has to be, to identify possible energy savings and to develop measures for an increase of energy efficiency. Especially the differences in energy consumption when comparing the single quarries and the reasons for these differences were of particular interest.

The aspects mentioned are of special importance in the field of mining. The characteristics of a mineral deposit are naturally given conditions and as such they cannot be influenced. The configuration of the production process has to be planned decisively based on these characteristics. It cannot be configured independently, as it is the case for many other industrial production processes. The result is a large number of parameters, which influence the energy consumption. In addition the influence that can be exceeded on these parameters differs from pit to pit. For example, the energy demand, which is necessary to produce one tonne of basalt rock, will typically vary depending on the rock properties, even if two quarries use the same mining method. The overall specific energy consumption is often used for the comparison of two mines. This has to be seen skeptically. The following table shows the overall energy consumption in the quarries that participated in the research project. The demand of diesel fuel and electric energy is given specifically per tonne of mined mineral.

Table 1 Specific energy consumption in the participating quarries

	type of rock	annual production [t]	energy consumption per tonne of mineral	
			diesel fuel [litres]	electricity [kWh]
quarry 1	basalt	205,000	0.52	n.a.
quarry 2	basalt	760,000	0.57	7.24
quarry 3	diabas	575,000	0.41	6.14
quarry 4	gneiss	600,000	0.48	n.a.
quarry 5	greywacke	200,000	0.58	n.a.
quarry 6	diorit	750,000	0.75	4.20
quarry 7	diorit	520,000	1.01	3.85
quarry 8	shell limestone	400,000	0.77	2.71
quarry 9	rhyolith	760,000	0.64	3.16

By comparing the different quarries, the fuel consumption varies between 0.4 and 1.0 litre per tonne of mineral. Even quarries that mine the same type of mineral (*cp. quarry 6 and 7*), differ by of 0.25 litres of fuel demand per tonne. But a statement regarding possible energy savings cannot be given based exclusively on this information. Neither it is recognizable in which sub-process a certain share of the total energy consumption occurs, nor it can be determined if the parameters that cause this consumption can be influenced. Thus, based on the overall specific energy consumption an optimization of the energy efficiency is not possible.

In some operations basic data were not recorded or to be more precise were not available. For example the electricity consumption in three quarries was only known to the central business administration of the companies, but no to the technical management. It has to be assumed that this data has not been checked in consideration of technical aspects. That also means that the energy efficiency of the electrically powered facilities in these quarries is not monitored so far.

The approach of this research project was to capture operational data from different levels of the operations. The data were classified according to the following system:

1. Operation level
   - Total energy consumption
   - Characteristics of the mineral deposit
   - Production rates (mass movement of overburden and saleable mineral)
   - Pit design, bench heights, haulage distances
2. Process level
   - Characteristics of the drilling and blasting operation
   - Equipment combinations and coordination in the loading and haulage process
   - Data of processing stages (primary and subsequent crushing stages)
3. Machinery and equipment level
   - Fuel and electricity consumption of particular machines
   - Machinery dimensioning
   - Cycle times of equipment
   - Equipment availability and utilization
   - Time demand for auxiliary operations
   - Maintenance intervals or inspection of certain parameters (e.g. tire pressure or track tension)

The collection of data was executed during on-site inspections and discussions. It has to be mentioned, that there were significant differences concerning the availability of data in the different quarries. This was already pointed out by the example of the overall electricity consumption. Plenty of verifiable information was available regarding the dimensioning of equipment and facilities as well as in the field of blasting techniques. In case of other aspects it became obvious that there has to be a significant improvement of data recording and analysis in the quarry industry prior to a successful optimization of energy use. This applies for example to the loading and hauling equipment. For these machines there was just little knowledge concerning cycle times and waiting periods as well as concerning equipment utilization. The same applies to the share of load and idle times of the primary crusher, when looking at the total operating hours. In this respect, the technical management of the quarries usually was usually able to indicate estimated values, but not to verify these data.

The following table shows the availability of certain data in the participating quarries. It was only evaluated, if and in which quantity data recordings were available. But type and use of these data are not rated in this table. The classification is based on the percentage of questions and parameters the technical management of the quarries was able to inform about. The classification set up includes the categories sufficient > 75 %, middle-rate > 50 % and problematical < 50 %, which refers to the amount of available information, respectively answered questions. If there was no information available, this is indicated by the abbreviation "n.a." (not available).

Table 2 Survey of the availability of data in the quarries by selected examples (*categories: sufficient +, middle-rate o, problematical -, not available n.a.*)

	Q 1	Q 2	Q 3	Q 4	Q 5	Q 6	Q 7	Q 8	Q 9	Ø
Fuel consumption										
operation level	+	+	+	-	o	+	+	+	+	+
machinery and equipment level	o	o	+	-	-	+	+	+	o	o
Electricity consumption										
pit level	n.a.	o	+	n.a.	n.a.	+	o	+	+	o
machinery and equipment level	n.a.	-	-	n.a.	n.a.	o	n.a.	-	n.a.	-
Drilling *(esp. annual drilling meters, drilling rate, etc.)*	o	-	+	-	-	+	+	+	-	o
Cycle times of loading and haulage equipment	o	o	o	-	-	-	-	-	-	-
Equipment availability and utilization *(esp. waiting periods)*	-	o	o	-	-	o	o	-	-	-
Time demand for auxiliary operations	-	n.a.	-	o	-	o	-	-	o	-
Primary crusher *(esp. load and idle times)*	n.a.	-	-	n.a.	-	o	-	-	-	-

As can be seen from the table, only little data has been captured in the past in those operations. Furthermore it turned out that the information available is often not sufficiently analyzed or that it is not made use of existing systems for the recording of energy data. This applies for example to options, the system control of processing plants usually possesses. Another example is represented by the systems for data storage, which dump trucks of newer production series are typically equipped with and which can be read out in cooperation with the manufacturer of the machine. In this context, also the amount of data has to be considered, which is generated by such recording systems. The effort required for the analysis of this information is only practical for major production machines in a quarry. Therefore a ranking of the machinery should be set up considering the energy consumption and operating time to identify the main energy consumers.

Irrespective of that, the table illustrates how little assured information and verified data concerning certain processes and parameters exist in a large number of quarries. This especially applies to the operational status of the machines and facilities regarding utilization. In case of the loading and hauling equipment, this includes for example the percentage of nonproductive time referred to the total operation time and hence the coordination of both sub-processes. This is also true for the load and idle times of the primary crusher and subsequent processing stages. Another example is the use of the main loading equipment for auxiliary operations, as the cleaning of the quarry face or the fragmentation of boulders in the blast rock. The time demand for these auxiliary operations and hence also the fuel demand of the loading process can be significantly reduced e.g. by a modification of the blasting method. All of these conditions attract just little attention and are rarely documented.

In the fields mentioned, the knowledge usually consists of empirical values. Those are not necessarily incorrect. Often they even represent helpful and relative precise reference values. However, these empirical values cannot be seen as a reliable basis for an optimization because of their insufficient verifiability. To increase the energy efficiency of particular sub-processes of quarrying or even of the whole operation, a more systematic and specific data capture is required, than it was carried out so far.

This includes among others a basic documentation of the fuel consumption of single machinery, as for example the loading and haulage equipment. In many quarries fuel logs are common but very rarely evaluated regarding possible fuel savings. For this purpose it is useful to combine the measured fuel consumption with information about loading and haulage capacities as well as operating and idle times. Waiting periods at the loading and discharge points should also be recorded. The same applies for the load and idle time of the primary crusher and the related energy consumption. For example, this can be realized by the installation of a wattmeter in the electric circuit of facilities, which have comparatively high driving powers or running times. So far, only a central data record of the overall electricity consumption is common in the quarry industry, which is not sufficient for an evaluation and optimization of the energy use.

4 Conclusion

As part of a research project at the Department for Surface Mining and Drilling the energy use in quarrying of natural stone was analyzed. Though quarrying is an energy-intensive process, the energy use is mostly not systematically monitored so far. Mainly political reasons have encouraged many companies of the quarry industry to start with the implementation of an energy management system in the recent past.

Hence the companies can derive economic benefit from these systems, the capture of energy data has to become more methodical and specific in the future.

Aspects evaluated were the availability of energy data and the fields of operation, in which there is need for improvement of data capture. It has to be mentioned, that there were significant differences concerning the availability of data in the quarries. Usually, information concerning sub-processes and machinery as well as their operational status and utilization is not sufficiently documented. This applies especially for the loading and haulage equipment and the primary crusher. As basic fields of interest load and idle times, waiting periods and the time demand of auxiliary operations were identified. These aspects are important regarding single equipment and also for the coordination of the different sub-processes of quarrying. Such information is crucial for a methodical identification and utilization of energy savings, which is of sustainable benefit because of the present and future importance of the industrial rocks and minerals industry.

References

[BMWI 2010] Bundesministerium für Wirtschaft und Technologie & Bundesministerium für Umwelt, Naturschutz und Reaktorsicherheit, Energiekonzept für eine umweltschonende, zuverlässige und bezahlbare Energieversorgung, Berlin, p. 12 (September 2010)

[EYMER *et al.* 2006a] Eymer, W., et al.: Grundlagen der Erdbewegung, 2nd edn., Kirschbaum Verlag GmbH, Bonn, p. 76 (December 2006)

[EYMER *et al.* 2006b] Eymer, W., et al.: Grundlagen der Erdbewegung, 2nd edn., Kirschbaum Verlag GmbH, Bonn, p. 114 (December 2006)

[MIRO 2013] Bundesverband Mineralische Rohstoffe e.V., Bericht der Geschäftsführung 2012/2013, Duisburg, p. 10 f (2013)

[STATBUNDESAMT 2011] Statistisches Bundesamt, Statistisches Jahrbuch, Wiesbaden, p. 372 (2011)

[STATBUNDESAMT 2013] Statistisches Bundesamt, Erzeugerpreise gewerblicher Produkte (Inlandsabsatz) – Preise für leichtes Heizöl, schweres Heizöl, Motorenbenzin und Dieselkraftstoff, Wiesbaden (December 2013)

Wirtgen Surface Miner – The First Link of a Simple Extraction and Materials Handling Chain in "Medium Hard"-Rock

Bernhard Schimm and Johanna Georg

Wirtgen GmbH, Windhagen, Germany

Abstract. The application of an IPCC system in an open pit operation can comprise a reduction of operating costs by 20 – 60 % in comparison to truck haulage but still the latter is the favored transportation method. However since the costs for e.g. fuel and labor increase, IPCC systems again come into focus. Therefore three theoretical approaches how to connect the Wirtgen Surface Miner to a conveyor system are discussed because the surface mining technology can offer additional benefits by merging the process steps drilling, blasting, loading and crushing to one step. The three approaches cover

- The continuous transportation by direct loading of the Surface Miner onto a fully mobile conveyor,
- The direct loading of the Surface Miner onto a truck and discontinuous haulage to a semi-mobile feeder and shiftable conveyor and
- The Surface Miner operating in the windrowing mode, where the material is picked up by a front end loader and discharged onto a mobile feeder connected to a mobile conveyor.

All three concepts offer a potential applicability within an open pit operation. It is concluded that a detailed evaluation needs to be carried out to gather more data to be able to compare the three approaches with IPCC systems.

1 Introduction

Within the mining industry Bucket-wheel excavators and downstream conveyor systems are the leading technology for continuous mining and transport in soft and free diggable rock mining since more than half a century. Due to the need of increasing the production, while mining deeper and lower grades makes it necessary to have high production rates. It is a matter of fact that conveying high quantities of bulk material over larger distances and inclinations is known as the most cost effective method.

However drilling, blasting, loading and transportation of the material by truck out of the mine is still the most common procedure used. Increasing costs, for example fuel and labor costs (see Figure 2), as well as longer transport distances

C. Niemann-Delius. (ed.), *Proc. of the 12th Int. Symp Continuous Surface Mining - Aachen 2014*,
Lecture Notes in Production Engineering, DOI: 10.1007/978-3-319-12301-1_44

and lifting heights force the mine planner and the accountancy to reconsider the discontinuous transport method.

2 In-Pit Crushing and Conveying (IPCC) Systems

In an open pit operation the focus on transportation is very important as approximately 50 % (see Figure 1) of the operating costs within the mine are haulage costs (including for example fuel, lubricants, tires and labor costs). These costs can be reduced by integration of an IPCC system into the mine. The costs for drilling, blasting and loading remain the same.

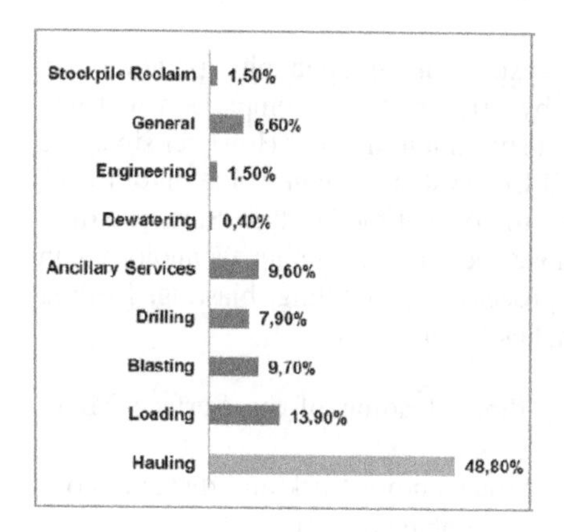

Fig. 1 Typical open pit operating costs (Tutton & Streck, 2009)

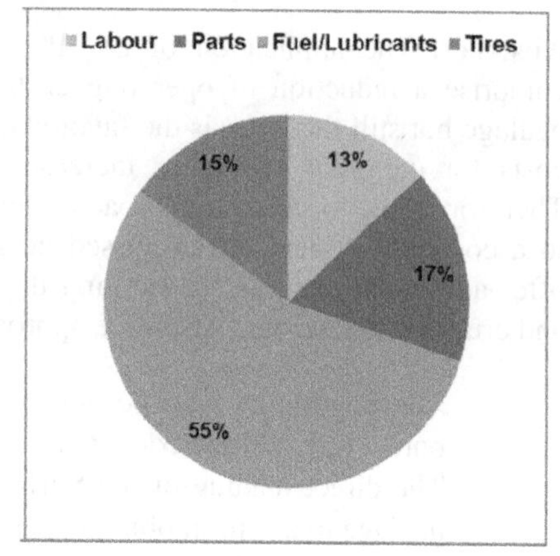

Fig. 2 Composition of hauling costs based on hard rock open pit mines (Tutton & Streck, 2009)

IPCC is defined as a system where the crusher is positioned in the mine and the material is transported by conveyor after leaving the crusher to its destination. Loading of material can either take place directly by excavator, by wheel loader or by truck. IPCC systems can be fixed, semi-mobile and fully mobile as can be seen exemplary in Figure 3.

Within the fixed IPCC system the crusher has been positioned in the center of mass movement over the long-term mine plan of the pit and connected to the conveyor. The material is hauled from the workface to the crusher by truck where it is unloaded, crushed and then transported to the processing plant. This eliminates the overland truck haulage and minimizes the truck haulage distance within the pit. The fixed in-pit crushing plant is stated to be 20 % less in operating costs compared to truck haulage.

The semi-mobile IPCC system will be relocated within the long-term plan. Therefore the semi mobile crusher needs special moving equipment. Since it can be relocated the truck haulage distance between workface and crusher can be adjusted. This system is stated to be 40 % less in operating costs compared to truck haulage.

Fig. 3 Exemplary application of a fully mobile IPCC system (Niemann-Delius, 2012)

Application of a mobile IPCC system allows the elimination of the truck haulage in the mine. The crusher is placed at the workface where the excavator loads the blasted material directly into the hopper. There it is crushed and fed directly onto the mobile conveyor. The mobile system follows the workface subsequently, which reduces the operating costs for transportation up to 60 %.

Concerning the planning of an IPCC system a step-by-step mine planning is indispensable. The dimensioning of the IPCC system within the planning phase is most important. It has to consider an increase of production over the years but may not be oversized since as well as the capital costs as the load factor have to be taken into count. Fixed IPCC systems need a detailed and confident mine planning considering the whole life-of-mine. This means the production has to stay more or less the same, capacity increments are hardly viable. Mobile and semi-mobile IPCC systems offer a more flexible application and can fit easier in material changing deposits. Anyhow a good mine planning for dimensioning of the equipment is necessary.

Comparing IPCC systems with truck haulage, the latter is still the most flexible system. But the following advantages and disadvantages need to be considered when selecting the transport equipment.

Advantages of the IPCC system:

- Shortening the truck haulage distance
- Operating costs associated with fuel, lubricants and tires are reduced
- Labor costs are reduced
- Safety risks are reduced
- Dependency on tires and fuel reduced

- Conveyors can traverse inclinations of up to 30°
- Continuous flow of material

Disadvantages:

- Short-term flexibility is reduced
- Upfront costs are higher
- Capacity increments are easier to achieve with trucks
- Lump size is limited within the feed of the conveyor
- Shutdown of the IPCC for maintenance results in no production in this time (need of stockpiles close to the processing plant)

3 Connecting Surface Miner with In-Pit conveying (IPC)

The Wirtgen Surface Miner unites and simplifies the process steps of the conventional mining method drilling, blasting, loading and primary crushing in one process step. As a result the use of a primary crusher to prepare the ROM (run of mine) for the conveyor transportation is obsolete. The Surface Miner is an areal working mining equipment. It mines layer by layer and each layer is cut path by path. There are two modes in which the Surface Miner can be applied. The first mode is the direct loading onto a truck here a discharge conveyor is attached to the machine. Another possibility is to operate the Surface Miner in windrow mode. In this mode the Surface Miner deposits the cut material behind the machine. In a following step the material can be picked up by a front-end loader.

Besides the simplification of the excavation process and the reduction in operating costs, Wirtgen GmbH has the objective to integrate the Surface Miner into an in-pit conveying (IPC) system. This additionally reduces the operating costs for transportation. Therefore three theoretical concepts are presented for discussion, on how the Surface Miner can be integrated in an IPC system:

Definition of terms used in relation to mine planning and Surface Miner:

Layer:	all cuts on the same level are referred to as layer.
Block:	one block is the amount of material that is mined in one operation, starting at the surface and ending at the designated depth of the pit.
Strip:	all blocks that are mined in one line are referred to as strip. The first strip is stated as box-cut.

1. Direct loading from the Surface Miner onto a shiftable conveyor belt (focus of this paper).
2. Loading material from a Surface Miner onto a truck which transports the material to a semi-mobile feeder within the mine.
3. Surface Miner working in windrow mode, the material is loaded with a loader and dumped onto a mobile feeder which is connected to a mobile conveyor system.

For all concepts the same scenario for the mining sequence has been defined. The Surface Miner mines blocks which have a working length of about 500 m, a width of 60 m and a depth of 50 m (see Figure 5). At one side of the block a ramp with an inclination of 1:10 (5.7°) for the Surface Miner as well as for the transportation equipment has to be constructed. The cutting width is about 4 m since a 4200 SM Wirtgen Surface Miner is applied. The average cutting depth of the Surface Miner is about 50 cm (max. 80 cm, min. 10 cm). In operation the Surface Miner will cut layer by layer, as can be seen in Figure 4, until a whole block is mined. At the end of each mined path a turning area of 30 m for the Surface Miner is required. After finishing the excavation of one block the Surface Miner is moved to the next block.

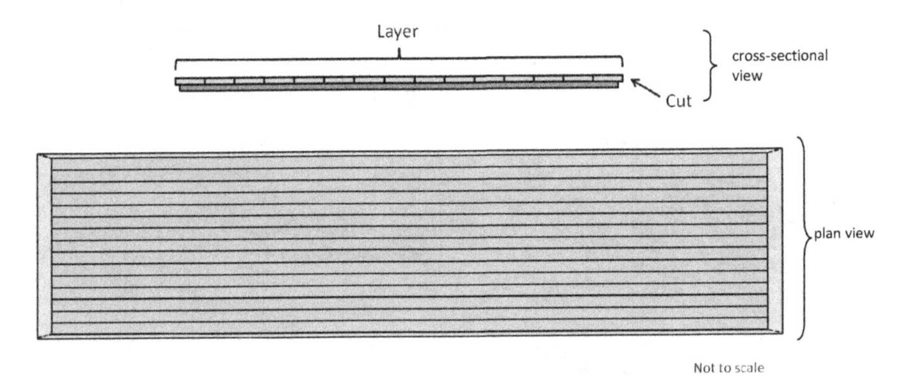

Fig. 4 Definition layer and cut (Niemann-Delius & Ranft, 2011)

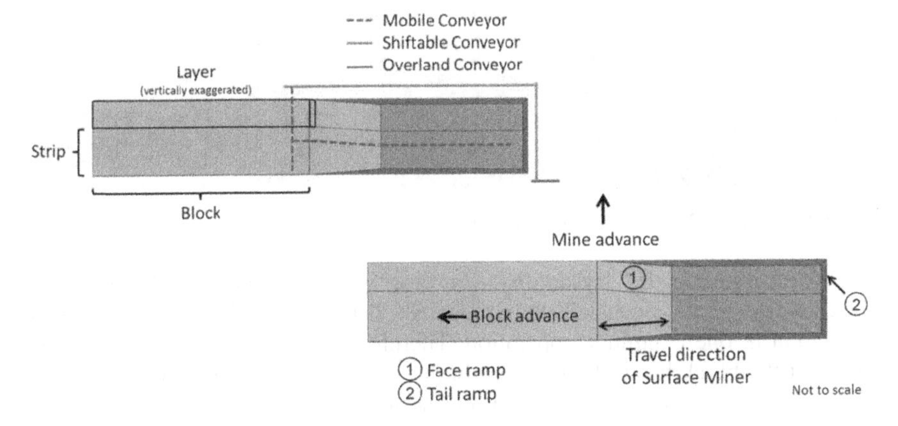

Fig. 5 Definition block and strip (Niemann-Delius & Ranft, 2011)

3.1 Direct Loading onto a Mobile Conveyor Belt

As part of a case study a fully mobile IPC system to remove overburden linked to a Surface Miner has been elaborated by Wirtgen GmbH. The idea is to connect the Surface Miner to an overland conveyor by a fully-mobile conveyor system. In this case the overland conveyor is the preferred solution to transport the overburden because the dump area is 20 km advanced from the mine. To connect Surface Miner and overland conveyor a in every direction mobile multiple module conveyor (e.g. Flexiveyor, see Figure 6) has to be implemented into the system and integrated into the mining sequence. Since the multiple module conveyor is still in the pilot stage, this concept is a theoretical approach. The approach has been based on the mobile multi module conveyor systems that have been introduced into the mining market for room and pillar mining operations under the expectation that they can be adapted to surface mining operations. There they have proven their applicability within the mining operation.

Fig. 6 Flexiveyor (Diversified Mining Services, 2013)

Linking a mobile IPC system to a Surface Miner has the big advantage that truck haulage in the mine will be eliminated. This enables the Surface Miner to mine a cut in one continuous operation. Therefore a mobile conveyor belt of the length of the block has to be installed in the mining area. At the top of the ramp another mobile conveyor is installed and beside the blocks a shiftable conveyor has to be implemented because this conveyor has to be shifted after four blocks have been mined (in this case 1&3 and 2&4, see Figure 7). The shiftable conveyor is connected to the overland conveyor, where the material is distributed to its destination. Two blocks will be mined parallel by two equipment units to achieve a higher utilization of the overland conveyor as well as a faster advancing exploitation.

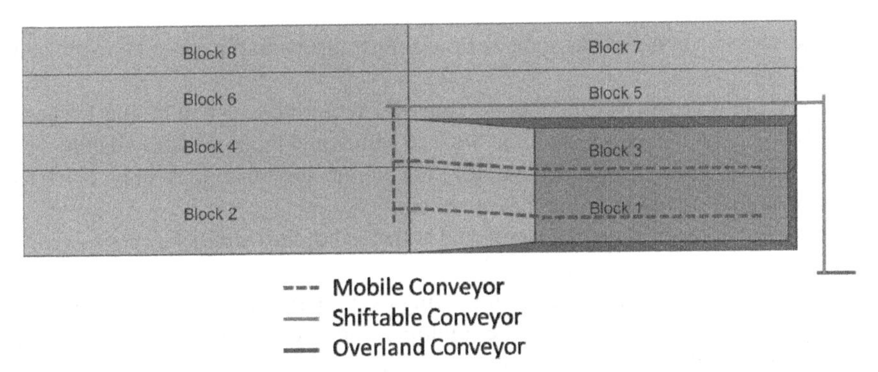

Fig. 7 Mining two blocks parallel (Niemann-Delius & Ranft, 2011)

Based on the before mentioned mining sequence of the Surface Miner the main challenge of the application of a conveyor system will be that it has to be moved every third cut perpendicular to the conveying direction. This maintains a continuous loading of the conveyor belt by the Surface Miner. To avoid unnecessary shifting of the conveyor or interference of the Surface Miner with the conveyor a detailed mining sequence has to be planned before the start of the mining operation.

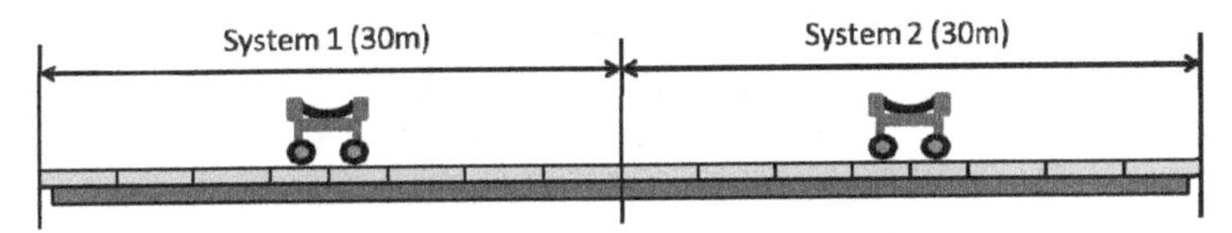

Fig. 8 Example of working widths (Niemann-Delius & Ranft, 2011)

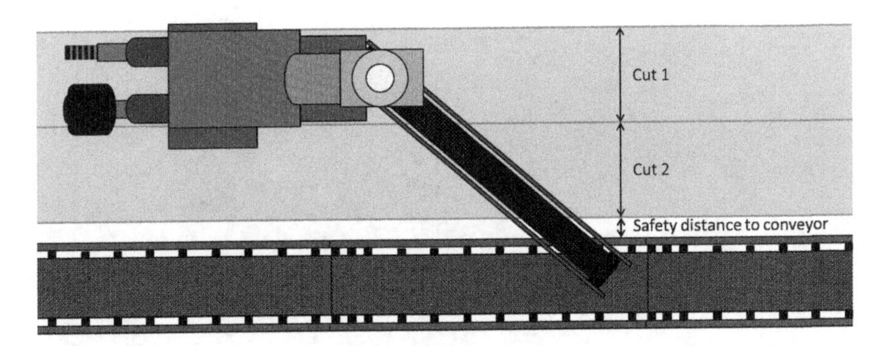

Fig. 9 Joining Surface Miner and Mobile Conveyor (Niemann-Delius & Ranft, 2011)

The mobile IPC system in combination with the Surface Miner offers great advantages once it has proven reliability:

- Less processing steps
- Lower operating costs
- Continuous production, once the IPC system is implemented

However, this concept has some serious constrains, which have to be evaluated in trials:

- Very often shifting of the mobile conveyer required, every third cut, because of the high advance speed of the Surface Miner and the low cutting depth.
- A mobile conveyor is needed for this application but does only exist as conceptual design and is no proven equipment.
- Material removal at the beginning and at the end of the cut/block is not solved.

For a realization of this concept, following exemplary issues have to be solved:

- Direction of movement of the multiple module conveyor: In conveying direction and perpendicular to the conveying direction
- Loading operation: Surface Miner has to be able to load the material on the multi module conveyor at any point of the conveyor line, a discharge conveyor has to be mounted on to the Surface Miner

3.2 Loading Material from a Surface Miner onto a Truck Which Transports the Material to a Semi-mobile Feeder within the Mine

The main idea of this approach is to use the benefits of truck and conveyor haulage and to eliminate the negative aspects of each transportation method. The truck haulage distance is shortened and the uphill transport minimized, since the costs for uphill truck haulage represent about 40-50% of the total transportation costs. Further this concept enables a high flexibility of the system and no major changes within the mining sequence have to be made.

Fig. 10 Exemplary integration of a Surface Miner into an IPCC system (Oberrauner & Ritter, 2013)

The Surface Miner will also mine blocks. Hence, the cut material will be discharged onto the trucks and hauled to the next installed semi-mobile feeder. The material is then transported with a shiftable conveyor and the overland conveyor to the distribution point.

This system has the advantage that it can be adjusted to changing deposit conditions and has the charm that it can be implemented in already existing mine layouts and mining sequences. Additionally the number of trucks and associated expensive labor can be reduced. This reduces operating costs. The disadvantage of this system is again the dependency on the truck availability.

3.3 Surface Miner Working in Windrow Mode, Loading the Material with a Loader and Dumping it onto a Mobile Feeder Which is Connected to a Mobile Conveyor System

Within this approach the Surface Miner is applied in the windrow mode. The cut material is deposited behind the machine and will be picked up by a front-end loader. The loader transports the material to a nearby moveable feeder where it is discharged onto a mobile conveyor.

Here the usual mining sequence of the Surface Miner can be applied too because the mobile conveyor is not connected directly to the Surface Miner. This results in an alternative design of the mobile conveyor as it does not have to follow the cuts of the Surface Miner. Accordingly an existing system like Lokolink from Metso can be applied.

Since the trucks can be eliminated within this approach the Surface Miner can cut with maximum speed. The system is now constrained by the front-end loaders which only have an economic travel distance of about 200 – 250 m. The total IPC system also needs a detailed mine planning to avoid unnecessary shifting of the mobile conveyors.

4 Discussion

The three concepts comprise the application of a Surface Miner in combination with a fully mobile or semi-mobile IPC system. In the following Table 1 the main issues of each concept are summarized.

All introduced systems are moveable and can be applied very flexible in the mining sequence. However the second concept offers an even more flexible application in the pit since it can be adapted to changing deposit conditions.

Nevertheless a detailed mine planning is essential when an IPC system should be deployed successfully. Therefore the production capacity over the life-of-mine has to be determined in the mine plan because a capacity increment of the IPC system is only viable to a limited extent.

Table 1 Summarizing the main issues of the concepts

	1. Concept direct loading onto conveyor	2. Concept truck haulage, semi-mobile IPC	3. Concept front-end loader, mobile IPC
Type of haulage and loading	continuous haulage and loading	discontinuous haulage and loading	continuous haulage and discontinuous loading
Mobility of the conveyors	fully mobile	semi mobile	fully mobile
Flexibility of the system	relocation of the system possible	relocation of the system possible, very flexible	relocation of the system possible
Equipment available	no multi module conveyor available	available	available, but has not been applied in combination with a Surface Miner

5 Conclusion

The three approaches of the applicability of the Surface Miner with an IPC system have shown that a detailed evaluation of the concepts should be accomplished. Therefore trials have to be performed to gather reliable data. Further the data can be used to compare the conventional IPCC systems with the Surface Miner IPC system. The researches indicate that using a Surface Miner in combination with an IPC system instead of an IPCC system can result in further cost benefits. This is because surface mining technology in combination with an IPC system offers benefits in transportation as well as mining and material processing by merging the process steps of the conventional mining method to one step.

References

Chadwick, J.: New IPCC ideas. International Mining, pp. 33–41 (June 2010)

Darling, P.: SME mining engineering handbook. Society for Mining, Metallurgy, and Exploration, Inc., U.S.A (2011)

Diversified Mining Services, Prairie Machine & Parts Flexiveyor System. Hunter Valley, New South Wales (2013)

Foley, M.: In-pit crushing: wave of the future? Australian Journal of Mining, 46–53 (May/June 2012)

metso (2014) metso, http://www.metso.com/miningandconstruction/mm_crush.nsf/WebWID/WTB-100112-2256F-E7688?OpenDocument (accessed January 22, 2014)

Niemann-Delius, C.: The role of mining models when comparing alternative vs. conventional mining technics. Miskolc, University of Miskolc (2012)

Niemann-Delius, C., Ranft, H.: Internal project study - Surface Miner and mobile conveyor - Development of a mining scheme. Wirtgen GmbH, Windhagen (2011)

Oberrauner, A., Ritter, R.: IPCC-Mine Planning - how it is supposed to work. Sandvik, Cologne (2013)

Saxby, P., Elkink, J.: Material transportation in Mining - trends in equipment development and selection. Australien Bulk Handling Review 27, 10–13 (2010)

ThyssenKrupp Robins, The IPCC challenge: putting all pieces together. Niagara-on-the-lake, Mining Magazine Congress (2009)

Tutton, D., Streck, W.: The application of in-pit crushing and conveying in large hard rock open pit mines. Niagara-on-the-Lake: Mining Magazine Congress (2009)

Zimmermann, E., Kruse, W.: Mobile crushing and conveying in quarries - a chance for a better and cheaper production. ISCSM Aachen (September 2006)

Concept for Applying the Continuous and Selective Mining Features of Surface Miners in a Strip Mining Operation

Claudel Martial Tsafack

RWTH Aachen University, Aachen, Germany

Abstract. Surface miners in the last years have experienced a remarkable development, from a technological point of view as well as in their area of application. The factors that determine the applicability of the machine are the hardness of the material to be cut and the dimensions of the working area. Additionally when different main equipment are teamed with the Surface miner (SM) in one operation the individual system constraints have to be merged. An example for this is the application of SM in Strip Mining operations. Furthermore, since SM have proven their advantages in stratified deposit with interburden, schemes are required for combined operation. This paper analyzes the systematic approach to merge the constraints of Strip Mining operations with those of a SM. Of utmost importance for the SM mining sequence of coal (and interburden respectively) is the way the DL continues after finishing one and starting the next strip. Accordingly this paper furthermore discusses the consequences of the different options for the DL and the SM operation.

1 Introduction

SMs nowadays are gaining more significance and are becoming in some operations the major mining equipment. The capability of mining material selectively and to operate continuously promotes them to become an alternative as well as a concurrent for conventional mining equipment. Both features together give them the advantage of being applicable in specific operations. SMs have proven their applicability as well as their productivity in some mining projects all around the world. They use to be considered as auxiliary equipment though they are becoming in numerous projects the main production machinery. They are applied in coal mining operation in India and in tabular iron ore deposits in Australia.

Nevertheless looking at the SM as principal mining equipment or as one of the major ones, additional constraints, advantages and disadvantages result from

C. Niemann-Delius. (ed.), *Proc. of the 12th Int. Symp Continuous Surface Mining - Aachen 2014*,
Lecture Notes in Production Engineering, DOI: 10.1007/978-3-319-12301-1_45

merging it with other machines or system of machines. One operation where fitting the SM may have upside potential is the coal recovery in a strip mining operations. The conventional coal removal here is up to now predominantly done by shovels and trucks or wheel loader and trucks. Especially when dealing with multi-seam deposits the use of the SM may lead to reduction of DL rehandling and the need for further DLs or machine fleet to deal with interburden.

This paper focusses on the development and the presentation of schemes for integrating the SM in strip mining operations. The development is based on SM and DL way of operation as well as some of their specificities. Furthermore the description and the presentation of the systematic for teaming DL and SM constraints are conducted.

2 Surface Miner (SM)

Surface miners are used for cutting material while traveling along a defined working path. The cutting drum is either attached in front, at the back or in the middle of the machine. The Wirtgen 4200 SM considered in this paper is equipped with a middle drum.

The SM combines almost all mining steps (drilling, blasting, loading and crushing) in one operation. As selective and continuous mining equipment SMs are capable to mine material with strength up to 80 Mpa economically. [Wirtgen SM Handbook, 2010]. The mined material can be handled in two ways. It can either be dumped after the cutting process that is the so called "cut to ground" or directly loaded on trucks, that is the so called "direct loading".

The application of SM holds, beside the capacity of doing several mining operation in one step, some other advantages. One is the production of good quality ROM due to the selectivity feature. Different materials are able to be mined separately this being a substantial advantage while dealing with multiple layers deposits and lead additionally to costs reduction in the processing plan. Furthermore a higher level of safety is achieved as no explosives are used for the material loosening process. Additionally avoiding negative blasting effects such as vibration, dust and noise has a positive effect on the acceptance of the neighboring communities in densely populate areas.

Nevertheless the SM also has some disadvantages. The performance of the machine for instance is extremely dependent on the material hardness. This leads to high operating costs due to outstanding fuel consumption or to relatively high bit pick consumption. Furthermore the decision of applying a SM is also dictated by the type of the deposit. They are applied in layered and tabular deposits such as coal where they reach high productivity. In addition, their effective production capacity also depends on the length of the chosen working area. Then the maneuvering operation occurring at the end of each cut, and the production losses during that time, force the minimum working area to be some 500 m long representing an effective cutting time of around 70 %. [Wirtgen SM Handbook 2010]

3 Dragline (DL)

Draglines were developed in the USA and in the former UDSSR to allow an economical mining of tabular or bedded mineral deposits with extremely thick overburden. Their comparatively low operating costs and high productivity make them to be unbeatable in waste removal operation and dumping when conditions like high production requirement are met.

They are used in so called Strip Mining operation. Basically draglines moves overburden from the current strip and places it in the former mined out one. They can handle up to 85 m of overburden depending on the class of machine. The resulting strip width varies between 30 and 100 m. [SME Book, chapter 10.3]

The working level and position of the dragline depends on the thickness of the overburden or the overall material above the deposit. The operation mode is chosen base on the geology, the material characteristics, the deposit specifications and the production rate for instance. For the purpose of this study the Simple Side Cast operation mode of the DL is applied. For simplicity reason only the applied operation mode used in this study is described next.

3.1 The "Simple Side Cast" Mode (SSC)

Here the dragline stands on the top of the overburden and strips down to the edge between the overburden and the valuable mineral (in our case coal). This is the dragline standard mode of operation. The stripped material is directly placed in his final position (previously mined out strip) after a swing of 90°. Basically for the appliance of the SSC mode the material needs to be unconsolidated. Otherwise fragmentation (blasting) is needed.

4 Applying the Continuous and Selective Feature of the SM in a Strip Mining Operation: Description of the Systematic Approach

4.1 Prerequisite

The constraints resulting from merging a DL and a SM are mainly due to their respective way of operation and the resulting mandatory working sequence when team together. Those are:

- The SM minimal working length,
- the safety distance to be kept to the DL,
- the maneuvering time at the end of every cut,
- the advance speed of both machines and finally,
- the possible production loses causes by production interruption due to additional standstill, are the principal constraints faced.

The latest constraint is by far the most crucial for DL operations.

At the beginning of each strip the DL first need to uncover a sufficient working area to allow the SM to perform efficiently. The time required for stripping a certain area will depend on the deposit characteristics, the DL operation mode, and the planned production rate. For the purposes of this paper the "simple side cast" mode is applied. Additionally the strip widths resulting from DL operation vary between 30 and 100 m [SME Book, chapter 10.3]. This means that the SM cutting operations can only be efficient when applied lengthwise. According to figure1 a widthwise (max. 80 m effective cutting length) cutting length results to an effective cutting time less than 40 %. Therefore the length of the working area should be some 500 m to allow the SM to use about 70 % of his effective cutting time. In addition to that taking into account the safety distance of 200 m to be kept between DL and SM, results in an initial length of some 700 m. This is the strip length to be uncovered by the DL before the SM can start with coal recovery.

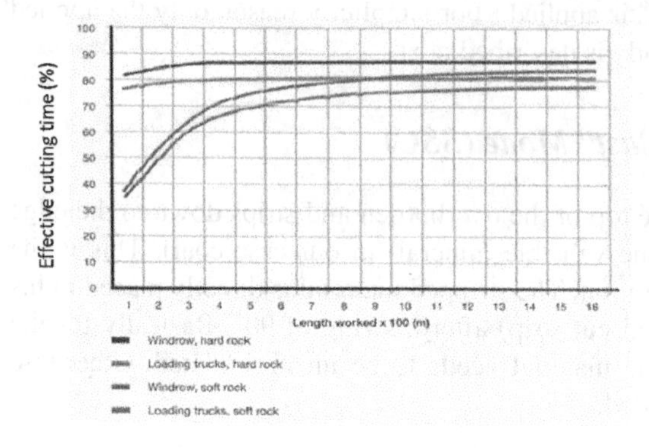

Source: Wirtgen SM Handbook 2010, showing the impact of the working length on the effective cutting time

Fig. 1 Effective cutting time (%) vs. working length (m)

4.2 SM Operation

At the begin of the strip the SM builds a "box cut" or "initial cut" in the coal layer to create a proper working area allowing a continuous coal mining operation in the direction of the DL. This is achieved by cutting a "V-shape" "box cut" consisting of two slopes having different angles as depicted in figure 2. The left side of the box cut has a wall slope of ~1:10. Additionally the figure shows the first access ramp which is used as haulage road at the beginning of the coal recovery and which is connected to the spoil side ramp. Coal haulage will be done initially via the end slope, the initial ramp and the spoil side ramp perpendicularly to the pit. The access ramp furthermore serves as maneuvering platform for the SM.

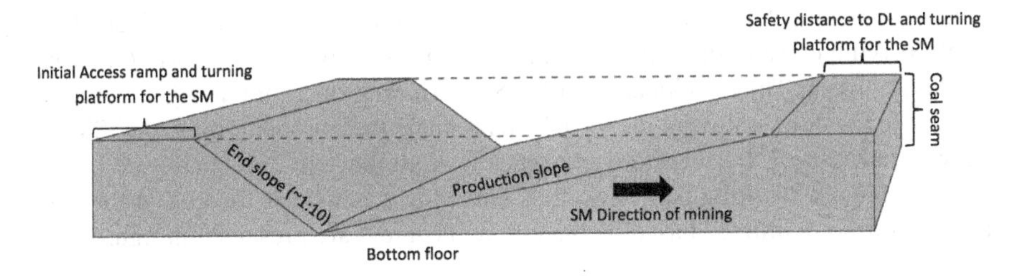

Fig. 2 Box cut in the coal layer (not to scale)

The right side of the "box cut" is characterized by a comparatively flat slop representing the "advancing slope" or "production slope" and 200 m safety distance. During SM box cut operation the production slope has to be extended up to 500 m minimum. (See figure 1). It is assumed that after the box cut, the production slope progresses in the direction of the DL with a constant distance.

After the box cut the SM now mainly works on the production slope which progresses towards the DL leading to an increase of the haulage distance compared to the position of the first access ramp. Therefore a sequence for the relocation and the optimal position of the access ramp along the strip has to be developed. Furthermore the SM, in his stripping activity, mines precise cuts next to each other till the working area is completely mined out. However, the SM always has to turn back (about 180°) after reaching the end of each cut in order to position itself in front of the next one for cutting in the opposite direction. Because the maneuvering time has to be kept to a minimum it is highly important to define the way the different cuts are going to be mine depending on the strip width available.

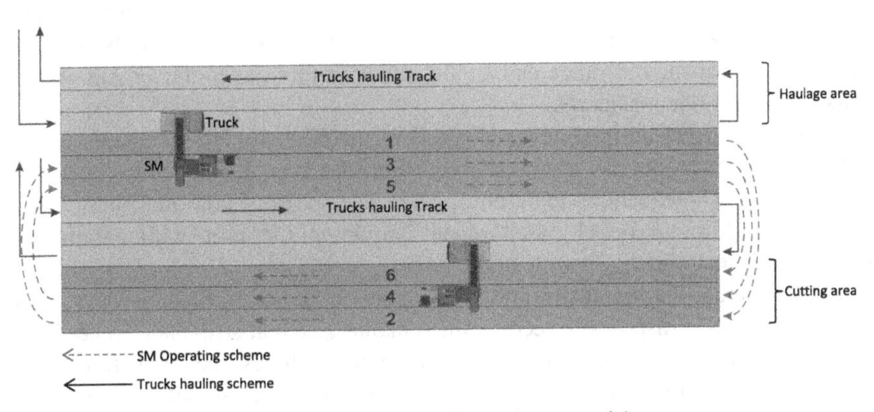

Fig. 3 SM cutting scheme and trucks hauling scheme (not to scale)

For the purposes of this paper a strip width of around 80 m is considered and the applied SM is the Wirtgen 4200 SM with a cutting drum width of 4.2 m and a cutting depth of up to 0.83 m. [Wirtgen Handbook 2010]. The proposed cutting scheme is depicted in figure 3. Moreover the stripped coal is directly loaded on trucks driving close to the SM. The direct loading is enabled by the SM 16 m long discharge boom. The developed path for SM stripping operation as well as trucks hauling ones for achieving to an efficient maneuvering pattern of the SM is showed in Figure 3. Furthermore an overview of all the activities conducted by the SM in a precise working path as well as a hauling scheme for trucks and the way access ramps are utilized based on the mining progress are presented in figure 4.

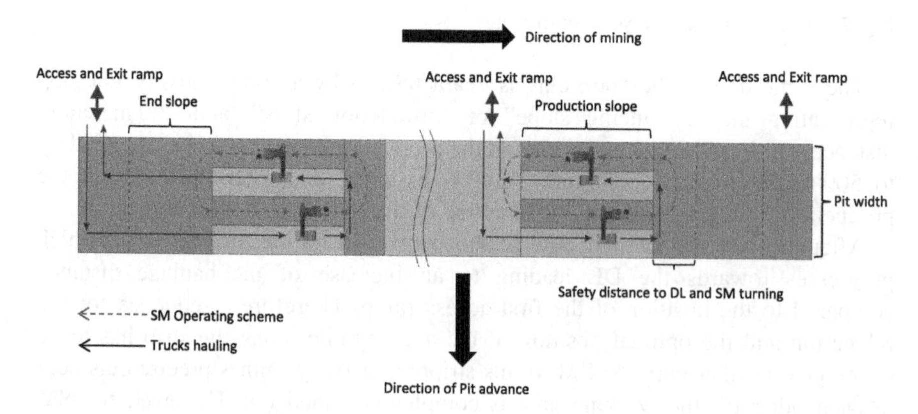

Fig. 4 Hauling scheme (not to scale)

The SM integration as well as it is way of operation in a strip has been described and presented above. Nonetheless the impact of such integration on the DL activities has to be assessed. The correlation of both machine advance speed is therefore conducted. After uncovering the coal the DL have either to wait for the SM to recover the remaining coal before starting the next strip or walk back to start a new strip at the other end of the pit. For this paper the possibility of walking back is not considered.

According to pre-defined advance speed of the SM, the DL standstill can be unlikely long. A model based on linear equations has been developed to overcome this situation. The equations represent the advance speed of both machines. They are integrated in an Excel based model providing the user with parameters influenced by teaming a SM with a DL.

The maximum DL standstill time is commonly assumed to be three days. This is the time the SM disposes to recover the remaining coal at strip end. To keep the DL waiting time to a maximum of three days, both machines linear equations need to be defined. Therefore the DL “SSC” mode DL is considered. After defining the equations the linear function representing the advance speed of both machines are set up.

Setting DL function

Equations

Daily volume flow: Q_d [m^3\d] = B*C_t*F_g*T_e*(1-S_f)*W_{hD}/h [1]

Face Advance speed: V_d [m\d] = Q_d\O*W [2]

DL safety distance to the SM: D_{safety} (t) [m] = V_d*t-l_{safety} [3]

Parameters

- Bucket size: B
- Cycle time: C_t : 60 s
- Fill grade: F_g : 90 %
- Swell factor: S_f : 20 %
- Time efficiency: T_e : 90 %
- Working hours: W_{hD} :6000 h\a, 16 h\d, 365 d\a
- Overburden Thickness: O
- Strip width: W
- Cycles per hour: C_h
- Excavating Time: t

Function

Linear function of DL advance: D (t) [m] = V_d*t [4]

Setting SM Function

Setting the SM function is more complex than the DL one due the starting "delay" of the SM operation compares to the DL one and to the safety distance between both machines.

Equations

Daily volume flow: Q_{sm} [m^3\d] = Q*CT*15.07 h\d [5]

Angle of inclination of the advance slope: α [°] = tan-1(M_c\L) [6]

Horizontal progress after one complete cut of the advance slope: P [m] = C_d\sin (α) [7]

Volume of a complete cut: V_c [m^3] = P*M_c*W [8]

SM advance speed: V_{sm} [m/h] = P*Q_{sm}\$V_{sm}$ = Q_{sm}\$M_c$*W [9]

Parameters

- Theoretical volume flow: Q ~ 900 [fm^3\h]
- Cutting time percentage: CT
- Working Hours: 5500 h\a, 15.07 h\d, 365 d\a
- Coal thickness: M_c = 1 to 10 m
- Strip width: W
- Angle of inclination of advance slope: α
- Cutting depth: C_d
- Advance slope length: L

Considering the starting "delay" of the SM, the resulting function starts at following coordinate:

- Abscissa: At time t_W needed for the DL to strip a sufficient area allowing the SM to start working add to the time for developing the box cut t_b. $\mathbf{t = t_w + t_b}$
- Ordinate: Initial strip length necessary to perform the box cut l_b

The resulting point combined with the SM advance speed allows the determination of t_w and the resulting.SM function is therefore:

Function

- $$SM (t) [m] = V_{sm}*t+(l_b-V_{sm}*t_w) \quad [10]$$

At time t = 0 the SM starts recovering the coal after the DL has uncovered enough space. The latest represent the horizontal length of the box cut minus the advance speed multiplied by the DL stripping time. After setting the SM and the DL function they are plotted in a Cartesian coordinate. The plotting operation leads to intersections or none of both linear functions depending on the SM advance speed and leading to the following three cases.

Case1

Intersection of both functions, figure 4, at a point (5) representing the maximal strip length. Additional stand is avoided.

Case2

Influence of a relative high SM advance speed. The two functions intersect "earlier" leading to additional standstill time T(SM).

Case3

Slow SM advance resulting in "additional" standstill time for the DL at the end of the strip.

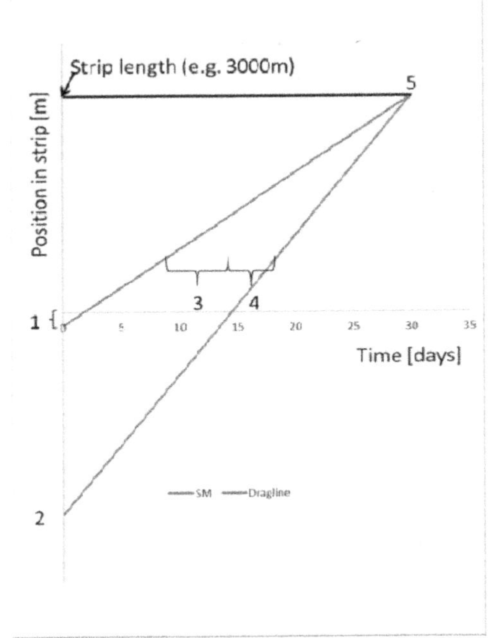

1 Safety distance
2 Y-axis intersection of SM function
3 Time span the dragline needs to clear a sufficiently long area for SM operation
4 Time span the SM needs to perform initial cut
5 Intersection of SM function and dragline function including the safety distance

Fig. 5 Functions intersection at strip end

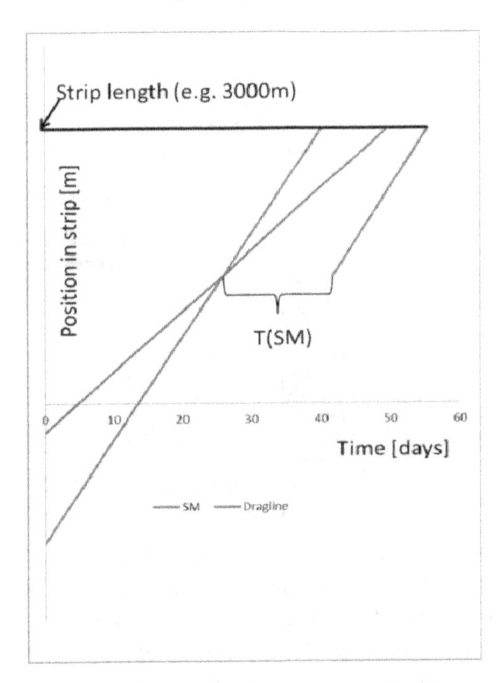

Fig. 6 Early intersection of the functions

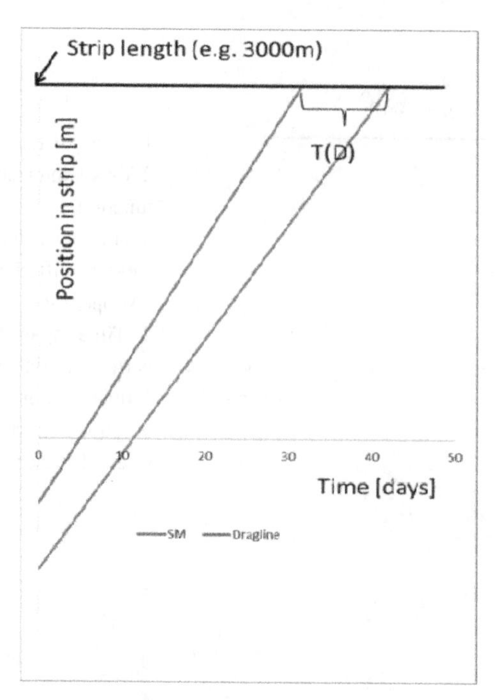

Fig. 7 Slow advance of the SM: no intersection of the functions

Table 1 Model: Data input window

Dragline Buckets		Cat 8750		
	Volume Flow	Advancing Speed		
96,00 m³	61.356 m³/d	26 m/d		
Dragline Parameters			**Strip Parameters**	
Cycle Time	60,00 s		Strip Width	60 m
Fill Grade	90%		Overburden Thickness	40 m
Swell Factor	20%		Coal+Intermediate Rock	1-10m
Time eff.	90%			
Working H.	6.000 h/a	16,4 h/d		
SM Parameters				
Effective Cutting Time	80%			
Working Hours	5.500 h/d	15,1 h/d		
Cutting Performance	900 bm³/h			
Actual Cutting Performance	10.849 bm³/d			
Cutting depth	0,83 m			
End wall slope length	100 m			
Safety distance to Dragl. Work. Face		200 m		

The input data used to develop both machine equations are variable. They are therefore settable according to the needs of the user. To do so the required parameters are given. Table 1 gives an overview on this part of the model. The output information on the oder hand indicates whether or not the input data (See table1) allows operating the chosen DL and SM. The results of the outputs part of the model base on the input data from table 1 is presented in table 2.

Table 2 Model: Data output window

96,00 m³	*1 SM*											
	y-axis intersection for SM function	-4.451,1 m	-4.451,1 m	-4.451,1 m	-4.451,1 m	-4.451,1 m	-4.451,1 m	-4.451,1 m	-4.451,1 m	-4.451,1 m	-4.451,1 m	-4.451,1 m
	Complete lag time	29,6 d	29,6 d	29,6 d	29,6 d	29,6 d	29,6 d	29,6 d	29,6 d	29,6 d	29,6 d	29,6 d
Table valid	Length all slopes	900 m	900 m	900 m	900 m	900 m	900 m	900 m	900 m	900 m	900 m	900 m
	Initial cut time	2,2 d	2,2 d	2,2 d	2,2 d	2,2 d	2,2 d	2,2 d	2,2 d	2,2 d	2,2 d	2,2 d
	Length of cutting slope	700 m	700 m	700 m	700 m	700 m	700 m	700 m	700 m	700 m	700 m	700 m
Thickness	Strip length	2000	2200	2400	2600	2800	3000	3200	3400	3600	3800	4000
1 m	SM-Dragline intersection	965,9 m	965,9 m	965,9 m	965,9 m	965,9 m	965,9 m	965,9 m	965,9 m	965,9 m	965,9 m	965,9 m
SM Advance Speed	Finish Time Dragline Safety Function	70,4 d	78,2 d	86,1 d	93,9 d	101,7 d	109,5 d	117,3 d	125,2 d	133,0 d	140,8 d	148,6 d
181 m/d	Finish Time SM Function	35,7 d	36,8 d	37,9 d	39,0 d	40,1 d	41,2 d	42,3 d	43,4 d	44,5 d	45,6 d	46,7 d
Min. Dragline	Additional Dragline Standstill	0,0 d	0,0 d	0,0 d	0,0 d	0,0 d	0,0 d	0,0 d	0,0 d	0,0 d	0,0 d	0,0 d
Standstill	SM Off-Time	34,7 d	41,4 d	48,2 d	54,9 d	61,6 d	68,3 d	75,0 d	81,8 d	88,5 d	95,2 d	101,9 d
0,55 d	% SM Off-Time	49,3%	53,0%	56,0%	58,5%	60,6%	62,4%	63,9%	65,3%	66,5%	67,6%	68,6%

5 Conclusion

The study presented in this paper shows the integration of a SM in a Strip Mining operation. It is worth using the SM as coal mining machinery as long as a clear and well-structured stripping scheme is applied. The main critical point during operation is the standstill time of the DL at the end of the strip. This can be fairly determined, forecasted and adjusted with the help of the presented model.

Although the study is a theoretical one it nevertheless reflects the potential of applying the SM in a strip mine operation for coal recovery. Further analyzes for assessing the outcomes of this paper base on a case study should be conducted next. Additionally an economical and operational model to compare conventional coal recovery operation (shovel, wheel loaders and trucks) with the application of SM in a strip mining operation will permit to define and capture the upside potential and downside risk of applying the SM in such an operation.

Development of the Cerovo - Veliki Krivelj Mining Complex at RTB-Bor

Dimča Jenić, Predrag Golubović, and Darko Milićević

Copper Mining and Smelting Complex Bor Group Bor, Serbia

Abstract. The strategic plan of copper production in Cerovo as part of the Veliki Krivelj mining complex is based on the certified balance sheet of copper ore reserves of over 511 million tons, on the possibility of more ore mining by purchasing new high capacity mining equipment, reconstruction and acquisition of new flotation equipment and the reconstruction of the smelter and construction of a new sulfuric acid plant, which will all provide for a better recovery and environment protection with high-tech environmental standards.

The paper contains a basic concept of mining developed and applied in this region.

Keywords: strategic plan, balance of ore reserves, high capacity equipment, reconstruction, flotation equipment, smelter, sulfuric acid plant.

1 Introduction

The techno-economic feasibility, perspective of work and particularly the development of mining and metallurgy at RTB Bor have been worked out in a document called the Copper Production Business Plan at RTB Bor for the period 2011-2021, thus confirming and justifying the investment in technological modernization of the entire production line and economic profitability of mining and processing of home copper resources.

Fluctuations of copper price in the last 10 years (from $1940 in 2000 the price rose to $8500 in 2011) with all forecasts in metal prices in world markets for the upcoming period, suggest there will be a serious economic and market basis for a significant increase in the physical volume of copper production.

RTB Bor has also installed facilities for the flotation and metallurgical processing of the total excavated copper ore, which will add value to its own mineral resources and production of finished products.

For these reasons, RTB Bor has planned a substantial increase in production from its large mining complex Veliki Krivelj - Cerovo, or from the mines of Cerovo and Veliki Krivelj as the concentrates produced in these two mines account for 55% of the total concentrate production at RTB Bor for the 2012 to 2021 period.

C. Niemann-Delius. (ed.), *Proc. of the 12th Int. Symp Continuous Surface Mining - Aachen 2014*,
Lecture Notes in Production Engineering, DOI: 10.1007/978-3-319-12301-1_46

2 General Geology

The mining complex of Veliki Krivelj - Cerovo is an area within the Timok Magmatic Complex which includes copper deposits of Cerovo and Veliki Krivelj.

These are large-scale porphyry deposits. Wider area of the deposits is shown in Figure 1.

Cerovo deposit is located ten kilometers northwest of Bor. There is a railway and asphalt road to Bor.

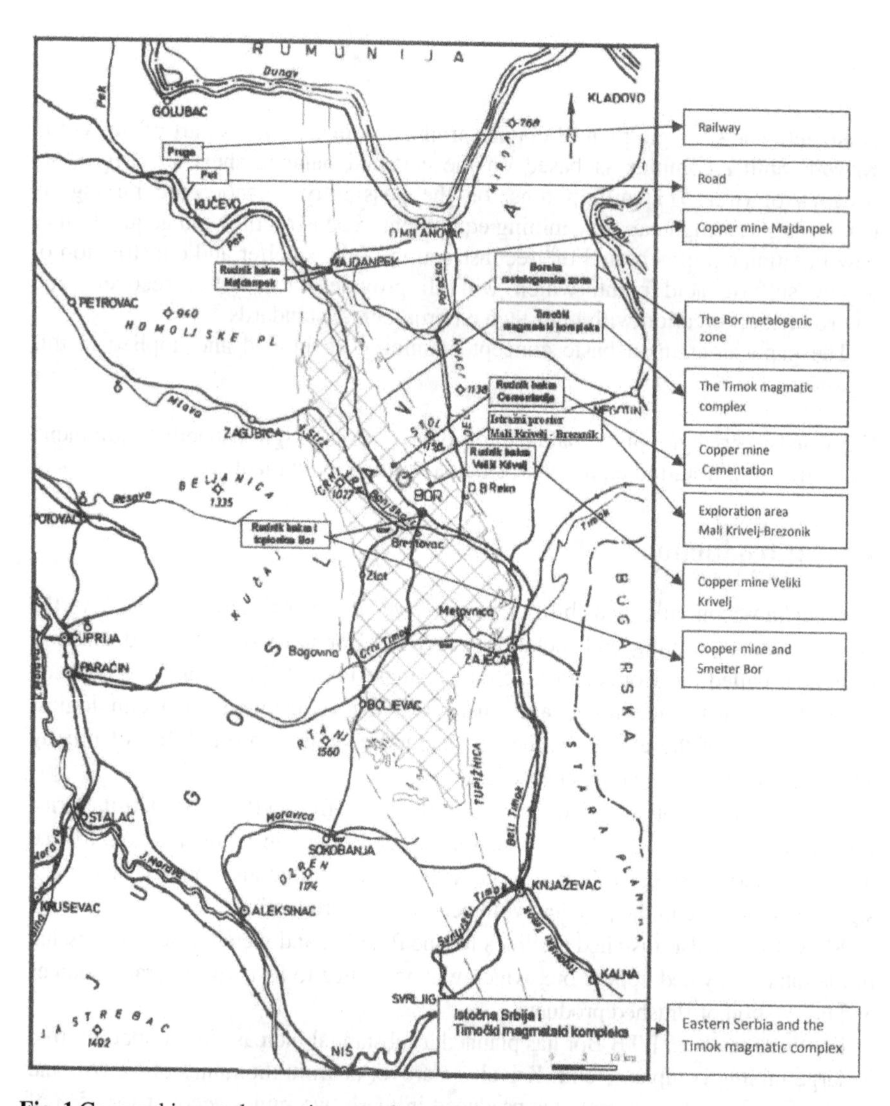

Fig. 1 Geographic-metalogenetic map of eastern Serbia

In the Cerovo deposit, porphyry mineralization is located in the zone of hydrothermal altered andesite rocks 1.5 km long with a maximum width of 600 m, which is dipping to the east - northeast. The deposit is in the horizontal projection of an oval shape, elongating in the NNW – SSE direction. In the vertical projection, it is irregular in shape and dipping below the level of -100.

Bearer of copper mineralization are chalcopyrite and bornite.

The Veliki Krivelj deposit is situated about 6 km north-east of Bor. There is an asphalt road to Bor.

Viewed in a horizontal cross section, porphyry copper deposit of Veliki Krivelj has direction NW - SE. Its longer axis is 1.5 km with a maximum width of 700 m. The deposit falls to the southwest, while we can not yet speak of its final depth as several wells of 800 m in depth are proof of mineralization.

The main copper ore barrier is pyrite and chalcopyrite as well as bornite at a small extent.

Surface areas are oxidized (30 - 50 m). Coating of limonite, malachite, azurite and tekorite is present too.

2.1 *Ore Reserves at Veliki Krivelj - Cerovo*

The development of mining at the Veliki Krivelj - Cerovo complex is based on the certified balance of ore reserves of 611 808 102 t with average copper content of 0.33 %. The balance is given in Table 1.

Table 1 Balance of geological reserves as of December 31, 2011

Deposit/ Ore body	Ore reserves category	Ore (t)	Cu (%)	Cu (t)	Au (g/t)	Au (kg)	Ag (g/t)	Ag (kg)
„Veliki Krivelj“	B+C1	461.669.785	0,326	1.503.091	0,056	25.854	0,248	114.494
„Cerovo“	B+C1	150.138.317	0,328	492.136	0,135	20.235	0,86	145.809
TOTAL RTB	**B+C1**	611.808.102	0,327	1.995.227	0,0955	46.089	0,554	260.303

3 The Basic Concept of Mining at the Veliki Krivelj - Cerovo Complex

Long-term production plan at RTB - Bor for the next ten to fifteen years will be based on:

1. Mass surface mining method in:

- Copper Mine Cerovo (deposits: Cerovo 1 and Cerovo 2, Cerovo Primary and Drenova), with a capacity of 2.5 x 10^6 of ore per annum in the first phase and a capacity of 5,5 x 10^6 t of ore per annum in the second phase.
- At the Veliki Krivelj copper mine - capacity of 10.6 x 10^6 tons of ore per year.

3.1 The Cerovo Open Pit Mine and the Processing Plant at Cerovo

The Cerovo copper mine is located about 25 km northwest of Bor. It was opened in 1993 and temporarily closed in December 2002.

Under the current circumstances, there is a real possibility to re-open the mine.

At the open pit mine of Cerovo, all the necessary infrastructure is already there: access roads, water and sewage networks, electric power supply to the substation and gas station, telephone network, drainage system etc.

There is also a facility for ore preparation and processing to the level of grinding along with the accompanying infrastructure.

While the mine was in operation, a 14 km hydro transport system was used to transport the pulp to the Bor Concentrator where froth flotation was applied and concentrate produced.

The annual production was 2,500,000 tons of ore.

The existing infrastructure will still be used for the future production along with the necessary reconstruction and rehabilitation to obtain the capacity of 2.5 Mt of ore per year and further expansion of facilities to reach the target of 5.5 Mt of ore per year.

To resume production, a new hydro transport system was constructed with a completely new technical and technological water system on the new route, having in mind that further processing will be done at the Veliki Krivelj Concentrator.

Getting ready for a re-launch of the mine, defects on the equipment were repaired, and other machines and devices installed on the technological ore processing line. Based on our inspection, underway now are reconstruction works and installation of equipment necessary for bringing the entire plant in operation for the installed capacity of 2.5 mt of ore per year, where the following will be included:

- Mechanical and electrical works at the primary, secondary and tertiary crushing and grinding with thickener;
- Measuring and control equipment for the primary, secondary and tertiary crushing and grinding.

3.1.1 The Concept of the Development of the Cerovo Mine

Our further plans for the development of mining in this deposit are based on continuation of production at the open pit Cementation 1 and opening of the new mine Cementation 2. In the next phase, our plan is to open Cerovo - Primary deposit and Drenova deposit.

Considering the possibility of further development of the Cerovo deposit, optimization was performed in the mining deposits of Cerovo 1,Cerovo 2, Cerovo Primary and Drenova.

Table 2 Summary with calculations for Cementation ½ Cerovo CPD mines

Elements	CEMENTATION 1 & 2	Cerovo CPD	TOTAL
Ore (t)	30911833	98 295 751.0	129,207,584
Overburden (t)	22151493	100,769,987.0	122,921,480
Excavated (t)	53,063,326	199,065,738	252,129,064
Cu (t)	94,698.30	327,968.000	422,666.3
Ag (kg)	37,191.12	106,517.000	143,708.1
Au (kg)	2,771.06	16413.7	19,184.8
Cu (%)	0.306	0.334	0.327
Ag (%)	1.203	1.084	1.112
Au (%)	0.090	0.167	0.148

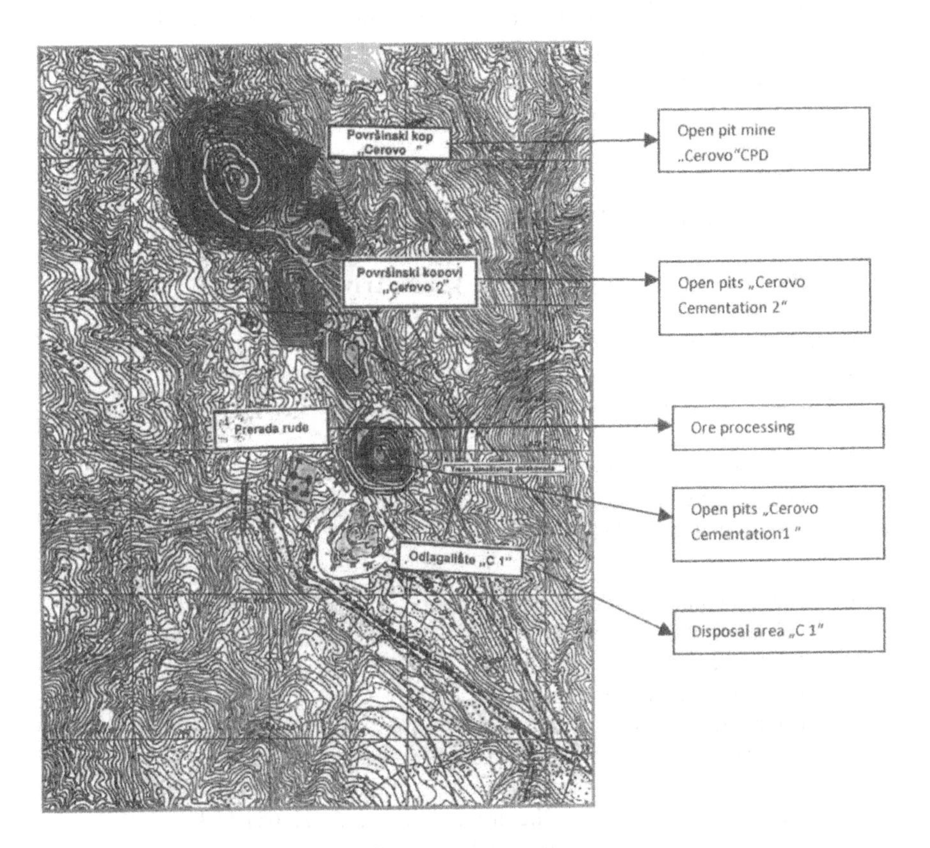

Fig. 2 Mining field of the Cerovo mine

In the first phase in 2012 and 2013, Cerovo 1, up to 2.5 Mt of ore annually will be mined and in 2014 mining of Cerovo 2 will begin with 5.5 Mt of ore per year. Practically, phase II will begin then with successive mining of Cerovo - Primary and ore body Drenova and will last until 2021 when mining will end without the relocation of the railroad.

Production up to the grinding level will be performed in the existing facilities of Cerovo, and the pulp will be further transported via a new hydro system to the Veliki Krivelj Concentrator.

In this procedure, without relocating the railroad, around 49 Mt of ore will be excavated with average copper content of 0,34%, 50 Mt of overburden wand the total amount of copper in the ore of 160,972 t or 140,000 t of copper in the concentrate.

The following table gives a summary of calculations of the Cerovo deposit until the end of its mining life, including the relocation of the railway.

The total mineable ore reserves can be increased by optimizing new mines, given the upward trend in metal prices in the world market.

Fig. 3 shows the Cerovo mining field.

3.2 *The Veliki Krivelj Copper Mine*

3.2.1 The Concept of Developing the Veliki Krivelj Open Pit

Based on verified reserves of A, B and C1 categories and requirements for the annual ore mining capacity of 10.6 million tons and the operation period of 20 years, the open pit was designed within its final contours of up to K-55 as shown in Figure 3.

The total amount of copper, silver, gold and tailings for the finally aproved contour surface of the Veliki Krivelj open pit mine and the mining life of 20 years, the following will be included for phase I:

- The total amount of excavated material, t 446 078 484
- The amount of overburden, t 233 481 038
- The amount of ore, t 212 597 445
- The marginal copper content in the ore,% Cu 0,150
- The average copper content in the ore,% Cu 0,316
- The average silver content in the ore, g / t Ag 0,245
- The average gold content in the ore, g / t Au 0,059
- The amount of copper in the ore, t 651 653
- The amount of silver in the ore, kg 50 523
- The amount of gold in the ore, kg 12 167
- Ratio of overburden, t / t 1,098

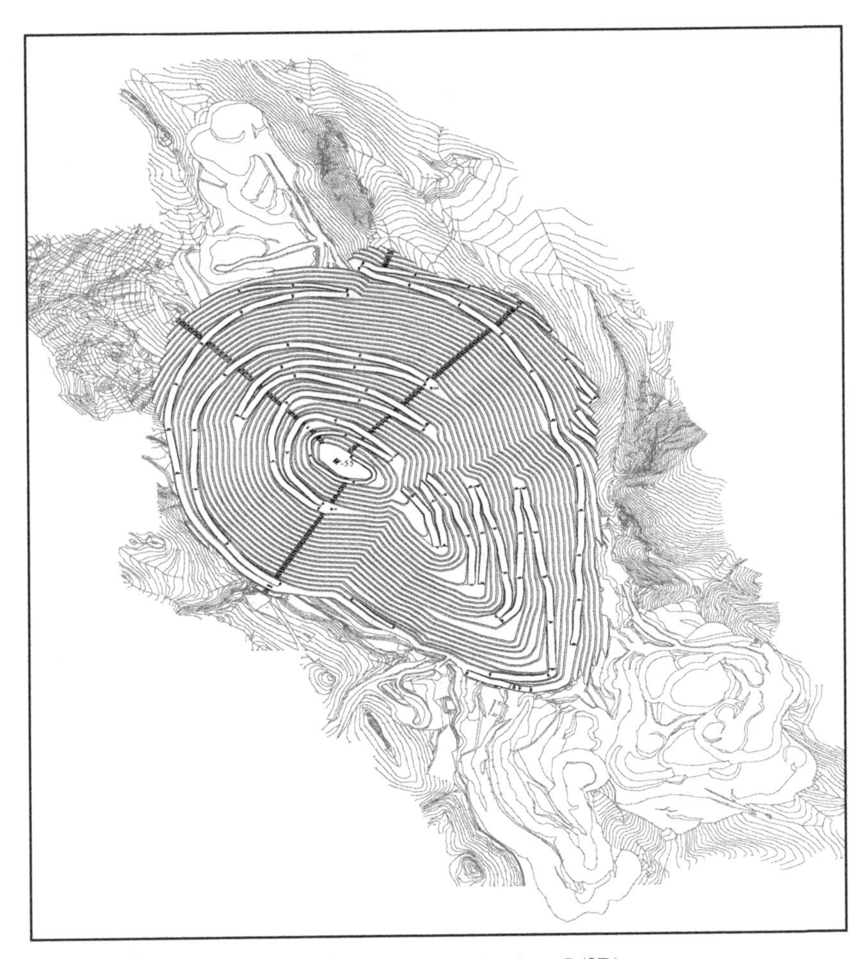

Fig. 3 The shape of the final contour of the open pit, phase I (2D)

4 The Concept of Ore Processing and Concentrate Production

According to the strategy adopted by the company of RTB Bor, increase in ore production capacity has been planned at the mining complex of Veliki Krivelj - Cerovo with copper ore of 16.1 Mt to be processed annually. At the Veliki Krivelj Concentrator, it will be possible to increase the processing capacity for the ore coming from Veliki Krivelj deposits after reconstructing the crushing, screening and grinding systems, as well as in the section of flotation concentration by replacing the existing flotation machines with larger ones. For the Cerovo ore, it will be necessary to install a new line of flotation cells.

During 1998 and the 1999, certain reconstruction activities on the crushing and screening systems were performed to reduce the size of crushed ore by 100 % - to 20mm.

However, it was found that the optimum coarseness at the inlet to the rod mill with F80 rods was from 12000μm - 14000μm, and that was the reason why it was necessary to reduce the size of the crushed product to 100 % - 16mm. It was the recommendation of Metso company which technology was implemented during the reconstruction of crushing and screening when new flotation equipment with an automated process was launched.

Along with the activities on the reconstruction of the crushing facility, we started installing new flotation cells of larger volume, such as those used in mines worldwide of similar capacity.

Basic engineering provided by Metso Minerals is the basis of the project of replacement old machines. The basic concept of replacing old flotation machines with larger ones is all about a better design, which will eventually lead to better technological results, savings in their maintenance and less power consumption per unit of processed ore.

Reduced coarseness of the crushed ore 100 % - 16mm will increase mills capacity and retain fineness of the ground ore of 58 % of class - 74μm in the hydrocyclone overflow (which will be measured by the latest PSM device). Mills will be able to process 424 t / h of dry ore for the adopted working index of Wi = 14 kWh / t. To achieve the newly designed capacity, the system has to be able to automatically control and regulate the grinding process.

According to the basic design, the entire hydrocyclones battery will be replaced with a new battery with hydrocyclones of a smaller diameter in order to obtain better and more stable fineness of the grind. A new battery will operate with a hydrocyclone pump and an attached rotation speed variator to provide a constant pressure at the inlet to hydociclones. In addition, all other pumps will have such variable speed rotation regulators.

Instead of 119 existing DENVER DR500 type machines for the basic Krivelj ore flotation, 18 RCS100 machines will be used, or the so-called tank cells.

Instead of 78 existing DENVER DR100 and DR300 type machines used for the primary Krivelj ore concentrate treatment, 12 RCS40 machines will be used. Additionally, instead of 8 existing DENVER DR500 type machines that were used for extended flotation, 4 machines RCS40 will be used.

In addition to installing new tank cells, a new flotation monitoring and controlling system will be installed too, with the latest equipment for copper content monitoring at the inlet and outlet of the flotation machine (X -ray analyzer). There will also be installed a part for the preparation and distribution of reagents in which two new mixers will be placed, as well as new tanks for the daily receipt of new reagents and their distribution. Other equipment for the preparation of lime, control of the pulp pH value, dewatering of the final copper concentrate and tailings, fresh and return water supply will not be subject of the aforementioned reconstruction, as it is believed that they can meet the demand of increased ore processing.

These changes will guarantee the technological recovery of 87 % and higher copper concentrate quality of 21 % Cu.

Guarantees are given by Metso Minerals, who is the supplier of the new equipment.

Overall Positive Effects of Larger Ore Processing Capacity at the Veliki Krivelj Concentrator

- With optimum size crushing of 100 % -16 mm, maximum grinding capacity will be achieved, power will be saved (about 5 %), as well as metal for grinding, rods and lining (10%).
- Better control of the hydrocyclone overflow, too much comminution of the raw material or at least insufficient release of the beneficial component will be avoided that will directly affect poor technological parameters.
- Before the pulp reaches the conditioner prior to the basic flotation and also at the output, copper content and the content of other components will be checked (modern X -ray analyzer), so that an optimum dose of collectors to be added can be determined, as well as the density of the pulp and the required amount of air in the machine, all of which will contribute to better technological parametrs.
- Flotation in new large volume tank cells which are less costly to maintain, will provide for much better results in terms of technological efficiency and the quality of concentrate.
- Fully automated flotation process means less physical strain for the operator, which will definitely lead to less risk of injury at work (better protection of workers).

Due to the fact that larger amounts of copper concentrate will be required for the new smelter, RTB has decided to reactivate the Cerovo mine to the level of grinding at Cerovo and hydro pulp transport to the Veliki Krivelj Concentrator.

Having in mind that funds needed to open any mine are extremely high, it was decided that in the first phase, the Cerovo mine should reach the previous capacity proven for many years while this plant had been in operation of 2.5 Mt / year, and after that the capacity of 5.5 Mt / year from 2014 onward will be be reached.

Phase One of 2.5 Mt / yr

The equipment to be used in the first phase of the Cerovo mine is the crusher now used at Majdanpek, after being generally repaired and made fit for the mine.

All available equipment that was used at the Cerovo mine will be used again (conveyers, bunkers, mills, thickeners, hydrotransport pumps) after a complete inspection and repair. For the hydro - pulp from Cerovo to the Veliki Krivelj Concentrator, two separate pipelines will be installed with a diameter of 340 mm each and a capacity of 2.5 Mt / yr as well as a necessary pipeline for the return technological and technical water all in order to reach the ultimate capacity of 5.5 Mt / yr.

In the first phase of processing of the Cerovo ore at the Veliki Krivelj Concentrator, old equipment will be used (DENVER type flotation cells) as follows:

For primary ore flotation of the Cerovo ore, 28 DENVER DR 500 machines will be used

  - Tertiery and secondary processing in 4 +8 machines DR100 and 4 +4 machines for primary and extended flotation
- Existing old conditioners for receiving and conditioning the pulp from Cerovo
- Necessary technical fresh water will be delivered from the lake water pool in Bor and the process water from the pumping station at the Krivelj tailings dump field I, first to the pool at the Veliki Krivelj and then to the pool at Cerovo.
- Concentrate will be dewatered and filtered with the equipment of the Veliki Krivelj concentrate as these two will join at the Concentrator and then taken to the thickener.

Second Phase of 5.5 Mt / yr

Second Phase of the Cerovo Mine Will Involve Investment in the New Equipment

Purchase of new crushers, mills (preferably autogenous and semi-autogenous) as well as buying new large volume flotation machines.

Purchase of new RCS 100 machinery is planned, which will only add to the series of three lines with three machines. This is how we have planned to increase the current capacity to 5.5 Mt / yr within the current expansion of the Veliki Krivelj Concentrator.

4.1 *Metal Balance of Production in the 2012 – 2021 Period at the Veliki Krivelj Cerovo Mining Complex*

Table 3.1a. Elements of production

No.	Name	unit	Open pit V.Krivelj	Open pit Cerovo	Total
1.	Wet ore	t	109.600.000	49.000.000	158.600.000
2.	Overburden	t	178.700.000	47.000.000	225.700.000
3.	Moisture in the ore	%	3,00	3,00	3,00
4.	Dry ore	t	106.312.000	47.530.000	153.842.000
5.	Copper content in dry ore	%	0,280	0,339	0,298
6.	Copper quantity in the ore	t	297.859	160.974	458.833
7.	Au content in dry ore	gr/t	0,060	0,146	0,086
8.	Au content in the ore	kg	6.349	6.944	13.293
9.	Ag content in dry ore	gr/t	0,229	1,084	0,493
10.	Ag content in ore	kg	24.383	51.515	75.898

Table 3.1b. Metal balance

No.	Name	V. Krivelj Concent.	
		unit	Total
1.	Dry ore	t	153.842.000
2.	Cu conc. wet	t	2.138.664
3.	Moisture in the concen.	%	10
4.	Cu concent. Dry	t	1.924.798
5.	Cu in dry conc.	%	21
6.	Cu in the concentrate	t	398.775
7.	Au in dry concentrate	gr/t	3.453
8.	Au in the concentrate	kg	6.647
9.	Ag in dry conecntrate.	gr/t	19.716
10.	Ag in the concentrate	kg	37.950
11.	Cu recovery	%	87
12.	Au recovery	%	50
13.	Ag recovery	%	50

5 Investment

In the observed period, investments are planned for the Cerovo mine processing facilities, for the purchase of new equipment for the mine, expropriation and relocation of power lines, as well as higher investments in the processing capacity from 2.5 Mt to 5.5 Mt of ore per year at Cerovo mine plants and the Veliki Krivelj Concentrator.

Table 4 The dynamics of investments

in `000 USD

No.	Sector	Year								Total
		2012	2013	2014	2015	2016	2017	2018	2019	
1	Veliki Krivelj open pit	1.500	-	4.396	5.483	10.990	6.594	11.590	-	40.553
2	Cerovo open pit	13.055	13.458	14.080	-					40.593
3	Veliki Krivelj Concentrator	4.500	5.400	-						9.900
Total		**19.055**	**18.858**	**18.476**	**5.483**	**10.990**	**6.594**	**11.590**	-	**91.046**

6 Conclusion

The planned development of the Veliki Krivelj - Cerovo mining complex as part of the copper production strategy will bring a positive impulse for the development of RTB Bor Group and the entire region. Strategic Plan to produce more copper is based on the confirmed copper ore reserves, possibility to increase mining capacity by purchasing new high productivity mining equipment, reconstruction and purchase of new flotation equipment, reconstruction of the smelter and construction of the new sulfuric acid plant. By implementing all these activities, more efficient technology will be purchased and environmental protection brought to the highest standard, and the respective economic performance will bring this significant copper producer to the highest world level.

References

1. Copper production business plan at RTB Bor for the period 2011 - 2021, RTB Bor (June 2011)
2. Feasibility study on the combined copper ore mining in the "Kraku Bugaresku" and "Cementation" copper deposits, IRM (December 2009)
3. Additional project of deposits mining for the Veliki Krivelj deposit with the capacity of 10.6 Mt of ore per year, IRM (March 2011)

“Exemptions” from the Management Objectives for Water Bodies Associated with Lignite Mining in North Rhine-Westphalia (NRW) Pursuant to the EC Water Framework Directive

Thomas Pabsch

District Government Arnsberg – Department of Mining and Energy in NRW

1 Water Management Situation

Currently, approximately 90 to 100 million tons of lignite are extracted every year up to a maximum depth of about 450 m in the active open-pit mines Garzweiler, Hambach, and Inden which are located in the tri-city area of Cologne, Düsseldorf, and Aachen. Most of the extracted lignite is converted into electricity in the power plants Niederaußem, Neurath, Frimmersdorf, Weisweiler, and Goldenbergwerk. Former open-pit mines in the western and/or eastern part of the Rhenish lignite mining district (the open-pit mines Zukunft, Frechen, Bergheim, and Fortuna-Garsdorf) as well as in the district's southern part (Ville) were recultivated and, predominantly, rehabilitated. The service lives of these open-pit mines will continue until 2030 for Inden and until 2045 for Garzweiler and Hambach.

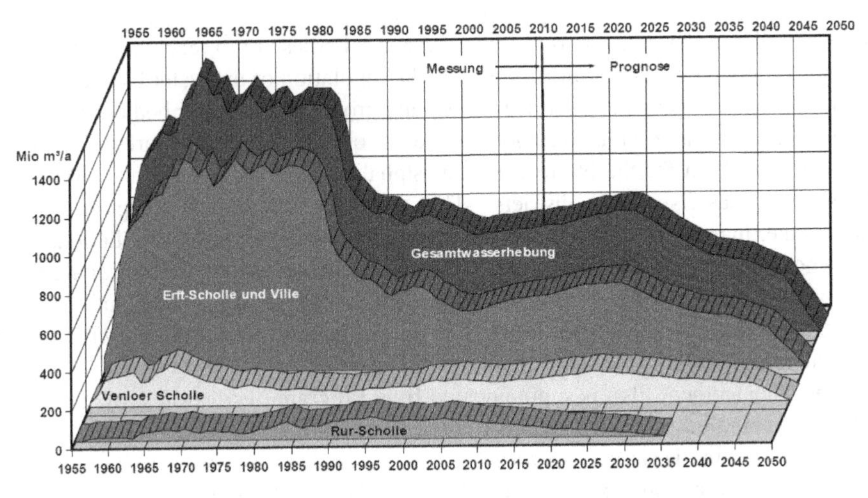

Fig. 1 The development of drainage water volumes between 1955 and 2050

C. Niemann-Delius. (ed.), *Proc. of the 12th Int. Symp Continuous Surface Mining - Aachen 2014*, Lecture Notes in Production Engineering, DOI: 10.1007/978-3-319-12301-1_47

In order to ensure secure lignite extraction in open-pit mines, the groundwater level has to be lowered as a preparatory measure and kept at a sufficiently low level during any open-pit mining operations. That is why the water level in the excavation slopes of the overlying stratum (above the coal) needs to be lowered to the bottom edge of the respective groundwater aquifer. Below the coal, the groundwater level needs to be lowered to the level of the operational underlying stratum (bottom edge of the coal) in order to avoid any hydraulic ground seepage.

The start of large-scale open-pit mining in Ville in the 1960s marked the beginning of extensive drainage measures in the Erftscholle region, and to a lesser extent also in the Venloer Scholle region located to the north as well as the Rurscholle region located in the west. How these drainage measures develop over time (please see Figure 1) is determined by the individual exploration points for open-pit mining, the pertinent extraction and rehabilitation phases (Forkel, 2011; http://www.gwz dresden.de/fileadmin/gwz/Downloads/Veranstaltungen/DGFT/ABSTRACT-2011-ge%C3%A4ndert.pdf).

2 Objectives of the EC Water Framework Directive

The EC Water Framework Directive issued in 2000 defines basic environmental objectives for all European water bodies, for brooks, rivers, lakes, groundwater, and coastal waters. The key objective of the European Water Framework Directive is to reach a "good status" for all rivers and lakes as well as the groundwater. The water quality is to be safeguarded and, if need be, further improved (http://www.umwelt.nrw.de/umwelt/umweltinformationen_umweltberichte/umweltbericht_umweltindikatoren/index.php). In order to maintain and/or improve biodiversity, the water bodies are to regain their good ecological status once again. Wherever this proves to be impossible because the surface waters were modified substantially and/or created artificially, the water bodies are to reach at least good ecological potential. An additional goal is the regulation of the water balance.

What the management of surface waters and groundwater (water bodies) seeking to accomplish these objectives looks like is outlined and summarized in the Management Plan for the North Rhine-Westphalian sections of the Rhine, Weser, Ems, and Maas river basin districts. This summary is based on comprehensive, basic water management data of the current situation and the causes of pollution, the existing utilization of the water bodies, and the available options for improving the status of such water bodies while also considering their utilization (http://wrrl.flussgebiete.nrw.de/Bewirtschaftungsplanung/index.jsphttp://wrrl.flussgebiete.nrw.de/Bewirtschaftungsplanung/index.jsp).

The Management Plan is supplemented by a Program of Measures, by profiles of the planning units, and by reports of the (inter)national Rhine, Weser, Ems, and Maas river basin districts.

These planning unit profiles include water body related information on the status of surface waters and the groundwater, the pollution of and management objectives for individual water bodies as well as the program measures intended for individual water body groups.

North Rhine-Westphalia is characterized by its high population density, its history of environmental burdens dating back to the era of mining and heavy industry, and its current utilization by industry, power generation, agriculture, and transportation routes.

The objectives of the Water Framework Directive can be achieved – even though, in many substantiated cases, this will not happen before 2027. So far, exemptions have become necessary in the lignite mining district, in the limestone quarrying sector, and for few water bodies affected by former ore mining.

According to the Official Notification by the North Rhine-Westphalian Ministry for the Environment, Nature Conservation, Agriculture, and Consumer Protection (MUNLV) of March 29, 2010 (Ministerial Gazette MBl. NRW., 2010 Edition, No. 12 of April 15, 2010, Pages 249 to 258), the Management Plan and the Program of Measures are binding for the public authorities and need to be observed with regard to the enforcement of water management.

The Background Document on Lignite substantiates the exemption made for lignite mining with regard to groundwater and the discharge of drainage waters particularly into the Untere Erft River, which is contrary to the management objectives laid down in the Water Framework Directive, and integrates it into the Management Plan/Program of Measures for the federal state of North Rhine-Westphalia (http://wrrl.flussgebiete.nrw.de/Dokumente/NRW/Anhoerung/Hintergrunddokumente/Braunkohle.pdf).

Please note:
Due to their legal and technical complexity, most of the information provided in Section 3 below has been taken from the original Background Document on Lignite.

3 Exemptions from the Management Objectives for Lignite

3.1 Exemptions from the Management Objectives for the Quantitative Status of Groundwater

For the affected groundwater bodies, any drainage measures undertaken on behalf of lignite mining represent an abstraction of water from the groundwater resources. Pursuant to § 47 (formerly § 33a Para. 1) of the Water Management Act (WHG), groundwater is to be managed in such a way that any detrimental change in its quantitative and chemical status is avoided, any significant and sustained upward trend in the concentration of pollutants resulting from the impact of human activities is reversed, a sound balance between groundwater abstraction and groundwater recharge is assured, and a good quantitative and chemical status in accordance with the WHG is maintained or attained.

Figure 2 shows the mining-related lowered groundwater section in the top aquifer as well as the section of the lowered confined groundwater level in deeper aquifers (the so-called underlying stratum).

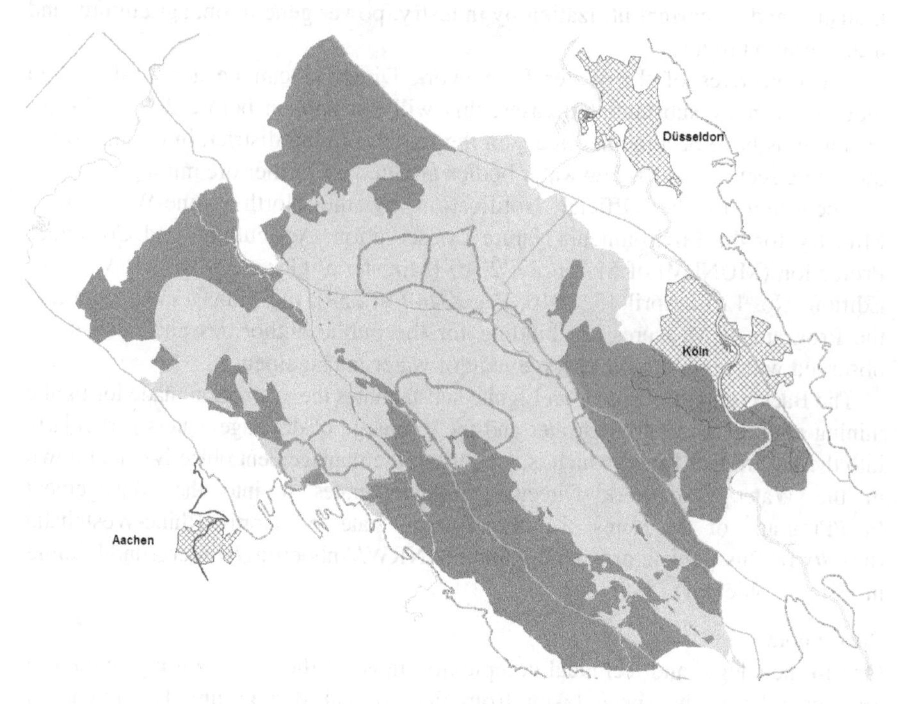

Fig. 2 Mining-related lowered groundwater in the top aquifer (light green) as well as the lowered confined groundwater level in deeper aquifers (dark green, partially superimposed by light green); (difference between 2006 and 1955)

Figure 3 provides an overview of those groundwater bodies which, according to current forecasts, are not expected to reach any good quantitative status in 2015 (and presumably also not by 2027) due to drainage measures carried out on behalf of lignite mining.

According to § 47 and § 30 of the WHG (formerly § 33a Para. 4 Sent. 3 of the WHG under the provisions of § 25d Para. 1 Nos. 1-4 of the WHG), any existing impairment of the groundwater management objectives resulting from the lowering of the groundwater level due to ongoing lignite mining operations constitutes an exemption from the management objectives for the quantitative groundwater status.

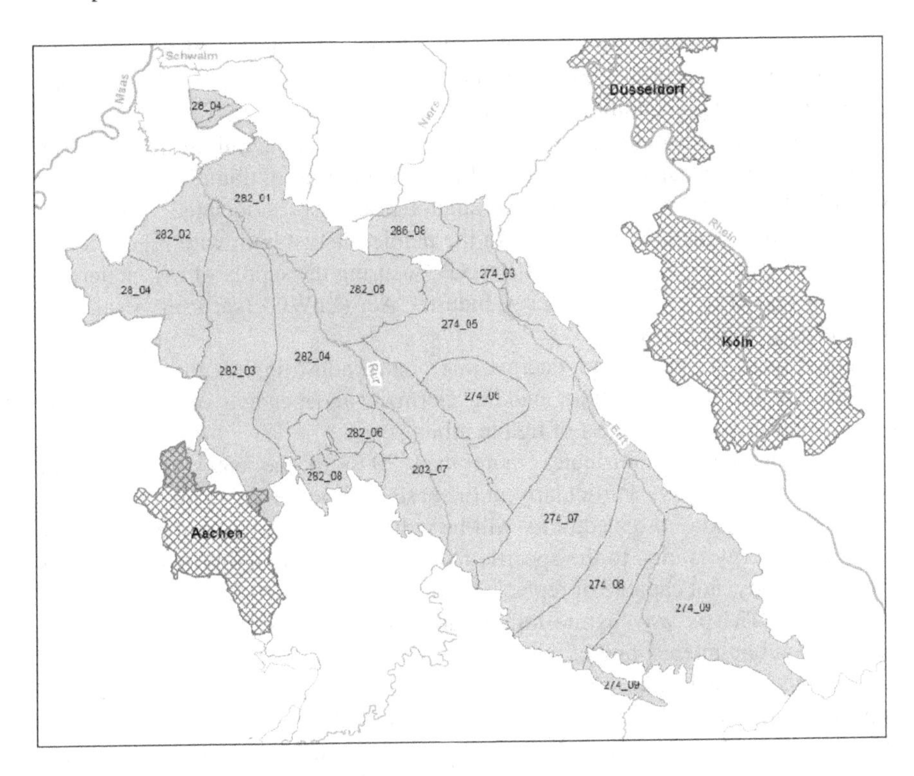

Fig. 3 Groundwater body which is not expected to reach a good quantitative status in 2015 due to the impact of lignite mining

It needs to be mentioned that due to the dynamic mode of operation and the dynamic progress of lignite mining, it will not be possible to clearly differentiate in the future between any previous interventions to the groundwater balance as a result of ongoing human activities (the consequences of which have, in part, either already occurred or will, to some extent, only appear for the first time in the future) and any new interventions and/or new changes associated with the future progress of the mining activities. Due to the spatial development of open-pit mines, some areas which have not been influenced so far will be newly affected by the lowering of groundwater whereas, in turn, initial groundwater resurgences already occur in the rear sections of open-pit mines. Another decisive factor is the delineation of groundwater bodies which can, of course, never be identical to the constantly changing boundary line reflecting the actual impact area of the drainage measures. That is why it was essential to also consider the exemption clause laid down in § 47 and § 31 of the WHG (formerly § 33 Para. 4 Sent. 2 of the WHG in conjunction with § 25d Para. 3 Nos. 1-3 of the WHG – pursuant to Article 4 Para. 7 of Directive 2000/60/EC dated October 23, 2000) in the Explanatory Statement in order to take into account any anticipated new changes. An exemption from the management objectives for the quantitative groundwater status may be granted provided that there are any recent changes in the groundwater level or in the physical properties of surface waters and, furthermore, provided that certain conditions are met:

Lignite is (in addition to regenerative energy carriers) the only primary energy carrier used in Germany which is available completely from domestic mining and which is, thus, free of any availability and/or price risks associated with import markets. The extraction of lignite is in the public interest both with regard to § 1 Item 1 of the Federal Mining Act (BBergG) (securing the supply of raw materials) and with regard to § 1 of the Energy Industry Act (EnWG) (securing a safe and reasonably priced energy supply).

With regard to the future energy mix, indigenous lignite continues to be available in the long run; thus, making Germany independent of imports from other countries (public interest of lignite mining).

Indigenous lignite contributes more than 20 % to the entire gross power generation in Germany. Particularly in the base load power generation sector, the conversion of lignite into electricity will be indispensable in the future since even renewable energy is, due to the specific features of its availability at the business venue Germany, not capable of replacing lignite which is used for base load power generation. That is why lignite mining and its conversion into electricity is of particular public interest and of overriding importance in Germany when it comes to securing a sustainable supply of energy with a specific focus on the energy supply of tomorrow. According to expert estimates, the costs of generating power from lignite in Germany are at the bottom end of the cost scale for the production of electricity in Germany. This demonstrates that the use of indigenous lignite for power generation not only makes a vital contribution towards a secure energy supply that does not depend on imports, but also substantially contributes towards an affordable supply of electric power (secure and inexpensive energy supply).

Public interest in lignite mining was already considered in the State Planning Decision which is part of the Lignite Mining Plans that had been prepared and approved for the individual open-pit mines. The objectives of this State Planning Decision are binding for the regional authorities in the federal state, and the aforementioned decision constitutes a binding framework specifying the necessity of lignite extraction for the duration of the scheduled extraction activities.

Public interest which is reflected by this State Planning Decision is also paramount in light of the fact that the conversion of Rhenish lignite into electricity actually secures numerous competitive jobs in and beyond the mining district (State Planning Regulations).

Even under the aspects of climate protection, there is and continues to be a primarily public interest in lignite extraction and its conversion into electricity in the Rhenish lignite mining district. Within the district, the RWE Power AG corporation operates lignite power plants, which are currently the most modern ones of their kind, at the locations Niederaußem and Neurath (BoA 2/3). In line with the provisions for older plants, existing blocks will be permanently shut down and replaced. Thus, RWE Power AG's power plant renewal program is making a vital contribution towards reducing specific CO_2 emissions resulting from the conversion of lignite into electricity (consideration of the climate protection aspects).

The energy management objectives, which are pursued in the form of groundwater abstraction with the above mentioned consequence of a deviation from the quantitative targets for groundwater management, cannot be achieved with other suitable measures having considerably less adverse impact on the environment. For lignite mining, the quantitative status in the affected groundwater bodies is, at the same time, impaired by the lowering of groundwater levels in such a way that no sound balance can be guaranteed between groundwater abstraction and groundwater recharge by 2015, and that it will be neither possible to maintain nor to achieve a good quantitative status of the groundwater by that time. In fact, maintaining the groundwater abstraction from the respective groundwater bodies is a mandatory prerequisite for a proper and secure continuation of lignite mining which, in turn, is necessary from an energy management perspective.

When considering alternative energies, the Background Document specifies that a safe and secure energy supply of the population requires an effective energy mix which also includes the conversion of lignite into electricity reaching well into the 21st century.

Experts attest that such potential alternative extraction methods as deep level mining, underwater extraction, and underground gasification are not alternatives applicable to lignite extraction in the Rhenish lignite mining district.

To ensure the secure operation of open-pit mines, complete drainage is indispensable from a geohydrological point of view.

Within the scope of the Explanatory Statement, all potential measures were tested for their applicability on groundwater bodies affected by drainage measures. These investigations also revealed that such technical measures as, for example, impermeable walls are not viable in the Rhenish lignite mining district.

This was followed by an analysis of those measures which had been identified as being suitable with regard to the requirement of altering the good quantitative groundwater status with the least possible deviation:

The State Planning Regulations included in the Lignite Mining Plan also define the actual mining site and its mining limits under the applicable water management and ecological aspects as was the case with the open-pit mine Garzweiler II, where an Executive Decision reached by the North Rhine-Westphalian State Government (1991) reduced the mine's initially intended total surface area by several kilometers due to, for example, water management and ecological issues (the so-called water management and ecological protection line) (considering the impact on the groundwater balance when determining mining limits).

In any case, groundwater is abstracted only in such quantities as are absolutely necessary to safeguard the stability of slopes and bottoms in open-pit mines. In addition, drainage waters are – as far as this is possible – assigned to the appropriate water uses (for example, drinking and industrial water, ecowater, and immission protection), and/or the requisite water abstractions for external uses are also included in the drainage strategy (minimum drainage).

Another option for reducing the adverse effects of groundwater abstraction on the quantitative status of groundwater bodies is to stabilize the groundwater table with infiltration and seepage measures. These measures essentially focus on maintaining the groundwater level in groundwater-dependent areas deserving protection (in other words: Groundwater-dependent terrestrial ecosystems and surface waters), and they are well suited for this purpose depending on diverse general conditions such as, for example, intensity of groundwater abstraction, extent of the areas deserving protection, and available water resources (in particular, drainage water).

Large-scale support was provided, for example, north of the open-pit mine Garzweiler, an area with widespread, interconnected groundwater-dependent wetlands. This is done, on the one hand, by direct infiltration of drainage water into the groundwater body – via seepage trenches, dry wells, and injection lances – and, on the other hand, by stabilizing the water balance through surface discharge into flowing waters and wetlands. Today, approximately 70 million m^3 of drainage water are discharged every year either directly and/or indirectly into the groundwater aquifers as so-called ecowater designed to stabilize the wetlands; by 2030, this amount is expected to increase to more than 100 million m^3 per year. Normally, these seepages occur in the top aquifer; but depending on the specific hydrogeological conditions, an infiltration into deeper aquifers might also be useful in order to avoid lowering the upper aquifers (large-scale groundwater replenishment through reinfiltration of the drainage water).

As is practiced in several wetlands, for example, along the inflows of the Rur River which originate in the Venloer Scholle region, it might also be reasonable to use groundwater abstracted from deeper aquifers as seepage water for the in-situ stabilization of the top aquifer.

If the receiving waters are sufficiently effective as is the case with the Erft and Rur Rivers, then it is common practice to carry out water-compatible abstractions from these receiving waters and to feed them back into nearby terrestrial ecosystems (please see Figure 4) and smaller surface waters and/or to construct drainage channels which are also designed to stabilize groundwater-dependent terrestrial ecosystems and surface waters. Additional specific individual groundwater conserving measures are coordinated with the respective public authorities in charge (in-situ groundwater stabilization and other in-situ measures).

In addition to the requirement for supporting the water supply with substitute supplies of drainage water or water obtained from other sources, the Explanatory Statement substantiating the exemption also focuses on the mass deficit, the so-called residual holes, which remain in the aftermath of lignite mining. So that raising the water level is accelerated in these residual holes (and/or then residual lakes) as well as in the surrounding groundwater bodies, water from external sources (often from efficient receiving waters in the vicinity) is fed into the residual lakes which helps achieve a good quantitative status within a considerably shorter period of time. Despite this infiltration of external waters, though, a sound water balance can only be expected decades after the lignite mining has ceased (substitute water supply, accelerated groundwater resurgence by filling residual lakes with water from external sources).

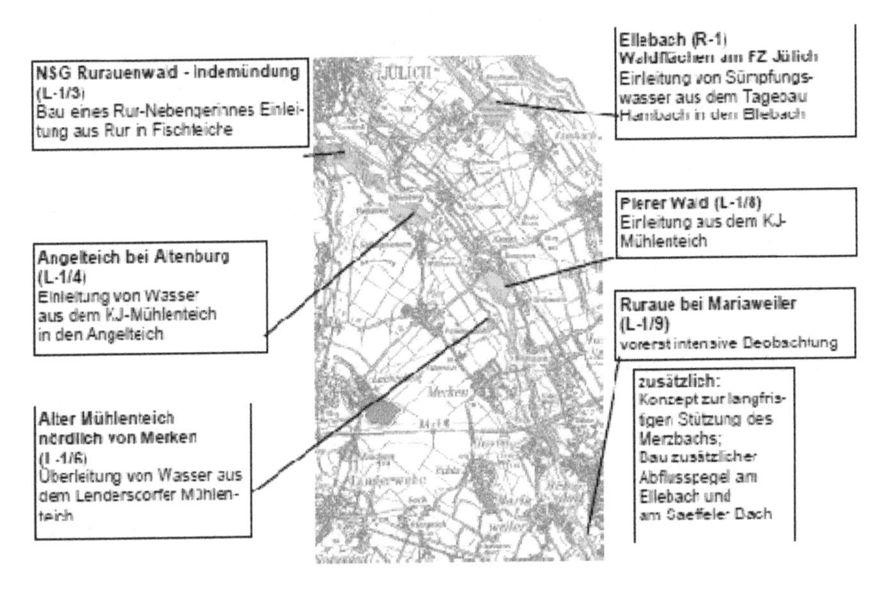

Fig. 4 Diverse in-situ measures to stabilize groundwater-dependent terrestrial ecosystems and surface waters along the Rur River east of the open-pit mine Inden

The measures described above encompass anything that is technically feasible and practically suitable for reducing the adverse impact on the quantitative status of the groundwater body and its affiliated surface uses (groundwater-dependent terrestrial ecosystems and surface waters).

According to a previously established categorization, the appropriate <u>suitable measures</u> were classified into six types and allocated to the affected groundwater bodies. These measures are implemented through the respective drainage, discharge, and seepage permits which are granted in accordance with the water law.

In addition to the actual implementation, supervising the lowering of the groundwater, determining suitable measures, and supervising the effectiveness of such measures are all subject to comprehensive monitoring carried out by monitoring work groups which were specificially established in this regard and in which all involved and/or affected public authorities and institutions can participate. For this purpose, the mining operator has all-embracing duties and obligations to cooperate and report (http://www.bezreg-koeln.nrw.de/brk_internet/gremien/braunkohlenausschuss/monitoring/index.html).

3.2 *Exemptions from the Management Objectives for the Chemical Status of Groundwater*

The lowering of groundwater and the resultant aeration of the rock mass as well as, above all, the shifting of soil materials, which are to some extent sensitive to acidification, which might occur in the course of the lignite mining activities and the resultant physical intervention into the water body create hydrochemical processes inside the mine dump body during which the pyrites (FeS_2) geogenically contained in the rocks are oxidized through aeration. Along with the resurging groundwater, sulphate as well as iron and hydrogen ions are then released into the dump bodies of open-pit mines which is also accompanied in some sections by acidification and the dissolution of heavy metals – this depends on the specific hydrogeological conditions on site. Locally, lignite residues in mine dumps also result in the formation of ammoniac nitrogen (NH_4-N). Any contamination with heavy metals, ammoniac nitrogen, and iron as well as any acidification are basically limited to the actual mine dump and/or its immediate runoff area. Only sulphate, which exhibits as a substance essentially a conservative behavior, leads to increased sulphate contamination even in extended groundwater runoff sections of overburden dumps and, thus, also to a deterioration of the groundwater quality in such areas. These contaminations inside the dump body and subsequently also in the groundwater runoff section result, in part, in a potential deviation from the management objectives pursuant to § 47 (formerly § 33 a Para. 1 of the WHG) according to which any detrimental changes in the chemical status of groundwater are to be avoided and any significant trends of increasing pollutant concentrations due to the impact of human activities are to be reversed.

Currently, this deviation specifically exists in so-called "old dumpsites" of former open-pit lignite mines (along the open-pit mines Ville and Zukunft) as well as in outside dumps (for example, Sophienhöhe, Vollrather Höhe, Glessener Höhe).

With regard to the existing groundwater contaminations, an exemption from the management objectives for the chemical groundwater status might initially be considered in accordance with the rules laid down in §47 and § 30 of the WHG (formerly § 33a Para. 4 Sent. 3 of the WHG under the provisions of § 25d Para. 1 Nos. 1-4 of the WHG).

Subsequent thereto, deviations may be made from the objectives to safeguard and assure a sound balance between groundwater abstraction and groundwater recharge as well as to maintain or achieve a good quantitative and chemical status of the groundwater by 2015; under certain conditions, less stringent management objectives may be determined as well.

When it comes to lignite mining, these prerequites are met with regard to the actions lowering the groundwater and shifting the material which both trigger pyrite oxidation.

Concerning the requirements of § 30 of the WHG (formerly No. 3 of § 25d Para. 1 of the WHG – avoidance of further deteterioration), the dynamic progress of open-pit lignite mining needs to be mentioned here because it does not permit any clear differentiation between previous interventions into the groundwater balance as a result of ongoing human activities (the consequences of which have, in part, either already occurred or will, to some extent, only appear for the first time in the future) and new interventions and/or new changes associated with the future progress of the mining activities.

Against this backdrop and in order to substantiate any deviations from the water protection objectives for groundwater, it was necessary to also take into acount § 47 and § 31 of the WHG (formerly § 33a Para. 4 Sent. 2 of the WHG in conjunction with § 25d Para. 3 Nos. 1-3 of the WHG – pursuant to Article 4 Para. 7 of Directive 2000/60/EC dated October 23, 2000) so as to keep abreast with the changes that are already foreseeable today.

An exemption from the management objectives for the chemical groundwater status may, under certain conditions, be granted if there is any change in the groundwater level due to drainage measures carried out in the course of lignite extraction as well as the requisite physical change to the water (shifting the groundwater body) and the detrimental changes which ultimately occur as a result of the aforementioned changes (in order to avoid repetitions with similar contents, only the new aspects are mentioned below):

Significant influencing variables for deviations in the chemical status from the management objectives as a result of pyrite oxidation are the presence of pyrite in the rock mass, the contact of pyrites with oxygen, and groundwater resurgence. Deviations from the management objectives are, therefore, avoidable if one of these three conditions can be prevented. The following options were examined:

The presence of pyrites is geogenically determined and inseparably linked to the lignite deposit; it can, thus, not be avoided. The contact of the pyrites with oxygen results primarily from shifting the overburden above the lignite and the associated lowering of the groundwater.

Ultimately, the resurgence of groundwater is supposed to be consistent with the objective of a good quantitative status, which is to be achieved again after lignite extraction has ended, and is unavoidable if this objective is to be met. That is why pyrite oxidation in the overburden dumps of open-pit lignite mines and the mobilization of its products along with the resurgence of groundwater cannot be avoided.

Initially, potential measures were determined; this was followed by an examination of their applicability to the affected groundwater bodies. Next, those measures were identified which are neither technically feasible nor productive such as, for example, mine dump sealing systems.

This was then followed by an analysis of all suitable measures which need to be undertaken to reduce any potential detrimental effects on the chemical status of the groundwater resulting from dewatering the rock mass and dumping the overburden.

When shifting the material during the extraction and dumping process, overburden masses with higher pyrite contents are primarily dumped into lower sections while upper dump sections are primarily built from overburden masses with lower pyrite contents. In places where it is useful from a geohydrological point of view and where it is possible due to the mass disposition of the pertinent overburden material, the dump section adjacent to the upper groundwater aquifer (also referred to as dump ridge) has to have as little pyrite as possible in it (selective dumping; the so-called A1 measure).

Another method of reducing pyrite oxidation is to minimize the air exposure of those layers which exhibit higher pyrite contents in such a way that the bottoms of open-pit mines – where in-situ layers of near-surface materials are exposed to air for a longer period of time – are relocated to areas exhibiting the lowest possible pyrite contents (optimized location of bottoms; the so-called A2 measure).

If the pyrite contents increase for geological reasons, then limestone is added to the overburden masses during dumping. Even though it is actually impossible to reduce pyrite oxidation by adding limestone, this addition accelerates the immobilization of the resultant secondary products and raises the pH value once again to almost neutral ranges (liming of dumps; the so-called A6 measure; please see Figure 5).

The specific on-site conditions prevailing at the open-pit mine Garzweiler provide the opportunity of arranging retention wells even retroactively in the dump's groundwater runoff section. With the help of these retention wells, sulfate-contaminated water can be removed, subsequently cleaned, and finally infiltrated into the groundwater aquifer again. However, a decision on this potential measure can only be reached decades after groundwater resurgence and groundwater runoff commenced (option of retention wells).

Depending on the degree of any potential qualitative impairment resulting from pyrite oxidation products running off from the dumps and depending on specific alternatives, it might be useful and/or necessary to relinquish the water supply at least partially at that location and to move it to a different location and/or to a groundwater horizon which is not and/or less affected by any dump water runoffs (adjustment of water supply locations and horizons).

In line with the previously established categorization, the appropriate suitable measures were classified into four types and allocated to the affected groundwater bodies. These measures are implemented through the respective approvals and permits granted by the District Government Arnsberg in accordance with the mining and water laws and subject to monitoring. The measures outlined above exploit all options that are technically feasible and practically suitable to reduce pyrite oxidation and its resultant secondary products, and they also help reduce the runoff volume of these secondary products into the top groundwater aquifers which, in turn, play an important role in water management.

Fig. 5 Conveyor belt collection point Garzweiler II with lime silos for the buffering of overburden masses – an A6 measure

3.3 Exemption from the Management Objectives for Surface Waters in Association with the Introduction of Drainage Water from Lignite Mining

Pursuant to § 27 (formerly § 25a) of the WHG, surface waters, provided that they are not classified as artificial or substantially modified, are to be managed in such a way that any adverse changes in their ecological and chemical status are avoided and that a good ecological and chemical status is maintained or achieved.

The drainage activities may affect surface waters in the river basin subdistricts of the Erft, Rur, Niers, and Schwalm Rivers.

When it comes to larger bodies of water (specifically the Rur and Erft Rivers), lowering the groundwater entails a loss of groundwater contact at least in some areas and, thus, reduces the runoff volume. In contrast, the runoff volume increases through the introduction of drainage waters into the Erft and/or Inde Rivers (and via the latter, also into the Rur River). It is, in particular, the Erft River into which so much drainage water is introduced – even though this drainage water is already used, to a large extent, directly and/or indirectly via the Erft River to supply the power plants along the Erft River – that it is currently overcompensated to a large extent as a result of the reduced runoff volume caused by the lowering of the groundwater which, in turn, changes the runoff system of this water body and has, at least during low water periods, a formative influence on runoffs occurring in the lower stretch of the Erft River.

Since the mid-20th century, the industrial extraction of lignite has required the large-scale drainage of groundwater as described above. In order to discharge this drainage water, but also to improve flood protection, the Erft River was expanded several times. Increased runoff volumes permit the extraction of cooling water for lignite power plants. Most of the accumulated drainage waters are fed into the Erft River near Bergheim. It should be noted at this point that diverse additional morphological changes were made to the Erft River which occurred independent of lignite mining (melioration and straightening for the draining of the Erftaue floodplain, flood protection measures). The information provided in the Background Document, however, focuses exclusively on mining-related topics (achieving the objectives of lignite mining and its conversion into electricity) and, thus, on the introduction of drainage water.

Even when it comes to its qualitative properties, the introduction of drainage water has an impact on the chemical and ecological status of flowing waters. Compared to natural flowing waters, drainage waters exhibit, geogenically determined, relatively low oxygen contents, increased iron and sulfate contents as well as higher temperatures. In conjunction with the drainage water volume that is introduced, the resultant high heat load causes the Erft River to warm up particularly in the winter and, when considering a period of one year, entails an unnatural homogenization of temperatures over the entire year.

Due to the dynamic progress in lignite mining and the associated changes in the volume and quality of the drainage waters that are introduced and due to the various new measures that need to be undertaken (particularly limiting the heat load supply and the associated medium-term and long-term reduction of drainage water volumes that are introduced as well as the ecological transformation of the water bed), the situation – particularly along the Erft River's lower stretch – is subject to changes which are designed to trigger the transformation of the Erft River's lower stretch into a good chemical and ecological status (and/or potential).

For example, a reduction of the drainage water volumes that are introduced, which is indispensable in the long run, results in an increased concentration of diverse water body contents originating from upstream riparians (for example, phosphate) because the thinning effect of the introduced drainage waters decreases. Furthermore, the reduced flow rate leads to reduced flow velocities and longer retention times in impoundment sections with the associated ecologically negative, reinforced oxygen consumption processes. Even though these negative trends are to be counteracted with parallel measures, it cannot be completely ruled out that the status of partial sections along the lower stretch of the Erft River will deteriorate temporarily due to the new changes revolving around the introduction of drainage waters resulting from the lowering of the groundwater as well as the intended physical changes to the Erft River's lower stretch.

Similar to the Erft River, but to a much smaller extent, the impact on surface waters also applies to the discharge into smaller water bodies (for example, the upper stretches of the Niers, Schwalm, and Mühlenbach Rivers) into which treated drainage water is introduced in order to prevent them from desiccation; this type of impact is, though, not relevant for the Rur and Inde Rivers.

In accordance with the provisions of § 30 (formerly § 25d Para. 1) of the WHG, a deviation from the management objectives for the ecological and chemical status of surface waters and/or the specification of less stringent management objectives for surface waters are initially taken into consideration for the current introductions to, utilizations, and/or impairments of surface waters which are necessary to safeguard a secure and effective lignite extraction in the Rhenish lignite mining district. And less stringent objectives can be determined thereafter provided that certain prerequisites are met.

Pursuant to § 30 (formerly Nos. 1, 2, and 4 of § 25d) of the WHG, any lowering of the groundwater on behalf of lignite mining and the resultant introduction of drainage water into the Erft River which is changed over time due to the dynamic development of open-pit lignite mines is also subject to § 47 and § 31 of the WHG (formerly § 33a Para. 4 Sent. 2 of the WHG in conjunction with § 25d Para. 3 Nos. 1-3 of the WHG in accordance with Article 4 Para. 7 of Directive 2000/60/EC dated October 23, 2000).

According to the aforementioned provisions and provided that certain prerequisites are met, any deviation from the management objectives for the ecological and chemical status of surface waters may be admitted, on the one hand, on account of any changes in the physical properties of surface waters and, on the other hand, on account of any changes in the groundwater level and any related, ultimately resultant adverse changes in surface waters (in order to avoid repetitions with similar contents regarding some of the prerequisites, only the new aspects are mentioned below):

The physical changes occurring in the form of ecological transformation, which the water body of the Erft River's lower stretch will undergo over the next few decades, are of greater public interest as was ascertained in the Concept for the Future on the ecological transformation of the Erft River region (http://www.erftverband.de/oberirdische-gewaesser/gewaesserbewirtschaftung/erftumgestaltung.html). This Concept for the Future also outlines in detail that the selected combination of measures cannot be replaced and/or altered by any other measures which would have had a significantly lower adverse impact on the environment and which would not have been associated with a disproportionately greater effort:

Due to the historic anthropogenic changes to water bodies and the extraction of lignite since the 20th century for the supply of energy and power, the lower stretch of the Erft River located in the Rhenish lignite mining district is affected specifically by the impact of lowering the groundwater to dewater the mining sections and the associated introduction of drainage waters in such a way that any adverse change to these water bodies cannot be avoided and that their good ecological and chemical status can neither be maintained nor attained by 2015.

It would only be possible to avoid the above mentioned changes in the thermal-physical properties of the Erft River's lower stretch if at least one of the two factors "increased temperature" or "introduced drainage water volume" was removed.

The water temperature of the discharged drainage waters is geogenically determined; drainage water is already used by the power plants located in the region, by industrial and commercial enterprises, for the public water supply, wetlands, and other ecowater introductions, and it also meets the on-site demand of open-pit mines. Any extended use of drainage water in remote regions would include more energy being consumed to transport the water as well as the landscape being impaired by the requisite construction of additional pipelines. In addition, longer pipelines also reduce the secure and sustained supply for water utilization. Expanding the water utilization is, thus, neither technically viable nor ecologically useful.

Another method of discharging drainage waters would generally be feasible from a technical point of view by transferring these drainage waters to the Rhine River. However, within the scope of the project "Transformation of the Erft River Region According to the Water Framework Directive" launched by the North Rhine-Westphalian Ministry for the Environment, Nature Conservation, Agriculture, and Consumer Protection (MUNLV) (today's Ministry for Climate Protection, the Environment, Nature Conservation, Agriculture, and Consumer Protection (MKUNLV)), it was ascertained in 2004 that the introduced drainage water also has a positive, i.e. thinning, effect on the concentration of nutrients and other harmful substances and pollutants. Any timely, significant reduction of the drainage water volume introduced into the Erft River would, thus, result in a substantial increase in the concentration of nutrients and pollutants. But it would also cause a considerable decrease in the flow velocities and, thus, increase the oxygen consumption processes in impoundments and slow flowing sections which, in turn, would have a significant, negative impact on the flora and fauna of the lower stretch of the Erft River. That is why any timely significant reduction of the drainage water volume introduced into the Erft River is counterproductive from an ecological point of view; it can only be implemented in conjunction with a long-term morphological transformation of the Erft River's lower stretch along with the respective reduction of the influx of nutrients and other harmful substances and pollutants from the Erft River's medium stretch.

In order to ascertain and implement suitable measures designed to reduce the adverse effects of both the mining-related hydraulic and thermodynamic stress on the lower stretch of the Erft River and the stress factors that are not related to mining, the MUNLV (today's MKULNV) had initiated the project "Transformation of the Erft River Region According to the Water Framework Directive" (or, more precisely: The lower stretch of the Erft River) in 2004.

Within the scope of this project, the decisive influencing factors on the chemical and ecological status of the Erft River were initially examined, and potential measures were investigated in order to avoid/reduce these influences.

As a key measure designed to improve the ecological status along the Erft River, a decision was reached in favor of transforming the Erft River region pursuant to the Water Framework Directive. This will be implemented with the participation of the RWE Power AG corporation in accordance with a general agreement that had been concluded between the MUNLV (today's MKULNV), the Erftverband water management association, and the RWE Power AG corporation in September 2008.

Another decision was reached in favor of additional drainage water aeration because of its basically positive ecological impact and comparably low costs. It should be noted in this context that RWE Power AG had carried out the aeration of drainage waters at two significant discharge points already before 2004.

Also with regard to iron, RWE Power AG already conducts diverse measures for the deferrization of drainage waters (operation of mine water purification plants and sedimentation basins, optimized control and monitoring of drainage and discharge systems).

Various measures designed to avoid/reduce the hydraulic and thermodynamic stress on the lower stretch of the Erft River have already been discussed above. These measures include, on the one hand, a reduction of the drainage water volume by minimizing the actual groundwater abstraction and, on the other hand, the extensive use of drainage water for water supply purposes. In addition, various other methods of discharging drainage waters were also analyzed in the above mentioned study with the result that any timely, significant reduction of drainage waters prior to implementing any other measures (morphological transformation of the Erft River region, treatment of precipitation water) would not only be extremely cost-intensive, but also ecologically counterproductive.

Due to the significant influence which the introduction of drainage water has on the Erft River's heat balance, the heat load, though, needs to be limited in the future – in line with the stipulations of the project "Transformation of the Erft River Region According to the Water Framework Directive" – in order to prevent any resultant deterioration of the ecological water body status. That is why the alternative of a "constant heat load" was selected from among a number of scenarios that had been examined for the introduction of drainage water (please see Figure 6). This alternative focuses on the introduction of drainage water until the end of the mining operations while also considering the ban on deterioration pursuant to the Water Framework Directive. This means that depending on the – geogenically determined – continuously increasing temperature of the raised groundwater level, the approved discharge volume, which currently amounts to 8.5 m^3/s, will decrease to 7.6 m^3/s over the next couple of years. Limiting the heat load, as mentioned above, also results in further limiting the iron load.

RS 0: Reference scenario "zero alternative" (complete discharge of the accumulated drainage water)

RS A: Reference scenario "complete decoupling"

a1, a2: Alternative "Fish Waters Ordinance" (permissible temperature rise $\leq 3°C$)

b1, b2: Alternative "temperatures over an entire year"

c: Alternative "constant discharge volume"

d: Alternative "constant heat load"

for a, b: Reduction by 2015 (a1, b1) and/or by 2035 (a2, b2)

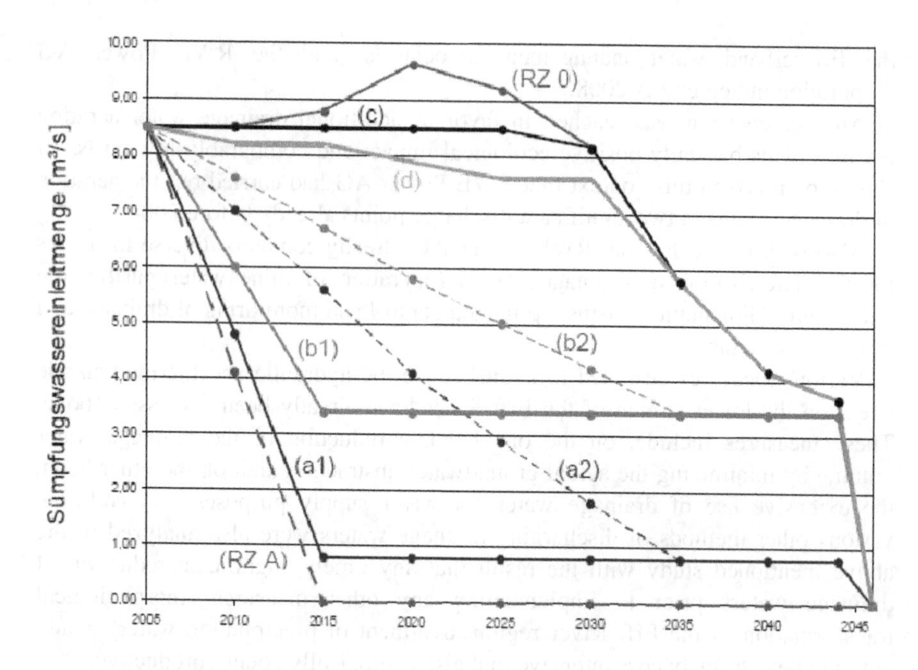

Fig. 6 Scenarios for the future development of drainage water discharge

The measures focusing on additional drainage water aeration and the idea of a "constant heat load" were already implemented. A period of approximately 40 years was mentioned in the Concept for the Future as the time horizon for the morphological transformation of the lower stretch of the Erft River which had also been part of the decision.

This period of time is based, on the one hand, on the scope and intensity of the transformation (transformation of a 40 kilometer long, technically developed river stretch with diverse dams into an approximately 53 kilometer long flowing water in line with the Water Framework Directive) but, on the other hand, also on the population density and utilization intensity of the affected region which requires an intensive, technically and legally substantiated coordination of individual measures with all the stakeholders and the public at large. It will, thus, not be possible to morphologically transfer and accordingly transform the entire lower stretch of the Erft River into a good ecological and chemical status for its surface waters so that it meets the requirements of the EC Water Framework Directive within the deadline set for 2015.

4 Management Plan and Program of Measures

The Management Plan for the North Rhine-Westphalian sections of the Rhine, Weser, Ems, and Maas Rivers lists the exemptions for lignite mining.

The current situation of lignite mining is described in Chapter 6 Status of Surface Waters and Chapter 7 Status of the Groundwater with reference to the Background Document. Additional details are included in Chapter 8 Analysis of Human Activities and the Significant Contamination of Water Bodies. Item 10.6.4.1 Lignite Mining of Subchapter 10.6.4 Less Stringent Management Objectives in North Rhine-Westphalia which is a part of Chapter 10 Management Objectives for Individual Water Bodies contains exemption clauses for lignite mining which are in accordance with Article 4 Para. 5 of the Water Framework Directive (exemption from achieving the objectives) as well as in accordance with Article 4 Para. 7 of the Water Framework Directive (exemption from the ban on deterioration). These exemption clauses are abbreviated as A2 and A4.

The Program of Measures (http://wrrl.flussgebiete.nrw.de/Dokumente/NRW/Bewirtschaftungsplan_2010_2015/Ma__nahmenprogramm/Ma__nahmenprogramm_NRW_Gesamtdokument.pdf) for the North Rhine-Westphalian sections of the Rhine, Weser, Ems, and Maas Rivers is basically limited to those water bodies which require mandatory reports to the EC, i.e. limited to all flowing waters having a catchment area of more than 10 km^2, to lakes having a surface area of more than 0.5 km^2, and to groundwater bodies. A summary of the Program of Measures is also included as Chapter 8 of the Management Plan.

In addition, the "planning unit profiles" contain tabulations listing the particular measures that are intended for each water body group. The requisite measures are described programmatically; namely, with the help of a "catalog name," a specific description, information on the supporting entity or entities of the measures as well as the respective implementation period (2012, 2015, or 2021/2027).

Reference to the exemptions for lignite mining was made in the Program of Measures. **Chapter 6 Measures to Reduce the Contamination of Groundwater**, for example, refers to Chapter 10 of the Management Plan which does not specify the planning of any additional measures in the lignite mining sector because comprehensive measures designed to reduce the impact of lignite mining are already implemented within the scope of lignite mining operations and because a detailed description of the correlations is already included in the Background Document on Lignite.

The **Concept for the Future of the Erft River Region** is mentioned in **Chapter 8.1 Measures for Specific Cases and Other Contaminations**. The renaturation and development of the Erft River is, thus, indispensable for the 50 km stretch which is influenced by drainage water. According to the timeline, this is to be gradually implemented in chronological phases; short-term measures by 2015, medium-term measures by 2027, and long-term measures by 2045. The time horizons 2015 and 2027 not only permit the realization of this transformation process, but at the same time also the objectives set by the EC Water Framework Directive and the requirements of the Water Management Act. The time horizon 2045 is dependent on the discontinuation of drainage water discharge after open-pit mining has ceased. The estimated total costs for the "Concept for the Future of the Erft River Region until 2045" add up to approximately 95 million euros.

5 Outlook

The requirements for water protection will be pursued with great effort by the mining operator and continue to be subject to intense monitoring by the public authorities. The transformation of the Erft River region is already underway. For this purpose, a Water Body Implementation Schedule was prepared in 2013 (http://www.erftverband.de/oberirdische-gewaesser/gewaesserbewirtschaftung/umsetzungsfahrplan/).

In those areas where drainage water is introduced along the Erft River, the requirement of a constant heat load will be consistently pursued in a strategic manner. The mining operator will make major investments in this regard.

Comprehensive data and information on the current situation of surface waters and groundwater resources in North Rhine-Westphalia were compiled in the updated Regional Survey of the Water Framework Directive which was completed in late 2013. A state-wide summary of this Regional Survey will be published in 2014 as a separate report.

With the publication of the "Significant Water Management Issues" in 2013/2014, the second phase of public participation in preparing the second Management Plan commenced. The report provides an overview of the current situation of the North Rhine-Westphalian water bodies and groundwater bodies and presents those aspects which are important in implementing the EC Water Framework Directive http://www.flussgebiete.nrw.de/img_auth.php/d/db/WWBF_2013_WEB.pdf.

Both the Management Plan and the Program of Measures will remain in effect until December 22, 2015, and need to be updated in accordance with § 84 of the WHG by the end of 2015. North Rhine-Westphalia's contributions to the Management Plans for the Rhine, Weser, Ems, and Maas Rivers are to be adopted by the State Government by December 22, 2015, and they shall become compulsory and binding for the public authorities in North Rhine-Westphalia. The described exemption clauses for lignite mining are adjusted therein.

Securing Final and Border Slopes in the Open-Pit Lignite Mines of the Rhenish Mining District from the Perspective of the Mining Authority

Rolf Petri

Bezirksregierung Arnsberg Abteilung 6 Bergbau und Energie NRW, Düren, Germany

1 Characteristic Features of the Rhenish Lignite Mining District

The Rhenish mining district in North Rhine-Westphalia has a deposit volume of 35 billion tons of economically extractable reserves; thus, making it one of the most significant lignite deposits in Germany and Europe. Over more than 100 years, a mining industry has evolved which is concentrated in the three large-scale open-pit mines Garzweiler, Hambach, and Inden of the mining company RWE Power AG today. Approximately 100 million tons of lignite are mined every year and used primarily to generate electricity.

The lignite deposits in the Lower Rhine Basin date back to the Tertiary Period. The overburden consists of loosely packed sedimentary rock mass originating in the Quaternary and the Tertiary Periods. It has a series of water bearing and water storing layers arranged in several groundwater horizons. For mining, the overburden has to be dewatered in several steps all the way to the seam floor. Up to three lignite seams with a total thickness of about 60 m are available for mining. The underlying stratum of the oldest seam is found at depths of up to 420 m below the terrain's surface. The deposit consists of several fault systems running from the southeast to the northwest in four blocks.

At some of the faults, slow tectonic shifts as well as earthquakes occur even today. The Lower Rhine Basin is one of Germany's most earthquake prone regions; although in a comparison with the rest of the world, Central Europe has to be classified as a region with few earthquakes. Yet more than 1,000 quakes have been registered since exact measurements and recordings were first taken in 1955; the strongest one occurred on April 13, 1992 with a magnitude of 5.9.

The region in which the lignite is mined used to be originally characterized by agriculture. Due to the high earnings resulting from the agriculture and the steadily expanding commercial and industrial sector as well as an excellent infrastructure, the population density is higher than 500 inhabitants per km^2 and corresponds to the median population density of North Rhine-Westphalia while the average for the Federal Republic of Germany is considerably lower at an average value of 226 inhabitants per km^2.

C. Niemann-Delius. (ed.), *Proc. of the 12th Int. Symp Continuous Surface Mining - Aachen 2014*,
Lecture Notes in Production Engineering, DOI: 10.1007/978-3-319-12301-1_48

Lignite mining in the Lower Rhine Basin faces primarily the following problems:

- Considerable water management efforts have to be undertaken to dewater the open-pit mines
- Thick overburden layers and, thus, considerable extraction depths
- Additional seismic stress
- High conflict potential with the people and the business community in the surrounding region due to several relocations and many points of contact between the mining operation and the neighborhood

It is due to the above mentioned issues that the safe and secure construction of open-pit border slopes and final slopes assumes a particularly important role.

2 District Government Arnsberg as the Mining Authority in North Rhine-Westphalia

The official approval and monitoring of the safe operation of lignite open-pit mines in the Federal State of North Rhine-Westphalia (NRW) are the duties and responsibility of the District Government Arnsberg, Department of Mining and Energy in North Rhine-Westphalia, which is the responsible mining supervisory authority for the entire state. Within the scope of its responsibilities associated with mining supervision, the governmental office regulates and monitors also the safe and secure construction of open-pit slopes.

3 Elements of State Lignite Planning as the First Step in Safety

Border slopes, which are created along the excavation boundary of an open-pit mine, as well as permanent final slopes are subject to a two-step safety concept in the governmental planning and approval process.

During the first step, the mining is delimited in the so-called lignite plan. Lignite plans are important elements in North Rhine-Westphalia's regional planning. They are created in a special regional governmental board, the lignite committee, to plan an open-pit mining project or relocations and are approved by the state government. In the lignite plans for mining projects, a safety zone is already fixed by determining the excavation boundary and safety line in accordance to the provisions[1] of the State Planning Act. The stipulations of the lignite plans have to be observed by all governmental offices in their respective approval procedures.

The excavation boundary specified in the lignite plan includes the excavation area within which the extraction of lignite has priority above and beyond other

[1] Ordinance to Implement the State Planning Act (State Planning Act DVO – LPlG DVO) of June 8, 2010 GV. NRW. P. 334.

utilizations and functional claims. During the later mining approval procedure, which is based on the lignite plan, mining may only be approved within the excavation boundary. The excavation boundary represents the inner boundary of the so-called safety zone while the outer boundary of this zone is formed by the safety line which is specified by parcels. The safety zone is the section between the edge of the excavation/dump site and the safety line whose width is based primarily on mining safety related aspects. Its width usually corresponds to either half the depth or even the full depth of the open-pit mine at the site in question, but at least 100 m. The safety line encloses and surrounds that area which would be affected directly by excavation and/or dumping activities taking place on the terrain's surface. It is in particular in this area, if it should prove to be necessary, that measures can be undertaken against hazards and any other issues associated with mining. This zone also provides room for operational measures and facilities, for example, wells and pipes, to dewater open-pit mines as well as installations which are used for emission control during the time of operation. When approving the general operating plan for mining, the mining authority makes the excavation boundary and the safety zone legally binding for the mine operator.

4 Approvals in Accordance with the Mining Law as the Second Step in Safety

While a first safety element comes into existence by determining a horizontal distance on the terrain's surface during lignite planning, the creation of a stable slope system comes into play during the second step of the safety concept. Geometrically, this applies to the design of the general incline of the slope which is defined as the angle of the line between the upper edge of the slope and the lowest part of the slope's toe above the horizontal. The general incline of the slope is determined by the inclination of the individual slopes and the corresponding number and width of the berms structuring and dividing the entire slope. This second safety element assumes a particular, decisive role in the practical application.

The mining authority demands proof of the stability attuned to individual cases for all relevant border and final slope systems of an open-pit mine in the course of the official mining approval procedure authorizing the general operating plan and also subsequently at regular intervals when submitting the special operating plans.

5 Guideline for Stability as a Generally Accepted Code of Practice

To examine and evaluate the stability of border slopes and permanent slopes in open-pit lignite mines and the requisite elevated dumps as well as the residual mining lakes, the generally accepted code of safety practice in North Rhine-

Westphalia is the District Government Arnsberg's Guideline[2] for Investigating the Stability of Slopes in Open-Pit Lignite Mines in its most recent version containing the first supplement dated August 8, 2013 – 61.19.1-2-1. It describes the requirements which need to be considered in the verification procedure. According to the stipulations of the Mining Ordinance for Lignite Mines, it is necessary to furnish proof of the stability of the border slopes and final slopes. This proof is to be submitted by the mining company already during the planning phase and has to be approved by the mining authority prior to its implementation.

The stability can be verified via:

- Geotechnical investigations (geological, hydrogeological, and geomechanical studies)
- Surveying documents
- Calculations of the stability
- Evaluations of the stability
- Results of monitoring measures

The mining authority generally determines already while examining the general and main operating plans for which slopes it is necessary to prepare stability calculations.

The requirements for stability need to be given sufficient consideration already while planning the slopes. That is why the preliminary exploration of the geological and hydrological conditions needs to be carried out in a timely manner.

6 A Requirement for Stability Calculations

Stability calculations are needed as essential evaluation criteria for verification in line with the Mining Ordinance for Lignite Mines if and when the stability cannot be taken for granted on the basis of previous experiences and evidence.

Stability calculations are generally to be prepared if and when:

- The stability cannot be considered as given after the geotechnical investigations have been properly assessed;
- any objects in need of protection (e.g. special buildings, industrial plants, traffic structures, supply and waste disposal pipes, flowing bodies of water, retaining structures, nature reserves) are found in the area bordering on the slope's edge;
- any ground conditions that favor landslides (e.g. tectonic stress zones, stratum boundaries or strata with little stability, adverse hydrological conditions, additional static loads or vibrations and tremors, for example, due to transportation infrastructures, old mining structures, coal pillars or permanent coal pillars, former dump sites) are present; or
- the mining authority demands it in individual cases.

[2] http://esb.bezreg-arnsberg.nrw.de/a_2/a_2_019/a_2_019_007/index.html.

Stability calculations are to be carried out while using the results of the geotechnical investigations which were conducted in procedures which are appropriate for the actual local conditions and conform to the state of technology. Generally speaking, calculations are to be carried out using both averaged as well as adverse soil mechanical parameters; the influence of the characteristic value approach on the stability is to be determined through comparison calculations.

7 Calculation Methods for Different Fracture Mechanisms

The applicable calculation methods and calculation programs along with the documents needed for the verification are to be submitted to the mining authority for approval. That is why the calculation methods were verified and officially confirmed for:

- Fracture mechanisms with circular slip lines (slice procedures)
- Composite fracture mechanisms with straight slip lines (rigid body method)

In many cases, it is often possible to realistically describe fracture mechanisms with the help of circular slip lines. Geological cross-sections depicting the strata, faults, and strata boundaries as well as miscellaneous weak zones, groundwater levels and/or dewatering objectives, and the slope's geometry are to be prepared before determining the potential slip lines for the slope area under investigation. The slope area under investigation is to be divided into perpendicular slices which correspond to the loose rock layers dissected by the slip lines. The calculation method is, thus, called the slice procedure; it is described in more detail in the DIN standard 4084.

The slip lines under investigation are to be specified while taking the instable geomechanical zones into account. The slip lines, which are decisive for adverse stability, are to be ascertained mathematically.

When residual water is present in the sector of the slope under investigation, then it is necessary to consider the pore water pressure with regard to the height of the water level and the specific weight of the water saturated soil. If present, the additional static loads influencing the slope body need to be also considered.

The stability is calculated according to BISHOP's method which is explained with examples in DIN 4084. It is also possible to carry out additional stability calculations which have been tested in open-pit lignite mining (e.g. the B.O.R. or JANBU methods). BISHOP's calculation method can only be used for plain cylindrical slip surfaces; for straight slip line sections and/or fracture sections, calculations need to be carried out using the methods described below.

Due to the tectonic conditions in the Rhenish lignite mining district, it is necessary to also consider for fracture mechanisms that reach deeper into the ground slip lines that run straight in sections due to instable zones which are predetermined by the rock mass. The individual fracture blocks can only move parallel to the external slip lines. The procedure that is also called the rigid body method considers fracture mechanisms which are kinematically possible and which can be clearly solved mathematically. This method is also described in DIN 4084. The procedure is to be used primarily for fracture blocks for which the location of the external and internal slip lines is clearly determined by geomechanically instable zones, such as, for example, thin clay horizons and faults. The investigation is to be carried out for the most adverse fracture mechanisms.

To ascertain the stability coefficient η according to FELLENIUS, a uniform reduction is carried out for the shear strength parameters of all slip lines.

The specific values which are used in the stability calculations to describe the soil mechanical properties come from the soil mechanical field and laboratory investigations. The mining company has been operating a certified geomechanical and soil mechanical test lab for many decades.

8 Assessing the Stability

When assessing whether the slope is stable, it is necessary to take the previous experiences into account while using the results of the geotechnical investigations, the surveying documents, and the calculation results.

If stability calculations need to be carried out, then the necessary stability coefficient needs to be specified and substantiated depending on the extent of the geotechnical investigations, the reliability of the applied characteristic geomechanical values while also considering the risk potential of the objects found in the sector of the slope's edge that need to be protected, the scheduled lifetime of the slope, and the protection of the deposit for each sectional plane under consideration.

The mathematically determined stability coefficient η of individual slopes and slope systems has to be above 1.0 for the worst case scenario.

For objects that need to be protected in the sector of the slope's edge and for permanent slopes, the stability coefficient has to be at least 1.3 for the slope system. When this value is not reached, then an explanation has to be given for each individual case.

When assessing the stability of slopes, it is also possible to take into account the spatial restraints of the slope's toe as well as the measures designed to maintain or increase the stability and to monitor the slope. And the intended utilization has to be taken into consideration as well.

9 Considering Earthquakes

Due to the seismic features of the Rhenish lignite mining district, the impact of potential earthquakes has to be also considered for permanent slopes of residual lakes and elevated dumps. This is investigated with the help of pseudo-static or dynamic methods.

The prerequisite for investigations applying these methods is the absence of liquefaction effects in the materials with which the slopes are actually constructed. That is why the permanent slope has to be designed and built in such a way that soil and/or ground liquefaction is not to be feared. The mining company has to properly demonstrate this.

When the pseudo-static method is used, the forces that come into play during earthquakes are either considered to be part of the gravitational acceleration (k being classified as the "seismic coefficient") or the peak ground acceleration caused by earthquakes (χ being the "pseudo-static coefficient"). The value domain of the seismic coefficient k varies from region to region and the more so, the stronger the earthquake is that is to be anticipated for a particular region. When using the pseudo-static coefficient χ, the magnitude of the earthquake which is to be anticipated is factored in using the peak ground acceleration PGA.

In order to validate the applicability of the selected pseudo-static method for the specific case of elevated dumps and residual lakes in the Rhenish lignite mining district, Prof. Dr.-Ing. Triantafyllidis was commissioned by the District Government Arnsberg to prepare an expert's report.[3] By using dynamic calculation methods, he carried out a retrograde calculation of the pseudo-static coefficient which is to be used in pseudo-static methods. It was, thus, possible to demonstrate that the pseudo-static calculation which was used in this approach can come up with results that are at least equivalent to those of the dynamic methods.

Determining the recurrence intervals and ascertaining the respective ground accelerations for the stability verification is to be carried out as follows:

- For permanent slope systems of residual lakes:
    - Until the final water level is reached (filling phase):
      Design basis earthquake 1 – recurrence interval T = 500 years
    - After having reached the final water level (final state):
      Design basis earthquake 2 – recurrence interval T = 2,500 years
- For permanent individual slopes of residual lakes and permanent slopes of elevated dumps:Design basis earthquake 1 – recurrence interval T = 500 years
  Design basis earthquake 1 – recurrence interval T = 500 years

The definition of these data is carried out as per DIN 19700-10[4] whereby large dams of the Category 1 are to be assigned a recurrence interval of 2,500 years for a global failure due to earthquake damage (probability of occurrence 4*10-4). Compared to Eurocode 8, slope systems of residual lakes actually get higher acceleration rates (recurrence interval Eurocode 8: T = 475 years). The calculation for

[3] T. Triantafyllidis: Gutachterliche Stellungnahme zu Standsicherheitsberechnungen mit Ansatz von „Erdbebenbeschleunigungen für Böschungen im Rheinischen Braunkohlenbergbau“ – Überprüfung des quasistatischen Ansatzes der Erdbebenbeschleunigung bei Standsicherheitsuntersuchungen und Bewertung der Rechenverfahren zur Böschungsstabilität, June 2013.

[4] DIN 19700: Dam Structures, July 2004.

the residual lake slopes is, thus, designed to take a possible later land use into account which is to meet the requirements of DIN EN 1998 (Eurocode 8).[5]

The peak ground acceleration (PGA) on the terrain surface is to be ascertained either via the site-specific data retrieval of seismic hazards from the German Research Centre for Geosciences Potsdam (GFZ Potsdam)[6] or via a site-specific seismic report of the Geological Service North Rhine-Westphalia (GD NRW), or any other competent agency. When the acceleration of the bedrock is to be ascertained, it is necessary to consider ground reinforcement due to the existing sediment cover.

The horizontal and vertical components of an earthquake's side effects are assumed to be acting simultaneously in the calculation. The acceleration is always considered to be a horizontal component moving in the direction of the open slope. For vertical acceleration, both possible directions are to be considered in different calculation steps. If information is only available on the horizontal seismic acceleration, then the vertical acceleration is to be ascertained from it with the factor 0.7.

When using dynamic methods, then the ground acceleration caused by earthquakes is to be calculated according to its elapsed time without any reduction.

If pseudo-static methods are used, then it is necessary to reduce the maximum acceleration PGA which is caused by earthquakes by applying the pseudo-static coefficient χ. The dimension of χ is to be determined in accordance with Prof. Dr. Triantafyllidis' expert report subject to recurrence intervals as well as the location of the fracture mechanisms (near surface or deep fracture mechanisms).

The impact of the earthquake acceleration on the pore waters in the slopes and the water level of the residual lake are to be taken into consideration with the help of the appropriate methods.

10 Official Verification and Approval Procedure

The calculations submitted by the mining company are to be verified on behalf of the mining authority by the Geological Service North Rhine-Westphalia with the help of comparative calculations. The Geological Service also investigates other possible fracture blocks and fracture mechanisms. The results of the soil mechanical field and laboratory investigations will also be verified for their plausibility and completeness. When specific problems exist, it is also possible to consult additional experts.

---

[5] Eurocode 8: Design of Structures for Earthquake Resistance - DIN EN 1998-1:2010-12; DIN EN 1998-1/NA:2011-01; DIN EN 1998-5:2010-2012; DIN EN 1998-5/NA:2011-07, 2010-2011.

[6] German Research Centre for Geosciences (GFZ Potsdam): Interactive Retrieval for Site-Specific Acceleration Response Spectra and Seismic Hazard Maps for Germany for Different Hazard Levels in Accordance with DIN 19700.

The test report of the Geological Service includes information and recommendations for the construction and monitoring of slopes. And finally, the mining authority verifies the stability verification in accordance with all the requirements of the Federal Mining Act and grants approval.

This procedure assures the verification and assessment of the stability by three different agents which is why it is also called the six eyes principle.

11 Constructing and Monitoring Slopes

The company has to construct stable slopes. Dump sites are, in particular, to be built in such a way that alternating strata of different materials are created which produces a special stability. The construction of the slope is accompanied by stability investigations. Surveyors have to record the slope's structure on behalf of the company. With the help of the surveying results, the mining authority can determine whether a slope has been built as planned.

Deformations of the slope's edge systems have to be monitored. That is why manual measurement methods and automated measurement systems with alarms are used. The proper operation of the measurement systems is monitored by the mining authority. Significantly higher accelerations in the overburden deformation can be the first indication of sliding surfaces forming in the overburden and can signal the beginning of a disproportionate deformation. In such a case, the automated measurement systems would trigger an alarm at a permanently staffed location. Past practice has shown that the speed of the overburden deformations in the Rhenish mining district initially starts slowly and then accelerates continuously. When there is a timely warning, then there is sufficient reaction time to initiate measures designed to maintain or strengthen the resistant forces in the overburden. This includes, for example, an immediate compensation of the inner overburden dump, support fillings at the slope's toe, or an additional reduction of the groundwater levels. In order to reduce the primary forces in the overburden, it is possible to carry out relief dredging at the slope's head or to undertake additional drainage measures.

12 Work Group Ground Mechanics to Improve the Codes of Practice

In the course of advanced professional training, regular joint discussions are held with experts in order to exchange information about the operating conditions and the ascertained measurements, on the one hand, as well as new developments, on the other hand. As a consequence of this exchange of expertise, the mining authority has founded together with the Geological Service North Rhine-Westphalia and the mining company RWE Power AG a work group for ground mechanics in lignite mining which systematically seeks to advance the generally accepted codes of safety practice in this field.

The organization consists of a coordination group which conducts the strategic planning, formulates the respective professional requests, and assures the general framework for the technical work. This technical work is performed by a smaller core work group which handles the tasks formulated by the coordination group by topic. The core work group is supplemented also by the work of external experts if so required.

Waste Management in the Rhenish Lignite Mining District

Peter Asenbaum

Cologne, Germany
peter.asenbaum@gmx.de

1 Introduction

Today, the national law in the member states of the European Union (EU) is primarily determined by higher-level decisions which are reached by the EU. When it comes to environmental protection, this also applies to the waste legislation guidelines. The fact that they are implemented differently into national law demonstrates the individuality of the member states. In addition to general waste handling provisions, the EU also specified its own guidelines for mining waste. National implementation follows these guidelines in the form of different legal standards without abandoning the uniform, standardized provisions.

Germany's waste legislation standards, for example, include specific regulations pertaining to approval procedures, requirements for dumpsites and landfills, public participation, and the necessary controls exercised by public authorities.

A set of rules, including the EU documents on the "best available technology," supplements the catalog of guidelines which have to be observed.

Essentially, any freedom in implementing these decisions is only to be found within the scope of the member states' national legal structures. EU Regulations are directly applicable and do not require any national implementation. In contrast, EU Directives and Council Decisions need to be transferred into national law.

It should be noted that the substatutory national set of rules is always relevant and applicable in the specific legal branch in which its authority is grounded.

2 EU Waste Legislation

2.1 Waste Framework Directive

The EU Waste Framework Directive, which has been updated several times since it became effective, establishes the European framework for waste legislation in the individual member states:

DIRECTIVE 2008/98/EC
OF THE EUROPEAN PARLIAMENT AND OF THE COUNCIL
of 19 November 2008
on Waste and Repealing Certain Directives

This Directive constitutes a legal framework for handling waste in the European Community. It pursues the objective of contributing towards environmental

C. Niemann-Delius. (ed.), *Proc. of the 12th Int. Symp Continuous Surface Mining - Aachen 2014*,
Lecture Notes in Production Engineering, DOI: 10.1007/978-3-319-12301-1_49

protection and human health by avoiding the harmful effects emanating from the generation and management of waste.

The Directive includes definitions for such important terms as waste, recovery, and disposal, and it creates the basic requirements for waste management, in particular, an obligation to obtain the official approval to register any plants, facilities, and/or companies undertaking waste management measures, as well as an obligation to establish waste management plans on part of the EU member states.

The Directive also encompasses such major principles as, for example, an obligation to handle waste in such a way that the environment and human health are not adversely affected, as well as a call for compliance with the waste hierarchy and, in line with the "polluter pays" principle, a requirement according to which the costs of waste disposal are to be borne by the actual waste holder, any former waste holders, or the manufacturers of the product from which the waste originates.

Fig. 1 Waste from cafeterias (cans, etc.)

Here are some key terms in conjunction with the aforementioned legislation:

- **Waste**:
  Any substance or object which the holder discards, intends to discard, or is required to discard
- **Waste Management**:
  The collection, transport, recycling, and disposal of waste, including the supervision of such operations as well as the follow-up monitoring of disposal sites, and including those actions which are undertaken by dealers or brokers
- **Waste Prevention**:
  Any measures which are undertaken before a substance, material, or product has become waste

- **Recovery**:
  Any operation of which the principal result is waste fulfilling a useful purpose
- **Recycling**:
  Any recovery operation through which waste materials are reprocessed into products, materials, or substances either for the original, intended purpose or any other purpose

2.2 *Landfill Directive*

The EU has also issued a number of additional regulations for dumpsites and landfills in the form of Landfill Directives which are defined more precisely by a Council Decision on hazardous substances.

Directive 1999/31/EC of the European Council of 26 April 1999 on the Landfill of Waste
as well as
COUNCIL DECISION 2003/33/EC of 19 December 2002 Establishing Criteria and Procedures for the Acceptance of Waste at Landfills Pursuant to Article 16 of and Annex II to Directive 1999/31/EC

By specifying stringent technical requirements for dumpsites, landfills, and waste, the European Union seeks to avoid and/or reduce as much as possible the harmful effects resulting from the storage of waste on the environment and, in particular, on the ground and surface waters, soil, air, and human health.

A distinction is made between various categories of waste (municipal waste, hazardous waste and/or non-hazardous waste, inert waste). These categories apply to all dumpsites and landfills defined as waste disposal facilities for the storage of waste above and below the earth's surface.

According to EU law, landfills are classified into the following three categories:

- Landfills for hazardous waste
- Landfills for non-hazardous waste
- Landfills for inert waste

An approval procedure for the operation of a landfill is specified in the Directive. The member states have to ensure that existing landfills only continue their operation if and to the extent that they fully comply with the provisions of this Directive.

The Council Decision regulates the criteria and procedures for the acceptance of waste in landfills pursuant to the guiding principles of Directive 1999/31/EC, with a particular focus on Annex II.

Threshold values and other criteria are established for the various landfill categories. The member states have to ensure that landfills accept only such waste which meets the acceptance criteria for the respective landfill category. In addition, harmonized standards are determined for samplings and test procedures.

Fig. 2 Ash landfill "Craiova", Romania

2.3 Mining Waste Directive

Mining waste ("waste from mineral resources") is excluded from the application of the Waste Framework Directive and from the supplementary set of rules both at a European and a national level; mining waste is subject to a special regulation.

The European Union has introduced measures which are designed to prevent and/or reduce any adverse effects on the environment as well as any resultant risks for humans derived from the management of waste (for example, residues and excavated material) in the extractive industry.

Directive 2006/21/EC
of the European Parliament and of the Council
of 15 March 2006
on the Management of Waste from Extractive Industries
and Amending Directive 2004/35/EC

The above mentioned Directive establishes uniform European standards for the handling of mining waste and complements the previous set of EU rules for waste in a consistent and harmonized manner.

Fig. 3 Mining waste (waste rock), Westkalk corporation, Warstein

The Mining Waste Directive provides the member states with a binding framework for the management of waste resulting from the extractive industry; however, numerous aspects of this framework still need to be specified in detail and further developed under national law.

The waste resulting from the extractive industry includes topsoil, overburden, waste rock, and residues accruing from the prospection, extraction, and processing of mineral resources. This waste represents the largest single waste stream in Europe, accounting for more than 20 % of the overall waste volume.

The impact caused by the management of such waste ranges from physical effects on ecosystems (for example, sedimentation on riverbeds) to the leaching of acidic seepage water all the way to the discharge of heavy metals and other hazardous substances which are used during ore beneficiation. The instability of facilities containing such waste (for example, waste heaps, tailings ponds, or dams) may have widespread devastating effects on humans and the environment for which national borders present no obstacle, as was demonstrated by the serious accidents that occurred in Aznalcollar, Spain, in 1998 and Baia Mare, Romania, in 2000 (both incidents resulted in severe damages to the freshwater ecosystems and had extensive socioeconomic effects).

As a new legislative instrument, the EU Directive has introduced the waste management plan which is to be established by the entrepreneur in order to safeguard and assure that the waste disposal concept is substantiated in detail and reported to the respective public authority already prior to the start of any mining activities.

2.4 European Waste Catalog

The European Waste Catalog serves as a tool for the description and classification of waste consistent with its specific monitoring requirements. Waste is classified according to its origin.

COMMISSION DECISION of 3 May 2000 Replacing Decision 94/3/EC Establishing a List of Wastes Pursuant to Article 1 (a) of Council Directive 75/442/EEC on Waste and Council Decision 94/904/EC Establishing a List of Hazardous Waste Pursuant to Article 1 (4) of Council Directive 91/689/EEC on Hazardous Waste (2000/532/EC)

In the European Union, the European Waste Catalog (EWC) classifies waste with the help of 6-digit waste codes. The initial classification is done by allocating the specific branch of origin, followed by a more refined itemization of diverse branch-typical processes, and finally by a numerical list. In this catalog, which is now standardized, hazardous waste (formerly also referred to as special waste) is characterized by adding an asterisk (*).

For example:

01 Waste Resulting from the Prospecting, Exploitation, and Extraction as Well as from the Physical and Chemical Treatment of Mineral Resources

01 01 Waste resulting from the mining of mineral resources

01 01 01 Waste resulting from the mining of metalliferous mineral resources
01 01 02 Waste resulting from the mining of non-metalliferous mineral resources

Fig. 4 Landfill for drilling mud (mineral oil)

2.5 *Free Access to Information on the Environment*

Access to environmental information held by public authorities is an essential prere-quisite for a more intense application and control of the Community's environmental law.

The European Union, for example, also specifies the rules which help guarantee both free access to environmental information held by public authorities and the dissemination of this information. And the EU also determines the fundamental prerequisites under which access has to be provided to such information.

COUNCIL DIRECTIVE of 7 June 1990
on the Freedom of Access to Information on the Environment
(90/313/EEC)

Information relating to the environment is considered to be any and all available information in written, visual, aural, and/or database form on the state of water, air, soil, fauna, flora, and natural habitats, and on any and all activities and/or measures adversely affecting or being capable of affecting this state, and on any and all activities and/or measures designed for their protection (including administrative measures and environmental protection programs).

This Directive pursues the objective of making environmental information available and publicly distributing this information in a systematic manner.

The differences in the laws of the individual member states regulating access to environmental information held by public authorities may result in unequal conditions for the citizens of the Community when it comes to information access and competition.

3 National Waste Legislation – Germany

3.1 *Closed Cycle and Waste Management Act*

Waste management in the Federal Republic of Germany has undergone substantial changes since the beginning. In Germany, the act on the disposal of waste (Waste Disposal Act, AbfG) issued in 1972 created the first standardized national regulation of the waste law. Today, the act on promoting closed substance cycle waste management and ensuring environmentally compatible waste disposal (Closed Cycle and Waste Management Act, KrWG) forms the core provision of the pertinent waste legislation.

Act on Promoting Closed Substance Cycle Waste Management
and Ensuring Environmentally Compatible Waste Disposal
(Closed Cycle and Waste Management Act – KrWG)
of 24 February 2012

With the KrWG, the requirements of the EU Waste Framework Directive (Directive 2008/98/EC) are transferred into national law. The closed substance cycle management is to be aligned even stronger than before with resource, climate, and environmental protection.

Fig. 5 Coincineration of waste wood

As a follow-up regulation, the KrWG maintains the key components of the Closed Cycle and Waste Management Act and of the Waste Disposal Act (KrW-/AbfG). The KrWG is supplemented and substantiated by numerous statutory ordinances.

Fig. 6 Recycling compost for immission protection purposes

3.2 Landfill Ordinance

Since June 1, 2005, the storage of organic, biodegradable municipal waste has no longer been permitted without pretreatment, and since July 15, 2009, landfills which are not compliant with EU standards may no longer be operated – except for such re-quisite measures as are needed to decommission such landfills. These are just two requirements of the Landfill Ordinance which has been available in its amended version since 2009.

Ordinance on Landfills and Long-Term Storage Facilities (Landfill Ordinance – DepV) of 27 April 2009

The DepV applies to landfills and long-term storage facilities of the categories 0, I, II, III, and IV. The DepV

- Contains requirements for the construction, operation, decommissioning, and follow-up monitoring of landfills falling under the above mentioned categories and, in addition, also the requirements for their supervision and control
- Regulates the treatment and storage of waste
- Regulates the use of waste for the production of recycled landfill materials used in construction as well as the recovery and treatment of waste as recycled landfill materials used in construction on above-ground landfills
- Contains requirements for long-term storage facilities falling under the above mentioned categories

This ordinance applies to waste holders and waste producers as well as to (also private) supporting organizations, owners, and/or operators of landfills and long-term storage facilities. It also applies to operators of plants and systems for the production of recycled landfill materials used in construction.

Fig. 7 Landfill for power plant residues, mono section for mineral construction waste

3.3 Waste Recovery and Disposal Records Ordinance

Germany's currently applicable Ordinance on Waste Recovery and Disposal Records (NachwV) determines the type and extent of the requisite documentation and recording of hazardous and non-hazardous waste recovery and disposal in electronic form.

Ordinance on Waste Recovery and Disposal Records of 20 October 2006

Since April 1, 2010, it has been a mandatory legal requirement for all stakeholders in the waste management industry to keep electronic records and, thus, furnish proof of any and all hazardous waste (electronic waste management system, abbreviated: eANV).

The NachwV is intended for

- Producers and/or holders of waste (waste producers)
- Collectors and/or carriers of waste (waste carriers)
- Operators of facilities and/or companies which dispose waste through a process pursuant to Annexes II A and/or II B of the Closed Cycle and Waste Management Act

The above mentioned ordinance does not apply to

- Private households
- The shipment of waste across state borders (this is subject to Regulation (EC) 1013/2006 and Germany's Waste Shipment Act (AbfVerbrG))

3.4 Waste Catalog Ordinance

This ordinance transfers the European Waste Catalog literally unchanged into national law.

Ordinance on the European Waste Catalog (Waste Catalog Ordinance – AVV) of 10 December 2001

In line with the EU requirements, the annex to this ordinance provides a complete list of all waste types with the appropriate six-digit waste codes, divided into chapters and groups. Those waste codes in the list which are marked with an asterisk (*) characterize hazardous waste types; this classification is linked to diverse legal consequences. Waste types not marked with an asterisk are non-hazardous waste.

3.5 *General Federal Mining Ordinance (§ 22a)*

On May 1, 2008, the Third Ordinance on the Amendment of Mining Law Ordinances dated January 24, 2008 (Federal Law Gazette I, p. 85) became effective. This ordinance transferred the EU Directive 2006/21/EC of March 15, 2006 on the management of waste resulting from the extractive industry into German law; specifically for those facilities and plants which are subject to supervision and monitoring by the Mining Authorities.

Mining Ordinance for All Mining Sectors
(General Federal Mining Ordinance – ABBergV)*)
of 23 October 1995
§ 22a – Requirements for the Disposal of Mining Waste –

In order to explain and ensure the standardized implementation of § 22a ABBergV, the Federal States Mining Committee (LAB) has prepared and compiled specific enforcement notes. These enforcement notes include seven annexes. Annexes 3 to 6 deal with decisions on part of the EC Commission which serve as supplements to the above mentioned Directive 2006/21/EC.

Fig. 8 Tailings pond near Oradea, Romania

All told, the residues resulting from mining operations can be assigned to one of the following three categories:

- Mining waste
- Waste which is subject to the Closed Cycle and Waste Management Act as well as the Waste Disposal Act
- Materials which do not constitute any waste

The term "mining waste" is defined in § 22a ABBergV on the basis of European legislation. Thus, mining waste is, without exception,

- Dead rock (waste rock) or
- Process slurries (tailings)

§ 22a exclusively regulates the disposal of such mining waste.

Fig. 9 Drilling mud from open-pit mining (lignite)

Mining waste which is disposed outside the mining operations is to be considered as waste pursuant to the Closed Cycle and Waste Management Act and is subject to its provisions, which also includes the substatutory set of rules (for example, the Waste Recovery and Disposal Records Ordinance) as well as the waste laws of Germany's federal states. This means whenever mining waste is set aside for its disposal outside a mining operation, this mining waste is subject to the general waste law.

Mining waste which is disposed either in one's own or in a third party's mining facility pursuant to mining law is at no point in time subject to the Closed Cycle and Waste Management Act. Furnishing the necessary proof of its shipment between different mining facilities needs to be specified and regulated by the operating plan.

New since May 1, 2008, are specific provisions for the disposal of mining waste in "waste disposal facilities" located within the mining facility, which means waste heaps, tailings ponds, etc.

In addition to the operating plan for waste disposal, a waste management plan specifying the disposal of mining waste needs to be submitted to the Mining Authority as well.

If an operating plan for waste disposal contains the required contents of the waste management plan, then this operating plan can, at the same time, also serve as a notification of the waste management plan pursuant to § 22 Para. 2 ABBergV.

Fig. 10 Settling pond in Baia Mare, Romania

Initially, the classification of disposal facilities according to their specific risk potential (for example, a Category A facility) is to be carried out by these facilities. The binding classification, however, is the responsibility of the Mining Authority which is to be provided with the requisite documentation, if need be.

Environmental Information Act (UIG) and Freedom of Information Act (IFG)

Germany's **Environmental Information Act** (UIG) pursues the objective of providing free access to environmental information and disseminating environmental information. It is directly applicable to those federal offices which are required by law to provide such information. The environmental information acts of Germany's federal states are applicable to the offices which are required to provide such information in the respective federal states, and they either refer to the UIG or they regulate the same issue independently.

The **Freedom of Information Act** (IFG) is a German law governing the freedom of information. So far, eleven federal states have introduced their own similar laws for their respective area of responsibility.

Freedom of information is a civil right which grants the public the right to inspect and review documents and files kept by the public administration. Within this scope, for example, governmental offices and authorities may be required to publish their files and cases (the principle of public access to official records) and/or make them accessible to citizens (administrative transparency) and, to this end, define binding quality standards for their access.

3.6 *Federal State Legislation*

State Waste Acts

The Federal Government's Closed Cycle and Waste Management Act is supplemented and substantiated by the individual **waste acts of Germany's federal**

states. Because of the Federal Government's concurrant legislative power for waste management (Art. 74 Para. 1 No. 24 of the Basic Law (GG)), state law regulations are only possible in those sectors which are not yet covered by federal law. That is why the state waste acts essentially refer to enforcement questions, for example, the designation of corporate entities which are subject to mandatory waste disposal and the designation of the responsible public authorities in the waste sector.

Waste Statutes
The collection and processing of waste associated with households is specified at the municipal level in the form of **waste statutes**. For example, waste statutes include provisions on the compulsory participation and utilization. And fees for using the waste disposal service are also charged on the basis of municipal waste fee statutes.

Landfill Self-Monitoring
The construction, operation, and decommissioning of landfills and dumpsites are subject to monitoring on part of the respective authority which is responsible for such monitoring. The fundamental requirement for the supervision and monitoring of landfills and dumpsites accrues from the Closed Cycle and Waste Management Act (KrWG). Additional requirements are specified in the Landfill Ordinance.

In North Rhine-Westphalia, the

Regulatory Ordinance on the Self-Monitoring of Above Ground Landfills (Landfill Self-Monitoring Ordinance – DepSüVO) of 27 August 2010

also regulates the self-monitoring of landfills. A specific focus of this ordinance is the submission of documents furnishing proof of this self-monitoring in form of a landfill annual report to be provided by the landfill operator in electronic form.

3.7 Sets of Rules

The above mentioned legal standards are supplemented by numerous sets of rules which document, for example, the current state of technology. These sets of rules do not have any direct, binding legal force so that their individual implementation is subject to a decision made by the public authorities.

Best Available Techniques Reference (BREF)

At the EU level, the BREF documents need to be mentioned here first. BREF is the English acronym for "Best Available Techniques Reference" or "Best

Available Techniques Reference Document," abbreviated: "BAT Reference" or "BAT Reference Document." In German, the term "BREF" is used in place of a BAT reference document.

A BAT reference document is an EU document which describes the best available techniques (BAT) for the prevention and reduction of adverse environmental effects in a specific economic branch and which needs to be considered when it comes to the approval of facilities by public authorities within the EU. BAT conclusions stipulate emission values which can be obtained with the best available techniques. Affected facilities need to comply with these emission values no later than four years after the BAT conclusions have been published in the EU Official Journal.

For example:

BREF Mining
Best Available Techniques Reference Document for the Management of Tailings and Waste Rock in Mining Activities
January 2009

Fig. 11 Dam of a tailings pond near Oradea, Romania

Joint Waste Commission of the Federal Government and the Federal States (LAGA)

LAGA is a working committee established by the German Conference of the Ministers of the Enviornment (UMK). Its objective is to safeguard and ensure the enforcement of the Federal Republic of Germany's waste law in the most uniform manner at the federal state level.

For this purpose, LAGA promotes the exchange of information and experiences between the Federal Government and the federal states. It maintains contacts to associations and interest groups. And it also develops proposals and provides impulses when it comes to further developing the statutory provisions and representing the interests of the federal states by defining Germany's position within international committees.

In order to solve waste management tasks, LAGA prepares reference documents, guidelines and directives as well as information material. Sample administrative regulations are compiled in order to enforce waste management legislation.

The LAGA Ad-hoc-AG work group, for example, determines the "Uniform National Quality Standard" for Germany's landfills in accordance with the Landfill Ordinance.

Joint Mining Committee of the Federal Government and the Federal States (LAB)

Germany's supreme authorities at the federal and state levels whose responsibility is the mining industry join forces in the Joint Mining Committee of the Federal Government and the Federal States (LAB) in order to discuss issues associated with their fields of activities, to work out solutions, and to make recommendations.

This includes, in particular:

- Providing reciprocal information and coordinating measures which require uniform administrative action on part of the Federal Government and the federal states; this also includes matters pertaining to the European Union
- Guaranteeing the uniform enforcement of the Federal Mining Act (BBergG) in the Federal Republic of Germany
- Providing reciprocal information and coordinating measures related to substatutory federal state law pursuant to the Federal Mining Act (BBergG)
- Providing reciprocal information and working out proposals designed to enforce the mining law and any other statutory federal provisions pertaining to mining in the federal states, particularly in the environmental protection and occupational safety sectors
- Discussing technical questions and providing expert advice as well as drawing up recommendations

As a standing committee appointed by the Standing Conference of the German Ministers for Economic Affairs, LAB reports on special events and activities. The Joint Mining Committee of the Federal Government and the Federal States has the following standing expert panels which convene as required:

1. Expert Panel on Technology in the Mining Industry
2. Expert Panel on Mining Law
3. Expert Panel on Environmental and Mining Affairs

As a working committee, the Expert Panel on Environmental Affairs (FAU) is a standing subcommittee of LAB. Its task is to inform the higher and supreme Mining Authorities of the federal states about environmentally relevant issues which have an impact on enforcement in mining facilities. On behalf of LAB, the expert panel develops interstate enforcement principles for this sector. Since representatives from the Joint Commissions of the Federal Government and the Federal States for Waste (LAGA), for Water (LAWA), and for Soil Protection (LABO) are also members of this expert panel, the outcomes and results of the work it carries out are developed in agreement with these commissions which are subordinate to the Federal Ministry for the Environment.

FAU focuses, in particular, on the disposal of mineral waste from non-mining and inherent mining sources in mining facilities. Since waste disposal plants are operated in numerous mining facilities, these facilities make a vital contribution towards safeguarding waste disposal in the Federal Republic of Germany. These facilities include above ground and underground landfills where waste is disposed and which are, in part, also significant for waste disposal at a European level. In addition, mineral waste is salvaged within the scope of rehabilitation and/or recultivation measures in numerous above ground and underground facilities. The vast majority of the open-pit mining areas which are rehabilitated as part of their recultivation are made available to agriculture, forestry, and/or nature protection.

The North Rhine-Westphalia State Agency for Nature, Environment, and Consumer Protection (LANUV NRW)

The North Rhine-Westphalia State Agency for Nature, Environment, and Consumer Protection (LANUV NRW) is a state agency which is subordinate to the North Rhine-Westphalian Ministry for Climate Protection, Environment, Agriculture, Nature Conservation, and Consumer Protection (MUNLV).

The state agency was founded in 2007 as a spin-off of diverse precursor institutions and builds on their competence and their many years of experience. LANUV NRW is a special technical-scientific authority which is active in the fields of nature protection, technical environmental protection for water, soil, and air, plant safety, climate change issues as well as animal health and food safety within consumer protection.

Ministerial Decrees

A **decree** is an official order handed down by an executive authority (usually ministries) to other state offices or to the population. In most cases, though, governmental authorities are only entitled to issue such decrees to their subordinate agencies, but not to citizens.

Letters from a superior federal or state authority, regardless of their content, to a subordinate authority or institution are also considered to be decrees; this includes, in particular, letters from ministries to authorities within their area of responsibility.

As a general rule, decrees unfold no third party effect; they are to be implemented and enforced by the public authorities (for example, requirements for approvals, etc.). Hence, decrees are guidelines and instructions for law enforcement authorities.

3.8 Information Systems

Today, the legal standards and sets of rules are amended by publicly accessible information systems in accordance with the EU Directive on the Freedom of Access to Information on the Environment and/or national requirements stipulated by the UIG and the IFG. Access to information is primarily provided via the internet. In addition, these systems are also used for supervisory purposes on part of the authorities and for environmental statistics.

ADDISweb
In conjunction with waste disposal sites, the

Landfill Data Information System ADDIS and/or ADDISweb of the North Rhine-Westphalia State Agency for Nature, Environment, and Consumer Protection (LANUV NRW)

is applied in North Rhine-Westphalia as a supporting tool for the self-monitoring of landfills and dumpsites and for their supervision by the authorities.

The ADDIS landfill data information system for the self-monitoring of landfills and dumpsites, which had been in place since 1998, was replaced by the new web-based information system ADDISweb in 2011. A major innovation is the direct access to the online data bank which is also available to the public at large.

Operators of those North Rhine-Westphalian dumpsites and landfills which require mandatory supervision are obliged to transfer their landfill self-monitoring data according to an interface specification to the information system or to use the LANUV NRW's available web user interface for entering their data.

ASYS (Waste Monitoring System)
In addition, hazardous waste is monitored and supervised through mass flow and fate verification via Germany's national waste monitoring system ASYS. The system permits the processing of those data which are required for monitoring the proper disposal of hazardous waste. And the master data of the participating facilities can be captured as well. All relevant catalogs are stored.

As a functional option, ASYS permits the automated, electronic exchange of data between public authorities and participating facilities. For the electronic verification procedure, the requisite data are exchanged via the Central Waste Coordination Office (ZKS-Abfall).

ENADA
ENADA is North Rhine-Westphalia's specialized information system for waste disposal sites. The system is designed to collect, forward, and evaluate the data provided by North Rhine-Westphalian waste disposal facilities; in particular, when it comes to assessing the existing plants, plant capacities, and plant technologies. And it also permits the identification of environmentally compatible disposal methods and plant-related mass flows to be applied in waste recovery and disposal.

AMEDA

AMEDA is LANUV NRW's waste volume data bank which permits the separate recording and evaluation of the waste volumes provided by all LANUV NRW information and enforcement systems (such as ISA, ASYS, ABILA, ERIKA, and ADDIS) for each and every waste holder. It is, thus, possible to get an overall view of as well as to compare and (if need be) rectify the waste volume figures provided by the participating data banks; each of which only covers partial aspects of the waste management business.

ABILA (Waste Balance for Municipal Waste)

ABILA manages the data obtained from North Rhine-Westphalia's municipal waste balances. The system permits the recording, management, and evaluation of municipal waste balance data from North Rhine-Westphalia's 53 public waste management organizations, and the LANUV is obliged to report these data to the Ministry for Climate Protection, Environment, Agriculture, Nature Conservation, and Consumer Protection (MUNLV) once a year.

ERIKA (Online Survey System for Sewage Treatment Plant Waste)

ERIKA is the online survey system for waste from sewage treatment plants. Since early 2006, ERIKA has been replacing the previous paper-based data query on the volume, quality, and fate of municipal sewage sludge, sand trap materials, and grit chamber screenings in municipal sewage treatment plants.

ISA (Information System for Substances and Plants)

The computer programs of the ISA information system for substances and plants are available to assist the district governments, counties, and independent cities in North Rhine-Westphalia in their multifaceted tasks revolving around immission protection rights and environmental protection.

4 Official Licenses

All official licensing procedures pursue the objective of safeguarding an environmentally compatible disposal of waste. Depending on the specific field of law, different administrative procedures are applied which are, to some extent, also associated with environmental impact assessments and public participation.

In addition to waste management legislation, the respective local laws are also applicable to certain recovery and/or recycling measures. Recovery and/or recycling measures in the mining industry are approved in accordance with the appropriate operational planning process as specified in the mining law. Likewise, the mining law is also relevant for the disposal of mining waste resulting from mining activities.

4.1 Landfill Law

According to the Closed Cycle and Waste Management Act, an approval procedure based on the waste law is required both for the construction and operation of facilities for the disposal of waste (landfills) as well as for any substantial modifications made to these facilities and/or their operations (§§ 35 et sqq. of the KrWG).

It has to be determined during approval procedures which are based on the waste law whether dumpsites and landfills can be built and operated in such a way that the protection of the environment, neighborhood, and general public is guaranteed and that the duties and obligations of the operators as stipulated by law are met.

These approval procedures are carried out in accordance with the provisions of the Closed Cycle and Waste Management Act (KrWG), the Landfill Ordinance (DepV), and the Administrative Procedure Act (VwVfG). The rules and regulations of the Environmental Impact Assessment Act (UVPG) are applicable, where required. The generally accepted technical regulations are also relevant.

Because numerous other public decisions are reached simultaneously whenever an approval is granted in line with the KrWG, the following legal fields may, in addition to the waste law, also be considered specifically in the evaluation of the admission requirements:

- Regional planning acts, zoning laws, building codes
- Nature protection laws, landscape protection laws, forest laws
- Soil protection laws, water protection laws
- Immission protection laws
- Hazardous substances laws
- Labor protection laws
- The Equipment Safety Act
- The Explosives Act

When it comes to approving dumpsites and landfills, the KrWG essentially distinguishes between the following two types of procedures:

- The planning assessment procedure as the usual revision procedure and
- the planning approval procedure.

If existing dumpsites and landfills are to be changed and/or modified, then it is also possible that under certain circumstances the notification procedure may be used which is based on waste management legislation.

As a basic prerequisite, the construction and operation of dumpsites and landfills as well as any major changes in such facilities and/or in their operations are subject to planning assessment, which means that a procedure needs to be followed which also involves public participation and an environmental impact assessment.

Instead of an official planning permission, a planning approval can also be granted if the scheduled preliminary examination proves that the planned project will not result in any substantial adverse effects on one of the above mentioned objects of protection.

Every planning assessment procedure which is based on the waste law requires an environmental impact assessment in accordance with the Environmental Impact Assessment Act. This does not apply to planning approval procedures.

An environmental impact assessment includes as a legally dependent part of the planning assessment procedure the identification, description, and rating of a dumpsite's or landfill's impact on the environment; and it is to be conducted with public participation. The requisite tests encompass the effects of the planned project including its operation and its on-site inventory after decommissioning (life cycle considerations) on humans, animals, plants, soil, water, air, climate, landscape, cultural assets, and other material assets as well as the respective interactions.

4.2 Mining Law

In mining, the operational planning process pursuant to §§ 51 et sqq. of the Federal Mining Act (BBergG) is a common and appropriate procedure when it comes to obtaining official licenses and approvals for activities to be carried out in conjunction with operational waste. This process is a binding administrative act, which means that as soon as the admission requirements are met in accordance with § 55 of the BBergG, the respective governmental authority must grant the approval. Otherwise, the authority has to ensure by virtue of incidental provisions that the respective admission requirements will be met.

The operational planning process as stipulated by mining law is to be applied to recycling waste, even to such recycling waste which is tendered to the mining facility by third parties and which is to be recovered in the mine. The responsible Mining Authority has to specify the requisite requirements for environmental protection and, if need be, also demand the provision of security measures.

This could be, for example, mineral construction waste which helps build immission protection dams as a preparatory clearance measure for the subsequent operation of open-pit mines, or supporting dams which are built during the actual open-pit mining operations. And this also includes the recycling of compost for immission protection purposes at the bottoms and on the slopes of open-pit mines.

The EU Mining Waste Directive was implemented at a national level by incorporating the new § 22a with its Annexes 5 to 7 into the Mining Ordinance for all mining facilities (General Federal Mining Ordinance – ABBergV) of October 23, 1995.

As a new legislative instrument, the EU Directive has introduced the waste management plan which is to be established by the entrepreneur in order to safeguard and assure that the waste disposal concept is substantiated in detail and reported to the respective public authority already prior to the start of any mining activities.

The requirements for waste management plans are defined in § 22a Para. 2 as well as in Annex 5 "Waste Management Plan" with reference to Annexes II and III of Directive 2006/21/EC. Additional requirements for the construction, operation, and decommissioning of waste disposal facilities for mining waste can be found in Annex 5 to § 22a Para. 3. Annex 6 contains additional requirements for the provision of security measures pursuant to § 56 Para. 2 of the Federal Mining Act for Category A waste disposal facilities.

Such a waste management plan must encompass the entire life cycle of the respective mine. Waste management plans pursue the objective of avoiding or reducing waste (specifically mining waste) and any harm emanating from such waste. The waste management plan is to be continuously updated and submitted to the responsible governmental authority on a regular basis.

Mining waste accrues either as waste rock or as sludge containing mill tailings and process slurries. If waste disposal facilities (waste heaps, slurry ponds, etc.) need to be built and operated for these wastes, then this requires in addition to a waste management plan also an official license for the respective waste disposal facility pursuant to § 22a of the General Federal Mining Ordinance (ABBergV).

In accordance with § 22a Para. 1 ABBergV, and irrespective of the provisions governing the obligation to establish and maintain an operating plan, entrepreneurs need to undertake appropriate measures for the disposal of mining waste when it comes to the construction, management, and shutdown of their plant so as to avoid or reduce as far as possible the environmental impact as well as the resultant risks for human health. In so doing, entrepreneurs need to consider the current state of technology with regard to the specific characteristics of their waste disposal plant, its location, and the environmental conditions on site. That is why the use of a specific technology is not mandatory. In this context, reference is also made to the EU BREF documents – or more precisely: The BREF document "Management of Tailings and Waste Rock in Mining Activities."

For waste disposal facilities, § 22a Para. 3 ABBergV specifies the compulsory application of the operational planning process as stipulated by mining law; depending on the specific risk potential, this might also include public participation.

In the Rhenish lignite mining district, the legal requirements of § 22a ABBergV are met by a special operating plan illustrating waste management in a comprehensive manner. This special operating plan represents all waste management activities in accordance with the waste and the mining laws and is continuously updated every year.

This special operating plan relieves the compulsory main operating plan for open-pit mines. The latter has, in addition to a general overview of waste management and the dumpsites and landfills in operation, only a reference to this special document.

5 Waste in the Mining Industry: Examples

5.1 Common Waste

Waste resulting from mining operations encompasses both that group of waste which accrues indirectly during the prospecting, extraction, and processing as well as during the associated storage of mineral resources in mining facilities (common waste) and mining waste as stipulated by the exemption clause of § 2 Para. 2 No. 7 KrWG.

Common waste is to be categorized into the following types according to its origin:

- Household waste and the like
- Commercial waste (from workshops)
- Mineral waste (excavated soil and building rubble)

Fig. 12 Mineral construction waste (building rubble)

Household waste and the like is waste coming from, for example, recreation rooms and cafeterias which usually also accrues elsewhere in comparable volumes and qualities. Recreation rooms and cafeterias are, by the way, not specific to the mining industry.

Machinery and vehicles are maintained and serviced in numerous workshops of the mining industry which also include maintenance and repair shops. Waste accruing from these activities includes, for example, oily cleaning rags and machine parts as well as used fuels and lubricants. That is why much of the waste has to be identified with hazard symbols and disposed separately. This waste group also exhibits no special characteristics which would be unique to mining.

Large quantities of excavated soil and building rubble in different qualities accumulate, for example, within the scope of preparatory clearance measures, which also include the relocation of entire villages, for the subsequent operation of open-pit mines. Comparable waste is also created in the course of large-scale projects in urban areas which is why they do not justify any exemption pursuant to § 2 Para. 2 No. 7 KrWG (which means they are not mining waste).

The provisions of the waste law are to be applied in full to the wastes mentioned above, in particular, if and to the extent that their due and proper disposal is not part of the actual mining operations. Exemptions only apply to specific recovery purposes applied in mining, for example, the recovery of mineral fractions for road construction.

5.2 *Mining Waste*

As defined by the EU, only dead rock (waste rock) and process slurries (tailings) are to be subsumed under the term mining waste. Generally, dangers and risks emanate from inadequately secured heaps and ponds as well as hazardous ingredients.

In the Rhenish lignite mining district, mining waste only accrues in the form of drilling mud within the scope of drainage measures carried out prior to the actual mining operations. The drilling mud is dumped together with the overburden, but in smaller quantities. The actual overburden is used for backfilling the open-pit mining holes. Following this definition, overburden is, thus, neither mining waste nor waste.

Fig. 13 Tailing pond in Baia-Mare, Romania

All other waste is subject to the provisions of the general waste law (KrWG) and is to be disposed accordingly.

5.3 *Waste Recovery*

According to the nomenclature of the waste law, the avoidance and recovery of waste takes precedence over its removal.

Recovery pursuant to the KrWG refers to any process within the plant or in the economy at large wherein the principal result is waste serving a useful purpose by replacing other materials which would otherwise have been used to fulfill a particular function, or waste being prepared in such a way that it fulfills that function.

Thus, the legal possibility and/or requirement have been created for using suitable waste prior to landfilling within the scope of recovery measures. That is why it does not matter whether it is one's own waste or the waste of third parties.

When it comes to official approvals for recovery measures, the respective local law is to be applied. In the mining industry, this includes in particular the use of soils for the reduction of extraction-related mass deficits or the use of soils in recultivation measures. But mineral construction waste is also suitable, for example, for road construction or the construction of supporting and immission protection dams. For immission protection purposes, suitable composts can be spread over the bottom of open-pit mines and on their slopes. The creation of green areas, thus, reinforces the desired purpose.

In industrial power plants of upgrading facilities, substitute fuels, sewage sludges, and other suitable wastes can be recovered and recycled within the scope of coincineration processes in accordance with the provisions of the 17th Federal Immission Control Ordinance (BImSchV).

The provisions of the Commercial Waste Ordinance (GewAbfV) also assign certain mandatory recovery quotas for common waste to mining facilities. Within the scope of a full service contract, the RWE Power AG corporation, for example, employs a certified waste management company (Entsorgungsgesellschaft Niederrhein, EGN) which separates the waste collected from mining facilities according to recovery and disposal in its waste treatment plants and which furnishes proof that the quotas required by the Commercial Waste Ordinance are met.

Fig. 14 Collection systems for commercial waste (EGN)

In addition, mining operators are also required to either observe any obligations that may exist regarding the tendering of waste to corporate entities which are subject to mandatory waste management pursuant to § 17 KrWG or, alternatively, to apply for an exemption.

5.4 Waste Disposal

Provided that waste cannot be recovered and/or recycled, it has to be disposed properly. Disposal means landfilling. Properly means in accordance with the existing rules and regulations and in such a way that the interests of the general public are neither violated nor infringed upon.

Fig. 15 "Vereinigte Ville" landfill for power plant residues

In this context, for example, power plant residues and/or ashes generated by lignite are to be mentioned which accrue in large quantities and consistent qualities at power plants located near mines and at customer facilities. For the disposal of these wastes, the RWE Power AG corporation, for example, operates a number of dumpsites and landfills in the Rhenish lignite mining district in accordance with the provisions of the Landfill Ordinance and, thus, in line with European standards. Currently, landfills for power plant residues are operated at the following locations:

- Garzweiler
- Fortuna
- Inden
- Vereinigte Ville

For the disposal of non-recyclable mineral construction wastes resulting from preparatory clearance measures prior to the actual mining operations, some landfills for power plant residues have established mono sections above the prospective groundwater level. This permitted the shutdown of "mine dumps" which had been operated in open-pit mines until then.

5.5 *Qualification*

§ 59 of the KrWG stipulates that operators of facilities and plants in which hazardous wastes accrue periodically, as well as operators of stationary recycling and/or waste disposal facilities and plants, are required to immediately appoint one or several official representatives for waste (waste management officers); provided that this proves to be necessary with regard to the type and/or size of these facilities and plants.

Waste management officers are trained and educated employees who are appointed in writing and act as advisors to the plant management. In addition, waste management officers exercise a control function and submit reports to the corporate management once a year.

According to this requirement, the RWE Power AG corporation trains and educates centrally and decentrally established waste management officers and notifies the Mining Authority of these appointees. Independent certifications as a special waste management facility supplement the qualification, for example, when it comes to the coincineration of waste in industrial power plants of upgrading facilities.

The full-service waste management provider, the Entsorgungsgesellschaft Niederrhein (EGN) corporation, also warrants and assures its qualification with the appropriate professional certification.

6 Official Supervision

The official waste management supervision in the lignite mining facilities of the Rhenish district is carried out centrally for the entire district by North Rhine-Westphalia's Mining Authority; specifically by Department 6 of the District Government Arnsberg at its Düren office. The experts in this office approve and supervise the waste streams close to the respective locations and in a very effective manner. For example, two waste controllers are available to the public authority for on-site samplings and additional specialists for operating the pertinent electronic waste management systems.

7 Conclusion

Waste management in mining facilities needs to be differentiated. In addition to mining waste, common waste is also generated there which has to be disposed in accordance with the rules and regulations of the general waste law. Waste management includes both recovery, which has priority, and disposal, which assumes a subordinate role.

The European Union has published binding requirements for both waste management methods. The Waste Framework Directive and the Mining Waste Directive have to be seen as the central rules here. This framework is further specified by national standards and supplemented by sets of rules which reflect, for example, the state of technology.

For such typical mining wastes as waste rock and process slurries or tailings, a specific approval instrument has been created in accordance with the provisions of § 22a of the General Federal Mining Ordinance (ABBergV).

Generally, administrative procedures associated with approvals are to be conducted with public participation and by taking into consideration the requirements for environmental protection as well as the life cycle of mines.

A special operating plan for individual mines ("presentation of the waste management") has proven to be successful in the Rhenish lignite mining district which documents both the common waste and the mining waste. It is, thus, at the same time also a "waste management plan" in accordance with European legislative requirements.

Depending on their suitability, internal and external wastes are recovered and recycled in open-pit mines and industrial power plants of upgrading facilities pursuant to the respective local law and/or the Federal Immission Control Ordinance (BImSchG). Residues resulting from lignite combustion in power plants are removed and disposed on specific landfills for power plant residues. And non-recyclable mineral fractions generated during clearance operations prior to the actual mining activities are also removed and disposed here in accordance with the provisions of the Landfill Ordinance.

RWE Power AG's centralized waste management guarantees the proper handling of waste under the supervision of North Rhine-Westphalia's Mining Authority.

Approval Procedure and Compensation Measures for the Removal of an European Nature Protection Area (FFH Area) by the Cottbus-Nord Opencast Mine

Christoph Gerstgraser[1], Hendrik Zank[2], and Ingolf Arnold[2]

[1] Gerstgraser Ingenieurbüro für Renaturierung,
Gaglower Str. 17/18,
03048 Cottbus, Germany
dr.g@gerstgraser.de

[2] Vattenfall Europe Mining AG,
Lead Engineer Ecology / Landscape Planning / Conservation,
Head of Geotechnics,
Vom-Stein-Str. 39,
03050 Cottbus, Germany
{hendrik.zank,ingolf.arnold}@vattenfall.de

Abstract. The Energy Company Vattenfall operates opencast lignite mines in Brandenburg. One of the approved opencast mine projects included the Lakoma pond area ("Lakomaer Teiche") which was registered as FFH area and claimed by the opencast mine. The European Commission was involved in the approval procedure. The approval for claiming the FFH area was granted under the imposition of the obligation of comprehensive compensation measures with regard to nature conservation. In accordance with the FFH directive the compensation measures had to be effective already at the time of the intervention. The approval of the plan according to water law was the first approval ever granted to completely remove a FFH area in Germany.

1 Introduction

In Germany, not far away from Cottbus in the Land of Brandenburg the Cottbus-Nord opencast mine is operated by Vattenfall Europe Mining AG. Since 1981, lignite has been mined in the opencast mine to supply the Jänschwalde power plant. The Lakoma pond area was within the approved mining field of this opencast mine. It consisted of 22 ponds with a total area of 69 ha. The ponds were use to breed carps. As a consequence of the pond farming practised for many centuries a high diversity of habitats and species developed around the waters. In the area totally 1320 species and 68 protected biotopes were identified.

C. Niemann-Delius. (ed.), *Proc. of the 12th Int. Symp Continuous Surface Mining - Aachen 2014*,
Lecture Notes in Production Engineering, DOI: 10.1007/978-3-319-12301-1_50

In May 1998 the mining company announced the project for abandoning the Lakoma pond area with the respective authority, the State Office for Mining, Geology and Raw Materials (LBGR). In 2002, the zoning application was submitted. As a result of the previously held discussions the creation of a new pond area was intended. This concept was refused by the Brandenburg State Environmental Office (LUA). The LUA alternatively proposed the renaturation of the Spreeaue (Spree floodplain) north of Cottbus as a complex compensation measure.

In January 2004 the revised planning documents were submitted. The Land of Brandenburg, however, proposed the Lakoma pond area as FFH area in December 2003 and in March 2004 the Federal Republic of Germany submitted the proposal to the European Commission. In the Flora-Fauna-Habitat Directive (92/43/EEC) the member countries of the EU have not only committed to establish protected areas for certain threatened habitats and species - the so-called FFH or Natura 2000 areas - but also to implement comprehensive compensation measures in the case of impacts in these areas to preserve coherence of the Natura 2000 network. End of 2004 it was necessary to involve also the European Commission in Brussels in the approval procedure and to obtain a position statement according to Article 6 Para. 4 because there was a suspicion that the priority specie hermit beetle *(Osmoderma eremita)* has settled in the trees (FREYTAG et al., 2007).

Fig. 1 Cottbus-Nord opencast mine

Resulting from the changed legal basic situation the compensation concept had to be revised several times. To preserve the coherence of FFH relevant habitats and species, species-specific compensation measures had to be planned in further compensation zones. In November 2006 the European Commission gave a written statement that the revised compensation concept maintains the overall coherence of the Natura 2000 network. In December 2006 the LBGR made the decision on the official approval of the plan. The abandonment of the Lakoma pond area was the first example for approving the complete removal of a FFH area in Germany (WIEDEMANN et al., 2008). Due to the explosiveness of this subject the whole planning approval procedure was accompanied by objections and legal disputes with nature conservation organisations.

2 Concept for Maintaining the Coherence of Natura 2000

Type and scope of the compensation measures mainly depend on the existing of relevant FFH species and habitats in the impact area. According to the FFH Directive the compensation measures shall be effective at the time when the damage occurs in the impact area unless it can be proven that simultaneity is not necessarily required (EC 2000). For claiming the Lakoma pond area it was therefore necessary to start implementation of the compensation measures as early as possible.

When exactly the impact will cause interference will mainly depend on the concerned FFH specie or habitat. An impact can cause an immediate or a delayed interference. If the Lakoma ponds would have been drained for example in summer the impact for amphibians had been effective immediately. Therefore the ponds were only drained after the animals have retreated into the places where they spend the winter season. For this reason the impact effect of drainage only started in the following spring time. When implementing the compensation measures it is also necessary to differentiate between the time of completion and active effectiveness. They can become effective immediately or with a delay in time (GERSTGRASER, 2009).

A timetable was established for all habitats and species concerned which showed the beginning of the impact and the time when the compensation measure should become effective. However, delays in time during the approval procedure in connection with the continuation of the opencast mine resulted in ever shorter time for implementing the compensation measures. It was therefore necessary to adjust both the technological realization and sequencing of the compensation concept several times.

3 Implementation of Compensation Measures

An area of 130 ha of the core zone of the FFH area Lakoma was claimed by draining the Lakoma ponds. Extensive investigations were carried out in the neighbourhood of the impact area to select suitable compensation zones. Altogether, the compensation measures were implemented in seven different zones with a total area of more than 530 ha (GERSTGRASER et al., 2008).

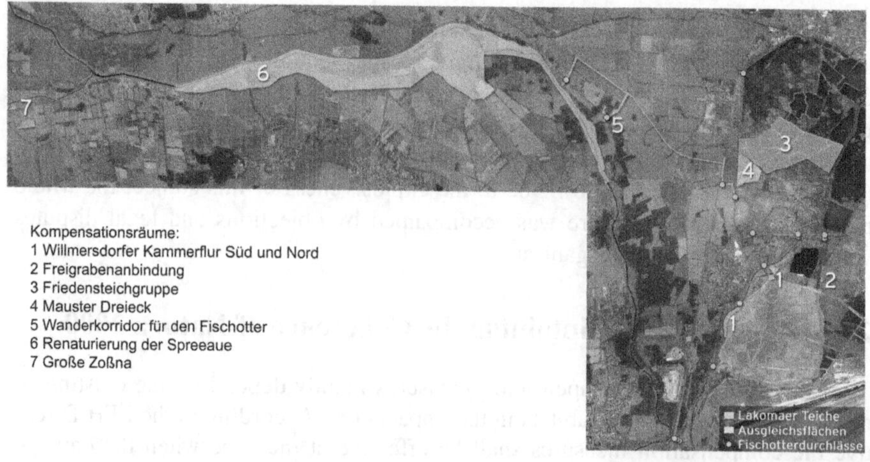

Fig. 2 Impact and compensation areas

The compensation measures were carried out within the period from 2007 to 2013 and included a number of single and complex measures with different functions and scopes. From the various compensation measures some single measures are shortly illustrated hereinafter.

3.1 Renaturation of the Spreeaue and the Spreeaue Ponds

The Spree River is the most important river in Lusatia. Over centuries the Spree and its floodplains were extensively changed by human activity. The Spree is diked on both sides and within the compensation area it was a monotonous river with lack of structural diversity. Missing discharge dynamics by storage dams in the upper reaches and missing longitudinal connectivity resulted in a reduction of variety of species. Over long distances the Spree was not cross-linked with the environs, wetlands were drained, riverine and floodplain forests were almost missing.

North of Cottbus, the Spree was restored to its near-natural state by comprehensive renaturation measures on a length of 11 km and within an area of 400 ha. The following measures were realized:

- Establishment of longitudinal connectivity by removal or redesigning of the ground ramps into river bottom slides
- Structural improvements with groins, islands and side channels
- Connecting of meanders, section wise diversion of the Spree and creation of new side streams
- Relocation of dikes on a length of 3 km for providing new flooding areas
- Improvement of variation in width by flattening of banks above mean water level
- Initializing of riverine and floodplain forests

- Improvement of connectivity of Spree and dike hinterland by installing a new river course on a length of 6 km
- Improvement of the water regime of the region by disconnecting existing drainage systems
- Conversion of forest and establishment of new marshland forests in the dike hinterland
- Creation of habitats for amphibians, otters and other species in the dike hinterland

Fig. 3 Left: 2006 the Spreeaue before renaturation, right: 2008 after renaturation

Another measure was the creation of the Spreeaue ponds. The ponds especially serve the securing of the Natura 2000 network for otter, fire-bellied toad and FFH habitats. The pond group consists of eight ponds with a water surface of 21 ha. They have wide shallow water zones and islands. Water is supplied from the Spree and is regulated by inflow and outflow structures.

Comprehensive landscaping measures were conducted for structuring the pond area and upgrading the environment. For a fast development of a pond vegetation reed rhizomes and pond ground from the impact area were settled in the ponds. Along the ponds trees, shrub groups and in the surrounding area marshland forests and reed were established.

The competent implementation in combination with a focused maintenance and cultivation of the ponds resulted in the establishment of an impressive countryside in the Spreeaue in a very short time. Therefore the renaturation of the Spreeaue is an important element for securing the Natura 2000 network. In addition, numerous

Fig. 4 Development of the Spreeaue ponds, left - before installation, right - 15 months after installation

other animal and plant species benefit from the renaturation of the Spreeaue. Ecological networking in the region is supported and a place providing a quiet and peaceful recreation for the local residents is created.

3.2 Resettlement of Amphibians

All amphibians existing in the impact area had to be resettled because there have not been available any suitable alternative habitats in the environment. During summer 2007 and 2008, amphibian larvae were collected in the Lakoma ponds. All animals were determined, counted and released in suitable new ponds in the Spreeaue.

The resettlement of old and young animals started in September 2007. Until June 2010, amphibians were resettled both in spring and autumn over a period of several weeks. For the collection of the animals 22 km of rescue fences for amphibians were installed in the Lakoma pond area. Altogether more than 180,000 larvae and animals, of it 76,220 fire-bellied toads, were resettled (gIR, 2012). About 150,000 amphibians were resettled in the ponds of the Spreeaue and more than 30,000 amphibian larvae were handed over to the Landesumweltamt Brandenburg (Environmental Authority of Brandenburg) for resettlement at other places.

For the resettlement of the old and young animals in the Spreeaue selected ponds with surrounding wet and forest areas were fenced. This measure prevented migration of the animals out of the pond area for the first two years. In the meantime all fences have been removed and the animals can move freely in the re-established waterlogged Spreeaue.

Fig. 5 Resettlement of amphibians

3.3 Creation of a Migration Corridor for the Otter

A 5 km long safe migration corridor for the otter was built between the Spreeaue and the Peitzer Teiche - as one of several compensation measures. Therefore, new ditches were built with river bank wood and 16 otter passages were built new or reconstructed at dangerous crossing points. Hence, the risk to otters at roads and trails was lowered and more distant natural areas were safely made accessible for the otter.

The Monitoring showed that the migration corridor and all other passage structures were accepted by the otter (ALKA-KRANZ, 2012). It was noted that the intensity of the use of the passages is quite different. Occasionally, the passages were used frequently and thereafter not used for a longer period of time. Nearby the ponds the migration behaviour also changed with the seasonally and spatially changing fish stock.

Fig. 6 Otter passage below a four-lane main road

3.4 Relocation of the Hermit Beetle Trees into the Grosse Zossna

The hermit beetle, as a priority FFH specie, is a beetle who settles in hollow trunks of old tree stands. In the impact area there were 5 of such trees with the suspicion that they were settled by hermit beetles. These 5 trees were endangered by the drainage of the Lakoma pond group. However, since the hermit beetle can complete the development from larval stage to beetle in deadwood the drainage of the water was not a direct impairment. But the massive abstraction of water supply results in the accelerated shrinking of the habitat of the beetle on a large area.

The suspected trees were carefully cut in winter after the imagos decease and brought to a nearby 2 hectare large area with mature stands of hornbeam trees and English oaks. The trunk parts were placed in form of tree pyramids at the new location. The larvae shall develop in the trunk parts. During the monitoring it was observed that one beetle flew from one of the relocated trees (BIOM, 2010).

The development of potential habitats for the hermit beetle within the forest stands in the renaturation area of the Spreeaue will be a further measure for the long-term. For this reason non-native forest stands were transformed into deciduous forest stands of different age in a nature oriented way. This way new habitats for the hermit beetle will be created in the long run.

Fig. 7 Suspected hermit beetle trunks were relocated and placed pyramid-shaped

4 Control of Success

Since 2007 the effects of the compensation measures have been controlled by a complex monitoring program involving various experts. Monitoring comprises investigations of flora and fauna as well as abiotic factors. Up to 16 different comprehensive reports per year are submitted to the authorities. Control of success is carried out over a period of 5 years. If the compensation targets will not be achieved during the monitoring period the LBGR can determine additional measures at any time.

The monitoring differs with regard to content, locations and times. A focused project management ensures that the single surveys are coordinated. All results of the experts are managed in digital form and entered in a GIS data base.

As a result of the monitoring executed so far it can be concluded that all compensation targets have been achieved and the Natura 2000 network was maintained (GERSTGRASER/ZANK, 2012) Altogether, unique conclusions about effects and development of the compensation measures can be drawn from the available monitoring data.

5 Conclusion

For the removal of the FFH area "Lakomaer Teiche" with approximately 130 ha of significant habitats, comprehensive compensation measures have been implemented in seven different zones covering totally 530 ha. The example of the claiming of the FFH area Lakomaer Teiche by the Cottbus-Nord opencast mine shows that despite unavoidable interferences with nature and landscape mining also offers chances for preserving and upgrading species, habitats and landscapes.

Early and comprehensive knowledge about the concerns of the habitats and species in the impact area are important for securing the FFH-relevant protection targets. Because the ecosystem is subject to natural annual fluctuations, sufficiently large and interconnected compensation spaces must be available for such sensible compensation measures to be able to balance natural fluctuations.

Professional implementation along with the monitoring of success of the measures essentially contributed to the success of the compensation measures. This ensures that in case of need improvement can be provided in short-term.

References

Alka-Kranz – Ingenieurbüro für Wildökologie und Naturschutz: Fischotter Monitoring, Report. Graz. 65 S. (2012)

BIOM – Martschei: Final report on Monitoring in "Große Zoßna", Jänschwalde. 14 S. (2012)

EC-European Commission: Natura 2000 – Area Management. The specifications of Article 6 of the Habitats Directive 92/43/EWG. European Union, Luxemburg, 71 S. (2000)

FFH Directive: Council directive on the Conservation of natural habitats and of wild fauna and flora (92/43/EWG) (May 21, 1992)

Freytag, K., Pulz, K., Neumann, U.: Cottbus-Nord opencast lignite mine – compensation measures outside post-mining landscape. Glückauf 143(10) (2007)

Gerstgraser, C., Arnold, I., Dingethal, H.: Implementation of compensation measures for claiming an FFH area by the Cottbus-Nord opencast lignite mine. Bergbau 8. 59 Jhg. Zeitschrift f. Rohstoffgewinnung, Energie, Umwelt. Hrsg. RDB Ring Deutscher Bergingenieure e.V. Essen. S.373–S.377 (2008)

Gerstgraser, C.: Implementation and effects of nature conservation compensation measures in practice. Cottbuser Schriftenreihe zur Ökosystemgenese und Landschaftsentwicklung. Band 9 Brandenburgisch Technische Universität Cottbus. S.35–S.53 (2009)

Gerstgraser, C., Zank, H.: Compensation of the removal of a FFH area. Naturschutz und Landschaftsplanung, Verlag Ulmer, Stuttgart 44 (10), S.293–S.299 (2012)

Wiedemann, B., Arnold, I., Zick, H.: Planning approval procedure for the Cottbus-Nord opencast lignite mine – special characteristics and solutions. World of Mining – Surface & Underground 60 (2008)

Build-Up and Support for Regional Initiatives around Opencast Mines

Michael Eyll-Vetter and Jens Voigt

RWE Power, Köln, Germany

Abstract. The relations between Germany's lignite industry and other regional actors is an important gauge for the regional anchoring of this branch of Germany's energy industry. Partnership at eye level is vital to these relations and has given rise to new forms of cooperation. This becomes especially apparent when it comes to shaping regional structures and further strengthening of the region's economic structure.

The *Indeland* and *Terra Nova* projects are good examples of the efforts to reshape regional structures in the post-mining phase. Within the scope of the *Indeland* project, Düren county and several towns and municipalities in the vicinity of the Inden opencast mine came together to take action. In the years that followed, the non-profit Aachen-based Kathy Beys foundation lent its support to the process with the aim of redeveloping the structure of the post-mine landscape and the rural area around the opencast mine.

The *Terra Nova* project network in the area of the Hambach and Fortuna opencast mines likewise consists of several modules and has worked out an overall regional concept with the support of RWE, and this concept is being implemented. In a next step, the neighbouring communities around the Hambach and Garzweiler opencast mines are to be included in the further development within the initiatives in the mines' vicinity. On account of the much longer mine operating times, the focus in Hambach and Garzweiler is also on the gradual implementation of measures for upgrading the townships on the mines' rim, this in addition to the planning of recultivation. Inter-municipal structural projects are also to be developed.

The Rhenish mining area has for many years now had an efficient and broad-base economic structure. Its further strengthening is a joint task for regional actors and the lignite industry. The region's economic clout is to be strengthened and diversified along these lines by providing further economic mainstays besides lignite. The lignite industry and regional entities have traditionally driven forward the development of industrial estates and a corresponding settlement policy on a joint basis. The focus here is above all on medium-sized business: 3,500 new jobs have been created by investment in the last ten years based on industrial estates at various locations in the mining area, which have either been implemented or are in the pipeline.

In the agricultural region, the food sector, too, plays an important part. RWE Power has been successfully working for some time now on the utilisation of the energy content of waste heat from the cooling-tower and sump water for agricultural production. A partnership-based approach underlies the cooperation between

C. Niemann-Delius. (ed.), *Proc. of the 12th Int. Symp Continuous Surface Mining - Aachen 2014*,
Lecture Notes in Production Engineering, DOI: 10.1007/978-3-319-12301-1_51

lignite companies and municipalities in the development of projects and also in the distribution of opportunities and risks.

1 Introduction

The regional structure in the Rhenish lignite-mining area bears the stamp of the lignite industry. At the same time, however, the region has also left its stamp on the mining and use of lignite. Shaping this interaction is key to the region's development (see Fig. 1: Paffendorf township and Fortuna opencast mine).

Fig. 1 Paffendorf township and Fortuna opencast mine

The extraction and use of Germany's valuable domestic lignite deposits – with approved reserves of 3 billion tons – is meaningful and necessary, and will be so for many decades to come. Lignite makes a contribution to a secure energy supply and, together with renewable resources, helps offset the phase-out of nuclear energy. Moreover, it does not drive up the price of electricity and, in the recent past, has gone far to enable the German economy with its intact industry to more than hold its own in times of crisis. This being so, it is an asset for the industrial State of North Rhine-Westphalia that must not be underestimated.

The term 'emancipation process' describes well the cooperation between lignite and region that has evolved across the past decades. Back in the 1980s, the lignite industry was financially strong and left its imprint on structures. The participation of regional bodies and individual citizens in decisions on investment and jobs was limited. Now, however, the rationalisation of the past decades with the necessary cutting of jobs has also led to losses in the structural rooting of lignite in the region. The region has logically diversified into a more varied and balanced economic and social system in the recent past. RWE today is one of many actors in the region and draws up concepts for future regional development together with a much more extensive circle of players. Today's forms of joint action have considerable implications for methods when it comes to tackling the issues involved in lignite extraction and utilisation.

2 Shaping the Regional Structures

Besides recultivation in the classic sense, re-shaping the landscape after the intervention of mining today takes account of a whole host of ecological, commercial and leisure aspects, and also of issues relating to the social and technical infrastructure. One prime example of this is the *Indeland* project to the west of the Rhenish lignite-mining area.

2.1 The Indeland Project

The question of shaping the follow-up structures for the Inden opencast mine, where operations will taper out toward 2030 was addressed early on. To this end, the start of the last decade saw the formation, with broad social support, of the *Indeland* initiative in the west of the Rhenish lignite-mining area in order to seize the opportunities offered for restructuring the entire, more agricultural space in the environs of the opencast mine. For this, it specially created an organisational framework in the guise of the "*Indeland Entwicklungsgesellschaft*" development company.

Members include Düren county, the towns of Eschweiler, Jülich and Linnich as well as the Aldenhoven, Inden, Langerwehe and Niederzier municipalities, plus RWE Power in an advisory capacity. With the private Aachen-based Kathy Beys foundation, it was also possible to win over a knowledgeable and neutral authority as moderator and driver behind the process.

Trigger for the new activities was above all the amendment to the lignite plan Inden II encouraged by important local-government actors. The bottom line is the creation of an 11-km² lake instead of the originally envisaged backfilling of the opencast mine using overburden, thus enabling new utilisation concepts. In producing these concepts, use was made of experience gained with the successful remediation of shut-down opencast lignite mines in Central and East Germany and with the rapid development of the water-related leisure and tourism industry in the area of the Dutch Maas river. In parallel, one focus in equal measure was on the further development of the economic structure.

Fig. 2 "Inde man"

The 36-meter viewing platform "Inde man" is the project landmark in this concerted effort (see Fig. 2: "Inde man"). For the vicinity of the opencast mine, "*Indeland Entwicklungsgesellschaft*" – with the support of RWE – is developing a regional master plan which defines joint planning and development goals for the municipalities involved, focusing mainly on projects in residential- and industrial-estate development, traffic infrastructure, renewable energy sources, leisure and tourism. The process of changing the rehabilitation of the Inden opencast mine has thus evolved into a new form of inter-municipal and social cooperation.

2.2 The Terra Nova Project

The towns of Bedburg, Bergheim, Elsdorf and Rhein-Erft county have together set up the joint *Terra Nova* planning network to shape and further develop the landscape produced by opencast lignite mining. The former long-distance conveyor-belt route, the northern edge of the Hambach opencast mine, the industrial estate and the power plant form the four modules at the core of *Terra Nova*. They provide impetus for shaping the landscape, while at the same time creating long-term visibility of lignite mining and use for the visitor. With the former overburden conveyor belt, as main axis measuring some 12 km, *Terra Nova* today links the towns of Bergheim, Bedburg and Elsdorf.

Fig. 3 Speedway *Terra Nova*

Fig. 4 Forum Terra Nova

The former long-distance conveyor-belt route today is a popular destination for outings by strollers, cyclists and skaters. (see Fig. 3: speedway *Terra Nova*). Following on from the route, given an ecologically ambitious form as "biosphere band" is the opencast-mine rim, which is being designed as a landscape garden, itself a piece of "terra nova". Part of the 'opencast mine rim' module is the *Forum Terra Nova* (see Fig. 4: *Forum Terra Nova*). As information and exhibition venue, its remit is to provide visitors with glimpses of ongoing opencast mining operations and, in a few years' time, recultivation progress. This building is outstanding, both visually and functionally, and is located right on the mine rim. It is equipped with future-geared technology, eg a heating system using sump water.

The Niederaussem power plant, another *Terra Nova* module, forms part of the learning landscape for energy management and, with the information centre, offers an opportunity to experience modern power-plant engineering and have a look at the new *BoAplus* power-station project with its manifold innovations round and about coal.

As fourth *Terra Nova* module, an inter-municipal industrial estate is to be developed. The aim is the settlement of scientific facilities, combined with industrial operations, stressing the innovative use of resources. Hence, the overall *Terra Nova* complex, like the *Indeland* project, is developing its special importance for the region due to the variety of the subjects included and the networking of many project partners to achieve joint goals.

2.3 The Hambach and Garzweiler Neighbourhood Initiatives

The future development of the Hambach opencast mine and its recultivation affects neighbouring townships to a high degree. Those concerned are six towns. They – together with the opencast-mine operator and the administrative districts in charge – wish to form the core of a planning community or community of interests for the Hambach post-mine landscape. Owing to the long term of the Hambach opencast mining operations, the various focuses and the urgency of the concrete arrangements for rehabilitation vary substantially. All the same – also as backing for the further development of the Hambach mine in the coming years – such a neighbourhood initiative is set to accompany the cooperation between mining operations and neighbouring municipalities on a regular basis.

Since the mining area is crossed by the boundaries of the two counties of Düren and Rhein-Erft, these two county councils in particular play an important role in coordinating the planning, in the moderation of possible conflicts and as provider of ideas and engine for future developments. Moreover, they input experience from other projects implemented across the township borders. After all, they must be crucially involved in the approval procedures, eg via the Lower Nature-Conservation Authorities.

The Cologne regional government as intermediate authority of the State of NRW in this process will mainly play a role as approval authority, inter alia, because it must examine the compatibility of the above planning with the stipulations of the lignite and regional plan.

In the north of the Rhenish lignite-mining area, too, some projects have already been implemented, especially in recent years, together with the neighbouring municipalities of the Garzweiler opencast mine, and these have highlighted the value created for the region by RWE Power beyond actual lignite mining.

These efforts are to be continued with the further development of the mine to the west. Hence, last year saw the constitution of an inter-municipal working group, consisting of representatives from the municipalities bordering the Garzweiler mine. The aim is to launch today already, early on and in a constructive and structured manner, a development process for economic valorisation geared toward the social interests of the mine-rim communities, also including the later post-mine landscape.

This planning for the future is taking place at different levels, so that people living in the vicinity of the opencast mine can input their short-, medium- and long-term ideas into the development process. In a first step, the current situation of the mine-rim communities must be analysed against the background of the existing spatial environment. Fields of action can then be derived that jointly affect the mine-rim communities and are mainly marked by social, property-law and infrastructural issues.

Finally, visions are to be developed from this that reflect the social consensus as regards the shape to be given to the locations. Here – depending on the specific situation of the municipality relative to the opencast mine – the options may range from a short-term enhancing of the attractiveness of recultivation and the immediate neighbourhood of the mine all the way to the valorisation of the regional economy. Development of the landscape, which is changing owing to the opencast operations, by providing appropriate link-ups with the settlement areas via new road concepts and green structures, taking account of the shape to be given to the residual lake, counts among the short-term measures. The pinpointed development of residential and industrial estates in the immediate vicinity of the Garzweiler opencast mine, by contrast, is part of the valorisation of the regional economy.

The further structured fine-tuning of the planning design is to take place via a master-plan process. On its basis, the detailed planning for the later shape given to the lake can then commence at the start of the 2020s.

2.4 *Shaping the Living Space Thanks to Modern Agriculture*

Traditionally, regional agriculture in the Rhenish lignite-mining area has had co-operative relations with the lignite industry. It is the most important area partner for lignite mining, even the entire energy sector. In addition to the operating surface of the opencast lignite mines and power plants, it frequently also provides the land for power lines. Adequate offset and compensation arrangements form a comprehensive basis for sustainable cooperation.

Especially the arrangements and agreements made in connection with lignite extraction ensure a largely frictionless sequence of land utilisation in the run-up to opencast mining and the land return after recultivation. Extraordinary developments, like surface-relevant amendments to lignite-plan procedures, eg the re-planning of agricultural rehabilitation in favour of a residual lake in the area of the Inden opencast mine or additional need for surfaces for species-conservation measures, lead to a scarcity of farmland. The tensions this inevitably entails can only be resolved in a relationship that is based on mutual trust in the partner's commitment.

Cooperation has a quite practical form within the scope of agricultural rehabilitation. RWE's own agricultural operations, called pioneer farms, act in the regional agro business as customers for merchandise and services, clients, suppliers of agricultural goods and members of the *Maschinenring* cooperative. After the land is returned, the farmers as owners or leaseholders continue developments on recultivated farmland commenced with the rehabilitation. But they also act as service providers for mining, be it for transport work or in air-pollution control and noise abatement.

The most recent example concerns the habitat-supporting measures for protected, forest-bound species, specifically bats. Round and about the Hambach mine, measures are being implemented on some 700ha of agricultural land. In the reshaping of the necessary green-land surfaces and their utilisation with grazing animals, preference goes to farmers affected by measures when it comes to earth-moving work, sowing and care during the growth phase, all the way to the later pastoral economy (see Fig. 5: Park-like, half-open landscapes with Glan cattle).

Fig. 5 Park-like, half-open landscapes with Glan cattle

Being aware as a mine operator that it is responsible to agriculture for land utilisation, the company is looking for novel cooperation possibilities. Worthy of mention here are the supply of the Neurath and Paffendorf biogas system to underpin the raw-material supply as well as the Agrotherm and Hortitherm systems which produce asparagus, tomatoes and decorative plants under intensive conditions.

3 Further Development of the Economic Structure

An economic structure marked mainly by lignite and the energy sector is to give way in future to wider, diversified economic activity. For a mining company the question arises of the specific areas in which opportunities can be found and promoted.

3.1 Agro Business

With this in mind, a 2010 Prognos study on the "*Potenziale der Industriestandortentwicklung an Braunkohlekraftwerks-Standorten*" (Potentials of the development of industrial sites at lignite-fired power-plant locations) pointed to the food industry, too, as target sector alongside the settlement of energy-intensive processing and manufacturing operations. Against this background, RWE Power has been successfully working for some time now on the use of heat for agricultural production in conventional greenhouses or, with the Hortitherm and Agrotherm systems, on the use of residual heat from cooling-tower and sump water. In addition, talks with numerous companies in the sector based in the mining area have shown that formation of a network could readily offer potentials for new settlement in a region marked by agriculture and the food sector.

Fig. 6 Gardenlands greenhouse park

Together with the *Landgard* cooperative and four garden centres, we took a joint decision to build NRW's biggest greenhouse at the Neurath power plant (see Fig. 6: *Gardenlands* greenhouse park). The first stage has already seen the implementation of 11ha under glass. A second stage is being planned. Heat is supplied by the Neurath power station. Together with local farmers, RWE also operates, on 10ha in the Elsdorf area, an asparagus-growing surface, the so-called Agrotherm system, where the asparagus harvest is brought forward thanks to sump water, so that regional produce can be marketed much earlier (see Fig. 7: Early asparagus harvest thanks to Agrotherm).

Fig. 7 Early asparagus harvest thanks to Agrotherm

The long-term goal is the development of an agro business supplied by lignite operations: extending from the cultivation of agricultural produce, via further processing, packaging and, eventually, delivery to people in the region. Apart from substantial CO_2 savings and shorter marketing channels, the idea here is to strengthen development in the rural area.

3.2 Trade and Industry

Small and medium sized enterprises (SME) based trade and industry, with its wide variety of settlements in the Rhenish mining area is another factor determining local economic strength in addition to agriculture. The further development of the SME-based trade and industry structure by way of the systematic and structured development of industrial estates requires in-depth support from many partners.

RWE Power is flanking these processes and is also using its competence in developing land within the scope of rehabilitation schemes.

As a case in point, the town of Kerpen, together with the mining company, has developed the "*Türnich*" industrial estate and business park on some 86ha of the recultivated Frechen opencast mine. Since 1998, about 60 companies have been successfully settled here, creating some 2,100 jobs (see Fig. 8: industrial estate *Türnich III*).

Fig. 8 Industrial estate *Türnich III*

In Bedburg, too, a showcase project has emerged in the shape of the *Mühlenerft* industrial estate for 34 companies covering a surface of over 100ha. With the extensions to the inter-municipal industrial estate in Grevenbroich/Rommerskirchen to include a further 12ha, and with the Nievenheim am Rhein estate located in the extreme north of the mining area, there are other projects in the pipeline. What all of these projects have in common is that they are being developed and implemented on a partnership basis.

For individual municipalities, job losses do occur in the wake of economic developments. However, these can be compensated in the efforts to re-design the post-mine landscape and other surfaces no longer needed for operations. RWE is actively supporting such schemes and has made a substantial contribution toward the creation of over 3,500 jobs in the last ten years. This is also reflected in an investment boost of approx. €25 million given within the scope of land development and a further €620 million in building construction, all triggered by the measures.

3.3 Energy Partner / Stakeholder Models

For the further development of the economic structure in the Rhenish mining area, too, expanding the use of renewable energy plays an important part. Besides risks, Germany's energy turnaround in close partnership with conventional energies offers, most of all, opportunities that are being exploited by the implementation of projects. Stakeholder models can constitute a future-geared concept for many of these projects. The term "energy partner" here refers to a partnership between municipalities and RWE in the field of energy generation and marketing based on renewables. The idea behind forming these project partnerships is to find a sensible use for the company's own land surfaces that will benefit the municipalities, but individuals as well. In 2012, for example, "Energiepartner Kerpen GmbH" was set up by the town of Kerpen and RWE. The firm has erected a photovoltaic system with 2MW output on a noise-abatement embankment in Kerpen constructed by RWE.

In January 2014, the Vereinigte Ville landfill for power-plant residues is seeing the commissioning of a photovoltaic system with 23,500 installed modules and a capacity of 3,000kWp, with an annual output of some 2.8 million kWh/a. This concerns a joint project by "Die BürgerEnergie eG" and RWE. Here, RWE is offering participation via citizen cooperatives (see Fig. 9: BürgerEnergie eG Ville).

Fig. 9 BürgerEnergie eG Ville

Together with several wind farms and in partnership with flexible lignite-fired power plants in the post-mine landscape, it is making renewable energy visible in the Rhenish mining area. These concepts are to be further expanded and, proceeding along these lines, a total of over 100MW in wind farms is to be erected in the Rhenish mining area.

4 Developing Local Structures

In parallel with the regional level, local government, too, has seen a future-geared planning and implementation process being established in order to shape new structures in the course of the changes produced by the lignite industry. Here, local government and mining company are jointly implementing major projects, thus ensuring that income flows to all those involved. The *Faktor X* concept is a prime example for this.

4.1 Faktor X

For more than a decade now, RWE Power and the municipalities have been partners in the development of building land in the Rhenish lignite-mining area and, hence, are shapers of an important part of the post-mine landscape. During this time, about 1,500 housing sites have been developed. This process is to be continued and further developed in the coming years with innovative projects, like the *Faktor X* model estate. In essence, the concept takes account of demographic change to which Germany will be exposed in the coming years that will no doubt lead to a contraction of the rural population and, in its wake, tend to mean pronounced falls in property prices. Long-term value retention for new properties will thus be possible using resource-sparing and, hence, cost-lowering construction (see Fig. 10: *Faktor X* housing concept).

Fig. 10 Faktor *X* housing concept

In the design of residential districts, special account is to be taken in the planning phase already of a fundamental reduction in resource consumption, of the future viability of the site-development and civil-engineering concept, and of the value and flexibility of the infra-structure and residential buildings. In this sense, the aim of the concept development is a reduction in raw-material consumption for residential areas by a factor X (= 10) and, via the improved economic efficiency thus obtained, an enhanced attractiveness of rural life in the long term as well. Cooperation on this is currently taking place with the Inden municipality and the town of Eschweiler. Construction sites each measuring 3ha are now to be designed with this in mind.

4.2 Village-Development Concepts

Structural changes make it important for rural communities in particular to identify their perspectives and potentials and to create the basis for structured and future-geared neighbourly assistance, alongside the development and implementation of specific building projects.

To this end, the State is offering the village-development planning instrument. Within this framework, RWE Power is currently sponsoring, among other ideas, a village-development plan in Bergheim Rheidt-Hüchelhoven, a township of some 1,800 inhabitants close to the Bergheim-Niederaussem power plant.

This offers the option of combining measures for reducing burdens with a structured process for shaping the future. In workshops on key subjects, like the structural development of the township, the design of playgrounds, but also on the creation of bridleways, cycle lanes and footpaths and at workgroup meetings, citizens are actively involved in the planning effort, so that they can exert a direct influence on the future of their township.

Qualities and deficiencies of a township are highlighted in this process and development opportunities identified. At the end of the process is a detailed plan of measures which serves as guiding principle for coming years. Thanks to the good experience gained, RWE Power aims to accompany village-development planning in other townships located in the vicinity of its operations as well.

5 Organising the Cooperation

A clear organisation at three levels of the cooperation ensures that planning does, in fact, become implementation (see Fig. 11: Partnership-based cooperation at different levels). An integral component of the overall concept for the Rhenish mining area here is the close networking of regional, local and economic structures.

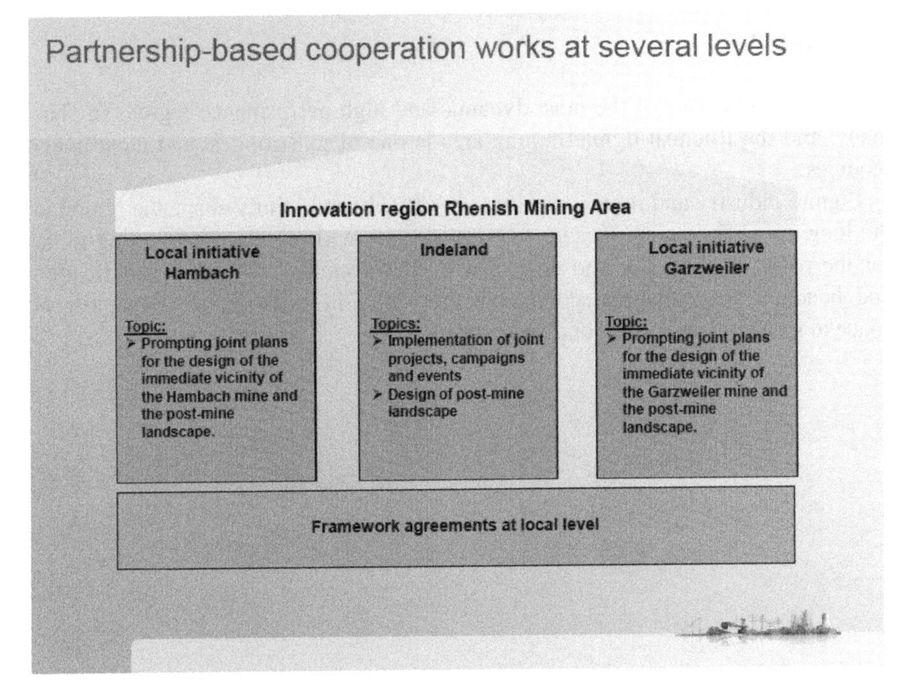

Fig. 11 Partnership-based cooperation at different levels

At local level, framework agreements are in place, with local government and mining company as partners. The agreement binds town and company to the joint goal, eg of implementing previously defined projects in accordance with defined rules. This framework agreement also provides for periodic meetings in which the status of processing, the deployment of funds and the next steps are addressed. This form of cooperation has proved extremely fruitful and is to be expanded further.

Inter-municipal planning associations, like the development company *Indeland Entwicklungsgesellschaft* or *Terra Nova*, are proceeding from the post-mine landscape and giving an important boost toward sustainable development in the region. They form the intermediate level of the cooperation. Beyond *Indeland* und *Terra Nova*, new planning associations are currently being formed with the neighbourhood initiatives Hambach and Garzweiler. Their aim is an optimal and joint design of the phase of opencast-mine development along the line of townships. Also in view of the lake to be created at a later date and other recultivation measures, the scene can be set correctly early on in this way.

The projects at municipal and inter-municipal level are also making a contribution to the initiative "*Innovationsregion Rheinisches Revier*" (Innovation region Rhenish mining area) created at State level. It represents the level of area-crossing approaches and is currently being realigned and, hence, geared even more closely to the concerns of the Rhenish mining area than in the past.

6 Upshot

The Rhineland is one of the most dynamic and high-performance regions in Germany, and the Rhenish lignite-mining area is one of its strongest and most lively modules.

Lignite industry and region want to help actively and jointly shape the region in the long term. For many decades to come, lignite will remain a supporting pillar for the regional economy and do its bit in the economy's further diversification and, hence, in strengthening economic activity, and, in this way, give new partners space to grow in importance and perspective.

State of the Lignite Rehabilitation and Current Challenges

Klaus Freytag and Hans-Georg Thiem

State Mining Authority of Brandenburg, Cottbus

1 State of the Lignite Rehabilitation and Current Challenges

In the times of the eastern block mining facilities and mineral deposits had been in the ownership of the public (German Democratic Republic state-property). After the political turn in the last period of the German Democratic Republic (GDR) and after the German reunification the viable part of mining industry became privatized, for example parts of lignite mining and quarrystone industry. The non viable parts of mining industry had been overtaken by special federal companies. Lignite mines and up-grading facilities was overtaken by the federal Lausitzer und Mitteldeutsche Braunkohlenverwaltungsgesellschaft (LMBV).

Before the German Reunification in 1990, almost 70 percent of the energy supply in the GDR originated from the mining, power generation, and upgrading of lignite. In comparison, this proportion only amounted to eight percent in the old federal states. Due to the lack of other energy sources, the GDR focused on the abundantly available lignite deposits. In fact, lignite mining increased fivefold from 1950 to 1990 in the GDR. However, as a consequence, the world's most extensive lignite mining and lignite utilization had a substantial impact on the landscape and environment (Figure 1).

Following Reunification, restoration of abandoned mining areas became particularly important in the area of the former GDR. At the end of the 1980s opencast mining there produced about 300 million tonnes of raw lignite per annum. Raw lignite covered around 70 per cent of the primary energy demand in the former GDR and supplied around 80 per cent of the fuel needed for power generation.

The land reclamation and derelict land clearance of those areas that still had been under mine supervision in 1990 had been licenced using al plan of operation by the mining inspectorate based on the federal mining act.

The restoration and rehabilitation of former lignite mining facilities is Germany's largest environmental project. This undertaking commenced immediately after the German Reunification in 1990 and has reached remarkable results during its more than 20 years of existence. This is best illustrated by the newly created post-mining landscapes with their expansive woodlands and wildlife preserves, with lakes reclaimed from former opencast sites that are now embedded harmoniously into the countryside as well as newly established and developed industrial and commercial zones.

C. Niemann-Delius. (ed.), *Proc. of the 12th Int. Symp Continuous Surface Mining - Aachen 2014*, Lecture Notes in Production Engineering, DOI: 10.1007/978-3-319-12301-1_52

Fig. 1 open-cast mine Greifenhain 1995 (LMBV)

1.1 Legal Aspects

The extraction of raw materials and natural resources found close to the land surface and mined in open-cast pits has a severe impact on the landscape. Proper landscaping after the shut-down of the mines has always been the responsibility of the mine operator. The "General Mining Act for the Prussian States", published as early as in 1865, contained already essential features of modern recultivation regulations. The rather generally worded regulations of the General Mining Act were then more clearly defined in a Mining Police Directive for the Rhenish lignite mining region issued by the then Superior Mining Authority in Bonn in 1929: "Overburden masses shall be deposited in disused open-cast mines in such a way that as much land as possible is reclaimed so that it may then be given over to forestry and farming."

Especially the "Practical proposals for cultivating dumpsites, fallow land, dunes and wasteland", published by Rudolf Heuson, the forest administrator of the Niederlausitzer Kohlewerke in the 1920s of the last century, are evidence for the long tradition and the commitment of the lignite mining industry in the Lusatian mining region to reclaim the post-mining land and recultivate it lastingly.

The 1980 Federal Mining Act (Bundesberggesetz - BBergG) stipulates the mine operator's obligation to take preventive recultivation measures and to complete these measures properly after the mining operations have come to an end.

The production of lignite in open-cast mines has assumed such dimensions that the land claimed and used in the lignite mining regions of Germany has become the subject of separate planning activities at state level. The instrument used for these regional planning efforts is the so-called "Lignite Plan".

As far as its function is concerned, lignite planning is a part of regional planning and must at least be allocated to the planning process at state level, as is the case in the federal state of Brandenburg. Special features of lignite planning are that the planning is closely linked both with the targets of the energy policy of the relevant federal state and with the mining law, This, in turn, will ensure that the regional plans and the mine operation plans are in accord with each other. This applies especially to the final operation plans, when it comes to land reclamation. However, it must also be pointed out in this context that the obligation of the mine operator to adapt the operation plans to the lignite planning, as it is laid down in the statutory provisions for the lignite planning, is legally disputed, since the BBergG (the Federal Mining Act) does not make any such reference. Only the "predominantly public interests", as referred to in Article § 48 II, p. 1 of the BBergG, would allow lignite planning finding its way into the mine operation planning procedure. Responsible for preparing the lignite plans is the state planning authority.

Lignite planning in the form of so-called island plans for the regions affected by lignite mining controls the resettlement measures as well as the relocation of roads and ways. The lignite plan outlines in particular the rough structure of the post-mining landscape. Prepared by the state planning authorities in consultation with the Lignite Committee, the lignite plan will then be put in force by the government of the federal state concerned and is thus legally binding. The mine operator is now sufficiently informed about the objectives of state planning in his mining region and will take these objectives duly into account during all other statutorily required permit procedures.

The mine operator indicates and identifies the long-term implications of his mining projects in the general operations plan that is required under mining law. Depending on the legal basis, this general operations plan (frame plan) will be dealt with and approved either as part of a simplified administrative procedure or in the course of a public works planning procedure, if the operations exceed environmentally relevant limits, which is usually the case with the lignite mining.

Due to the provisions contained in the Energy Contract (Unification Treaty of September 23, 1990), operations in the open-cast lignite mines of the new federal states were deemed to be procedures in progress and continued on the basis of frame plans in accordance with Article § 52, paragraph 2, of the BBergG (simplified administrative procedure). In accordance with the provisions contained in the State Planning Act of the federal state of North-Rhine Westphalia (LPLG NRW), the environmental impact assessment (UVP) will be carried out as part of the lignite planning procedure. The mine operator includes a schedule for the land reclamation measures in the frame plan which has to be drawn up in line with the mining law. This schedule cannot be very detailed as yet, since the measures to be taken refer to a period some decades ahead.

The mine operator is only entitled to start operations, after the main operations plan has been approved. The approval usually covers a period of two years. The main operations plan includes early statements about land reclamation. This means for the mining operations to determine the way how the overburden is dumped (overburden management), to produce substrates that are suitable for cultivating the reclaimed land (creation of loess reserves) as well as to take the general profile of the landscape into account.

The mine operator will have to submit a final operations plans as required by the mining law, when the entire company or parts thereof cease operating and when the land is to be returned to any other use but mining. In the sense of an early integration of the main operations plan and the final operations plan, it has been a practical advantage to prepare a final operations plan also for those "units of the mining company which are still active". The preventive land reclamation measures, as contained in the main operations plan and implemented accordingly, can then be smoothly transferred into measures for shaping the final land surface.

The final operations plan contains the details of how the more general requirements concerning the "proper land reclamation", as enshrined in the Federal Mining Act, will be satisfied.

The BBergG defines the term "proper" (land reclamation) only partly. A more detailed definition as to what land reclamation is supposed to mean can be found in the state-specific mining directives (such as the Mining Directive for Lignite Mines/BVOBr of the Superior Mining Authority of the federal state of North-Rhine Westphalia) and in guidelines issued by the state mining authorities.

The Land Reclamation Directive of the state mining authority of Brandenburg (LBGR), which has been prepared in close cooperation between the mine operators and the environmental and mining authorities of the federal state of Brandenburg, contains details as to how land can be reclaimed which is intended for forest and agricultural use, i.e. the main branches of land utilization in post-mining areas. Based on the soil substrates prevalent in the Lusatian region, the directive makes reference to the technical suitability of these substrates for soil cultivation measures. A separate section of the directive deals with land reclamation under the aspect of nature conservation. The directive demands that the reclaimed land is integrated as speedily as possible into the existing landscape and connected with the neighboring biotopes that have existed before and are intact. Biotope systems linked with each other, special areas of land as well as integration measures taken in the forest and agricultural sector are named and shown as a suitable approach to promoting this integration process. The positive effects of a successive use of large land areas are also mentioned.

Apart from the mining standards in their own right, the final operations plan is the vehicle for implementing further statutory requirements concerning land reclamation. Take the Federal Soil Protection Act (BBodSchG) as an example: It does not contain any independent provisions as to mining projects requiring approval, but it stipulates the physical requirements how to treat the land and how to handle the soil, i.e. provisions which will have to be observed during the permitting procedure for the mine operations plan. Likewise, these requirements will also have to be satisfied when materials, such as soil improving substances or waste materials, are deposited in the ground or on the land surface. The specific requirements as regards soil protection usually cover a rooting depth of about 2 m for any subsequent vegetation.

Apart from the impact the soil protection provisions have on the land reclamation process, as has been outlined above, the provisions contained in the Water Management Act (WHG) and in the state water laws are also of great importance. The mass deficit created by the extraction of raw materials in the mining area will usually result in the formation of lakes, for which a permitting procedure under the water law is required, if they are to be developed further.

In the process of implementing all the different statutory regulations referred to in the mining license, which is undoubtedly a necessary requirement, one will have to bear in mind that these mining regulations must be carefully weighed against each other. All these requirements that are meant to regulate the mining operations from outside must not lead to a situation in which the mining privilege would eventually be strangulated and any vested rights be disregarded. The statutory requirements will have to keep within the clearly defined framework created and limited by the state planning regulations or by the mining license.

The proper implementation of the land reclamation process is closely connected with the end of the supervision through the mining authorities. In accordance with Article § 69 of the BBergG, "the supervision through the mining authorities ceases after the final operations plan has been implemented ... i.e. at the time when, according to common knowledge and experience, detrimental effects with negative consequences for the general public … are no longer expected". A period of 5-7 years will usually pass between the land reclamation and the end of the supervision through the mining authorities. This time is required to establish the fundamental success of the land reclamation efforts. It is quite possible to make statements about future positive developments as regards the reclaimed land after some time of intermediate land use for forest and agricultural purposes. Although the end of the supervision through the mining authorities is stipulated by law, the mining regions usually take measures, such as final inspections or planning notifications, that refer to the end of the mining authorities' responsibilities in this respect.

The success achieved in the field of land reclamation in all mining regions during the recent decades has shown that the lignite mining industry creates new landscapes that do not leave any scars in the countryside and merge very well with it. The increasing commitment of the lignite industry in this respect has reached a very high level by now, and the success thus achieved has contributed considerably to a further improvement of the image of lignite mining.

1.2 The Company LMBV

The Lausitz and Central-German Mining Administration Company (LMBV) was established on 9th August 1994 by the trust institution as a company for the decommissioning and rehabilitation of mining pits in the lignite regions of Lausitz and Central Germany. It is owned by the Federal Republic of Germany, represented by the Federal Ministry of Finance. The last production sites of LMBV lignite mining were taken out of operation at the end of 1999. Since then the focus of activities lies in the rehabilitation and recovery and increasingly also in the reclamation of the areas claimed for lignite mining. With this the foundations are laid for a comprehensive restructuring and recovery of the former mining regions.

Approximately 107,000 hectares of the areas claimed for mining in Germany lay originally in LMBV's area of responsibility. Up to now more than 23,000 hectares of agricultural and forest areas have already been reclaimed and several thousand hectares are left to develop naturally. So far over 10,200 hectares of wooded areas have been newly established and for this over 100 million trees have been planted. The reclamation rate for LMBV lies at 60 percent in Lausitz and at 74 percent in Central Germany.

The extraction of lignite had a massive effect on the water balance in both regions. Therefore until 1990 a depression in ground water level arose which reached to a depth of 80 metres. For restoration of balanced water resources that are extensively capable of self-regulation, it is in the long term necessary to recharge approx. 12.7 billion cubic metres of water into the aquifers and the mining lakes of the mentioned areas. Through replenishment of the aquifers and flooding from outside water sources up until now more than half has been rebalanced. So far ten of the larger 46 openmining lakes have reached the final water level while another 28 are still being flooded (Figure 2).

Fig. 2 open-cast mine Greifenhain 1995 and 2004 (LMBV)

However, there were also the less positive aspects of this intensive lignite mining and use: firstly, a leading position with regard to the per capita load of typical lignite pollutants such as SO_2 or dust, and secondly, extreme ecological damage, for example:

- ❖ The devastation of large areas (around 120,000 hectares) needed for the opencast mines and the lignite processing plants
- ❖ Severe interventions in the hydrological regime of the lignite mining region;
- ❖ The filling of worked-out open casts with both residues from lignite processing and industrial and municipal waste.

Therefore the intended rehabilitation of abandoned lignite mining sites in former East Germany comprised many tasks:

- ❖ Securing, rehabilitation and final landscaping of 215 worked-out open casts in 31 opencast mining areas.

  Of the 215 worked-out open casts, 163 are to be landscaped as lakes, 28 have to be filled and the slopes of 24 dry ones have to be secured.
- ❖ Around 1,063 km of slopes have to be secured, around 535 km of them facing potential liquefaction slumping,
- ❖ Demolition
  - o of 57 briquette factories
  - o of 48 industrial power plants and boiler houses
  - o of 2 coking plants, 2 low-temperature carbonisation plants and 1 gas works

The rehabilitation of the large scale industrial wasteland resulting from the demolition of the facilities, which in general covered large areas.

Work on these tremendous tasks was started at the beginning of the 1990s. The Federation and the so-called 'lignite federal states' in the former GDR, i.e. Saxony, Brandenburg, Saxony-Anhalt and Thuringia, decided to allocate massive funds to the rehabilitation project. As the owner of the areas and facilities to be rehabilitated, Lausitzer und Mitteldeutsche Braunkohle-Verwaltungsgesellschaft (LMBV – Lausitz and Central-German Mining Administration Company), acted as project leader. It was responsible for planning the rehabilitation, issuing invitations to tender and contracts and for controlling the execution of the work.

In dealing with these tremendous tasks various technical, economic and ecological targets had to be taken into account. In addition, there were social, in particular labour market aspects to be taken into consideration, especially due to the high number of redundancies resulting from the extensive closures of mines.

1.3 Rehabilitation of Opencast Mines

The rehabilitation of the opencast mines comprised various tasks, in particular:

- ❖ Securing the steep slopes of the worked-out open casts which were in danger of slumping

- Acceleration of the slow natural refilling of the worked-out open casts with water by using water from nearby rivers as well as draining water from opencast mines still in operation
- Recultivation of the artificial waste heap bottoms by soil improving measures, using the areas for agriculture and forestry and by making areas available for natural development and successions with substantial nature conservation potential.

Facilities for lignite coal upgrading which had been in operation for decades had to be decommissioned at more than 100 sites. This comprised briquette factories, power plants, heating stations, low-temperature carbonisation plants, coking plants, gas works and related facilities. They were technically outdated and very often contaminated by residue and pollutant depositions. In addition, soil and groundwater pollution had to be coped with. On the other hand, infrastructure was well developed in many cases, offering considerable development potential for such sites.

The rehabilitation of such sites comprised mainly:

- Dismantling the facilities by gutting and scrapping, demolition of the buildings including removal of underground floors (depending on the pollutant load), recycling of non-contaminated material and its reuse as construction material during the rehabilitation project.
- Removal and disposal of contaminated soil. This is particularly complex with tar products which cannot be pumped, desulphurisation products and carbonisation gas residues in installations, containers and pipes. In individual cases the volumes amounted to several thousand tonnes.
- Reuse or disposal of residues and pollutants amounting to several 100,000 tonnes, which were stored in large industrial settling basins or landfills that did not meet the technical security requirements for such installations.
- Assessment of groundwater contamination and purification of groundwater.

1.4 Restoration of a Largely Self-regulating Hydrological Regime

Groundwater deficits amounting to 12.7 billion cubic metres, distorted surface water discharge conditions in large areas and lakes forming due to rising groundwater in the worked-out open casts have severely affected the hydrological regime and will continue to do so in the future. They are considered the main ecological damage caused by lignite mining. Restoring a well balanced, largely self-regulating hydrological regime is therefore central to the rehabilitation of abandoned lignite mining sites. In view of the tremendous water deficits and the danger of water acidification resulting from the pyrite oxidation in the waste heaps, it is vital to master the problems with regard to water volumes and water quality.

The most efficient solution to this problem is to flood the largest worked-out open casts with both water from nearby rivers and draining water from opencast mines still in operation. This enables the worked-out open casts to refill in only a tenth of the time natural refilling would take and also speeds up the replenishing of aquifers. In addition, neutral external water helps to fight acidification in the opencast mining lakes. Where this is not achieved, as for example in the Lausitz region, neutralising treatment of the water is necessary.

1.5 Rehabilitation of Existing Polluted Sites

In the area of the rehabilitation project more than 1,200 areas were documented as potentially polluted sites as per the 31 December 2004.

The most common problem at the polluted sites is waste deposited in the worked-out open casts and small scale soil pollutions, so-called leakages, in production areas with machinery and vehicle fleets. These were dealt with in a standardised way and disposed of within the framework of the geotechnical rehabilitation of the slopes and the demolition of facilities. The need to trim landfills due to rising groundwater levels is a specific problem occurring when rehabilitating abandoned lignite mining sites.

Soil and ground water rehabilitation in the areas of former upgrading facilities such as coking plants and low-temperature carbonisation plants calls for a more complex rehabilitation concept. It is more time consuming and costly due to the typical pollutant load arising from upgrading raw lignite: petroleum-derived hydrocarbons, phenols, monoaromatics and polycyclical aromatic hydrocarbons. The best known sites in this context in Germany are Espenhain and Deuben in former East Germany and Schwarze Pumpe und Lauchhammer in the Lausitz region.

1.6 Financing

Between 1991 and 2013 more than 9.4 billion Euro were earmarked for the rehabilitation of abandoned lignite mining sites. The Federation contributed the greatest part of this, 75 per cent. The federal states concerned, i.e. Brandenburg, Saxony-Anhalt, Saxony and Thuringia, contributed 25 per cent to the funding, particularly in view of the fact that rehabilitating the abandoned lignite mining sites also improves the conditions for the economic development of this former lignite mining area. About eight billion Euros were used by LMBV for the securing of decommissioned open pits and further necessary work for rehabilitation. The company LMBV is aware of its huge responsibility towards the people of both regions. The company's 24-year experience in the matter of rehabilitation and reclamation of landscapes affected by mining will also be used over the coming years to continue the previous success and thus support the structural change of the regions.

1.7 Results of the Rehabilitation and New Challenges

The basic mining engineering rehabilitation has for the most part been concluded. Almost all of the slopes have been stabilised - a total length equalling the distance between Paris and Berlin. More than 50 per cent of the waste heaps have been recultivated for agricultural or forestry purposes.

The last phase of rehabilitation, the complex restoration of a well-balanced hydrological regime in an area which is almost one and a half times as large as the Lima Metropolitana region is under way at present and will also be the focus of future work.

In order to restore a well-balanced, almost self-regulating hydrological regime the aquifers and opencast mining lakes have to be replenished with 12.7 billion cubic metres of water. By replenishing the aquifers and using external water sources, around 6.5 billion cubic metres of water have already been achieved.

Marketing the newly landscaped areas is well under way. They are purchased by local authorities, nature conservation associations, foundations and private individuals. By the end of 2005, a total of around 53,500 hectares of former mining sites had been transferred to new owners.

New ecologically sustainable landscapes with economic potential are emerging from the largest landscape construction site in Europe - from Wulfersdorf near Helmstedt via Nachterstedt and the valley of the river Geisel up to the very heart of the East German mining area in the triangle formed by the cities of Halle, Leipzig and Bitterfeld. This also holds true for the Lausitz region, from the Spreewald to Dresden and further on to Görlitz and Zittau.

The major nature conservation project Lausitzer Seenland (Lake District Lausitz) is the high value centrepiece of the chain of lakes coming into existence which will constitute a water landscape of transregional importance. In addition to the Lakes Senftenberg, Knappen and Silbersee which arose many years ago from abandoned mining sites and are now popular tourist destinations, step by step another 15 lakes will come into existence in Saxony and Brandenburg thanks to the flooding of former opencast mining areas. In the central area of the Lake District Lausitz alone, with all its connecting navigable channels, a water surface area will be created which in Germany will be second only to Lake Constance in size.

In their final stage these new landscapes will be dominated by around 27,000 hectares of water and almost 18,000 hectares of nature conservation areas. They will improve the quality of life in the former mining areas considerably and offer substantial tourism potential.

In addition, the new industrial parks with their excellent infrastructure offering several thousand new jobs are no longer just a vision but have already started to become reality.

And there is also the technical and expert knowledge which was gained in the course of the rehabilitation works and which should not be undervalued. Companies like LMBV as project leader in the project presented here stand for high levels of competence with regard to the technical know how and management skills needed for such a complex rehabilitation project. With its experience and knowledge it is an attractive partner - in Germany as well as abroad. German experience in the field of rehabilitating abandoned mine sites is sought after.

Preventive care pays off. It saves a lot of money later on during rehabilitation. Therefore rehabilitation aspects should be included at a very early stage, i.e. when opening up a new mine.

Rehabilitation can be expensive but it pays off as well. Making use of specific know-how and appropriate technologies can help reduce costs in the medium and long run.

Do not postpone rehabilitation for too long. It should be tackled as soon as possible and in an effective way. This will bring about excellent result at relatively low costs.

The rehabilitation reached in 2012 a high level of good results in land rehabilitation and new use of former mining used areas.

But there are still challenges in Geotechnical and water issues. The refill of Groundwater reached good results but because of this we had Acid Mine Drainage (AMD) and instable dumps. The instable areas be located suddenly in old rehabilitated areas in 2006; these areas were in forest used since over 30 years. We know the problem of instable slopes and the focus was from the beginning to stabilize them, but not the areas behind. The philosophy was, that more than 3 – 5 meters of dry overburden is enough for stable land use. But now we are smarter and the universities are looking for new methods of stabilization.

The AMD in the new post mining lakes are solved with Inlake-Treatment (lime ship). By this method we bring lakes of a dimension of ha in good stable conditions.

In the years of 2008 and 2011 we had large rain fall in the Lusatia. This had a high rate of refill of groundwater, so large areas came early in contact with drains as we expected. By this process we had a large drainage of FeO_2 . So the rivers in the old mining areas get brown colored. Our strategy now is to build a protective wall to protected the naturally heritage of the Spreewald.

Because of the clogging of the river spree the LMBV together with the LBGR had developed a 10-point emergency program (Figure 3).

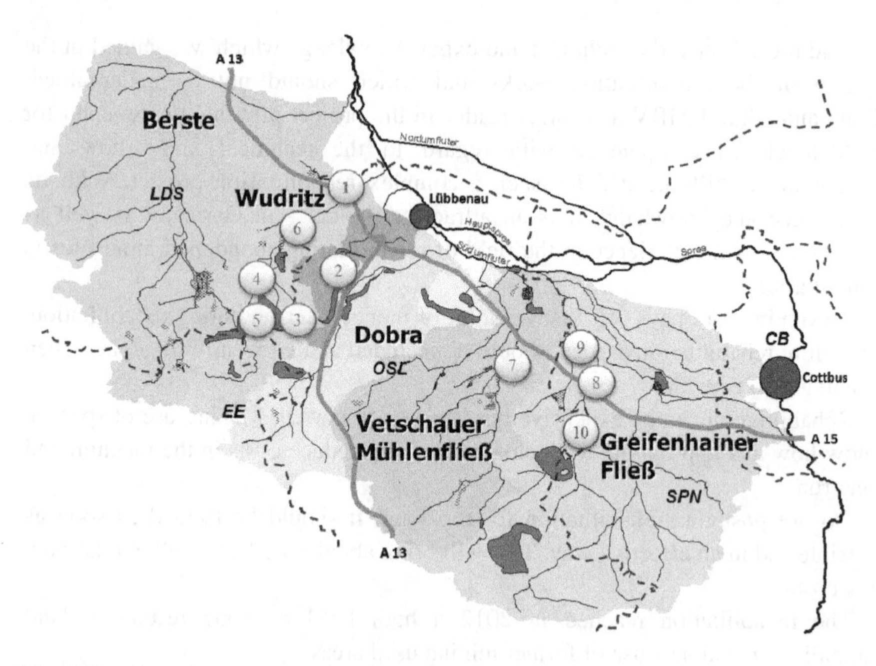

Fig. 3 10-point emergency program (LBGR)

This program takes the following measures:

1. Sludge site clearance
2. Neutralization of the lake Lichtenauer
3. Temporary Reconciliation from the lake Schlabendorf to the lake Lichtenau
4. Establishment of a temporary Conditioning plant at the flowing water Lorenz
5. Inlake-Treatment at the lake Schlabendorf
6. Temporary constructed wetland
7. Reactivation of the Pitwater Treatment plant
8. Sludge site clearance at Greifenhain
9. Reconstruction of the water purification plant of the former power station Vetschau
10. Diversion of water

Author Index

CPSIA information can be obtained at www.ICGtesting.com
Printed in the USA
LVOW02*1725090415

433931LV00010B/106/P